IDENTIFICATION AND MITIGATION OF LARGE LANDSLIDE RISKS IN EUROPE

IMIRILAND PROJECT

Identification and Mitigation of Large Landslide Risks in Europe

Advances in Risk Assessment

Ch. Bonnard, F. Forlati & C. Scavia, editors

European Commission
Fifth Framework Programme

IMIRILAND Project

A.A. BALKEMA PUBLISHERS Leiden / London / New York / Philadelphia / Singapore

of large landslides in the mountain environment: identification
d by the European Commission between 2001 and 2003 within
Environment Generic" (Contract No EVG1-CT-2000-00035),
fice for Education and Science (Contract OFES No 00.0361),

Management of the Project

Administrative Coordinator: Vincenzo COCCOLO – ARPA Piemonte, Torino.
Scientific Coordinator: Prof. Claudio SCAVIA – Politecnico di Torino.
Editorial Board: G. Amatruda, Ch. Bonnard, M. Castelli, G. Moletta, M. Morelli,
                             L. Paro, F. Piana and G. Susella.

ACKNOWLEDGEMENTS
Sincere acknowledgements have to be expressed to all the persons who significantly contributed
to the IMIRILAND Project as authors, as well as to those whose names do not appear as authors
in this book:
Technical assistants: L. Beccani, E. Fusetti, L. Mallen, H. Sarri, F. Tamberlani (ARPA Piemonte);
A. Allodi, S. Marello (Politecnico di Torino).
Research assistants: J. Moreno Lapiedra, R. Rojas Fuentes (Ecole Polytechnique Fédérale de
Lausanne); R. Russo, A. Stéphan (CETE Lyon).

Special thanks are also expressed to Prof. Oldrich Hungr and to Dr. Pedro Basabe who contributed
to chapter 10, to K. Barone who mastered all the difficulties of the final editing, as well as to
L. Martinenghi Stuckrath who revised the English form of the texts. The administrative staff in charge
of the coordination at ARPA Piemonte and Politecnico di Torino also deserves a vote of thanks.
    We are equally grateful to all the communal authorities and to their consultants, in particular
those of Ceppo Morelli, Rosone, Séchilienne and Sedrun, who provided interesting information
for the development of this research.
    Finally, the editors as well as the participants to the IMIRILAND Project are deeply indebted to
Marta Castelli of Politecnico di Torino and Noemi Giordano of ARPA Piemonte for their fundamental part in the success of the project. Marta, by her continuous involvement in the scientific
coordination tasks as well as in the planning and management of the different phases of the
research and the diffusion of its results, and Noemi, by her careful and precise implication in the
administrative aspects of the project that have to be assumed in order to ensure its completion.
Noemi also fulfilled the difficult task of translating some chapters.

Published by: A.A. Balkema Publishers, a member of Taylor & Francis Group plc
             www.balkema.nl and www.tandf.co.uk

ISBN 90 5809 598 3

The following institutions participated in the IMIRILAND Project:

AGENZIA REGIONALE PER LA PROTEZIONE AMBIENTALE DEL PIEMONTE (ARPA PIEMONTE)/ REGIONAL AGENCY FOR ENVIRONMENTAL PROTECTION OF PIEMONTE (Turin, Italy)
SETTORE STUDI E RICERCHE GEOLOGICHE, SISTEMA INFORMATIVO PREVENZIONE RISCHI

Arpa is a leading centre of studies and applied research in the field of natural risk. Its aim is the development of methodologies and tools to assess, manage and minimise the geological risk. It has a wide experience in data collection, organisation and elaboration with GIS techniques. It produces and disseminates scientific and technical knowledge, addressed to public and private end-users. Moreover it coordinates national and international projects.

POLITECNICO DI TORINO/TECHNICAL UNIVERSITY OF TURIN (Italy)
DIPARTIMENTO DI INGEGNERIA STRUTTURALE E GEOTECNICA

The activities of the Department follow three main directions: Structural Mechanics, Structural Design and Geotechnics. Within each of these areas, focused research is conducted, including fracture mechanics, earthquake engineering and rock mechanics. Theoretical, experimental and numerical aspects are addressed in order to determine solutions of problems related to structural and geotechnical phenomena affecting, for example, the statics and dynamics of bridges, large containers, buildings of strategic importance and historical monuments. In particular, the research in slope stability problems is dedicated to the geomechanical modelling of rock masses and the application of numerical methods to the definition of hazard scenarios.

ECOLE POLYTECHNIQUE FÉDÉRALE DE LAUSANNE (EPFL)/SWISS FEDERAL INSTITUTE OF TECHNOLOGY (Lausanne, Switzerland)
DEPARTEMENT ENVIRONNEMENT NATUREL, ARCHITECTURAL ET CONSTRUIT (ENAC) – LABORATOIRE DE MECANIQUE DES SOLS (LMS)

LMS has developed several methodologies for the hazard mapping and modelling of large zones affected by landslides. Furthermore, it has developed a method for the prediction of landslide behaviour based on neural network analysis. LMS also works on coupled thermo-hydro-mechanical modelling of saturated and unsaturated soils, including constitutive and thermodynamic aspects, as well as on geothermal structures and heat exchange problems. Finally it deals with soil reinforcement by cement injection and heat effect.

CONSIGLIO NAZIONALE DELLE RICERCHE (CNR)/ ITALIAN NATIONAL RESEARCH COUNCIL (Italy)
ISTITUTO DI GEOSCIENZE E GEORISORSE (IGG) (SEZIONE TORINO)

> CNR is involved in basic geological, structural, petrographical and mineralogical researches. A relevant activity of field survey for the 1:50000 scale National Geological Map of Italy, represents the basis to develop approaches for the analysis and mitigation of geologic risk. This topic has been worked out in the last years in co-operation with some Italian Institutions.

TECHNISCHE UNIVERSITÄT WIEN (TUW)/ VIENNA UNIVERSITY OF TECHNOLOGY (Austria)
INSTITUT FÜR INGENIEURGEOLOGIE

> Institute for Engineering Geology is involved in research activities concerning problems of slope stability and mass movements, site exploration and engineering geological mapping, numerical modelling, tunnelling (rock and soil mechanics, hydrogeological and grouting aspects), technical petrography (esp. for conservation of historical monuments).

UNIVERSITAT POLITÈCNICA DE CATALUNYA (UPC)/ TECHNICAL UNIVERSITY OF CATALONIA (Barcelona, Spain)
DEPARTAMENT D'ENGINYERIA DEL TERRENY, CARTOGRÀFICA I GEOFÍSICA

> This UPC Department forms a leading centre of research in the field of Soil and Rock Mechanics, in landslide hazards and landslide modelling, in the numerical analysis of coupled thermo-hydro-mechanical problems, and in laboratory and fieldwork. The Department has been considered as a Centre of Excellence for Research within the Catalonian Research Plan framework.

LABORATOIRE CENTRAL DES PONTS ET CHAUSSÉES/ CENTRAL LABORATORY OF BRIDGES AND HIGHWAYS (Paris, France)
CENTRE D'ETUDE TECHNIQUES DE L'EQUIPEMENT DE LYON (CETE)

> LCPC is a State research organisation working for the State and the local authorities in connection with professionals involved in civil engineering, transport, urban engineering and environment. One of the original characteristics of LCPC resides in its role as technical network co-ordinator for infrastructure, as far as natural hazards are concerned. The LCPC is relying on the regional CETE's to carry out the main project studies and some field research.

List of contact persons and participants in the IMIRILAND Project:

Arpa Piemonte – (ARPA)
Settore Studi e Ricerche Geologiche – Sistema Informativo Prevenzione Rischio

Contact person: **Dr. Ferruccio Forlati**
Corso Unione Sovietica 216
10134 Torino – Italy

Tel: +39-011-3169335
Fax: +39-011-3169340
e-mail: settore.20-3@regione.piemonte.it
URL: www.arpa.piemonte.it

Participants: Gianfranca Bellardone, Stefano Campus, Barbara Coraglia,
Ferruccio Forlati, Ermes Fusetti, Lidia Giacomelli, Noemi Giordano,
Luca Mallen, Luca Paro, Manlio Ramasco, Gianfranco Susella,
Herbert Sarri, Ferdinando Tamberlani, Carlo Troisi.

Politecnico di Torino (POLITO)
Dipartimento Ingegneria Strutturale e Geotecnica

Contact person: **Prof. Claudio Scavia**
C.so Duca degli Abruzzi 24
10129 Torino – Italy

Tel: +39-011-5644823
Fax: +39-011-5644899
e-mail: claudio.scavia@polito.it
URL: http://www.polito.it/ricerca/dipartimenti/distr/

Participants: Andrea Allodi, Gaetano Amatruda, Marta Castelli, Stefania Marello,
Marina Pirulli, Erminia Quacquarelli, Claudio Scavia.

Consiglio Nazionale delle Ricerche (CNR)
Istituto di Geoscienze e Georisorse (IGG) – Sezione di Torino

Contact person: **Dr. Riccardo Polino**
Via Accademia delle Scienze 5
10123 Torino – Italy

Tel: +39-011-530652
Fax: +39-011-530652
e-mail: r.polino@csg.to.cnr.it
URL: http://www.igg.cnr.it/

Participants: Michele Morelli, Fabrizio Piana, Riccardo Polino.

Laboratoire Central des Ponts et Chaussées (LCPC)
Division de Mécanique des Sols et des Roches et de Géologie de l'Ingénieur

Boulevard Lefebvre 58
75732 Paris – France
Tel: +33-01-40435246
Fax: +33-01-40436516
URL: www.lcpc.fr/

Centre d'Etudes Techniques de l'Equipement (CETE)

Contact person: **Dr. Jean-Louis Durville**
CETE de Lyon
25 Av. Mitterand CSE 1
69674 Bron CEDEX – France

Tel: +33-04-72143215
Fax: +33-04-72143035
e-mail: jean-louis.durville@equipement.gouv.fr
URL: http://www.equipement.gouv.fr/

Participants: Jean-Louis Durville, Laurent Effandiantz, Pierre Pothérat, Rosalba Russo, Ariane Stéphan.

Contribution: Philippe Marchesini (DDE-Isère).

Universitat Politècnica de Catalunya (UPC)
Departamento de Ingeniería del Terreno

Contact person: **Prof. Pere Prat**
Jordi Girona 1-3, Building D2
08034 Barcelona – Spain

Tel: +34-93-4016511
Fax: +34-93-4017251
e-mail: pere.prat@upc.es
URL: http://www.etcg.upc.es/

Participants: Jordi Corominas, Marcel Hürlimann, Alberto Ledesma, Pere Prat.

Technische Universität Wien (TUW)
Institut für Ingenieurgeologie

Contact person: **Prof. Rainer Poisel**
Karlsplatz 13
1040 Wien – Austria

Tel: +43-1-5880120-319
Fax: +43-1-5880120-399
e-mail: rainer.poisel@tuwien.ac.at
URL: http://www.tuwien.ac.at/

Participants: Rainer Poisel, Alexander Preh, Werner Roth, Ewald Tentschert.

Ecole Polytechnique Fédérale de Lausanne (EPFL)
Faculté Environnement Naturel, Architectural et Construit (ENAC) – Laboratoire de Mécanique des Sols (LMS)

Contact person: **Christophe Bonnard**
                         EPFL – ENAC-LMS
                         1015 Lausanne – Switzerland
                         Tel: +41-21-6932312
                         Fax: +41-21-6934153
                         e-mail: christophe.bonnard@epfl.ch
                         URL: http://lmswww.epfl.ch/

Participants: Christophe Bonnard, Xavier Dewarrat, Jordi Moreno, Rafaël Rojas.

Contribution: Francis Noverraz, Consulting geologist.

Contents

Preface

Throughout history, all mountainous countries have been exposed to natural hazards that often reached disastrous proportions. It is undeniably a fact that the public awareness for natural disasters is generally low and, therefore, progress in risk mitigation is mostly based on the upsetting experience of the people concerned. This is not only true for investments in risk mitigation measures but also for the fields of research, legislation and standardisation. A significant increase in damaging effects has been seen over the last two decades. The reasons for the steady increase in damage due to natural hazards are often ascribed to the supposed impact of climate change. Additional and perhaps less spectacular reasons, such as the increase in our living standards, in population density, and in infrastructures and goods in economically privileged but hazardous places, the development of settlements in disaster-prone regions, the enormous increase in mobility on road and rail, the use of hazardous technologies in industry and the increased susceptibility to problems of our overall living space, have to be considered as well (Fig. 1). In any event, to be able to develop efficient risk prevention and mitigation strategies, it is important to know the true reasons for this risk increase.

The disastrous events of the last few years have also clearly shown that attainable security in natural hazard management remains limited despite effective preventive measures. This is largely due to the vulnerability and the sensitivity of our infrastructures (Fig. 2), as well as our society and the environment which we live in. In addition, it is more and more questionable to distinguish between natural risk and that induced by human activity.

In the mountainous environment, landslides of various kinds, defined in a general sense as "the movement of a mass of rock, debris or earth down a slope" (Cruden, 1991) are frequently responsible for considerable losses of both money and lives, and the severity of the landslide problem worsens with increased urban development and change in land use (Programme INTERREG 1, 1996). Given this understanding, it is not surprising that landslides are rapidly becoming the focus of major scientific research, engineering study and practice, as is land use policy throughout the world, mainly in rugged topographic areas.

Figure 1.   Encampadana-Canillo landslide foot, Andorra – 2002, a region of intense touristic development.

When speaking of mountains, it should be realised that 14% of the EU and 11% of the Accession Countries' land area is formed of mountainous relief (not including Switzerland!) (Source: EEA, 1999). In Europe, major road and rail tunnels, high bridges and dams are concentrated in the mountains, and are prone to widespread, frequent and expensive damage. Furthermore, expansion of tourism in mountainous villages has spread accommodation and infrastructures into risk areas.

Risk assessment, prevention and mitigation require technical and socio-economic answers to some basic questions. The question "what risks can be considered as acceptable?" has to be answered at a socio-cultural level and needs to be compared with the questions "what could happen? how often? what are the consequences?", which require expert technical know-how. Science, education and politics have to face this challenge and to support this difficult decision-making process with adequate knowledge, research, and the development of practical tools as well as through insurance and legislation strategies. As a consequence, only a multidisciplinary approach by specialists from different countries using specific methods can increase confidence in risk assessment. Therefore, the problem needs to be addressed not by single countries, but at the European level.

This book is specifically dedicated to the impact of large landslides in the mountainous environment in terms of identification and assessment of risk. Several definitions exist for large landslides, for instance the following one presented in the world landslide inventory of the IUGS:

– volume of more than 1 million m^3
– causes one victim or more
– induces a loss equal to the work of one man/year.

Figure 2.   View of Cassas landslide, Italy – 1983 where a major motorway and railway are located at the toe.

This definition mainly focuses on sudden events of fairly large magnitude but does not at all reflect the situation of complete mountain slopes (in some cases affecting areas larger than dozens of km^2) in slow movement or threatened by huge rockfalls. Therefore, the present text deals with another kind of large landslide in which large communities or infrastructures are likely to be exposed to the eventual extreme development of a phenomenon characterised by a very slow basal movement induced by its geological features.

A methodology carried out for the determination of hazard, vulnerability and finally, risk, is presented, with some practical applications to large landslide cases located in the Alps. The context, derived from the first results of the project IMIRILAND, is supported by the EU within the fifth framework programme.

The editors:
Christophe Bonnard, Ferruccio Forlati, Claudio Scavia

1
Introduction

M. Castelli[1], F. Forlati[2]
[1]Politecnico di Torino; [2]Arpa Piemonte

Large landslides affect many mountain valleys in Europe. They are characterised by a low probability of evolution into a catastrophic event but can have very large direct and indirect impacts on man, infrastructures and the environment (Table 1.1). This impact is becoming more and more pronounced due to increasing tourist development and the construction of new roads and railways in mountainous areas. Methodologies for the assessment and mitigation of risk are, therefore, a major issue. Since very large slope movements are quite often directly or indirectly implicated in disasters (not only large landslides but also secondary slides or debris flows), their early identification, which is not always easy, is essential to an adequate risk assessment of the zones involved.

Table 1.1. Some major landslides which have occurred over the past centuries in the Alps.

Landslide	Location	Year	Lives lost
Mont Granier	French Alps	1248	About 3000
Salzburg	Austria	1669	250
Goldau	Swiss Alps	1806	457
Elm	Swiss Alps	1881	115
Vaiont	Italian Alps	1963	2100
Val Pola	Italian Alps	1987	27
La Salle en Beaumont	French Alps	1994	4

What do we know about landslide risks? A first theoretical definition was given by Varnes in 1984: "Risk is the probability of an event of a given magnitude multiplied by its consequences".

Since then, many technical and scientific papers have been written on this topic (i.e. Brabb, 1984; Brand, 1988; Bunce et al., 1997; Carrara, 1992; Carrara et al., 1992; Cherubini et al., 1993; Cruden & Fell, 1997; Cruden & Varnes, 1994; Einstein, 1988; Fell, 1994; Fell et al., 2000; Finlay & Fell, 1997; Hungr, 1981; Hungr et al., 1993; Hutchinson, 1992; Leroi, 1997; Mulder, 1991).

However, the evaluation of hazard is still made through different approaches, both qualitative and quantitative, which lead to differing results. Besides, only a few studies on vulnerability have been carried out (i.e. Leone et al., 1996). Finally, no well-defined procedure to include the results of risk analyses in land planning from a legal point of view exists. As a matter of fact, many experiences during the period of critical development of landslides (Bonnard, 1994a; Vulliet & Bonnard, 1996) have shown a lack of methodology and, above all, a nonsystematic approach of interpreted risks. In practice, risk management is carried out by local and regional authorities only during the actual critical event in a necessarily improvised way. This "reactive" approach brings negative consequences to the identification procedure and to preventive actions. For example, very expensive monitoring systems have been installed on several large landslides, but in spite of that:

- available data are often very dispersed (data banks, reports, publications ...) and their reliability is not always verified;

– landslide studies are isolated, limiting their interpretation to local factors;
– a well-established methodology linking the understanding of the mechanisms of deformation and the interpretation of the data obtained through the monitoring system to the practical questions concerning the management of the risk does not exist (Fig. 1.1).

Figure 1.1.   View of Ruines de Séchilienne landslide, France – 1994, where prevention works have been carried out but a global risk management policy has not yet been adopted.

A literature review (IUGS/WGL/CRA, Working Group on Landslides, committee on risk assessment, in Cruden & Fell, 1997) related to these specific aspects has pointed out a lack of knowledge in the following topics:

– experience obtained at a large scale to include the notion of actual risk in land planning;
– comparative analyses of different hazard prediction techniques carried out by multidisciplinary teams;
– application of risk assessment procedures to large landslide zones by considering the potential damage in a quantitative manner and the cost of protective measures in a complete economic balance including direct and indirect costs;
– comparative assessments of the applicability of such techniques in different frameworks;
– direct involvement of end users (regional and national technical administrations);
– application of risk assessment and mitigation methods to real cases, in which the administrative and legal aspects of the final results can be checked.

Furthermore, risk can extend well beyond local damage (for instance, risk of river damming which may induce major hydrological hazards: floods, inundation of sewage plants, loss of drinking water resources), so that it must be considered from a wider perspective. In this sense, a multidisciplinary approach by specialists from different countries using specific methods (i.e. tectonic analyses, photogrammetry, numerical analyses, etc.) at some well-monitored sites can improve the determination of the main landslide mechanisms and thus the assessment of the hazard level, while the contribution of administrators and end users in general can ensure practical checks of the reliability and suitability of the developed techniques.

Thus, the present text intends to examine the hazard analysis of different types of landslides more closely by the application of geological, geomechanical and statistical methods. Particular attention is given to the understanding of the structural setting of the landslide sites and adjoining areas. Relations between "on site" and "around site" structural features have been sought in order to define the actual boundaries of the potential landslide areas.

These procedures are based on the application of basic techniques of structural analysis (Hancock, 1985 and references therein), with a special emphasis on the recognition of fundamental structural associations according to Hobbs et al. (1976) (see also Forlati & Piana, 2000). Furthermore, extensive use of satellite images has been made to search for relations between the distribution of local and regional structural elements (Morelli & Piana, 2003). Structural models obtained for each landslide provided the basic inputs for the geomechanical numerical models. Particular attention was given to numerical models, where some of the most current finite element, boundary element and finite difference methods were applied in three-dimensional conditions on the same slope stability problems (Castelli et al., 2001). A comparative analysis of different hazard prediction techniques can then be carried out.

Finally, vulnerability and risk analyses have been developed for the same landslides considering short and long-term perspectives, direct and indirect consequences, as well as technical and social impacts (Fig. 1.2). The combination of the results obtained in relation to hazard analysis and through the risk approach will allow the development of a new practical and quantified risk assessment program which will be applied to several sites.

The book presents the results obtained by the IMIRILAND project, supported by the EU from March 1st, 2001 to December 31st, 2003.

The project was born of the common interest of many European countries in exchanging information and experiences in the field of the use and development of risk assessment procedures. In particular, the countries involved in the consortium (Italy, Austria, France, Spain, Switzerland) are aware of the importance of these problems and have been working on them for several years. The consortium involves different and complementary expertise, including some of the most scientifically advanced universities and research infrastructures dealing with natural hazard problems as well as two end users which must face practical land management problems every day.

Figure 1.2.   Dams on the Oselitzenbach landslide, Austria – 2002, needed to protect the valley from erosion phenomena which might trigger this slide.

The project allowed the selection of large landslide sites located in various mountainous environments and legal contexts for the development and illustration of risk assessment methodologies. These areas ranged from the Pyrenees (Spain, France) to the West, to the Austrian Alps to the East, passing through western (France, Italy) and central (Switzerland) areas, in order to grant a high degree of generality to the methodologies and to impart a European dimension to the problem. In this way, the results can also be diffused and applied outside the countries directly involved in the project. The partnership, therefore, created a remarkable critical mass for EU socio-economic development.

Eight landslide sites (Fig. 1.3) were selected (slides or rockfalls, volume $>10^6 \, m^3$) in the countries involved (Italy, France, Spain, Austria, Switzerland), where geological, geomechanical and monitoring data already exist. The selected cases (already occurred or in a pre-failure phase, see Fig. 1.4) imply a complex risk situation for which no reasonable stabilisation works can be executed. The partners involved in the project already have some knowledge about the sites, the data of which are already available but very dispersed (data banks, reports, publications). In this way,

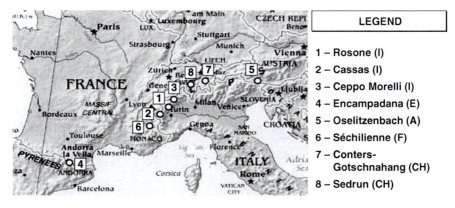

LEGEND

1 – Rosone (I)

2 – Cassas (I)

3 – Ceppo Morelli (I)

4 – Encampadana (E)

5 – Oselitzenbach (A)

6 – Séchilienne (F)

7 – Conters-
     Gotschnahang (CH)

8 – Sedrun (CH)

Figure 1.3.   Landslide site locations.

Figure 1.4.   Uphill-facing scarps in Sedrun landslide, Switzerland – 2002 whose movement has been significantly increasing over the last 20 years.

4

it has been fairly easy to collect a large amount of information without expending excessive amounts of time and energy.

A general scheme for data description was created and compiled for each site (Table 1.2). Each partner was then able to access data and to manage them in a homogeneous manner. The objective was also to check the reliability of the available data and to choose which were to be considered as significant for the risk analysis.

The principal aims of the project are to:

– carry out a review of hazard analysis methodologies,
– compare the reliability of hazard assessment in different situations according to various criteria,
– develop practical risk analysis methods considering direct and indirect potential damage,
– apply the developed methods to real situations in different countries,
– test the applicability of such approaches through close interaction with administrators.

Table 1.2.   Selected sites.

Site	Country	Approximate volume of main landslide [m³]/globally affected area [km²]	Main elements at risk
ROSONE	Italy	$50 \times 10^6 \, m^3/5.5 \, km^2$	Rosone village and main road SS 460 Hydroelectric power plant Valley downstream
CASSAS	Italy	$15 \times 10^6 \, m^3/0.9 \, km^2$	Highway A32 (Frejus) International Torino-Modane railway Valley downstream
CEPPO MORELLI	Italy	$5 \times 10^6 \, m^3/0.5 \, km^2$	National Road N. 549, only link to the Macugnaga tourist resort Prequartera and Campioli villages
ENCAMPADANA	Spain	Several tens $\times 10^6 \, m^3/1 \, km^2$	Important ski resort El Tarter, small village Main road Andorra – France Communities downstream
OSELITZEN-BACH	Austria	$0.5–2 \times 10^6 \, m^3/2.9 \, km^2$	Villages of Tröpolach and Watschig downstream Main road "Naßfeld – Bundesstraße", important link to the "Gailtal" Valley
SÉCHILIENNE	France	$25 \times 10^6 \, m^3/0.9 \, km^2$	National road RN 91 Valley downstream
CONTERS-GOTSCHNAHANG	Switzerland	$2 \times 10^9 \, m^3/25 \, km^2$	New A28 national road Town of Lanquart Valley downstream
SEDRUN	Switzerland	$100 \times 10^6 \, m^3/1.5 \, km^2$	Regional railway line National road Sedrun tourist resort

Note: The volumes indicated above refer to the modelled active landslide masses, whereas the areas indicated refer to the globally involved instability phenomena.

To achieve such objectives, the following phases have been foreseen:

1. Data collection. A data base has been created containing all available data on the selected sites.
2. Development of risk assessment methodology. In order to compare the results of different methodologies in terms of landslide scenario evolution, the following approaches have been

considered: (a) Field analysis, (b) Mechanical modelling. For each approach, state-of-the-art, critical analyses of the methods and applications to selected sites have been carried out. Vulnerability and risk analyses were also made.

3. Application to management. This concerns to the application of the developed methodologies to the management of endangered landslide zones. The actions are defined in relation to the problem urgency, the legal framework and the power of local and regional authorities, as well as the relevant economical conditions.

4. Dissemination of risk management methodologies.

The global and multidisciplinary approach which characterises the project represents the first important result for the entire consortium. In fact, it allows the testing of a new method of collaboration at the European level, gathering various researchers, technicians and officials from different countries and making them study similar problems (natural hazards in mountainous environments) in different contexts (from a socio-political point of view). By analysing and interpreting all of these experiences, a unified problem-solving approach was developed. In particular, for universities and academic organisations, the project represented a very useful opportunity for developing new methodologies or applying existing ones to several real situations, enabling a test of their applicability and reliability. As far as the other members of the consortium were concerned, who were also end users, these persons had the opportunity of establishing a direct relationship with the academic world, in order to work together to find practical solutions for the problems they must face as a part of their administrative tasks.

Another important positive consequence of the project is the improvement in the perception of scientific and technical problems by officials and public administrators working in the field. A specific work phase has been dedicated to this aspect; the comparison among different approaches to similar phenomena within different legal frameworks produced very interesting and useful results.

This special book is organised into eleven major sections, written by several authors:

1. Introduction
2. The meaning of risk assessment related to large landslides
3. A key approach: the IMIRILAND project method
4.–9. Integrated quantitative risk assessment applied to the six selected landslides
10. Specific situations in other contexts
11. Suggestions, guidelines and perspectives of development.

2
The meaning of risk assessment related to large landslides

Ch. Bonnard
Ecole Polytechnique Fédérale de Lausanne

2.1 SPECIFICITIES OF MOST LARGE LANDSLIDES

Most observed or inferred landslides (falls, slides, flows) generally affect a very limited area; therefore, their management requires a well-defined analysis and local action only in terms of prevention or remediation. On the contrary, large landslides which cover more than 1 km^2 or involve huge masses of rocks with extremely high potential kinetic energy require a global consideration in the context of land planning procedures, in which the aspects of risk assessment within the prevention action play a major role.

Thus it is important and even essential to identify the specificities of the majority of large landslides in terms of mechanisms, preparatory and triggering causes and potential development. On the basis of the case studies analysed within the IMIRILAND project (see Table 1.2), four different typical situations related to large landslides may be specified and analysed in order to understand the different circumstances and conditions which rule the assessment of the risks they involve.

2.1.1 *Slow permanent movement of extensive areas on which dwellings and infrastructures exist*

This typical situation is often encountered on Alpine or Pre-alpine slopes which have been affected by glacial retreat approximately ten thousand years ago. The observed movements, provided they have been identified and monitored over long periods, are generally fairly regular, but some occasional accelerations may occur which last for a short period of several months to a few years. Their specificity is mainly their surface extension, from 1 to 40 km^2, and their gentle morphology, thus inducing the population to use this apparently favourable land for settling villages on slopes with a shallow gradient and developers to build infrastructures such as roads and cable cars (see Fig. 2.1 showing the Lugnez landslide).

In many cases, the velocity pattern of the different sectors of this type of landslide is fairly homogeneous, and differential velocities are only recorded in specific zones, so that these zones are more recognised by the population as local shear bands, which display local damage to roads and buildings, rather than as the limits of extensive moving areas.

The risks induced by such continuous phenomena are especially related to serious climatic events of high intensity or long duration, which may bring about a momentary acceleration and an increase of differential movements, thus causing a disruption of the road network and serious damage to houses located near the shear zones (see Fig. 2.2 showing the acceleration of one point A located on the toe of La Frasse Landslide (Canton of Vaud/Switzerland), whereas another point B nearby does not show any acceleration) (Bonnard & al., 1995).

2.1.2 *Significant potential movement of large areas inducing major cracks in slopes, change of geometry at the toe and consequent damming of the valley*

This critical case is often ignored or improperly understood as the main movements of such a landslide are quite limited in terms of average velocity and as the evidence of an activity of the mass

Figure 2.1. Lugnez Landslide, located in Eastern Switzerland, displays a gentle and regular slope over an area of more than 30 km² on which several villages have developed since the Middle Ages, despite a regular movement of several centimeters per year (Noverraz & al., 1998).

is often limited to a localised zone near the scarp or at the toe. In the case of a marked reactivation, the movement of the mass may evolve into a rockfall phenomenon or a debris slide near the toe, which, according to the size of the blocks, might be able to dam the valley and cause flooding first upstream, and then downstream if a sudden release of stored water occurs following a dam break event. In this case, the impact of the phenomenon is much larger than its own surface extension and may affect hundreds of km² (see the Sechilienne landslide commented on inChapter 9). Such a case occurred in 1993 in southern Ecuador at La Josefina as presented in Chap. 10.2 below.

2.1.3 *Possible rapid or very rapid mass movement (rockfall, rock avalanche) causing a major threat to the communities and infrastructures below the source area*

This case is even more difficult to detect in many locations and geological settings, as the preliminary movements are hardly perceptible and the masses affected by typical preparatory mechanisms, such as toppling, may be quite extensive. If the unstable masses in one part of the slope experience a sudden acceleration, sometimes related to a specific climatic event, the induced rockfall may be massive and widespread, and can therefore seriously threaten the villages or communities below, especially if the slope is steep. As far as transportation infrastructures which cross these unstable zones at the toe of the slope are concerned, they are very difficult to protect due to the high energy of the falling blocks and to the uncertainty of their trajectory, so that the only way to handle this problem often implies the construction of a tunnel (see the case of Ceppo Morelli commented on in Chapter 7).

The suddenness of such phenomena implies that any monitoring program may be inefficient to trigger an alarm in due time and to protect the users of the transportation routes located below the source area.

2.1.4 *Large debris flows induced by the presence of an active slide on the side of a stream, causing damage in the valley downstream*

A critical situation in terms of risk is also induced by the occurrence of debris flows of large magnitude which are "fed" by erosional processes touching active slides on one or both banks of the valley.

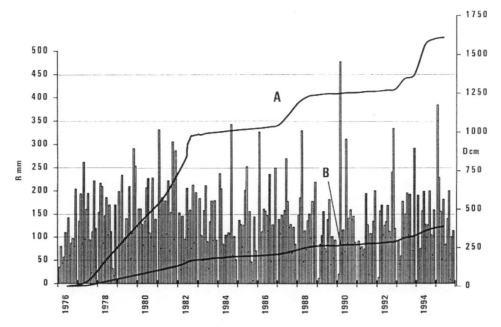

Figure 2.2.   Evolution over 25 years of the displacements of two points located on the lower part of La Frasse Landslide (right scale [cm]), as well as the monthly rainfall amounts (left scale [mm]). The crisis of 1981–1982 seriously affected a main road crossing the landslide (point A) whereas the movement at point B nearby was not appreciably modified.

This "in-chain" mechanism, which may lead to a momentary damming of the river at the toe of the slope, has serious consequences on the population and infrastructures downstream, as the extension and magnitude of the flow may far exceed previously observed phenomena (see Oselitzenbach case commented on in Chapter 6).

2.2   FACTORS INFLUENCING THE HAZARD ASSESSMENT FOR LARGE LANDSLIDES

Considering the specificities of large landslides which involve huge masses of soil or rock affected by gravity-induced movements and taking into account that their behaviour is never repetitive, it is quite difficult, if not impossible, to assess their hazard on the basis of statistical monitoring data as is commonly done in flood prediction. Therefore, hazard analysis must use particular criteria and consider various factors in order to assess the hazard level for different scenarios in a semi-quantitative way, so as to specify relative orders of magnitude of intensity and probability as explained in the methodology developed in Chapter 3. This approach, implying a combination of both para-meters in order to quantify the hazard level, will nevertheless allow a quantitative risk assessment which can be used to manage the exposed areas.

The proposed hazard qualification through a matrix system can be adapted in a flexible way to quantitative values when relevant information exists concerning either intensity ranges, obtained for instance by mechanical modelling, or probability ranges, based on the analysis of statistical data.

Of course, the corresponding hazard value only has meaning as a range which will qualify a certain surface extent on a map; such an analysis cannot reasonably be done for a specific point or individual object exposed to a certain danger. The significance of these hazard zones on a map, which must not be too detailed or fragmented, will then be considered in the vulnerability of the exposed objects, taken in a general sense, in order to produce risk maps.

Finally, the main factors influencing hazard assessment, in the case of large landslides, are first the intensity and probability of different events selected within reference scenarios, but also the development or propagation of such events in a regional context, and finally the combination of several events (extreme rainfall and related flood plus landslide causing the damming of the river, for instance). These catastrophic combinations, even if quite rare, represent the determinant scenarios for most tragedies and the cause of major disasters. Therefore, a comprehensive analysis of hazards related to large landslides must be carried out in order to assess the relevant risks properly.

2.3 EXPRESSION OF THE CONSEQUENCES INDUCED BY LARGE LANDSLIDES

In order to assess the risks induced by large landslides, it is necessary to determine in which way the hazard situations investigated in a dangerous area may affect the population, the buildings and infrastructures, as well as the environment. Indeed, risks do exist only if a potential damage of a given magnitude may be determined, whatever its characteristics are. Therefore, the notion of direct and indirect impact has to be defined first.

2.3.1 *Direct and indirect impact*

Any natural phenomenon modifying the geometrical and geomechanical conditions of a slope may have an impact on the population, buildings and activities which can be classified as direct or indirect consequences. Under direct consequences are grouped all of the physical impacts leading either to destruction, damage (cracks in buildings, failure of pipes) or even excessive deformation (inclination of a house due to the movement of a landslide, without any damage to the structure of the building, see Fig. 2.3).

These direct consequences include infrastructures hit by blocks and buildings on a sliding slope affected by differential movements, but also constructions in the valley downstream a slide which might be destroyed by a flash flood induced by the failure of a dam caused by a landslide blocking the valley floor. The impact on animals is also included in direct damage.

Figure 2.3. Inclined chalet on the Chloewena landslide, following the disaster of 1994; the building is considered as destroyed even though it is not structurally damaged (Vulliet & Bonnard, 1996).

10

The direct consequences include, of course, persons killed, wounded or affected by the loss of their houses and properties due to the landslide phenomenon. This even includes those persons that need to be evacuated before the occurrence of the landslide, because they might be affected by the mass itself or by the consequences of the induced phenomena, such as the flash flood mentioned above.

The direct consequences can also relate to the environment, in the way that a landslide may destroy forests, animals or a habitat zone for a certain type of protected animal; it may also affect the springs of water available at the toe of a rockfall. The method of assessing all of these direct consequences will be detailed in the following paragraphs.

As far as indirect consequences are concerned, these deal with all disruptions of or hindrances to the economic activities induced by the considered landslide, mainly due to the interruption of traffic, the reduction in tourist demand and other indirect consequences related to the modification of the state of the landslide surface. These indirect consequences are fairly difficult to assess because they may imply a spatial extent much larger than the landslide zone itself. Their nature may also change with time because when a hindrance is suddenly introduced in an economic system other solutions are found by all the partners to limit the negative consequences of this perturbation. Thus, the magnitude of the potential impacts for a given phenomenon must be expressed by different aspects, mainly social, environmental and economic, as presented in Chapter 3 below.

2.4   THE MEANING OF RISK FOR LARGE LANDSLIDES

The main specificity of large landslides, such as those considered in the IMIRILAND project, is that the probability of a critical global movement of the whole mass is very low, whereas the consequences may be catastrophic. It is, thus, indispensable to assess several scenarios, and not only the worst case, with their respective potential consequences, in order to quantify the corresponding risk components in comparable terms, and to propose preventive measures on a wide scale, so that such an extreme event does not hit the site by surprise.

The comparison of the impacts of different scenarios computed on parallel bases also allows the comparison among different prevention strategies.

Such a risk assessment gives a correct and comprehensive overview of the potential impact and thus provides a refined tool for decision makers, which would otherwise either ignore the risks as a politically incorrect perspective, or exaggerate the risks in order to call for subsidies and often carry out inappropriate actions.

2.4.1   *Possible management of risk*

In the same way, the meaning of risk is difficult to express; the management of the zones or regions exposed to landslide risk is politically complex. First, nearly no reference situation in the past can provide a guideline to adopt an adequate political and technical action, as even similar situations which happened some centuries ago (for instance in the case of rockfalls) are not always well documented and occurred in a very different socio-economic context. Then, it is certainly inappropriate to prohibit any development on the largest potential impact zone, especially as the people living in the mountainous areas in which large landslides occur have already many hindrances to face and must be encouraged to remain in these regions as much as possible. From the other extreme perspective, it is also inappropriate to ignore the potential hazards totally, even if they are quite rare, and to favour the construction of any buildings and infrastructures in exposed areas, without analysing alternative solutions or more careful layouts.

The ultimate goal of such research is thus to provide guidelines for the management of the exposed zones in order to favour sufficient attention to the risks without impeding any perspective of development.

11

2.4.2 *Impact of risk assessment in the case of large landslides*

It has often been observed that the consideration of landslide risks by local authorities is either too optimistic (the risk is nearly denied) or too pessimistic (major prohibitions are pronounced, major protection works are carried out without a reasonable risk analysis). It is, therefore, difficult to measure the real impact of a proper risk assessment on local development, because the analysis and consecutive political measures are not founded on a systematic approach. This often leads to the execution of apparently efficient protection works, which neither solve the stability problem nor limit the hazard in case of rare events but rather aim at preserving the good conscience of the authorities.

After a serious event with direct impact on exposed structures, even if it implies only a small portion of the whole landslide phenomenon, the authorities tend to favour extremely safe solutions and adopt preventive measures which may be excessive with regard to the real risk analysis results. This perspective is inappropriate for two reasons; first, it leads to an unnecessary investment with respect to the real necessity of protection of the exposed objects and, second, it creates precedents in the risk management policy so that the need of safety seems to grow unnecessarily. Thus, a rational approach is needed, especially from a European perspective which must aim at dealing the similar problems in a proportionate way.

3
A key approach: the IMIRILAND project method

G. Amatruda[1], Ch. Bonnard[2], M. Castelli[1], F. Forlati[3], L. Giacomelli[3], M. Morelli[4],
L. Paro[3], F. Piana[4], M. Pirulli[1], R. Polino[4], P. Prat[5], M. Ramasco[3], C. Scavia[1],

G. Bellardone[3], S. Campus[3], J.-L. Durville[6], R. Poisel[7], A. Preh[7], W. Roth[7], E.H. Tentschert[7]

[1]Politecnico di Torino; [2]Ecole Polytechnique Fédérale de Lausanne; [3]Arpa Piemonte; [4]CNR-IGG sezione Torino; [5]Universitat Politècnica de Catalunya; [6]Centre d'Études Techniques de l'Équipement de Lyon; [7]Technische Universitat Wien

3.1   INTRODUCTION

In order to supply quantitative data related to landslide risk to land planners, it is necessary to consider the existing available tools first.

Quantitative Risk Assessment (QRA) is a method of quantifying the degree of risk through a systematic examination of the factors contributing to landslide hazard and affecting the severity of the consequences, as well as the definition of probabilities for the individual factors. The QRA technique was developed, and is now used world-wide, to estimate risks from industrial plants, such as petro-chemical, etc. QRA can provide a framework for assessing the cost-effectiveness of risk mitigation options, formulating risk management strategies and facilitating decision-making on resource allocation to reduce risks posed by different types of landslide hazards.

In simple terms, the following questions are addressed in a QRA of landslides:

– What can cause damage?
– Where?
– How often?
– What could happen?
– How severe would the impact be?

The logical progression of landslide risk assessment has been discussed by many authors (see, for instance Varnes et al., 1984; Cruden & Fell, 1997). Most notable is the procedure developed by Varnes et al. (1984); UN (1991) and Fell (1994). This consists of key steps (Table 3.1), which can be covered quantitatively or qualitatively.

Table 3.1.   Steps in a hazard and risk assessment procedure for landslides (Fell, 1994).

Action
1. Recognition of hazard (i.e. landslides of a certain type)
2. Estimation of magnitudes (volumes)
3. Estimation of the corresponding occurrence probabilities
4. Determination of elements at risk
5. Estimation of vulnerabilities
6. Calculation of specific risks
7. Calculation of the total risk
8. Assessment of risk acceptability
9. Mitigation of risk (if necessary)

It should be noted that steps 8 and 9 deal with risk management rather with than risk assessment, as explained in the definitions given in § 3.2.

The evaluation of risk involves the notions of hazard, vulnerability and cost of consequences. Theoretically, it is defined as follows (Leroi, 1997):

$$R = \sum_i A_i \times \left(\sum_j V_{ji} \times C_j \right)$$

where, R: risk; A_i: hazard i; V_{ji}: vulnerability of object j exposed to hazard i; C_j: cost or value of the object j.

Recent works (Cruden & Fell, 1997) propose the following multidisciplinary procedure for the quantitative risk analysis of slopes as a standardisation of the previous definitions:

– Hazard analysis: analysis of the probability and characteristics of the potential landslides;
– Identification of the elements at risk, i.e. their number and characteristics;
– Analysis of the vulnerability of the elements at risk;
– Calculation of the risk from the hazard, elements at risk and the vulnerability of the elements at risk.

In the same reference, the Working Group on Landslides, Committee on risk assessment, discussed the state-of-the-art on QRA for slopes and landslides and detailed issues requiring research and development. On the basis of these outcomes, the IMIRILAND team addressed the theme aiming at the following proposed improvements:

– *"The process (i.e. QRA) would be carried out by suitably qualified and experienced persons"*
  The researchers and technicians involved (representing different countries) are aware of the importance of these problems, as they have been working for several years on landslide risk analysis. They believe that a good methodology could be derived from a multidisciplinary approach taking into account the different know-how, knowledge and results obtained by the different disciplines. Another important aspect of the project lies in the implication of end-users who have to face every day practical land management problems;
– *"QRA does not have to be based on large quantities of objective data, it is equally valid to use subjective, as well as objective, data"*
  A first project step consists in defining a logical organisation of all site-specific available data (historical and technical-scientific material) to facilitate the activities of data collection, data analysis, data processing updates and the creation of different outputs; keeping in mind all information which could be involved in a QRA process. A special effort was made to link subjective information (or best interpreted data) with measured data and to create a new working relationship between two fundamental disciplines: Geology and Geotechnics.
– *"The inability to recognise a significant hazard and thus underestimate the risk [has to be considered]"*
  A very delicate step in the QRA process is related to the danger characterisation and hazard evaluation. From a simplified point of view, the calculation of risk is to be considered as a mathematical computation, which is relatively easy when the hazard corresponding to each danger, the elements at risk, their vulnerabilities and values are known. The main objective taken into account in the IMIRILAND project consists, therefore, in improving the hazard analysis, either by specifying the role and contribution of the different disciplines involved in such an investigation, or by defining criteria for the field observation and modelling, aiming at the quantification of the different scenarios.

Chapter 3 gathers the contributions of the IMIRILAND team on risk analysis and explains the practical risk assessment framework developed. Particular attention is given to pointing out simple but rigorous tools devoted to risk analysis, and detailing the relevant steps.

3.2 DEFINITIONS

Several definitions of the terms used in QRA exist, sometimes with differing meanings. On the basis of the results obtained by the IMIRILAND project, the following definitions of the most important terms, as used in this book, are given below, including the Italian, French, German and Spanish terms. Some concepts proposed by Cruden and Fell (1997) have also been developed:

Hazard (Pericolosità, Aléa, Gefährlichkeit, Peligrosidad)

For a given danger, the characterisation of hazard has to include the probability of occurrence within a specific period of time, the potentially involved area and the intensity of the damaging phenomenon.

Danger (Fenomeno o Processo naturale, Phénomène ou Danger, Gefahr, Peligro) is an existing or potential natural phenomenon, in this case the landslide, which may induce negative consequences. Danger characterisation (unlike danger identification) includes: a relative and qualitative non-temporal probability of detachment (predisposition), geometrical characterisation, geostructural and geomorphological constrains, mechanism description. It, thus, provides the basis for the construction of geomechanical models.

Intensity (Intensità, Intensité, Intensität, Intensidad) is a measure of the destructive potential of a landslide, based on a set of physical parameters, such as downslope velocity, thickness of the landslide debris, volume, energy and impact forces. Intensity can be expressed qualitatively or quantitatively. Intensity varies with location along and across the path of the landslide and, therefore, it should ideally be described using a spatial distribution function or an appropriate map.

Occurrence probability (Probabilità di accadimento, Probabilité d'occurence, Eintretenswahrscheinlichkeit, Probabilidad) is the probability that a certain danger will occur over a specified period of time. It can be evaluated in both quantitative and qualitative terms and it contributes to risk evaluation. Sometimes a relative probability is used as a non-temporal concept. In this case, it should be clearly specified as such by the term "predisposition".

Hazard Scenario is the spatial representation of a specific hazard situation and its development.

Vulnerability (Vulnerabilità, Vulnérabilité, Verwundbarkeit, Vulnerabilidad)

It is the degree of loss for a given element at risk, or set of such elements, resulting from the occurrence of a natural phenomenon of a given magnitude. It is usually expressed in relative terms, using words such as "no damage", "some damage", "major damage" and "total loss", or by a numerical scale between 0 (no damage) and 1 (total loss). It can be more significant if, instead of assuming only one vulnerability coefficient, its different specific components (i.e. physical, social, economic, environmental) are considered.

Elements at risk (Elementi a rischio, Eléments à risque, gefährdete Elemente, Elementos expuestos a riesgo) include any land, resources, environmental values, buildings, infrastructures, economic activities and/or persons in the area that are exposed to natural and technological hazards. The elements at risk can be quantified by a monetary value or by a relative value scale.

Consequence is the resulting loss or injury, or the potential loss or injury. It is the product of the value of the elements at risk and their vulnerability, and can be quantified if the element at risk is expressed as a value and if the vulnerability is expressed numerically. When a consequence is expressed qualitatively, it is sometimes referred to as a "consequence rating".

Risk (Rischio, Risque, Risiko, Riesgo)

Landslide risk considers all the consequences of the landslide hazard. Simply stated, risk is the product of the occurrence probability and the consequences.

Physical Risk, Social Risk, Environmental Risk, Economic Risk are the risks connected to each specific component of vulnerability. They are separately computed for each scenario.

Total Physical risk is the sum of the respective risks for all the scenarios, whenever relevant.

Total Economic risk is the sum of the respective risks for all the scenarios, whenever relevant.

Total Social risk is the sum of the respective risks for all the scenarios, whenever relevant.

Total Environmental risk is the sum of the respective risks for all the scenarios, whenever relevant.

Risk Analysis is the processing of all available information to estimate the risk to individuals or populations, property or the environment, from hazards. Risk analyses generally contain the following steps: definition of danger, hazard identification and risk estimation.

Risk Estimation is the process used to produce a measure of the level of health, property, or environmental risks being analysed. Risk estimation contains the following steps: frequency analysis, consequence analysis and their integration.

Risk Assessment is the process of risk analysis and explicit or implicit judgement, considering the various components of risk (physical, social, economic, environmental), to relate the risk perception with the results of risk analysis.

Risk Management is the iterative process of risk assessment and risk control, including active and passive mitigation measures or strategies, incorporating the risk acceptability whenever possible or politically acceptable.

3.3   QRA IN THE IMIRILAND PROJECT

To deal with the problem of risk evaluation from a quantitative point of view (QRA), we presuppose that, rigorously speaking and according to the QRA definitions:

– each risk component should be numerically expressed;
– each risk component should be spatially represented.

In the light of this assumption, the methodology elaborated for the IMIRILAND project is based on the development of the whole process by means of integrated and consequential phases, referred to as a simple mathematical operation with a matricial approach and a spatial representation of risk through GIS techniques. The use of GIS techniques can be considered as a relatively new approach in slope risk assessment, with a logical structure in independent layers of geo-referred information. Different layers related to a single risk parameter can be easily combined and cross each other in order to derive a spatial quantitative risk representation (Fig. 3.1).

Each component of hazard and risk, represented by quantitative values, as objectively as possible, can be considered as a numerical attribute of a specific spatial representation.

The risk assessment process can be summarized in three steps:

1. Hazard analysis
2. Consequence analysis
3. Risk computation.

The most important question to be asked in each landslide hazard and risk analysis relates to danger individuation and its spatial representation (present or potential instability zones). A preliminary step for the complete process consists, therefore, of a detailed available data collection in order to identify the dangers. It is important to mention that the majority of large landslides are characterised by interacting specific mechanisms relating preparatory and triggering causes. Each mechanism can evolve in a specific way and, therefore, can represent a specific danger. In other terms (Fig. 3.2), a large landslide could generate many dangers and, as a consequence, the relation between a large landslide and the induced dangers can be single or multiple. Indeed, one or more hazards may correspond to each danger. This complex correspondence derives from the fact that, in the hazard assessment phase, several methods (e.g. geomechanical run-out modelling) can be applied and, therefore, different results may be obtained in terms of the affected area and intensity. Such results must be critically examined in order to select the most representative hazard scenario.

The proposed methodology is based on the consideration of four hazard components:

– Danger characterisation,
– The area affected in relation to a possible landslide evolution,
– Process intensity,
– Occurrence probability.

16

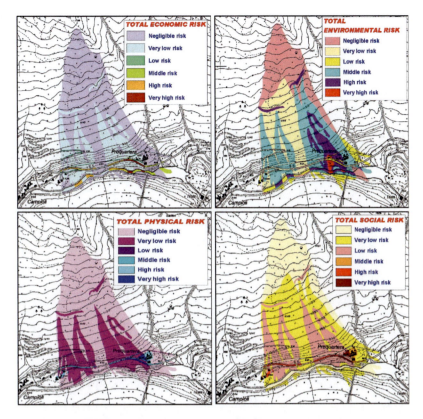

Figure 3.1.   Example of representation of different risk maps related to the specific vulnerability factors considered obtained using GIS techniques.

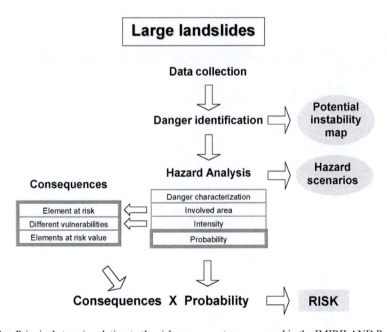

Figure 3.2.   Principal steps in relation to the risk assessment process used in the IMIRILAND Project.

17

The danger characterisation points out the most prominent geostructural and geomorphological constraints, allowing the construction of a geomechanical model. The affected area describes the limits of the possible evolution of a landslide of a given intensity.

The process intensity is a measure of the potential destructive impact of the landslide.

The occurrence probability is the probability that a certain danger will occur over a specified period of time. The hazard specific components supply, moreover, their contributions to assess the consequences of the landslide, in a specific way:

– The affected area permits the spatial identification of the elements at risk.
– The process intensity contributes to the determination of the vulnerability of the elements at risk.
– The occurrence probability, together with the consequences (the product of the value of the elements at risk and the vulnerability), defines the risk.

In order to perform the quantitative risk analysis, the process must first be divided into consecutive steps, then a matrix is used in which the row values increase vertically downwards and the column values increase horizontally to the right (Fig. 3.3).

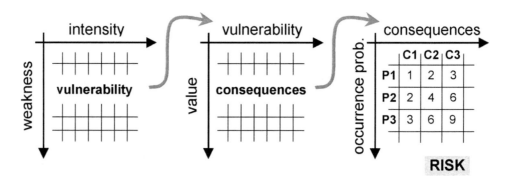

Figure 3.3.   Matricial risk computation scheme.

The matrix is built by multiplying each row value by the corresponding column value; thus a new risk-derived component is obtained. In order to simplify the computation for some risk parameters (i.e. the value of the exposed object), a relative index can be used.

Theoretically, the value of risk is computed by multiplying the probability that a certain phenomenon of a given intensity might occur by the vulnerability and by the value of the assets in question or of the exposed persons. This simple definition, introduced by Varnes (1978), can be refined, first, by considering that a significant probability can be established for each reference hazard scenario on the basis of the recurrence of historic or similar past events or, second, by considering the probability of triggering causes such as rainfall.

Different vulnerability components (i.e. physical, social, economic, environmental) are considered (see § 3.5.3).

The decision to resort to these different components arises from the difficulty in allocating comparable values to all of the different categories of goods and persons and, thus, in obtaining comparable risk scenarios. The presentation of risk related to several scenarios with a separate consideration of the different vulnerability factors allows a matricial vision of the different risk components (Fig. 3.4).

Several sums may thus be computed. One possibility is the sum of any risk for each scenario, connected to each specific component of vulnerability. This sum expresses the global risk value of the considered scenario, but requires the specification of a homogeneous scale, namely by the expression of risks in monetary terms. A second possibility is the sum of the respective risks for

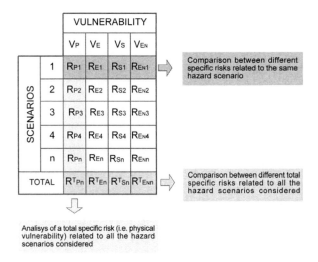

Figure 3.4. Matrix of individual risk values expressed as a function of different specific vulnerabilities and scenarios (P: physical, S: social, E: economic, En: environmental).

all scenarios considering specific components of vulnerability. This sum (total risk) is justified, provided that the objects affected as a consequence of one scenario will be repaired or rebuilt before the next scenario occurs.

This type of sum is not appropriate when one scenario is conditioned by the occurrence of another one (for example when a large rockfall is triggered by a preliminary boulder fall).

As previously mentioned, an extensive use of GIS has been made to perform such analyses and assessments. One of the advantages of using a GIS technique is the possibility of producing an iterative advanced risk mapping. The spatial overlapping of different risk components/factors permits a quite simple and useful evaluation of results. The possibility of organising the information into geo-referenced vectorial information layers allows the association of several attributes with each spatial entity.

In the following pages, a detailed description of the complete IMIRILAND methodology with its different phases and tools is presented.

3.4 DATA COLLECTION

Systematic collection, analysis and organisation of available data allows a better hazard and risk assessment including their scientific, technical, economical and political components and represents a first logical and necessary step in the consequential phases devoted to risk mitigation strategy.

The partners involved in the project generally have some knowledge concerning the sites which permits an appropriate danger identification, but the information is often very dispersed and available from different government sources; it may include maps, reports, aerial photographs, historical documents, etc. As spatial and temporal comparisons of the data are of great interest for landslide investigation and comprehension, it has been decided to define and compile a general scheme for data collection and description for each site.

A specific database structured in various sections to manage alphanumerical data, eventually coupled with GIS software to manage geographical data, has been developed. After the database compilation, landslide data organised in such a homogeneous manner are available for consultation. Data organisation is aimed at collecting all the information related to each landslide in order to allow an easy and fast comparison among:

- the geological framework (geology, structural setting, morphology, water conditions, hydrology, climate at regional and local scales) in which landslides have formed, where high risk conditions have been created,
- the type and rate of movement, landslide identification, history, morphometry, morphology, damage, etc.,
- the different investigations carried out to study each site, monitoring activities and instrumentation planning,
- the modelling approaches to understand the landslide kinematics and its possible evolution, either to forecast triggering/failure (if and when) or run-out (where) after failure,
- the stability analyses carried out to define the site "hazard",
- the vulnerability coefficients for goods and activities affected,
- the assessment of risk scenarios,
- the landslide hazard and risk reduction approaches.

In each of these sections, there is the possibility to insert and describe a great amount of data, according to different degrees of detail, using either look-up tables, numerical fields or textual fields.

The database structure adopted allows the following actions:

- to check the reliability of the available data, to individuate which items should be considered in the subsequent phases, as much attention has been paid to data quality (such as source type, scale, date) and data format (for example: a distinction has been made for in cartographic data between traditional maps, available only as images, and numerical ones, sometimes created with GIS software);
- to compare landslide studies and approaches easily, underlining differences, lack of information or redundancy in characterising different landslide typologies in different countries;
- to analyse available data in detail, to interpret and process information easily according to different representations or different aims;
- to extract all the information necessary to prepare exhaustive reports containing a concise description of the main features/factors identified in the studied sites, including information on land use and elements at risk (either in qualitative or quantitative terms, i.e. number and characteristics), a preliminary vulnerability analysis, a preliminary (often qualitative) evaluation of the expected scenarios, an indication of adopted regulations, and a critical analysis of what has been done and what could be done further.

3.5   HAZARD ANALYSIS

In order to assume a rigorous risk analysis which will supply useful elements to decision-makers, a complete and detailed hazard analysis must be carried out. With this aim, the IMIRILAND project develops a methodology based on a multidisciplinary approach for the hazard analysis, at the same time showcasing the valuable contributions of several professional experts (engineering geologists, geologists, geotechnical engineers.) The project has also benefited from experiences from different social, cultural and political contexts (Spanish, French, Italian, Swiss and Austrian).The different types of analysis used in the present methodology are detailed below; each one supplies elements of different weight aiming at the determination of the various hazard components (Table 3.2).

Table 3.2. Contribution of the various disciplines to the hazard component evaluation.

Analysis tools	Aim	Know-how	Contribution to quantitative risk analysis
Geological – structural model	To understand the basic hierarchical and geometrical relationships between the main regional structural features (*around site*) and geostructural landslide discontinuities (*on site*).	– Definition of landslide geometry and discontinuity roles. – Definition of typology and lithology of the materials involved. – Understanding of the kinematic mechanisms.	– Basic outline of the situation. – Input data for geomechanical models (triggering and run-out).
Geomorphological model	Qualitative evaluation of the typology, geometry and mechanisms of evolution of unstable areas.	– Landslide typology classification. – Spatial-temporal identification of the landslide evolution. – Characterisation of the slope in its initial and present stages, according to different dangers (sliding mechanisms, mass qualitative behaviour, dimensional parameters, qualitative nontemporal probability: a sort of predisposition).	– Outlines a first definition of hazard scenarios. – Adds information to the geological-structural model. – Input data for geomechanical models (triggering and run-out).
Historical analysis (data collection)	To collect information about damages, reactivation frequency, general instability evolution.	– Information about movement typology, occurrence, effects and damages. – Definition of the relationships between instability causes and related phenomena (e.g. rainfall and landslide).	– Outlines the landslide phenomenon. – Establishes a spatial-temporal frequency. – Refines the results obtained through other surveying or monitoring methods.
Monitoring investigation	To collect quantitative data about displacements, affected areas, trend of the movements.	– Spatial-temporal representation of the landslide activity. – Identification of the sensitivity of measured parameters with respect to the variation of boundary conditions. – Comparison of different measurement typologies.	– Refinement and calibration of geomorphological and geomechanical models.
Geomechanical modelling: triggering	To carry out slope stability analyses, starting from the geometrical and physical-mechanical characteristics of the involved mass (through different numerical methods).	– Failure mechanism representation. – Sensitivity analyses. – Back-analyses of the recorded movements.	Validation of the previous qualitative hypotheses (geological and geomorphological approaches) from a mechanical point of view, with the help of displacement data.
Geomechanical modelling: run-out	To evaluate the run-out areas and the intensity of the mobilized rock mass.	– Post-failure behaviour of the rock mass. – Sensitive analyses. – Back-analyses.	The results permit the definition of the mobilised mass distribution and the phenomenon magnitude (in terms of mass velocity or energy) for several scenarios.

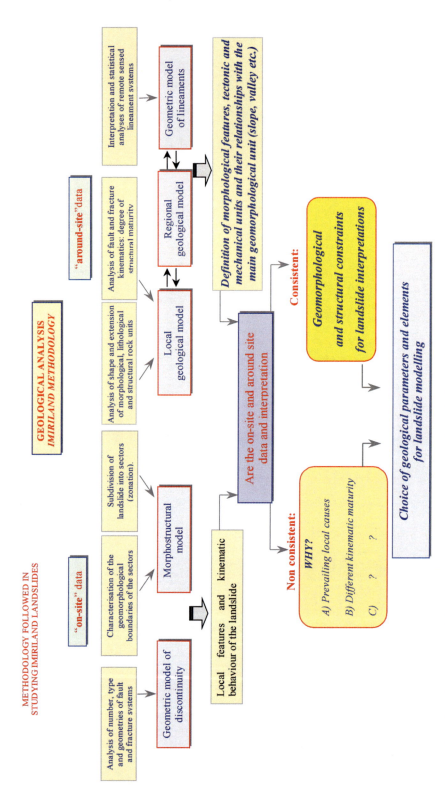

Figure 3.5.   Methodology followed in studying the geological aspects of the IMIRILAND landslides.

22

3.5.1 *Geological analysis*

In the framework of IMIRILAND project, the first fundamental step in the definition of hazard (Fig. 3.2) is the danger characterization, which leads to a connotation of the problem analyzed either from a geometric point of view or from the point of view of the qualitative understanding of the possible failure mechanisms. This phase contributes, moreover, to qualifying the susceptibility to instability, i.e. the potential instability phenomena. In other words, the danger characterisation supplies a first schematisation of the studied phenomenon, aimed at orienting and supporting the numerical modelling. In particular, the danger characterisation of large landslides requires the consideration and processing of the following items:

– geometry of the potentially moving mass: volume, area, depth etc.;
– geomorphological constraints;
– geostructural constraints;
– comments on the assumed mechanism of deformation at or before failure, with its boundary conditions;
– relative and qualitative non-temporal probability of failure (predisposition).

In order to treat the aspects mentioned in a rigorous way, a specific approach is required, focusing proper attention not only on the landslide body itself, but also on the local and regional geological context. The local analyses provide essential information, but they are not really useful for establishing relations between the "on site" and "around site" geological configurations. The regional analyses can give a knowledge of these relations, which are necessary even for a tentative assessment of landslide hazard conditions, based on the lateral extrapolation (for instance over an entire mountain slope) of the recognised geological features. Geological analyses of landslide sites cannot, thus, concern only the "on site" geomechanical properties of rocks and fractures, but relatively large surrounding areas ("around site") should also be investigated to allow the understanding of the basic geometrical relations among the main structural units and discontinuities. In particular, the around site analysis becomes a necessity when the on site rock mass is so damaged as to prevent the reconstruction of the original structural framework of the landslide. This will be compared with those geomorphologic elements that are thought to be strictly controlled by the structural setting of the bedrock.

This is an integrated methodological approach, where meso and macroscale data, regional models and remote sensed data are linked. This approach is summarized in a flow chart (Fig. 3.5) and will be described in the following paragraphs.

3.5.1.1 *"On site" analysis*
Geometric model
This results from the local characterization of the geometry and physical properties of discontinuity systems (type, number of systems, orientation, persistence, type of termination and cross-cutting relations), in order to define which are the most recurrent shapes and dimensions of rock blocks, probable kinematics and failure mechanisms. Exhaustive data collection is not always possible on landslide sites due to logistic difficulties in sampling, paucity of safe rock masses etc. In this case, representative sites will have to be chosen within the landslide body to describe the fracture network. If representative zones cannot be found, a synthesis of single observations has to be carried out to understand the fracture network.

Morphostructural model
In recent years, investigations on large landslides have more frequently concerned the whole slope area or even the entire valley on which they are located, in order to discover common (ubiquitous) elements that could have driven the gravitational instability.

A morphostructural model of the landslide is then defined, where the correspondence between morphological and structural elements is mainly sought based on the assumption (indeed very often verified, in the Authors' experience) that the morphological features of the landslide are the

23

surface expressions of the structural subsurface elements (such as schistosity, faults and fractures, etc.) of the bedrock.

However, the morphostructural approach also takes into account the geomorphologic context due to physical and chemical erosion agents.

This approach consists of:

- Photo-interpretation of aerial photos at different scales and/or analyses of multi-temporal and multi-spectral satellite images, in order to achieve a general knowledge of those morphological features that are peculiar for each slope or valley studied (for instance those related to deep rock slope deformation or glacial erosion). Successively, more detailed interpretations (scales 1:10'000; 1:20'000) are carried out to investigate the landslide site properly at a scale where peculiar features such as scarps, debris, trenches and undulations can be discerned.
- Field work in order to verify the relations between the features observed on aerial photos and the structural setting of the site, as well as to improve the early work hypotheses with new field data. This allows the definition of a morphostructural model of the landslide, i.e. the location of the main boundaries, the definition of landslide sectors and the definition of distinct kinematic behaviour for each sector (zonation of the landslide).

The integration of geometric and morphostructural models allows the definition of the local features and kinematic behavior of the landslides.

3.5.1.2 *Around site analysis*
These allow a better definition of the local geological model (Fig. 3.5) that refers not only to the landslide site, but also to the surroundings (slope or valley) and which is mainly based on the knowledge of:

- shape and extension of morphological, lithological and structural rock units. These must be defined, since differences in their lithological properties induce different mechanical behavior of the rock units. This is the case, for example, of the Rosone landslide where white mylonitic schists seem to control the triggering of the gravitational instability (see Chapter 5). Tectonic units can also be defined to recognize sectors subjected to instability, on the basis of their different fracture intensity. The landslides that have been studied in the IMIRILAND project are, in fact, placed in regional geological settings characterized by distinct tectonic units which have controlled the gravitational instability in a specific way.
- fault and fracture kinematics and degree of structural maturity. Peculiar features such as thickness, type of fault rock, persistence and subsidiary fracture networks correspond to each kinematic group of faults. The knowledge of these features provides constraints for defining rock block shapes which can be compared with the morphostructural features of the landslide body. Structural data are also organised into genetically coherent groups (structural associations[1]). These represent key elements to model the geometry of faults and fractures. This procedure leads to the identification of one (or more) basic shapes of rock units, each one characterized by distinct cross-cutting relations of their bounding surfaces. This geometry can be reproduced graphically or by numerical modeling. An example of the application of basic rock unit has been applied to the bedrock of the Cassas landslide to infer the type of instability mechanism (see Chapter 4).

The local geological model has to be compared with more recent understanding of the regional geological framework (regional geological model) in order to verify if the local interpretations can be placed in a wider sphere.

Extrapolation of local data to a larger scale (and vice versa) is also allowed by using remote sensed lineaments (interpretation of geometry of lineament: orientation, density distribution, statistical analysis). The geometrical and hierarchical relations of the lineament systems (geometric

---

[1] the structural association is a "geological tool" for the correlation of the different discontinuity elements with different geometry, kinematics and size into genetically coherent groups (Vialon et al., 1976; Davis and Reynolds, 1996).

model of lineaments) can thus be compared (qualitatively and quantitatively) to the architecture of the regional faults (Morelli, 2000; Morelli & Piana, 2003). Only when this comparison shows good similarity, can the lineament systems be used as geological structures. The lineaments present the advantage of being more homogeneous than field data, of being suitable for use to greater advantage in an automatic and computerized way for statistical and numerical analyses, and of being characterized by a number of elements that are statistically more representative and refer to a context of wide extension requiring relatively short processing times. This methodology has been applied on the Ceppo Morelli and Cassas landslides, where geometrical and hierarchical relationships, observed at the outcrop scale, are also confirmed by the statistical distribution and geometric characteristics of the lineament systems at both the regional and slope scales. This correspondence allows the extrapolation of the geometry and hierarchy of the structural elements to different scales (see Chapters 4 and 7).

The integration of local and regional geological models, as well as the geometric model of lineaments, allows the definition of tectonic and mechanical units and their relationships to the main geomorphological unit (slope, valley, etc.) to which the landslide belongs. Consequently, the landslide can be subdivided into sectors, each one characterised by distinct morphology, dimensions, fracture density and geometry. The danger intensity could thus depend on: the position of the sectors with respect to the slope's shape and drainage network, the lithostratigraphic setting, fracture density and intensity, structural maturity and kinematics along the sliding and/or failure surfaces.

Comparison between "on site" and "around site" data and interpretations

The geometric and kinematic features of the landslide are compared with the main morphological and structural features (tectonic and mechanical units) of the slope, watershed or valley to which the landslide belongs, in order to verify if the inferred geologic context, to be related with the peculiar instability conditions within the landslide body, is recurrent (or not) over the whole slope.

When these conditions are satisfied, it can be assumed that the "on site" and "around site" interpretations are consistent. In this case, there is the possibility of extrapolating the "on site" interpretations to broader areas or vice-versa.

In case they are non-consistent, it can be suggested that the gravitational instability could be triggered by peculiar features of the landslide body, for example by individual faults, mechanically weakened layers, geomorphological features or, in the case of a very mature landslide where the kinematic evolution is no longer controlled by pre-existent structural features.

Contribution to the quantitative hazard analysis

The aim of the geological and structural approach is to understand the basic hierarchical and geometrical relationships between the main regional structural features (*around site*) and faults and fractures on the landslide site (*on site*). This understanding leads to a zonation of the landslide body where each zone could have distinct mechanical features and kinematic behaviour. The contribution to quantitative risk analysis consists in providing constraints for geomechanical models as well as a qualitative interpretation for linking different kinds of data, such as geomorphological, geophysical, geotechnical and historical. The aim of the integrated geological and morphostructural approach is to evaluate, in a qualitative way, the typology, geometry and evolution mechanisms of the unstable area. This approach leads to a classification of landslide typology, a spatial-temporal identification of the landslide evolution and a characterization of the slope in its initial and present stages, according to different aspects (sliding mechanisms, mass qualitative behavior, dimensional parameters, relative probability of occurrence). The contribution to quantitative risk analysis consists of a preliminary definition of hazard scenarios and of inputs for the geological-structural and geomechanical models (triggering and run-out).

3.5.2 *Historical analysis*

Two basic sources of documents are available for the reconstruction of past events. First, the facts which can be found in the archives of public or religious authorities, as well as the synthetic analysis available in some textbooks (Heim, 1932; Abele, 1974; Eisbacher & Clague 1984). Then, aerial and ground photographs and ancient maps allow a detailed interpretation of the evolution of the slope over the period separating the two documents related to same area (Fig. 3.6).

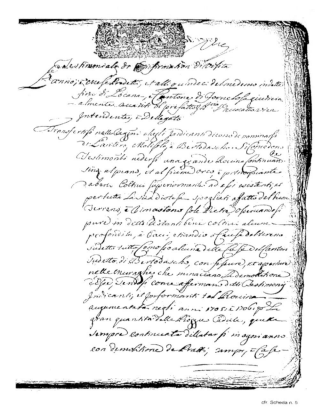

cfr. Scheda n. 5

Figure 3.6.   Document dated 1706 mentioning the occurrence of a landslide in Bertodasco during autumn 1705 in the area of the Rosone site (Italy).

Ancient cadastral maps, defining the limits of the plots of land which indicate the location of old houses, also help establish past rates of movement, as in the case of the La Frasse landslide in Switzerland, where ancient maps date back to 1768 (Bonnard, 1984).

The data obtained from archives are often overemphasised and reinterpreted, especially if they are ancient. Furthermore, the reading itself of archives requires specific practice to understand the context and exact circumstances of the past disasters. At the same time, such investigations cannot be made only by historians, as the interpretation of the recorded stories requires a detailed understanding of the landslide characteristics. Finally, the exact dates of the event are not always correct, so that it is difficult to correlate the gathered information with other events related to the disaster.

The comparison of aerial photographs allows a very precise identification and quantification of displacements, provided some well-identified fixed points exist around the landslide. However, the exact date and circumstances of the observed disaster cannot be determined. Moreover, such information of sufficient quality has only existed for about 50 years. In very particular situations, related to recent events, some television archives are very useful to understand the mechanism and to calibrate the modelling of the trajectories. Finally, the interviews of people who have lived near the disaster area and who have a good memory of major past events are useful to obtain details of the landslide sequence, but the testimony has to be thoroughly cross-checked with the in situ observations to be sure that the information is reliable. As an example, in the case of the Ceppo

Morelli landslide (Valle Anzasca, Piemonte, Italy), historical analysis of archives and recent field observations supplied the following results:

- Rockfall: 1940, Oct. 1971, Apr. 1977, Oct. 2000, Jun. 2002
- Movement of the entire mass (or of a large part of it): 312, 843, 1816, 2000

The reliability of all of this information is good, even if it is based on very ancient documents, as the affected valley was heavily used in the Middle Ages.

3.5.3 *Monitoring investigation to assess long term behavior*

Only past information collected by several monitoring techniques may be significant to assess the long term behaviour of large landslides. Through the new measurement of old topographical landmarks which have been surveyed many times since the beginning of the twentieth century (at least in Switzerland), for instance, it is possible to reconstruct the total movement of some landslides, as well as the evolution of the displacements of various points with time. The most prominent case, related to the Lumnez landslide (Canton of Graubünden, Switzerland), in which the spires of the churches in five villages have been surveyed regularly since 1887, has permitted the confirmation of the regularity of the movements for most of the surveyed points for over than a century (Bonnard & Noverraz, 2001). In the case of Sedrun (Switzerland), as detailed in Chapter 8, the regular check of the topographic points, first for maintenance objectives and then for safety reasons, permitted the observation of a marked and progressive acceleration from 1973 onwards.

The precision of the displacements obtained depends on the quality of the initial determinations, as well as on the structure of the survey net, but even for small velocities (<1 cm/year) measured over a period of more than 50 years, the results are quite significant.

The ancient cadastral maps still available in archives, indicating the limits of plots of land as well as the location of buildings, can also be compared to current maps, as well as to recent field surveys in landslide areas. This technique, used for the La Frasse landslide, allowed the determination of total horizontal displacements between 1768 and 1861 and from 1861 to 1981. Such a procedure requires that several fixed points can be identified on the cadastral maps; these have to be digitised and adjusted one by one or through global compensation, as they generally cover only a limited area.

The precision checked by the comparison of the relative position of fixed points indicates that the maximum error encountered in XVIII century documents is 2–3 m, whereas it reaches 1 m on maps of the end of the XIX century. As the movements over more than two centuries reach 25 m the zone of the village, the accuracy is quite acceptable.

A final source of information is related to the periodic check of the position of infrastructures or transportation systems, such as roads, railway lines or posts for cable cars. These data are often very significant because they refer to a complete transversal (e.g. road) or longitudinal (e.g. cable car) profile of the landslide.

The long term movement data, as well as piezometric readings (which are rarely available over a long period), are fundamental to establish the input parameters for the mechanical modelling and its calibration.

3.5.4 *Geomechanical modelling: numerical triggering models*

Numerical triggering models, when used in a hazard assessment framework, are often supposed to allow the determination of the occurrence probability in time related to a catastrophic landslide, on a mechanical basis.

The probability of failure in time of an existing slope is nil, unless a variation in time of the existing situation can be envisaged, for instance, the decrease in time of shear strength parameters or the increase in pore pressure in the slope. As such variations cannot be predicted in a quantitative way, only a non time-related probability is usually computed (see, for instance Einstein, 1988 and Scavia et al., 1990).

Unfortunately, the evaluation of such a probability is impossible when large landslides are studied, as it depends on the knowledge of a number of both geometrical and mechanical parameters, including their variability and that of the boundary conditions (for instance pore pressure distribution in the slope, initial state of stresses), which are very approximately known in reality. For these reasons, numerical triggering models are used in the framework of the IMIRILAND project as valuable tools to verify, from a mechanical point of view, the hypotheses made through the geo-structural and geomorphological analyses, and as well as the monitoring results, about the mechanism and the volume of the landslide. As a consequence, if the numerical triggering results do not match the geological hypotheses, a new drafting of the geo-structural and geomorphological models is required. If the numerical results are in good agreement with the geological hypotheses, on the basis of the triggering analyses, more information about the shape and the dimensions of the potentially unstable volume can be obtained and used as further input for the run-out analyses (see examples in Chapters 5, 6, 7 and 9).

In order to obtain significant results from these models, it is necessary to incorporate the main geo-structural data precisely in the construction of the model, to choose to work in 2D or 3D conditions according to the scope of the analysis and to refer to a discontinuous or equivalent continuum approach, as appropriate.

Numerical triggering models can indeed represent the rock slope as:

1. a discontinuous medium, where localized discontinuities (such as joints or interfaces) are introduced as specific elements into the geometrical model, allowing them to be taken into account for the analysis;
2. an equivalent continuum model where the discontinuous medium is substituted for the analysis by a continuous one, the material properties of which are modified, so that its behavior is equivalent to the discontinuous, real medium.

The choice of one of these methods depends on the size of the structure in relation to spacing, orientation and strength of the discontinuities and on the imposed stress level.

The main problem with the discontinuous approach is to determine the location and geometry of the natural discontinuities; the main problem with the continuum equivalent approach is the evaluation of the mechanical characteristics of the rock mass, which cannot be determined in the laboratory.

The most widely used numerical methods are the limit equilibrium method (LEM), the finite element method (FEM), the boundary element method (BEM) and the finite difference method (FDM).

Limit equilibrium methods do not allow the evaluation of stress and strain in the slope, so they are not able to reproduce the crucial role played by deformability in large landslides. Finite element methods and finite difference methods are very suitable when the slope is made of different materials, but the time of computation can be cumbersome in 3D conditions. Boundary element methods are not convenient when more than one material must be taken into account, but are very useful in 3D conditions. In the IMIRILAND project, three numerical codes were applied to the study of the triggering in the selected sites: DRAC (FEM), MAP3D (BEM) and FLAC3D (FDM).

The combined use of these 3 methods has permitted carrying out analyses in both 2D and 3D conditions, on the basis of both discontinuous and continuous equivalent approaches and adopting elastic, viscous and plastic constitutive laws.

A brief description of the three numerical codes is presented in the following sections.

3.5.4.1 *Code DRAC*

The computer package DRAC (Prat et al., 1993) is an analysis system based on a finite element program designed for the analysis of rock mechanics and other geotechnical problems.

It is a very flexible package, which can be used with a great number of boundary conditions and different types of materials. The program has an open structured architecture, allowing new modules serving both future engineering needs and new computational techniques.

DRAC can be used to solve two and three-dimensional problems. The program can perform linear and non-linear analyses, for which several techniques, such as line-search, are available, as well as finite displacement analyses. Time-dependent analyses are also possible. The element library has been built with the needs of the problems specific to rock mechanics in mind, for instance, with the development of a zero thickness joint element to model discontinuities.

This interface model, a key constitutive model for joints, interfaces and other discontinuities, is a simple but general model for normal/shear cracking based on fracture mechanics recently developed at UPC (Carol et al., 1997). It is defined in terms of the normal and shear stresses on an average joint plane, and the corresponding normal and shear relative displacements across the discontinuity.

Besides the finite element program itself, there are a number of auxiliary modules that complete the package: these are the pre- and post-processing programs and the libraries (mathematical, finite elements, material models, and solver). An external module allows the analysis of water flow problems within the rock mass. The package includes a post-processing graphics program specifically designed to handle the output of the finite element program in two and three dimensions, including a module to estimate the safety factor for instabilities due to the presence of discontinuities in the rock mass.

3.5.4.2 *Code Map3D*
The commercial code Map3D is based on BEM and in particular on the Displacement Discontinuity Method (DDM), an indirect boundary element methodology. Map3D, developed and supported by Mine Modeling Pty Ltd. (Victoria, Australia) was first devoted to simulate underground excavations and mining sequences, but has recently been applied to slope stability analyses in 3D conditions.

The code allows the easy simulation of natural discontinuities in the rock slope. Requiring the discretisation of the problem boundary, it lends itself very well to analysis in 3D conditions, even when a visco-elastic behavior of the discontinuities is adopted.

3.5.4.3 *Code Flac3D*
FLAC^{3D} (Fast Lagrangian Analysis of Continua in 3 Dimensions) is a powerful 3D continuum code for modeling soil, rock and structural behavior, and can be used interactively or in a batch mode (ITASCA, 1997). It is a general analysis and design tool for geotechnical, civil and mining engineers. FLAC^{3D} can be applied to a broad range of problems in engineering studies. The explicit finite difference formulation of the codes makes it ideally suited for modeling geomechanical problems that consist of several stages, such as sequential excavation, backfilling and loading. The formulation can accommodate large displacements and strains and non-linear material behavior, even if yield or failure occurs over a large area or if total collapse occurs. FLAC^{3D} uses a mixed discretisation scheme to provide accurate modeling of plastic collapse and flow. Materials are represented by polyhedral elements within a three-dimensional grid that is adjusted by the user to fit the shape of the object to be modeled. Built-in primitive shapes allow the generation of a variety of complex geometries. FLAC^{3D} has a built-in macro programming language called FISH that allows users to customize their analyses to suit their needs. Constitutive models may be defined using FISH, as can loading patterns, servo-control of test conditions, and grid generation sequences.

3.5.5 *Geomechanical modelling: numerical run out models*

In the previous chapter, it was shown how a volume and a qualitative probability of failure can be assessed for a certain danger through geomorphological, geo-structural, historical and triggering numerical analyses.

In order to provide a scenario for the danger under consideration, the analysis of the run-out of the rock mass volume down to the valley must be carried out. The results of such analyses are the area of propagation of the landslide and its intensity (see examples in Chapters 5, 6, 7 and 9).

29

As a function of the triggering mechanism and of the entity of the involved rock mass volume, rock mass movements can be subdivided into three broad categories:

– Rockfalls, blocks of rock fall freely from cliffs or mountainsides and accumulate at the base in talus slopes. Chemical and physical weathering plays a large role in loosening the blocks. Freeze-thaw cycles separate blocks along joints and the expansion of the water when it freezes to ice is enough to trigger the fall.
– Rockslides, large masses of bedrock slide as a coherent material along downward sloping joint planes or bedding surfaces.
– Rock avalanches, large masses of broken rock (millions of cubic meters) flow as a viscous fluid at high velocities (tens to hundreds of km/h).

In rockfall cases, the run-out can be simulated through the computation of the trajectory of a single boulder, jumping and rolling down to the valley bottom. Several non-interacting boulders can be considered.

Many computer codes are used for the evaluation of boulder trajectories, based on various simplifications of the phenomenon (see, for instance, Scioldo, 1999).

As the IMIRILAND project mainly refers to the study of large mass movements, the run-out methodologies described in the following present a possible approach for the analysis of rockslide and rock avalanche phenomena. Attention is particularly given to rock avalanche behavior, where the rock mass can be simulated as a massive, flow-like motion of interacting fragmented rock (Hungr, 2001).

3.5.5.1 *Classification of the run out models*

Run out prediction methods can be grouped into two broad classes (Hungr, 2001):

– Empirical methods
– Analytical methods

Empirical methods provide a quick estimate of the total run-out distance. They are based on simple correlations between the volume of the mass involved in the movement and the geometrical parameters of the slope. Empirical laws are then worked out through back-analysis on documented cases. As a function of the aspect taken into consideration, these methods can be further subdivided in relation to:

– Path profile
– Deposition area
– Mass balance

Analytical methods should be able to reconstruct the motion of every portion of the landslide mass in space and time, depending on path topography, characteristics of the material, both within the sliding mass and along the path, internal stresses within the mass and entrainment or deposition of material. We may distinguish among:

– Lumped mass methods
– Continuum mechanics methods
– Granular methods

Lumped mass models are unable to account for internal deformation: while they may provide a reasonable approximation for the movement of the center of gravity of the landslide, they cannot simulate the motion of the flow front, which is often the most important aspect in a run-out analysis (Evans et al., 1994).

Continuum mechanics models use techniques developed for the analysis of fluid flow in open channels. There are, however, important differences between fluids and earth materials, even when the latter are imbued with water and highly disturbed. It is necessary to find an "equivalent" fluid, whose rheological properties are such that the bulk behavior of the flowing body is able to simulate the expected bulk behavior of the prototype landslide.

Continuum mechanics analyses adopt an Eulerian or a Lagrangian approach to the study of landslides. The Eulerian approach correlates the calculation to a reference frame fixed in space and is particularly suitable for slow movements. Given the highly unsteady nature of landslide movements, it is preferable to use a moving Lagrangian reference frame (Potter, 1972).

Granular models consider the moving mass as the result of the interaction of single particles.

To model the possible run-out behavior of a landslide phenomenon, two methodologies have been applied:

- The Particle Flow Code (PFC) from the ITASCA (1999a and 1999b) Consulting Group in cooperation with FLAC (Itasca) as the Ball Wall model and stand alone as the All Ball model;
- The computer code DAN (integrated with the rockfall program ROTOMAP – Geo & Soft International, 1999 and 2003) based on a continuum mechanics approach.

3.5.5.2 *The Particle Flow approach*
FLAC is a continuum mechanics finite difference code that, used with the shear strength reduction technique, allows the estimation of the failure mechanism, the distribution of displacements and the critical parameters of the slope in three dimensions. The deduced failure surface of the slope is the basis for the run-out analyses carried out with PFC and DAN.

PFC is also a finite difference code which models the mass as an assemblage of circular (2D) and spherical (3D) elements among which exist bonds and friction; after the breaking of the bonds, propagation is caused. To generate displacements and rotations, the laws of motion take into account both the contacts of the particles and the boundary conditions. In general, this is not a rockfall program but it can be adjusted appropriately in order to generate the run-out mechanisms in a more detailed way.

The computer program PFC^{2D} and PFC^{3D} have been used as an All Ball and a Ball Wall model; in the PFC Ball Wall model the detached rock mass is modeled by balls, while the bedrock is simulated by linear (2D) and planar (3D) elements. Therefore, in the Ball Wall model, an estimate of the failure mechanism is needed as an input parameter. As previously explained, these estimations were carried out with the aid of FLAC.

In contrast to the Ball Wall model, in the All Ball model, the relatively stationary bedrock, the failure mechanism of the slope and the detachment mechanism are modeled by balls. This allows the all in one calculation of failure mechanisms, detachment and run-out. For calculating the failure mechanism, however, quite a demanding calibration of materials is necessary; this is particularly due to the fact that the modeling of the bedrock by balls determines a rather rough surface. When comparing the volume of the detached rock with FLAC, both methods correspond closely.

One limitation of PFC is due to the fact that it is not able to model the influence of water (e.g. pore pressure). On the other hand, as the balls are supposed to model simple rock blocks, and as the number of balls is limited by the computation capacity, a certain restriction exists in the modeling of large landslides of dozens of millions of m^3. Finally, the All Ball model allows an all in one calculation but the computational effort required obliges the modeling of the detached mass with less and bigger balls than those used in the Ball Wall model.

3.5.5.3 *The DAN approach*
The dynamic analysis model DAN (Hungr, 1995), was especially developed to simulate the motion of flows, flow-like slides and avalanches. In particular, it implements a one-dimensional Lagrangian solution of the equations of motion, formulated in terms of an open rheological kernel, in which the type of material (plastic, frictional, Newtonian laminar, turbulent, Bingham, Coulomb viscous, Voellmy) can vary either along the path or within the sliding mass.

The sliding mass is represented by means of a number of blocks (mass blocks) in contact with one another that can deform freely and retain fixed volumes of material in their downward path.

The main data used as inputs for the DAN code are the slope profile and the top profile of the initial mass. In particular, to perform a three-dimensional analysis, the width of the path must be determined at each path profile point by assuming that the entire envelope considered is enveloped by the moving mass.

Hungr's approach to width definition stems from a visual analysis of slope morphology and hence the lateral boundaries of the area involved are defined on the basis of experience with real cases.

In the proposed methodology, this evaluation has been replaced with an analysis conducted with ROTOMAP (Geo & Soft International, 1999), a three-dimensional rockfall program which makes it possible to define envelopes as a function of the trajectories of the rock blocks released from the toe of the unstable area.

The final outputs are: run-out distance and run-up on the opposite slope, landslide velocity and expiring time, limits of propagation and depth of the displaced mass once it stops.

The main limitation of the Hungr model is due to the fact that it is approximate, as it involves the reduction of a complex and heterogeneous three-dimensional problem into an extremely simple formulation. The simplicity of the model is, however, an advantage in making possible an immediate and rapid numerical simulation of real cases.

3.5.6 *Occurrence probability assessment*

Risk analysis requires the assessment of occurrence probability in time; however time is of great uncertainty in landslide hazard/risk analysis. Geomechanical models do not help in this case.

Occurrence probability, which can be defined as the chance or probability that a landslide will occur, can be expressed in relative (qualitative) or probabilistic (quantitative) terms. A rigorous, quantitative, procedure could include the application of mechanical-probabilistic methods, taking into account the uncertainty of all of the geometrical and mechanical parameters (slope geometry, shear strength, piezometric pressure, etc.). In the case of large landslides, however, a similar procedure cannot be applied because the uncertainty in the definition of the parameters involved is too large and it is not possible to define any reliable variability for them; finally, the number of events is not sufficient to obtain extreme values by statistical extrapolation.

An alternative procedure takes into account the relationship between the intensity of triggering events, such as rainfall or earthquakes, and the landslide occurrence, through the use of empirical correlations or more sophisticated methods (i.e. neural network methods in Mayoraz et al., 1997). In any case, it is still difficult to relate a given magnitude of a landslide phenomenon to a given intensity of the triggering factor. In particular, in the case of rainfall-induced slides, the real cause is often a combination of long antecedent rainfall and intense spells of rain. Often, the lack of data makes this approach inapplicable. Furthermore, from a rigorous point of view, it is not always possible to determine a statistically-computed probability value for most large landslides due to their non-repetitive behaviour.

The probability of sliding can be estimated using semi-quantitative methods related to the assessment of the historic record of sliding in conjunction with detailed geological and geomorphological characterisation of the investigated area and considering the past experience of similar events.

If several historical records exist on different instability process activities (different dangers) and if, in the studied area, a non-negligible hazard exists, the historical approach seems much more suitable to obtain information related to the frequency of failure. In its simplest form, this method consists in recording any type of sliding which occurred in the past and in its detailed description (type, volume and area involved, possible localization, etc.). Generally a long representative period of recording is needed in order to consider all phenomena types and a sufficient amount of data. Historical approach results can be quantified using analyses of aerial photographs.

A systematic multi-temporal interpretation of aerial photographs (it is often possible to obtain photographs taken for the same area 10, 20 or 50 years earlier, allowing a stereoscopic viewing) indeed permits an analysis of the factors leading to instability. It allows a subdivision of the investigated area into "homogeneous" zones by using interrelationships among geology and geomorphology. Morphometric multi-temporal analyses of landslides, combined with geological field investigation and surface observation, provides:

– surface and temporal characterisation of instability processes;
– quantitative movement displacement evaluation (between two different dated stereo pairs);
– temporal activity identification and interpretation of historical slope evolution;
– possible future evolution;

- semi-quantitative volume estimation;
- possible landslide-prone areas;
- principal sliding mechanism interpretation;
- reasonable semi-quantitative estimation of the probability of sliding and occurrence frequency for some repetitive instability processes.

This mixed method (historical and geomorphological approaches) can be a useful way of estimating the annual average probability of sliding, despite its limited accuracy. It also depends on the skill and experience of the persons carrying out the studies.

For instance, as mentioned previously (see Chapter 7 concerning the case of the Ceppo Morelli landslide/Valle Anzasca, Piedmont, Italy), an historical analysis supplied the following results:

- Rockfall: 1940, Oct. 1971, Apr. 1977, Oct. 2000, Jun. 2002
- Movement of the entire mass (or of a large part of it): 312, 843, 1816, 2000.

At this point, it is possible to link together each danger, characterised by geologic-structural and morpho-dynamic features and an occurrence probability class. An occurrence range of time has been obtained for each of the three scenarios (danger 1: rock fall, danger 2: movement of approx. $1 \times 10^6 \, \text{m}^3$, danger 3: movement of the entire mass, approx. $5 \times 10^6 \, \text{m}^3$) considering the minimum and maximum time intervals between two consecutive recorded events. On the basis of their mean value, a frequency has then been calculated in terms of events per year (Table 3.3). The computed values are to be considered as relative values.

Table 3.3. Example of computation of frequency on the basis of historical information as applicable to the Ceppo Morelli landslide.

Danger and scenario No	Recorded events	Occurrence range (years)	Average (years)	Frequency (event/year)
1 Rockfall	1940 Oct. 1971 Apr. 1977 Oct. 2000 Jun. 2002	0–30 (2–31)	15	1/15 = 0,066
3 Rock avalanche (B) $5 \times 10^6 \, \text{m}^3$	312 843 1816 2000	200–1000 (184–973)	600	1/600 = 0,0016
2 Rock avalanche (A) $1 \times 10^6 \, \text{m}^3$		(assumed intermediate value) 30–200	115	1/115 = 0,009

3.5.7 *Relevance and choice of significant scenarios*

The main characteristics of large landslides, such as those considered in the IMIRILAND project, is that the probability of a critical global movement of the whole mass is very low, whereas the consequences may be catastrophic. The majority of large landslides are characterized by a complex sort of interacting specific mechanisms, as well as by preparatory and triggering causes. Any mechanism can evolve in a specific way and, therefore, could produce different potential dangers and, consequently, different scenarios. It is, thus, indispensable to assess several scenarios, and not only the worst case, with their potential consequences, so as to quantify the respective and indirect risk components in comparable terms, and to propose preventive measures on a wide scale so that such an extreme event does not hit the site by surprise.

On the basis of the multidisciplinary application of the various methods mentioned above, a synthetic model of the landslide behaviour has to be determined for each danger identification. This process requires a confrontation and critical assessment of results that may not be in complete agreement. An assessment of the reliability of all of the results produced must be carried out, so as to give more weight to the more significant data.

The object of such a scenario selection is to qualify clearly the different danger situations leading to different consequences, so that each scenario represents an intrinsic phenomenon, for instance rockfall scenarios vs. rock avalanche.

For each scenario, the probability of occurrence has to be set up on the basis of historical information or past movements. This selection process is certainly the most delicate operation within the risk analysis. It can be related to the probability classes that are proposed in several documents or recommendations set up in different countries (Switzerland: OFAT, OFEE & OFEFP, 1997; France: PPR, 1995). For example, in Switzerland, high probability means a 1–30 year return period, medium probability 30–100, low probability 100–300.

The interpretation of historical data gives a justification for the selection of return periods, even though it is not always certain that the reported events have the same magnitude. For very low probabilities, for instance 1/1000 years, the probability of the scenario has to be considered as more qualitative than quantitative.

Finally, there is a unique correspondence between a specific danger situation and the related scenario. Similar mechanisms of different magnitude will induce different scenarios.

3.5.8 *Evaluation of the contribution of the various disciplines to hazard analysis*

On the basis of the experience gained by IMIRILAND project and the knowledge of the partners of this project, the applicability of the different disciplines proposed can be rated according to several considered aspects: intensity definition, definition of the area affected, relative probability of failure, occurrence probability evaluation. Table 3.4 points out that all disciplines contribute to the

Table 3.4. Hazard analysis approaches: contribution to hazard component evaluation.

Analysis type	Contribution to intensity definition	Contribution to definition of area affected	Contribution to occurrence probability evaluation (temporal concept)	Contribution to relative failure probability (non-temporal concept)
Geological – structural	•	•	•	•••
Geomorphological	••	••	•	•••
Historical (data collection)	••	••	•••	•
Monitoring investigation	••	–	•	•
Geomechanical modeling: triggering	••	•	–	••
Geomechanical modeling: run-out	••••	••••	–	–

intensity definition, but an appropriate geomechanical modelling of the run-out does provide outstanding quantitative results. The historical analysis definitely contributes in an essential way to probability evaluations. Finally, run-out models are indispensable to determine the spatial impact of the considered scenarios. It must be recalled that no significant results can be obtained without comprehensive geological and geomorphological studies.

3.6 CONSEQUENCE ANALYSIS

3.6.1 *Introduction*

Once the hazard scenarios have been identified, each one being characterised by an area in which the distribution of the process intensity is known, as well as the corresponding probability of occurrence, it is possible to proceed with the true and real consequence analysis. The necessary steps to allow the definition of consequences are the following:

- identification of elements at risk,
- evaluation of the value of elements at risk,
- vulnerability evaluation,
- consequences (element at risk value × vulnerability).

3.6.2 Exposed elements: identification and value

Superposing the involved area identified by the run out/morpho-dynamic model for each scenario with territorial data (land use according to town and country planning, population distribution according to statistical demographic studies, strategic elements defined by e.g. transport administration or electric power supplier, etc.), elements at risk are recognised. The elements at risk are classified in different categories (see below), and counted. The actual state is either taken into account in the classification or assessed through a depreciation coefficient expressing the difference of value between the present and initial cost estimate of each exposed object. The necessary inventory of the exposed elements also considers potentially developed zones and buildings under construction.

The elements at risk can be monetarily quantified or globally assessed in order to obtain a consequence (expected impact) calculation.

For the value computation, several approaches exist (Crosta et al., 2001):

1. Computation of a specific value for the individual elements.
2. Use of a utility function.
3. Use of empirical formulas.
4. Qualitative assessment of the global value for a certain area.

3.6.2.1 Computation of a specific value for the individual elements

In such an approach, the value of the elements at risk is expressed as the sum of the intrinsic value of each separate element. In most of the published cases, a distinction is made between the material goods/economic activities and human life. The evaluation of human life is, in each case, a difficult problem; as a simple guide Table 3.5, drawn from the project PER (DRM, 1990), is given

Table 3.5.   Scale of relative costs of human victims exposed to risk considered in France (PER project, DRM, 1990).

Human victims	Relative cost
Dead persons	1
Wounded persons	2–3
Homeless persons	0.2–1

and puts forward the paradoxical aspect in which wounded persons represent a higher value than dead persons because of the high social costs related to medical care and rehabilitation of handicapped persons.

In the assessment of the real value of buildings, the reference prices needed, depending on the local situations, have been taken into account (see, for example, the Sedrun site in Chapter 8).

3.6.2.2 Use of utility function

In this approach, the elements at risk are characterised by a utility function u(x) in which the social or individual cost induced by the loss of a specific element is expressed as a function and not as a single value. Therefore, for each element, it is necessary to specify a variation of the social or

35

individual utility (e.g. linear, logarithmic, exponential variation, etc.). The use of the utility functions imply a better flexibility and adaptation to the complex conditions which determine the global cost of the losses. In Figure 3.7, the utility functions used to express, first, the cost of the interruption of a road depending on the time in hours and, then, the cost related to the number of wounded persons, are presented; in the first case, the variation is logarithmic because after a certain number of hours of interruption of the road, alternative routes are identified and/or re-established; in the second case, the variation is exponential because with the increasing number of wounded persons the cost of rescue and medical care tends to increase dramatically.

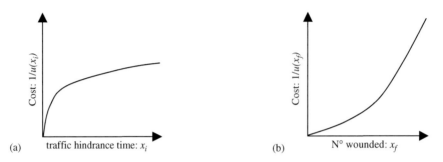

Figure 3.7. (a) Variation of cost of traffic hindrance as a function of time; (b) Nonlinear variation of cost of rescue and medical attention as a function of the number of wounded persons (Crosta et al., 2001).

3.6.2.3 *Use of empirical formulas*

The third approach foresees the use of empirical formulas for the computation of a global quantitative value of the elements and risk. As an example, the following formula is proposed (Del Prete et al., 1992):

$$W = [R_m(M_m - E_m)] * N_{ab} + N_{ed} * C_{ed} + C_{str} + C_{morf}$$

R_m = average income of the inhabitants of the area;
M_m = average age of death of the inhabitants within the exposed area;
E_m = average age of the inhabitants within the exposed area;
N_{ab} = number of inhabitants within the exposed area;
N_{ed} = number of buildings present in the area;
C_{ed} = average cost of existing buildings;
C_{str} = cost of existing structures and infrastructures;
C_{morf} = cost of morphological changes induced.

3.6.2.4 *Qualitative assessment of the global value for a certain area*

Such an approach appears to be somehow useful for particularly vast areas in which it is complicated to analyse the value of individual elements. In the methodology proposed here, this method of assessment is selected by proposing a scale of values based on a subdivision of the exposed areas according to land use and/or to the categories determined in the local management plan.

The parametric value appraisal of the elements at risk, generally, is carried out according to separate classifications (material assets and persons), therefore, calculating separate risk both for the assets and for the population. Moreover, in this methodology, environmental and economic values (related to interruption of economic activity) are individuated. In order to simplify the evaluation of the value elements at risk, it is sometimes possible to determine a relative value scale using several indices (in order to use indices correctly, a detailed analysis is necessary).

In the following tables (Tables 3.6 and 3.7), relative values related to some elements at risk categories are shown. For each element at risk, a "relative value index" for assets, economic activity interruption and environment has been indicated. The human lives relative "value" has been applied on the basis of the persons involved (the ranges can vary according to the area studied).

Table 3.6.   Assessment of relative values for exposed objects.

Element at risk	Assets relative value	Relative value of interruption of economic activity	Environmental relative value
Densely built modern cities (with high-rise buildings)	4	4	1
Historical city centers	4	2	4
Residential areas	4	1	1
Productive or industrial areas	4	4	1
Strategic services and facilities	4	4	2
Extra-municipal infrastructures and plants	4	4	2
Valuable buildings or valuable rural centers (historical, architectural, artistic and/or cultural value)	3	1	3
Tourist accommodation – buildings	3	3	1
Valuable environmental areas – buildings	3	2	2
Local infrastructures and plants	3	3	2
Transportation facilities (highway and international railways with operating services)	3	4	1
Strategic lifelines	3	4	1
Heavy traffic or strategic roads	3	4	1
Tourist accommodation – camp grounds	2	2	1
Valuable environmental areas – no buildings	2	1	4
Parks, sport and parking areas	2	1	3
Local services	2	2	1
Secondary roads	1	2	1
Secondary lifelines	1	2	1
Rural areas – farming and domestic animals	1	2	2
Forests (private and public properties)	1	1	3
Special risk objects			
hospitals	3	4	1
schools	2	4	1
garbage dumps	2	3	4

Table 3.7.   Assessment of relative values for exposed persons.

Human lives (number of persons involved)	Relative value of human lives
0 ÷ 1	1
2 ÷ 9	2
10 ÷ 19	3
> 20	4

When elements (assets and economic activities) can be monetarily evaluated, the indices represent a relative cost. In these tables, the relative values have been evaluated in an arbitrary way (using indices 1 to 4), and, therefore, have only a relative meaning in each of the considered categories. In one of the examples presented below (see Chapter 8), the assessment of the value of all buildings has been carried out in monetary terms, using average market values of similar objects as a basis.

3.6.3   *Vulnerability evaluation*

The degree of loss to a given element at risk, or set of such elements, resulting from the occurrence of a natural phenomenon of a given magnitude, is defined by a vulnerability coefficient (V). Vulnerability is usually expressed in relative terms as defined in the quotation below, using words such as "no damage", "some damage", "major damage" and "total loss", or by a numerical scale between 0 (no damage, 0%) and 1 (total loss, 100%).

"Although the state of the art for identifying the elements at risk and their characteristics is relatively well developed, the state of the art for the assessment of vulnerability is in general relatively primitive.

The vulnerability is affected by the nature of affected facilities, whether it is uphill, on, or downhill of the landslide, and the nature of the elements at risk. The velocity of movement also affects the vulnerability, with higher velocities usually leading to greater vulnerability. This can lead to different degrees of damage on or in the travel path of a landslide. For structures and persons onto which a landslide travels, the greater the depth of slide material, generally the greater the damage and vulnerability.

For structures, the assessment of damage and, hence, vulnerability, depends on modelling the landslide-structure interaction. This is relatively well documented for rockfalls (where structures have been designed to withstand impacts), and to a lesser extent for debris flows and extremely slow and very slow movements. For higher velocity slides, spreads and topples, there is very little guidance available. Often, it is necessary to use judgement. In any case, the notion of direct and indirect impacts has to be defined first" (IUGS/WGL/CRA, 1997).

As previously pointed out, the adopted methodology foresees the assessment of the different vulnerability factors, which include physical, social, environmental and economic components. In the investigation work carried out, two approaches have been considered for the assessment of the exposed objects, applying either monetary values or coefficients expressing their relative importance. The amount of the physical vulnerability coefficients considered takes into account the intensity of the hazard and the quality of the constructions.

3.6.3.1 *Physical vulnerability*

This term expresses the degree of loss or potential damage to a given element or set of elements at risk when they are affected by the behaviour of an unstable mass, which is qualified by a given magnitude or intensity. Such an impact must be assessed first in terms of structural failure, by the analysis of the effect of differential settlements or movements on a structure crossing a scarp, or by the impact of rock blocks hitting a building. It also has to be evaluated in terms of operational failure, for instance, when the tilting of a house or a road exceeds an acceptable value, even though no cracks are observed. The physical damage also depends on the quality of the materials used in the building or the infrastructures under study, as well as on the maintenance of these structures, especially when wooden buildings are concerned.

No theory has been elaborated up until now to model the damaging effects due to all types of landslides, with the exception of the sudden impact of a block on a wall; however, the number of variable parameters for this case (velocity, block mass, impact angle, position of the wall impact point, potential deformation of the whole structure, detailed geometry of the wall, strength of the material) does not allow the obtaining of a significant result in terms of global risk analysis for a large landslide area. On the other hand, in most situations where limited damage has occurred, no systematic monitoring procedure has been carried out, which would have allowed a proper evaluation of the effect of the landslide movements.

The main criteria to determine the value of the physical vulnerability coefficients are, in decreasing order:

– the phenomenon intensity (velocity for slides, energy for rockfalls);
– the types and function of the structure, as far as its strength is concerned;
– the state of maintenance and capacity of deformation.

The classification of these criteria may vary when networks are considered (electric lines, water pipes, sewage ducts) instead of buildings and roads.

3.6.3.2   Social vulnerability

This term expresses the rate of impact due to a landslide on the exposed population. In many rock-fall cases, even of limited intensity, this rate is 100% as it may cause death. Serious wounds leading to a permanent important handicap are also assessed as nearly 100%, because the long-term costs they induce for the society are high. Temporary wounds (such as a broken leg) are considered, on the contrary, as a minor vulnerability, as the persons can recover after a short time.

The coefficient of social vulnerability also includes the psychological consequences of the loss of a home, as it is often observed as a more difficult situation for the victims to lose their effective roots than to be wounded. Even the fact of being provisionally evacuated, for a certain duration, constitutes an impact on the population that deals with social vulnerability.

Thus, the main criteria to determine the value of the social vulnerability coefficients are, in decreasing order:

– the phenomenon intensity (which is in relation to the warning time);
– the population sensitivity, depending on its age and capacity to anticipate a landslide;
– the capacity of understanding the phenomenon and to move away from the exposed zone.

3.6.3.3   Environmental vulnerability

In order to face all the direct impacts of a landslide, it is necessary to add a third category after the physical and social domains, which deals with the natural environment. Indeed, it is often the first "target" of landslides, such as livestock, as the slopes where they occur are generally not densely populated or crossed by major infrastructures. On the other hand, the damage to these natural components cannot be evaluated in monetary terms, especially as far as forests (which now have nearly no merchant value), wild animals and rare plants are concerned. For the forests, this impact may be important as several zones are considered as protective forests against rockfall and avalanches, so that their potential value has to be assessed through the analysis of their global function (productive, protective, recreational).

As far as impacts on water resources and flow conditions in rivers are concerned, it is also important to analyse the potential loss of springs due to a rockfall, even if they are not yet equipped for water consumption, and the potential loss of fish if the river flow is disturbed by a sliding mass obstructing its course.

Thus, the main criteria for determining the value of the environmental vulnerability coefficients are:

– the intensity of the phenomenon, in relation to its disrupting effect on nature,
– the function of the forest or of the endangered animal and plant species,
– the sensitivity and rarity of these species.

3.6.3.4   Economic vulnerability

Beyond the potential destruction of assets on or near landslides, such phenomena may cause indirect economic impacts when they block a road or a railway line, destroy an electric line or a water pipe, or dam a valley causing a lake that represents a danger downstream, so that the economic activity in the valley below must be stopped or reduced, but perhaps without final destructive consequences.

It is important to realise that, in most real situations studied, the indirect economic impacts are much larger than the direct impacts. As an example, for the avalanches of the 1999–2000 winter in Switzerland, it was assessed that the indirect impacts were 4 to 5 times higher than the direct ones. In the case of the La Josefina landslide in Ecuador, the threat to the Paute hydroelectric plant located downhill, which might have been seriously affected by the overflow of the dam caused by the landslide in 1993, would have represented the loss of 70% of all the electrical energy production in the country; happily only limited damage was registered when the flow reaching $10'000\,\mathrm{m^3/s}$ filled the previously emptied reservoir, but this flood brought the equivalent of 2 years of silting in the reservoir, which meant several hundreds of thousands of US dollars for its pumping and evacuation in order to maintain the regulation capacity of the lake (see Chapter 10.2, below).

The economic impact on a road blocked by a landslide depends, of course, on the average expected traffic, but also on the existence of an alternative route. For example, when the Ceppo Morelli rockfall (Valle Anzasca, Italy) reached the access road to the Macugnaga resort station, it impeded any traffic to and from this important station for several weeks; an alternative road now exists on the other side of the valley, but its position does not guarantee its operation in case of large rockfall (see Chapter 7, below).

In other cases, an alternative road exists, but requires a very long bypass of several dozens of kilometres, as it is the case for the La Frasse landslide, in Switzerland, so that a large part of the tourist traffic coming in the area for one day will give up and visit other resorts, especially in winter and spring. If a threat of rockfall is known and publicised, many people will not drive to the concerned area, even if the risk is effectively limited. As far as transportation of goods is concerned, some may be stockpiled for a certain time, such as logs of wood, but other products, such as milk or cheese, need to be transported continuously to the factories or warehouses, so that there is a direct economic loss in case of disruption of traffic, or an indirect economic loss if a longer transportation distance must be foreseen by an alternative route.

The main criteria to be considered in the assessment of the economic vulnerability are, thus:

- the types of services implied (industry, tourism, transport);
- the types of economic activities affected;
- the traffic and the cost of blocking a route (which is very high for railways);
- the possibilities of alternative routes.

Finally, this aspect must include an analysis of the potential damage due to the secondary consequences of a landslide, such as the formation of a dam in a valley, inducing flooding upstream and a potential catastrophic flow downstream.

3.6.3.5 *Vulnerability computation*

The vulnerability of elements at risk depends on the typology of the element (T) (and, therefore, on its shock resistance features for assets) and on the process intensity (I):

$$V = f(T, I)$$

From a practical point of view, two approaches can be followed in order to evaluate vulnerability: (1) simply on the basis of effects on the element at risk (see Table 3.8), and (2) on the basis of the process intensity.

In the first case (1), 5 classes of percentage loss (0, 0.25, 0.5, 0.75, 1) have been applied to each vulnerability category. Through detailed studies it could be possible, however, to attribute various weights at various percentage classes based on "utility curves" [e.g.: in the road interruption case – economic vulnerability – some studies have evidenced a logarithmic type curve (Fig. 3.7a), as the vulnerability diminishes with the passing of time for the opening of the alternative roads (e.g.: variable indices – 0, 0.5, 0.8, 0.95, 1 – according to elapsed time)].

In the second case (2), process intensity may be indicated as energy thresholds (J) obtained from the run-out model. When no data models are available, it is possible to "translate" the intensity processes into qualitatively described effects (Fig. 3.8).

Several scales have been developed in order to qualify the intensity of the different types of landslide phenomena. For instance, the Swiss recommendations published in 1997 provide orders of magnitude in order to differentiate low, medium and high intensity for block fall, rockfall, slides and flows. They are expressed, either in terms of impact energy, or in terms of average velocity, or in terms of depth of the potentially erodible soil zone (OFAT, OFEE & OFEFP, 1997). For large rockfalls, it is evident that the intensity level in the exposed zones is always high to very high, whereas for slides, many large phenomena display low to medium intensity, according to the proposed scale.

Table 3.8.   Assessment of vulnerability factors values for physical, social, environmental, economic aspects.

Physical vulnerability			Social vulnerability	
Vulnerability description	Loss range	Index	Vulnerability description	Index
Intact structures	0	0	Non-affected persons	0
Local damage	1÷25%	0.25	No physical damage, evacuated persons	0.25
Serious damage (possible to repair)	26÷50%	0.5	Physical damage (persons may continue their activities)	0.5
Mostly destroyed structures (difficult to repair)	51÷75%	0.75	Seriously wounded persons (up to 50% disability)	0.75
Total destruction (out of use; e.g. >5% inclination)	76÷100%	1	Fatalities, 51–100% disability	1

Environmental vulnerability			Economic vulnerability	
Vulnerability description	Loss range	Index	Vulnerability description	Index
Intact element	0	0	No interruption	0
Local loss	1÷25%	0.25	Short temporary interruption (hours to day)	0.25
Serious damage (possible to repair)	26÷50%	0.5	Average temporary interruption (days to week)	0.5
Mostly destroyed elements (difficult to repair)	51÷75%	0.75	Long temporary interruption (weeks to months)	0.75
Total destruction	76÷100%	1	Permanent interruption	1

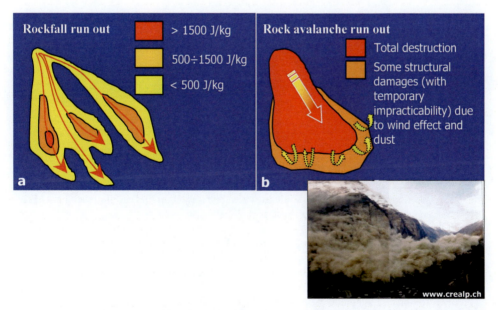

Figure 3.8.   (a) Quantification of impact in terms of unit mass energy; (b) Qualification of impact based on the gradation of damage. In the photograph: Randa rockfall in 1991 (Switzerland).

A more detailed analysis implies, however, the consideration of the acceleration phase, or the volume of potential unit blocks of a rockfall, as well as their possible interaction. But in any case, the limit values proposed for the intensity classes are indicative and have to be considered as qualitative references rather than as quantitative parameters.

3.6.4 *Assessment of consequences*

The assessment of consequences induced by a certain scenario is based on a theoretical computation, simply the product of the elements at risk value (V_E) and the vulnerability (V):

$$C = V_E \times V$$

In this methodology, each element at risk value is multiplied by a relative vulnerability category (e.g. physical vulnerability index $\times$ assets value index = physical consequence index). The operation is repeated for each hazard scenario, considering each specific process intensity class as far as it influences the vulnerability level (V) of the elements at risk. It is also repeated for each type of vulnerability (physical, social, environmental, economic).

Therefore, the matricial approach on which the proposed methodology is based foresees a simple multiplication, at this point of the process, between value classes, the result of which (consequence index) will enter in the next matrix for the computation of the risk (Fig. 3.3 and 3.4). In this way, the assessment of consequences results from an automated objective process. For this reason, and in order that the consequence assessment be significant, all the initial parts of the process (danger definition, evaluation of the value of the elements at risk, vulnerability assessment) which will inevitably suffer from a certain rate of subjectivity, require very careful analyses and assessments.

3.7 RISK AND TOTAL RISK ASSESSMENT

As previously defined, risk is the product of landslide hazard and its consequences:

$$R = H \times C$$

In the methodology analysed, hazard (H) has been defined by its three components: involved area, process intensity and occurrence probability. The involved area (used for the identification of elements at risk) and the process intensity (necessary for the vulnerability identification) have already been used and included in the consequences. So, risk is the result of the product of occurrence probability (P) and its consequences (C):

$$R = P \times C$$

In the same way as in the consequence computation, the matricial approach used in the risk asessment foresees a simple multiplication between two value classes the result of which is the risk index defined for each vulnerability category, thus obtaining the physical, social, environmental and economic risks. This operation must also be repeated in this case for each hazard scenario, or at least for each specific occurrence probability considered. In this way, four risk maps are derived for each scenario.

The problem of the management of numerous data and of a large quantity of resulting maps is solved thanks to the use of performing GIS techniques which allow the superimposition of various levels of spatial information and the development of mathematical operations between each implied term in an automatic way.

In order to obtain a synthetic assessment of the risk (total risk) for the studied area, it is sufficient to sum up the various risk indices obtained in the different scenarios for each category of impact, namely the physical, social, environmental and economic total risks.

This process of summing different risk index values corresponding to several scenarios implies the hypothesis that each considered scenario may occur independently without having the first occurred scenario modify a subsequent scenario. This is clearly the case when the considered scenarios are characterized by very different magnitudes. This assumption is also justified by the fact that "frequent" and less intense scenarios can occur before a very rare scenario. Moreover, it is also assumed that a certain lapse of time will exist between one scenario and the next one. Finally, this methodology supposes that the value of the exposed objects will not significantly change over time.

3.8   SCOPE OF APPLICATION OF THE PROPOSED METHODOLOGY

The methodology proposed by the IMIRILAND project has pointed out the importance of a correct determination of the intensity and probability for different scenarios in order to reach an appropriate hazard and risk assessment in the zone of direct impact of large landslides. However, these potential dangers have to be assessed in a regional context with all their consequences, especially as they are able to induce "in-chain" phenomena such as debris flows following a slide reactivation, a secondary rockfall with a trajectory modified by a first event, or above all the possible failure of a landslide dam releasing a flash flood due to the water accumulated behind the dam (see Chapter 10.2).

All of these complex scenarios must include the consideration of the simultaneous occurrence of rare events such as extreme rainfall and related flooding; but of course the probability of such combined effects decreases, as expressed by the conditional probabilities.

The fundamental specificity of a landslide dam failure is that its impact extends far beyond the landslide zone itself, either upstream with the progressive inundation of extensive areas if the river course dammed presents a reduced slope, or downstream with the sudden and violent flooding following the dam failure due to overflow or seepage. This final aspect requires a detailed study, not only of the extension of the flooded area, but of all the infrastructures and exposed objects threatened by such an event, especially as, in these cases, the vulnerability coefficients of all kinds are very high.

When international roads or railways following the river course downstream from the landslide site are damaged, when lifelines are affected even locally, interrupting traffic, the indirect consequences of even a very rare event may be huge on economic or tourism activities and, thus, represent non-negligible risks.

The analysis of such "in-chain" phenomena therefore requires a multiple assessment of the induced risks, as presented above, which makes it even more difficult to compare different risk situations and to evaluate their respective degrees of severity.

4
The Cassas landslide

F. Forlati[1], M. Morelli[2], L. Paro[1], F. Piana[2], R. Polino[2], G. Susella[1], C. Troisi[1]
[1]ARPA Piemonte, Torino, [2]CNR-IGG Sezione Torino

4.1   INTRODUCTION

4.1.1   *Landslide description*

The Cassas landslide is located on the right side of the middle Susa Valley, in the Salbertrand Municipality, in the "Gran Bosco" natural park (county of Turin, Piedmont, Italy). The Cassas slope is deeply touched by instability phenomena; the word "cassas" (probably from French "casser" = to break), used over the ages to indicate this zone, refers to debris accumulations (Brovero et al., 1996a).

The top of the landslide is located near a minor watershed at about 2'000 m a.s.l., while the landslide foot is placed at about 1'000 m in the wide Salbertrand valley flat that was formed by the filling of a landslide dammed lake in the locality Serre la Voute (Capello, 1941).

In this part of the Susa Valley, which is an ancient commercial and military route, the A32 motorway that connects with the Turin to Frejus Tunnel, the Turin-Modane international railroad and the national road SS 24 "Montgenèvre Pass" are located. In particular, a service station for the A32 motorway is located just at the foot of the landslide (Fig. 4.1).

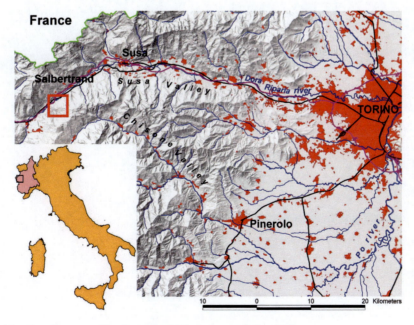

Figure 4.1.   Location map of the Cassas landslide (red square), Susa Valley.

4.1.2 *Historical background*

The first records of the slope's instability in this area are found in documents from the 18th century, in which the disastrous effects of the May 1728 flood are described (Bogge, 1975). In the 19th century, the instability of the right slope of this part of the Susa Valley caused several problems to the Turin-Modane railroad (Baretti, 1881, 1893; Sacco, 1898). These authors wrote that the causes of the instability were the intensively fractured rocks and the "dissolution of calcareous formations", outcropping at the slope's foot.

At the beginning of the 20th century, additional information about the instability phenomena in this area were extracted from documents of the Corps of Engineers and forest authority offices: in the former documents, the deformations of the railroad tunnels and the bypass channel of the Chiomonte power plant (Segré, 1920) are fully analysed; in the latter documents, a "huge landslide" of about 10 hectares is mentioned (Consorzio Forestale di Oulx, 1955; Corpo Forestale dello Stato, 1956). Such evidences of instability are also visible in the first available aerial photographs dated July 1954.

Between 12th and 14th June 1957 in the Susa Valley, a huge flood caused great damage to the entire valley. On that occasion, the western area of the Cassas landslide had a paroxystic phase: millions of m^3 of rock mass slid and caused a debris flow that reached the valley floor. Re-activations of the landslide were recorded until the middle Sixties (Peretti, 1967, 1969; Brovero et al., 1996a) (Fig. 4.2).

Figure 4.2.   General overview of the Cassas landslide in a photo from 1965, at the conclusion of the paroxystic phase of the period 1950–1965 (Brovero et al., 1996a).

In the Seventies, the Cassas landslide appeared much wider than the instability phenomena of the 50's and 60's, as stated in a survey about the "hydrologic instability" of the Susa Valley (Ramasco & Susella, 1978).

In the Eighties, most of landslide accumulations on the right slope of the middle Susa Valley, including the Cassas landslide, were interpreted as associated to the evolution of the so-called "deep-seated gravitational deformations" by Puma et al. (1984, 1989) and Mortara & Sorzana (1987). These phenomena involved the Susa-Chisone watershed for some kilometres and the slopes for hundreds meters of depth.

In the early Nineties, in order to mitigate the risks on the motorway and at the service station at the Cassas landslide foot, a monitoring system was installed on the landslide to verify its movements. The inclinometer data indicate a slow but constant movement up to the present.

4.2 REGIONAL FRAMEWORK

4.2.1 *Climatic and water conditions*

The climatic features of the Susa Valley are variable, with characteristics unique among the Piedmontese mountain valleys. Such variability is due to the articulated valley floor morphology (more than 70 km long and few metres to some kilometres wide) and the slope's altitude (more than 3'000 m a.s.l.). These morphological features control to a large extent wind direction, solar radiation, temperature distribution, humidity, etc. In Figure 4.3, the Piedmont climatic map is shown, prepared using the Thornthwaite method (Biancotti & Bovo, 1998a), based on the determination of evapotranspiration (real and potential) and of its comparison with precipitation amount. The map in Figure 4.3 shows valley areas where special aridity conditions have been observed (xerothermic areas of the Susa-Bussoleno valley floor and of the Bardonecchia basin).

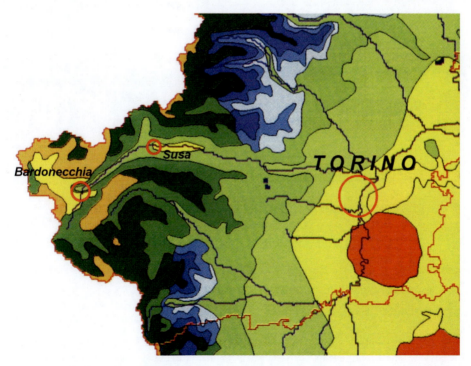

Figure 4.3. Map of Piedmont climatic regimes prepared using the Thornthwaite method (Biancotti & Bovo, 1998a). Simplified legend of climatic types: blue = perhumid; green = humid; yellow = from humid to subhumid; red = from subhumid to subarid.

47

On the basis of the rainfall data analysis, a bimodal trend is observed, showing the two highest values in spring and autumn: spring is the most rainy period of the year, while autumn records the greatest rainy water heights over a few days (showers). In summer, intense but short precipitations (storms) take place over small areas (Peretti, 1969; Biancotti & Bovo, 1998a). Through the updating of the regional weather monitoring net, it is possible to supply precise indications about the climatic trend in the valley (Fig. 4.4). In Table 4.1 some climatic data that refer to a time span between 1990÷1999 are listed. These data come from weather stations located in the upper parts of the Susa and Chisone Valleys (Fratianni & Motta, 2002). In Figure 4.5, a histogram of the annual precipitation data at the Salbertrand–Graviere weather station, located at the Cassas landslide foot, are reported.

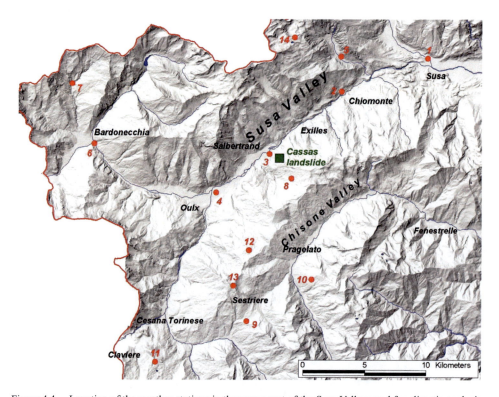

Figure 4.4.   Location of the weather stations in the upper part of the Susa Valley used for climatic analysis.

4.2.2   *Hydrological data*

The main historical floods of the Dora Riparia have been recorded since the 18th century; they are sometimes associated with disastrous events (Brovero et al., 1996a; Fratianni & Motta, 2002) listed as follows:

–   XVIII century: 11/1705, 05/1728, 1748 (?);
–   XIX century: 05–06/1827, 10/1839, 05/1856, 1863 (?), 06/1876, 05/1879, 07/1896;
–   XX century: 09/1920, 05/1948, 06/1957, 08/1972, 05/1977, 04/1981, 09/1993, 06/2000, 09/2000, 10/2000.

Analysing the flood events seasonally, a greater frequency is observed in the spring (April÷June), although in the autumn, rainfall is generally more frequent. Therefore, the annual maximum flood of Dora is the result of the spring snow melting and rainfalls as confirmed by the analysis of the events of May 1728 and of June 1957 (Bogge, 1975; Brovero et al., 1996a). During

Table 4.1.   Climatic data in the upper Susa Valley.

No	Weather station	Municipaliy	Alt. [m]	T_m [°C]	H_r [%]	R_m [mm]	S_m [cm]
1	Pietrastretta	Susa	520	12.2	59.2	710	–
2	Fniere	Chiomonte	813	10.5	63.1	813	–
3	Graviere	Salbertrand	1010	8.7	71.2	723	69
4	Gad	Oulx	1065	7.9	68.4	602	–
5	Val Clarea	Giaglione	1135	8.7	66.4	–	–
6	Prerichard	Bardonecchia	1353	7.3	66.2	786	98
7	Camini Frejus	Bardonecchia	1800	4.9	63.2	916	199
8	Le Selle	Salbertrand	1950	4.5	62.6	722	276
9	Sestriere	Sestriere	2020	1.5	65.1	–	300
10	Clot della Soma	Pragelato	2150	3.7	47.8	–	330
11	Colle Bercia	Cesana T.se	2200	2.9	61.7	–	451
12	Lago Pilone	Sauze d'Oulx	2320	2.3	–	499	456
13	Monte Fraiteve	Sestriere	2701	−0.8	67.3	–	–
14	Rifugio Vaccarone	Giaglione	2750	−0.7	55.4	–	–

T_m: Average of the monthly temperature average (1990–1999)
H_r: Annual humidity average (1990–1999)
R_m: Annual rainfalls average (1990–1999)
S_m: Sum of the monthly average hights of snow (1990–1999).

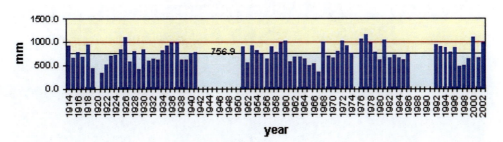

Figure 4.5.   Salbertrand-Graviere station annual rainfall (columns) and annual average (756.9 mm) for the period 1914–2002. The lack of data is due to instrument failure or to lack of measurement. Since 1992, a new pluviometric station has been functioning.

these events, the Cassas landslide immediately accelerated, as also occurred during the October 2000 heavy meteorological event which involved the whole Dora Riparia valley and a large part of the Piedmont region.

4.2.3   Water condition

No surface waters have been observed within the landslide area but there is an aquifer hosted in the debris cover of the landslide that lies on the calcschistic bedrock. Groundwater height varies according to rainfall regime. Waters related to rainfall and snow melting mainly infiltrate in the upper part of the landslide, as high fractured bedrock enables deep water circulation.

In the lower part of landslide, two springs run (estimated flow on July 1997: from 2×10^{-4} to $8 \times 10^{-4} \, m^3/s$). There are also some springs along the scarp that cut across the debris cover between the altitudes 1'350 and 1'550 m a.s.l. (Brovero et al., 1996a; Paro, 1997).

49

4.2.4 *Regional morphology*

The Susa Valley morphology is due to fluvial-glacial modelling and to gravitational instability of the slopes (Fig. 4.6). On the right side of the valley, the modelling process due to landslide deformations is evident, whereas glacial morphologies and deposits are almost absent (Carta Geologica d'Italia in scala 1:50'000). Also, the wide Salbertrand valley flat was formed by a landslide dammed lake in Serre la Voute (Capello, 1941). Furthermore, along the Susa–Chisone watershed (Fig. 4.7), there are trenches, sliding steps, depressions and spreading double-crested ridges related to deep-seated gravitational phenomena (Puma et al., 1984; Mortara & Sorzana, 1987; Puma et al., 1989).

The Cassas slope is not entirely affected by gravitational instability as there is a sector on the left flank of the landslide, termed hereafter as "1957 Cassas", where polished rock steps due to fluvial-glacial modelling and glacial deposits are preserved (Paro, 1997; Carta Geologica d'Italia in scala 1:50'000, 2002).

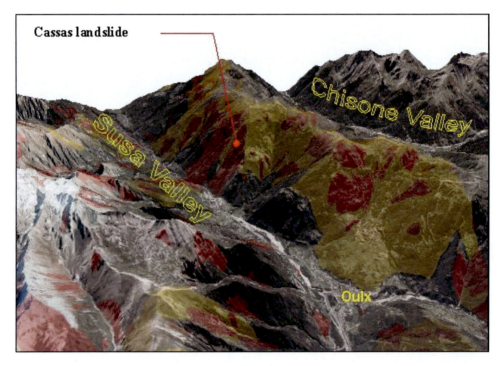

Figure 4.6.   General overview towards the East of the Susa and Chisone Valleys (2001 aerial photographs with overlapping Digital Terrain Model). Gravitational phenomena deriving from IFFI Project (Gravitational Phenomena Inventory in Italy. ARPA Piemonte, 2004) are shown: yellow = deep-seated gravitational slope deformation; red = gravitational phenomena of other typologies.

The sub-sectors of the Cassas slope will be described from East to West in the following section (Fig. 4.8):

– on the eastern part, a landslide accumulation called here "Sapé d'Exilles" (A) is distinguished; it extends from an altitude of about 2'350 m to the valley floor for about 2.5 km² (height more than 1'000 m). This landslide accumulation and that of Eclause in front of it (on the left side of the valley) formed the valley dam originating the Oulx-Salbertrand plain.
– To the West of "Sapè d'Exilles", other accumulations can be recognized (B and D); these are smaller and lobe-shaped, and gently overlap the alluvial deposits on the valley floor. The largest one (D) is particularly articulated as it covers another accumulation to the East (C), while to the

50

Figure 4.7.   Double-crested ridge (1 km long and 300 m wide) along the Susa-Chisone watershed (Testa dell'Assietta Mount) related to the deep-seated gravitational phenomena (to the right).

West it is partially covered by the 1957 Cassas accumulation, which partly re-mobilized this deposit. The main scarps of these accumulations (C, D and "1957 Cassas") are joined together and aligned to the sub-rectilinear East-West direction.

The Cassas slope is also characterized by many scarps that bound or split the accumulations described above. In particular, at the foot of Sector A there is a scarp whose origin is related to the erosion of the Dora Riparia river; the foot destabilization due to the erosion is probably the cause of the deformations verified along the A.E.M. bypass channel (the water of the Dora Riparia is transported to hydroelectric plant of Chiomonte through this channel), located at the foot of Sector A.

Also within Sectors B and D, two aligned scarps are recognized at an altitude of about 1'300 m. The alignment of such scarps is probably due to the presence of a previous (pre-landslide) morphology due to fluvial modelling, that had been successively covered by B and D accumulations and then partially exhumed.

Another evident scarp, some tens of meters high, is located at western part of the Cassas slope. The origin of this scarp is probably due to past fluvial erosion. At the foot of this sub-vertical scarp, a large talus and the motorway service station are located.

From a morpho-evolutionary point of view, the more interesting escarpments are observable in "1957 Cassas": one scarp is placed in the upper part of the sector, while another is cuts across the sector with a North–South direction (see § Geo-morphological analysis).

51

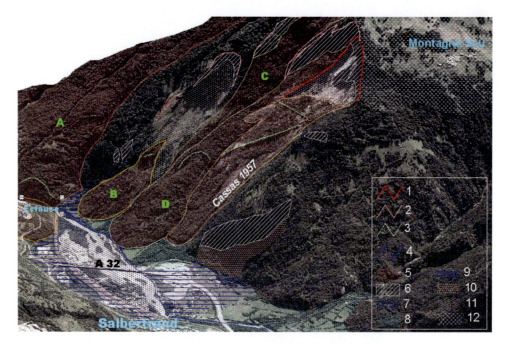

Figure 4.8. View of the Susa Valley right slope, from Salbertrand to Serre la Voute, with indications of the main morphological elements: 1. Cassas landslide main scarp, June 1957 event; 2. main scarps of gravitational phenomena; 3. secondary scarps of gravitational origin; 4. scarps by fluvial-torrential erosion; 5. landslide accumulations; 6. main rock outcrops; 7. actual fluvial-torrential deposits; 8. alluvial fans; 9. fluvial torrential terraced deposits; 10. talus; 11. landslide scar and transition area; 12. area characterised by morphological features related to deep-seated gravitational deformations.

In the valley floor, some alluvial fans are observed, such as the large fan formed by the Gran Plenei stream (right tributary of the Dora Riparia river). Some of these, of smaller dimensions, are located at the feet of landslide accumulations. Among those, the fan placed at the foot of the "1957 Cassas", was involved in the paroxystic event of June 1957. Successively, this fan was modified by the construction of the service station for the A32 Motorway (C in Fig. 4.15), which also rests on the alluvial deposits of the Dora Riparia river.

4.2.5 *Regional and structural setting*

In the middle Susa valley, several tectonostratigraphic units (Dela Pierre et al., 1997; Polino & al., 2002) of different paleogeographic origin have been distinguished (Fig. 4.9):

A. oceanic units consisting of meta-ophiolites and calcschists (Piedmont-Ligurian units);
B. continental margin units made up of Mesozoic dolomites and calcschists;
C. polymetamorphic basement units consisting of gneisses, quartzites, micaschists and metabasites (Ambin and Clarea complex) and relative Mesozoic sedimentary covers (Briançonnais units).

These tectonostratigraphic units are juxtaposed by west-dipping thrust planes and displaced by high-angle strike-slip or normal faults (see § local geology and structural analysis). These faults define the geometric framework of the middle Susa valley and gave origin to huge fault breccias and strongly fractured rock masses. This allowed the triggering and evolution of the large landslides of the middle Susa valley (Carraro & al., 1979; Giardino & Polino, 1997).

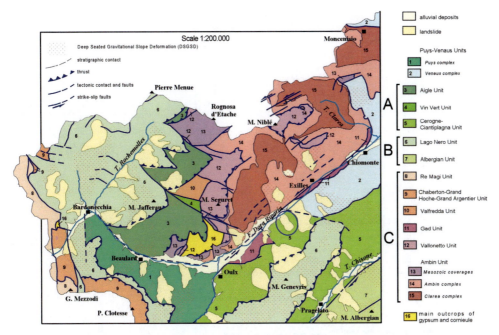

Figure 4.9. Geological structural map of the Bardonecchia sheet n. 153 of the CGI at the scale of 1:50'000 and legend of the tectonostratigraphic units (Carta Geologica d'Italia in scala 1:50'000, 2002). Blue circle: Cassas landslide.

4.3 HAZARD ANALYSIS

4.3.1 *Geomorphological evolution*

Since detailed historical documentation is lacking, other morphological information has been deduced from topographic maps.

The currently used topographic map (CTR 153110 Salbertrand – 1991) has been compared with the previous official map of the Military Geographic Institute (IGM sheet N°54, II SE, Oulx – 1934) in order to underline the main morphological changes suffered by the Cassas slope during the 1934–1991 time span.

The 1934 official map of the Military Geographic Institute already shows features that can be interpreted as instability clues on the Cassas slope. In this area (Fig. 4.10), small scarps (A), a trench (B) and a small landslide (C) are underlined.

The morphological evolution has been also reconstructed on the basis of the interpretation of aerial photos made in different years (1954, 1957, 1963, 1979, 1986, 2001). The revision of these documents gave new important elements for the interpretation of the Cassas landslide evolutionary processes.

4.3.1.1 *Aerial photographs dated 1954*

In these photos, some important features are shown (see Fig. 4.11):

- the upper part of the Cassas slope is already involved in incipient sliding (L);
- a sliding surface is shown at the top of this area (Cassas main scarp, F), with a vertical dislocation of some tens of meters;
- two sectors of active landslide (complex rock slide-rapid debris flow, D);
- accumulation of large blocks on the eastern side of this slope, probably derived from landslides of different age that reached the valley floor; this accumulation is partly remodeled in its western sector, where a scarp of about 80 m high is shown (S);

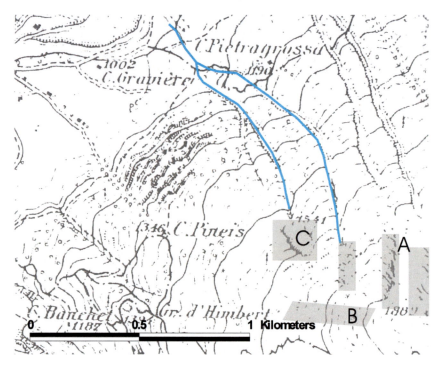

Figure 4.10. 1934 map at scale of 1:25'000 (IGM sheet n.54, II SE, Oulx). The shaded zones indicate the main instability features.

- below the S scarp, a wide wooden zone (from about 1'450 m to 1'200 m in altitude, A) is laterally bounded by two streams that produced a fan at the bottom of the slope. It has been hypothesized that the wooden zone is also a gravitational accumulation that controlled the onset of a new hydrographical pattern and that still shows evidence of instability;
- the accumulation area situated on the eastern side of Cassas slope (C) also presents some scarps due to movements at the foot.

4.3.1.2 *Photographs and surveys from 1962 to 2002*

The comparison between the images of the Cassas area before and after the event of June 1957, emphasises some important elements about the kinematic evolution of the gravitational phenomena, whose paroxystic phase ceased in the mid 60's in the past century. In Figure 4.14, the Cassas landslide appears at the end of the 1954÷1965 paroxystic phase: the debris accumulation is at its greatest extension, since it covers 2/3 of the fan surface in the valley floor.

Through the historical data and present-day morphology, it is possible to describe the landslide of June 1957:

- during the event of June 1957, the upper part of the Cassas slope (L in Fig. 4.11) underwent a total displacement of about 250 m, mostly by sliding. A sub-vertical scarp, tens of meters high, was formed on the right flank from which a part of the existing debris flowed (S in Fig. 4.12).
- The movement involved the bedrock (slide surface), that is still visible at present in the upper part of the slope. The strongly fractured and released bedrock induces several metric open fractures along the crown zone (Fig. 4.13).
- Most of the displaced rocks probably accumulated in the wide depression just below Sector L, where they covered the S scarp (Fig. 4.11). The displaced rock mass is not excessively disarticulated, as indicated by many preserved trees and by a large remnant of the soil cover on the

54

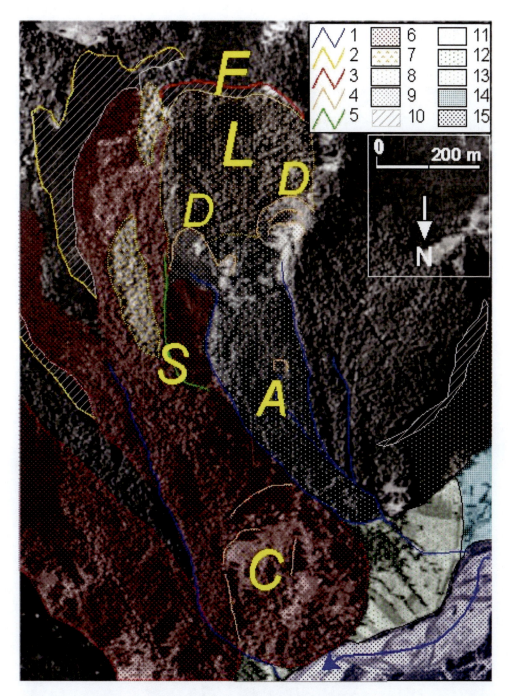

Figure 4.11. Geomorphological map based on aerial photographs dated July 1954 (1 – hydrographical pattern; 2 – gravitational scarps; 3 – Cassas main scarp; 4 – secondary gravitational scarps; 5 – erosional and gravitational scarps; 6 – landslide accumulations outside the Cassas area; 7 – accumulation of large blocks with no vegetation; 8 – sector of the Cassas landslide with evidences of movement; 9 – complex slide-rapid flow landslides; 10 – bedrock main outcrops; 11 – presumed landslide accumulation; 12 – alluvial fan; 13 – talus; 14 – terraced fluvial-torrential deposits; 15 – actual fluvial-torrential deposits).

accumulation body. Part of the displaced material reached the valley floor by debris flow processes, and lies on the alluvial fan at the slope foot with a maximum thickness of 1.5 m.

– Instability evidences, activated during the June 1957 event, indicate a movement of the whole slope, also involving an area located outside the Cassas landslide (Fig. 4.14).

Movements on the Cassas slope continued after the paroxystic events of 1957÷1965. In Figure 4.14 it can be observed that inside the accumulation body, some important scarps are delineating (N and R in Figure 4.15). The largest one (R) cuts across the landslide area, separating the upper

Figure 4.12.   Overview of the main scarp on the right flank. Heterogeneous debris accumulation is still visible today.

Figure 4.13.   Metric open joints along the Cassas landslide crown zone (photo by L. Paro).

zone, characterised by prevalent sliding, from the lower one, in which debris flows are prevalent. The scarp N is located in the same position as the dip slope change shown in the 1954 aerial photographs, indicating that this one has been covered for some years by the sliding mass.

Figure 4.14. Cassas landslide evolution (1965 and 1982: Brovero et al., 1996a; 2002: photograph by L. Paro).

The morphological evidence indicates that:

– the whole landslide accumulation is in movement (as confirmed by monitoring data);
– the scarps separate zones with different kinematics;
– the scarps correspond to morphostructural elements buried below the gravitational accumulation.

Finally, Figure 4.15 also shows an uphill-facing scarp (T), already visible in photographs of 1954, placed at the western side of the landslide, at an altitude of 1'600 m.

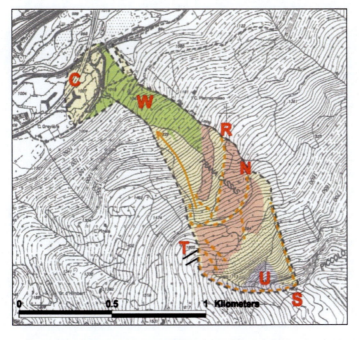

Figure 4.15. Synthetic sketch map of the main morphological elements of the Cassas landslide (by aerial photos 1963 ÷ 2001). S: main scarp; R and N: minor scarps; T: ancient up-hill facing scarp; U: bedrock; W: recent forest colonisation; C: Lower position of debris in the 1963 paroxysm.

4.3.2 Local geology and structural analysis

The Cassas landslide has been intensively studied in recent years. "On site" structural analyses have been carried out to characterise the main discontinuity systems. The results have been compared to the regional geological context ("around site" analyses) in order to achieve a better knowledge of the physical and geometrical characteristics of the main structural elements that concern the landslide.

- "On site" geological data consist of available geological data mainly collected on the landslide.
- "Around site" geological data consist of:
  - regional geological and structural data and related mesostructural and geomechanical data;
  - mesostructural data collected in the area surrounding the Cassas landslide;
  - lineaments detected from aerial photo analysis;
  - remotely sensed data.

4.3.2.1 "On Site" geological data
Data comes from geotechnical studies (Epifani, 1991; Brovero et al., 1996a; S.I.T.A.F., 2000) and the geological survey for the Bardonecchia sheet (Carta Geologica d'Italia in scala 1:50'000, 2002).

The Cassas landslide is placed in the Cerogne-Ciantiplagna tectonostratigraphic unit bounded at the top and the base by tectonic contacts (Fig. 4.9).

This unit consists mainly of:

- prevalent calcschists;
- carbonatic massive calcschists with interlayered, cm–mm thick micaceous schists;
- metabasites with local basaltic metabreccias;
- rare massive micaceous quartzites.

The available mesostructural data of the upper and lower part of the landslide permitted the observation of six main fracture (*s.l.*) systems (Fig. 4.16a):

- the main schistosity (**S1**) dips to the S-SE with an inclination of 40–50°. In some places, the schistosity gently dips to the N; in this case it is labelled **S2**;
- **K1** system: sub-vertical (80°–85°) persistent fractures dipping N350° and N170°;
- **K2** system: fractures dipping N220° and N240° with an inclination of 50°–70°; they are less persistent than K1 and K2;

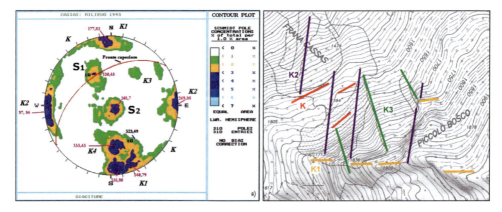

Figure 4.16. a) Stereographic net of the discontinuity systems; b) morphostructural lineaments observed in the upper part of the Cassas landslide. S1 and S2: schistosity; K1, K2, K3, K4 and K: joint systems (Brovero et al., 1996a).

58

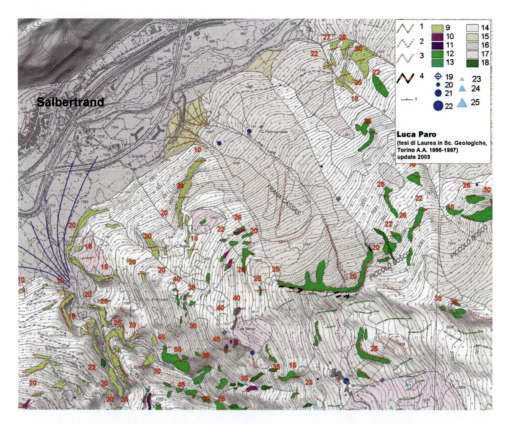

Figure 4.17. Extract of the geomorphologic and geological map of the right slope of the Susa Valley between Salbertrand and Exilles (Paro, 1997) updated in 2003. 1 – trenches; 2 – gravitational-structural scarps; 3 – scarp of fluvial-torrential erosion; 4 – metric open-joints; 5 – strike and dip of main schistosity; 9 – grey calcschists; 10 – quartzitic intercalations; 11 – olistoliths of metabasites and serpentinites; 12 – brown-yellow calcschists; 13 – phyllitic schists; 14 – landslide accumulations; 15 – fan of mixed origin; 16 – alluvial deposits; 17 – glacial deposits s.l.; 18 – travertines. Hydrology: 19 – water extinction point; spring (20 – <0.2 l/s; 21 – 0.2–0.8 l/s; 22 – >0.8 l/s; carbonate spring (23 – <0.2 l/s; 24 – 0.2–0.8 l/s; 25 – >0.8 l/s).

– **K3** system: sub-vertical (80°–90°) fractures dipping N90° and N270°;
– **K4** system: fractures dipping N350° with an inclination of 40°–50°.
– **K** system: sub-vertical fractures dipping to the SE and NW.

These fracture systems correspond to major morphostructural elements of the Cassas landslide, also confirmed by aerial photo analyses (Fig. 4.16b and Fig. 4.17). They are:

– the E-W striking lineament system (K1) that corresponds to the main scarp (western part) and open trench sub-parallel to the crown of the landslide;
– the N150° striking lineament system (K2) that corresponds to fractures and scarps within the landslide body and on the top of the landslide. These systems are sub-parallel to the lateral boundaries of the landslide.
– the N-S striking lineament system (K3) that corresponds to minor rock scarps within the landslide body;
– the N45° striking lineament system (K) that corresponds to main scarps (eastern part) and fractures within the landslide body and in the upper part of the landslide.

➤ Rock mass classification

Available rock mass data have also been analysed to give information on the geomechanical properties of the substratum of the Cassas landslide.

These data refer to some measurement sites in the surroundings of the landslide; these were chosen among those that were set up for the survey for the Bardonecchia sheet (Fig. 4.18).

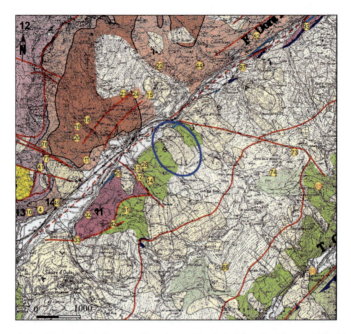

Figure 4.18. Portion of the geological map of Bardonecchia sheet (see also Figure 4.9) with location of the some measurement sites (yellow circle) collected during the geological survey of Bardonecchia sheet. The sites 12, 13, 14 and 28 are sites used for rock mass classification. Blue circle: Cassas landslide.

These data permitted the classification of the rocks of the slope according to the SMR method (Slope Mass Rating[1], see Table 4.2; Romana, 1990) according to the I.S.R.M. (1978) rules.

Table 4.2. Description of the rock mass quality classes (Romana, 1990).

Class	V	IV	III	II	I
SMR	0–20	21–40	41–60	61–80	81–100
Description	Very bad	Bad	Normal	Good	Very good
Stability	Quite unstable	Unstable	Partially unstable	Stable	Quite stable
Failure types	Translational slide	Translational slide and/or large wedge	Translational slide and/or small wedge	Some blocks	No one

---

[1] This method permits the classification of the slope in classes of quality by the following equation:

$$SMR = RMR + (F1 \cdot F2 \cdot F3) + F4$$

The RMR is the Rock Mass Rating (Bieniawski, 1974); F1, F2, F3 and F4 are corrective factors that define the type of slope failure resulting from geometric relationships between discontinuity systems and the slope (Romana, 1990).

The collected data are summarised in Table 4.3 and in the relative histogram (Fig. 4.18), where the RMR, SMR and RQD (rock quality designation) are reported.

The comparison between the SMR value and the Romana table put the rock mass of the Cassas landslide between classes of quality V and III.

These data underline the great variability of the geomechanical rock mass conditions as the measurement sites, although close to each other, give very different SMR values.

Table 4.3. Geomechanical data collected at some sites in the area surrounding the Cassas landslide (Lapenta & Trombetta, 1996).

Measurement Sites	Lithotypes	RQD	RMR	SMR
12	Carbonatic Calcschists	59	39	33.6
13	Calcschists	82	48	20.5
14	Carbonatic Calcschists	75	39	45
28	Calcschists	32	44	50

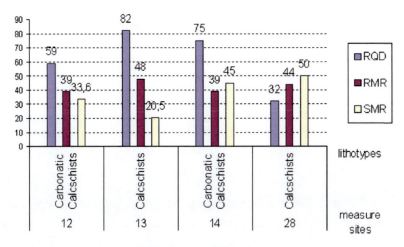

Figure 4.19. Histogram of the geomechanical data reported in Table 3.

4.3.2.2 "Around site" geological data

To improve the geometric and kinematic characterisation of the geological model, a comparison between the on site structural data and the around site geological and structural analyses has been made with the aim of finding relations between the morphological and tectonic elements of the landslide.

The amount of available regional data was high; those used here have been critically revised, since they were originally produced for very different purposes from those of the IMIRILAND project.

➢ Regional structural and geological data

The Cassas landslide is located in the joining zone of two main regional fault systems that control the morphology and geometry of the drainage network and bound some regional tectonos-tratigraphic units (Fig. 4.9).

These fault systems are:

– the N60° striking system (F1) that consists of some kilometre-scale extended pervasive faults and fractures. These are conjugate normal or left lateral faults dipping NW and SE, largely developed over the whole Susa-Chisone watershed;

61

– the N120° striking system (F2) that consists of some kilometre-scale extended right lateral and normal faults clearly visible in the Cassas landslide area.The geometric and kinematic features of these faults have been studied at mesoscale in the surroundings of the landslide.

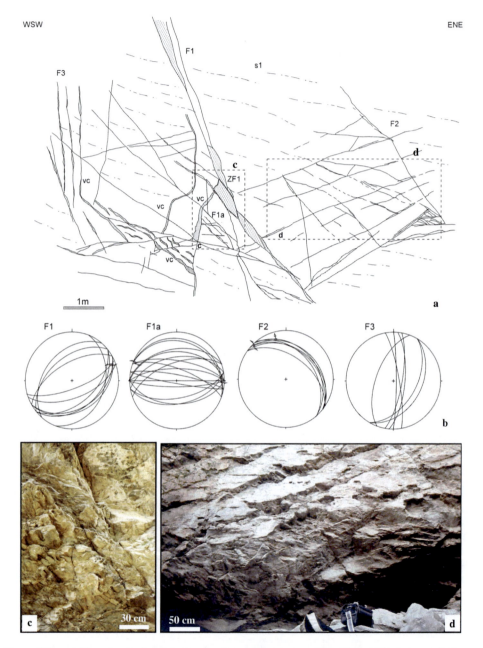

Figure 4.20.   a) Mesostructural sketch showing the peculiar cross-cutting relationships between fault systems (western boundary of Cassas landslide); b) Schmidt diagram (lower hemisphere) average cyclograph trace of the main discontinuity systems; c) and b) detailed pictures of the mesostructural sketch. s1: schistosity; F1, F1a, F2 and F3: fault systems; ZF1 shear zones (see text for further explanations); vc: calcite veins. In the sketch a) the location of c) and d) photographs is indicated.

The F1 and F2 fault systems have been analysed at 6 sites on the left side of the landslide where the rocks are relatively sound. These analyses allowed the definition of the structural associations[2], described in the following paragraphs (see Fig. 4.20, 4.21 and 4.22).

The **F1** mainly consists of transtensive conjugate faults dipping at high angle towards the SE and NW. The F1 system is the most represented at the outcrop scale; it defines metric to decametre-scale fault zones and disjoints the rock mass for several tens of metres (Fig. 4.21).

In some places, these faults are associated with steep SE-dipping metric shear zones (ZF1, in Fig. 4.20a), characterised by anastomised shear planes. Decimetre-scale to metric lenticular-shaped blocks and tectonic breccias and gouges are aligned on these surfaces. The thickness of these fault rocks, ranges from a few centimetres to some decimetres. The geometry of the F1 system is indeed very complex since a minor fault system (F1a, in Fig. 4.20a) is associated to the main one. The F1a consist of normal and left lateral faults dipping from low to high angle to the N and S (Fig. 4.20b), interpreted as synthetic Riedel shear planes (*sensu* Ramsay & Huber, 1987) of the F1. The F1 system roughly corresponds to the K, K1 and K4 fracture systems observed within the landslide (see "on site" geological data, Fig. 4.16a).

The **F2** mainly consists of discrete sinistral strike-slip faults dipping about 40° to the NE and decametre-scale shear zones dipping to the SW-SSW. The discrete faults have regular decimetre-scale spacing (Fig. 4.20), whilst the larger shear zones (F2) have a spacing of several metres and are usually associated to fault rocks (Fig. 4.21 and 4.22). The F2 system roughly corresponds to the fracture system (**K2**), observed within the landslide (Fig. 4.16a).

A third structural association, less evident at the regional scale but widespread around the Cassas landslide, mainly consists of high angle N-S to NNE-SSW striking normal faults (**F3**, Fig. 4.20). They are mainly discrete faults with associate sub-vertical centimetre-scale calcite veins (**vc**). The F3 system roughly corresponds to the fracture system (**K3**), observed within the landslide.

Figure 4.21. Decametre-scale fault zones associated to subparallel F1 fault systems (dashed red line) observed at the western boundary of the landslide.

Figure 4.22. Decametre-scale shear zone associated to the F2 fault systems observed at the western boundary of the landslide.

The schistosity (**s1**) dips at a low to medium angle to the SSE, both within and outside the landslide area. The schistosity also dips to the NE in some places near the landslide, probably due to

---

[2] The structural association is a "geological tool" for the correlation of the different discontinuity elements with different geometry, kinematics and size in genetically coherent groups (Vialon et al., 1976; Davis & Reynolds, 1996).

the presence of drag folds or tilting of rock blocks. Since the s1 has a relatively constant orientation, it can be easily extrapolated along strike and at depth for at least several hundreds of meters.

➤ Aerial photo lineaments

The geometric relations and the geomorphological characters of the discontinuity systems observed on the Cassas landslide have also been analysed at the scale of the slope by available photo-lineament analyses (Fig. 4.23) in the geotechnical study (S.I.T.A.F., 2000) and at the regional scale by means of interpretation of Landsat Thematic Mapper (Fig. 4.25). The integration of these different data led to a clearer understanding of the distribution of morpho-structural elements around the Cassas landslide.

Four main photo-lineament systems have been recognised (Fig. 4.23):

– a NE-SW system mainly consisting of regular vertical short scarps that show kilometre-scale lateral continuity and are clearly apparent on the left side and at the top of the landslide slope. These photo-lineaments are sub-parallel to the eastern part of the Cassas main scarp and trenches on the landslide body;

Figure 4.23. Geomorphologic map obtained from analysis of the aerial photographs (S.I.T.A.F., 2000) where is shown: the main landslide bodies (red colour); alluvial valley floor (light green colour); fluvial incision in erosion (brown colour); slope due to glacial modelling at low and height altitude (respectively yellow and light blue colour); lineaments (red line) Cassas landslide (blue dashed line).

– a E-W system consisting of aligned fractures, depressions, double-crested ridges and scarps with kilometre-scale lateral continuity. Some of these alignments are well developed in the upper part of the Cassas slope and they are also sub-parallel to the western part of the main scarp of the landslide (see also Fig. 4.17);

– a NW-SE system consisting of longer morphostructural elements that correspond to lateral boundaries of the Cassas landslide;

– a N-S system consisting of scarps (e.g. R in Fig. 4.15), trenches and fractures well developed in the upper and middle parts of the Cassas landslide.

➢ Remotely sensed lineaments

On the map (Fig. 4.24), the lineament distribution points out a paucity of lineaments on the Susa-Chisone watershed in comparison to the adjoining areas. It is worth noting that this paucity could be due to the uniform and wide distribution of fractures and faults that characterise the whole Susa-Chisone watershed. This high fracture intensity reduces the chance of detecting major spectral and morphological lineaments at the scale of satellite images.

Although the lineaments are not evident on the watershed, conversely, they show a good correspondence with regional geological structures in the surroundings of the landslide as also resulting from field observations.

Consequently, geometric relations between the regional fault systems and the satellite lineament systems have been sought to verify the statistic representativeness of the field structural data (Morelli, 2000; Morelli & Piana, 2003). Morphostructural and spectral criteria have been used for the detection of lineaments.

The identified lineaments were first analysed statistically and then compared with photo-lineaments and structural field data. The statistical analysis was carried out to characterise the relationships between the azimuthal frequency distribution and lineament length both for cumulative lengths and length classes. The statistical description is illustrated in the rose diagrams of Figure 4.25.

The frequency and cumulative length diagrams show a similar distribution (Fig. 4.25a, b). Both diagrams show a maximum of frequency and cumulative length on the N60°–N70° direction and another maximum on the N120°–N130° direction. Two other minor clusters are on the N80° and N170° directions.

The lineaments have been distinguished in terms of their length to determine the relations between frequency and length. Frequencies were divided into three length classes (Fig. 4.25c, d, e, and f).

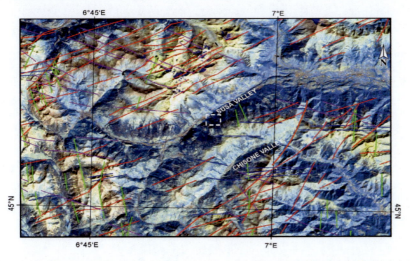

Figure 4.24. Landsat TM image processed (false colour composite 7-5-4 RGB,) with lineaments; the colours show the different lineament systems: LnS1 (red), LnS2 (violet), LnS3 (green) and LnS4 (blue). The blank dashed line is the landslide area.

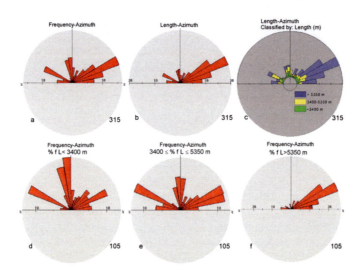

Figure 4.25. Landsat lineament diagrams showing the azimuthal frequency and cumulative lengths (a and b) and azimuthal frequency divided into classes of lengths (c-f). L: length of the lineaments; f: azimuthal frequency of the lineaments.

The frequencies for the classes of lineament length are more widely distributed and permit the identification of four lineament systems (see Fig. 4.24):

– LnS1 (NE-SW) lineaments are uniformly distributed in the three length classes. The LnS1 system consists of regularly spaced lineaments, uniformly distributed, especially on the left side of the Susa valley where they show geometric patterns similar to those of the F1 fault system.
– LnS2 (NW-SE) lineaments are mainly distributed in the intermediate and short classes. The LnS2 lineaments consists of regularly spaced lineaments, well exposed on the left side of the Susa valley where they cut the LnS1 lineaments or end against them. The cross-cutting relations of the LnS1 and LnS2 systems define lozenge-shaped domains and have geometric and hierarchic relations similar to those of the F1 and F2 regional faults.
– LnS3 (N-S) lineaments are better represented in the shorter class. The LnS3 system consists of minor lineaments particularly clearly observable in the eastern and western parts of the Susa-Chisone watershed. These show scarce evidence in the adjoining area of the Cassas landslide although widespread at the mesoscale.
– LnS4 (E-W) lineaments are better represented in the longer class. The LnS4 system consists of a cluster of lineaments mainly aligned along the middle Susa Valley. This system does not show cartographic evidence, but it is well exposed at the mesoscale in the surroundings of the landslide.

4.3.3 Investigation and monitoring

The monitoring system of the Cassas landslide consists of different instruments installed since 1991.

Five drilling surveys have been carried out until the present:

– 1991, 4 boreholes in the lower part of the accumulation;
– 1995, 11 boreholes in the middle–upper part of the landslide;
– 1998, 2 boreholes in the lower part of the accumulation;
– 1998, 2 boreholes in the upper part;
– 2002, 2 boreholes in the middle part of the landslide.

In total, 21 drillings were carried out to a depth between 25 and 80 m, nine of them instrumented with inclinometers, eight with piezometers and five with a geophones (piezometers and geophones measure continuously in real time). Some boreholes were supplied with fixed inclinometers. The last inclinometers installed were supplied with TDR systems.

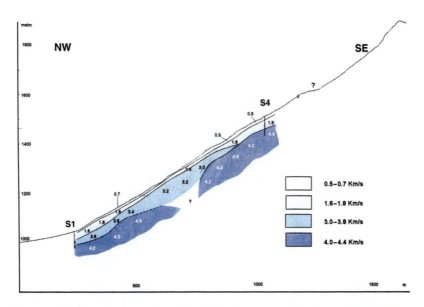

Figure 4.26.   Longitudinal seismic profile of the middle–lower area of the Cassas landslide (Brovero et al., 1996a).

At present, only one inclinometer is working (installed in 2002), as the other instruments were cut by the landslide movements.

Starting from 1997, near the crown, 13 extensometers and tiltmeters were installed: they measure the main fractures of the crown.

The rock wall at the left flank of the landslide is monitored by 12 joint gauges and by 1 geophone.

Begun in 1994, a microseismic network is available to monitor the rock noise, composed of 8 geophones (some of them located in superficial positions), both triaxial and uniaxial ones.

Some refractive seismic surveys have been carried out (Epifani, 1991) showing four different reflective horizons (Fig. 4.26).

4.3.3.1   Monitoring results

Inclinometers placed in the middle–upper part of the landslide body show a sliding surface at a depth of about 50–62 m. The stratigraphic logs indicate that the moving mass consists of a chaotic and heterogeneous material floating in a sandy-muddy matrix. Near the sliding surface, breccias and fine-grained materials are found. In the period 1995–2000, the average movement velocity was about 20 mm/y. Starting from the heavy meteorological event of October 2000, the movements have accelerated up to 70–90 mm/y.

The inclinometers located at the landslide foot show another sliding surface located at a depth of about 20 m, with an average movement velocity of about 30 mm/y. These data indicate that the middle–upper sliding surface of the mass is not related to the lower one. It seems that the upper surface intersects the slope at an altitude of 1'250–1'300 m, where a spring alignment is observed.

67

The surface instruments located on the Cassas landslide crown and on the rock wall at the left flank have not recorded significant movements (C.T.M., 2003).

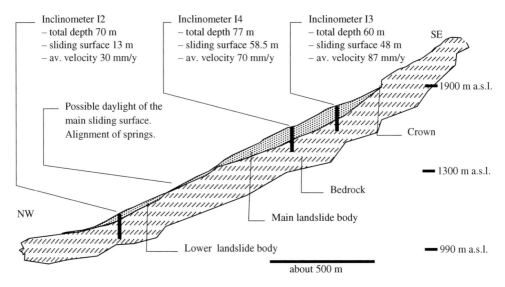

Figure 4.27.   Schematic section showing the most relevant monitoring instruments and results.

4.3.4   Danger characterization

4.3.4.1   Introduction
The definition of the dangers for the Cassas landslide is difficult, since it is complicated by the slow and continuous deformation at depth. The different studies carried out over the last ten years have supplied important elements for a correct understanding of the kinematics and dynamics of the landslide, leaving, nevertheless, several problems unsolved.

In this study, a detailed data analysis has been carried out; it supplies some indications about the evolutionary features of the phenomenon. Considering the difficulty of obtaining realistic results due to the complexity of this study, no numerical modelling has been carried out.

To obtain concrete conclusions about the dynamic and kinematic evolution, the data are analysed in detail in the following chapter.

4.3.4.2   Morphostructural constraints
In this study, the Cassas landslide has been divided into zones and particular attention has been given to the paroxystic phase of movement in the period 1954–1965 to obtain a better danger characterisation. Each of these zones is characterised by distinct kinematics.

Referring to Figure 4.28, it is possible to define the following zones:

A – main scarp zone: intensely fractured and released calcschists, with unstable blocks of variable volumes; in the crown zone many deep fractures are apparent;

B – active landslide accumulation: made up of rock blocks of variable size (max 15 m^3) floating in a sandy-muddy matrix; it is bounded at the base by a sliding surface at a depth of about 50–60 m, approximately at the contact with the bedrock as inferred on the basis of instrument data;

C – complex landslide accumulation: made up of rock blocks of variable size (max 5 m^3) in a sandy-muddy matrix; in this area the kinematics are complex because of the combination of sliding and slow-to-rapid flow movements;

D – alluvial fan area: area of debris flow expansion and accumulation originating in the middle and upper parts of the slope;

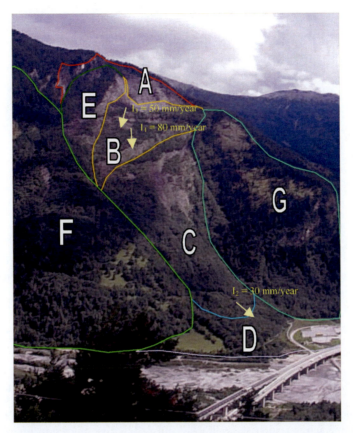

Figure 4.28.   Subdivision of the Cassas landslide to identify danger. Some inclinometers and the related movement data are located.

E – landslide accumulation in meta-stable equilibrium: block accumulation and mega-blocks (20–30 m^3) of older landslides, partly mobilised during the event of 1957;

F – ancient landslide accumulation: old accumulation body reaching the valley floor, partially reactivated by recent movements, probably still active at present.

To the West of the Cassas landslide a stable area may be seen (G) where superficial landslides are located; the monitoring data of the sub-vertical rock wall above the motorway service station do not indicate movement.

4.3.4.3   *Structural constraints for kinematic behaviour and displacement mechanisms*
The local discontinuity systems that were taken into account for the slope stability analysis (Fig. 4.29) belong to the main regional fault systems described above (F1, F2, F3). At the Cassas site, they correspond to:

–   decametre-scale to metric NE fault zones (F1) and associated F1a faults that displace the rock mass for several tens of metres. They are sub-parallel to the scar zone of the landslide and to some trenches observed within Zone B.
–   NW strike-slip faults (F2) and shear zones (F2sz) parallel to the lateral boundaries of the land-slide that correspond to alignments of many scarps more or less extended (i.e. the lateral boundary of Zone C).
–   N-S minor faults (F3) widespread all around the landslide; they are sub-parallel to the trenches and scarps in the middle and upper parts of the landslide. The most extensive scarp among those observed in the landslide body separates Sector B from Sector C.

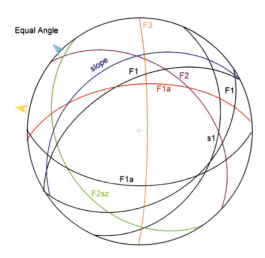

Figure 4.29.   Schmidt diagram (lower hemisphere) of the average cyclograph trace taken into account for the slope stability analysis. Legend: F1 fault zone with associated conjugate minor fault system F1a; conjugate F2 fault system and F2sz shear zone; F3 minor fault systems; s1 regional schistosity; the arrows shown the sliding movements toward WNW and NW (see text for further explanations).

These tectonic features permit the drawing of a schematic geological section of the bedrock (Fig. 4.30), where the structural framework is depicted by graphic repetition, below the entire slope, of the main structural associations observed in the field (a prototype structural site is sketched in the inset, where the actual configuration and cross-cutting relations of the features are reported). In this interpretation, the bedrock is affected by first order structures mainly dipping uphill, whilst second order structural features gently dip toward the valley.

Moreover, a concave discontinuity has been traced at the base of the shallower part of the bedrock, represented here as a strongly disengaged rock mass (debris). This discontinuity could correspond to the sliding surface inferred at the depth of about 50–60 m on the basis of monitoring data; this surface constrained the estimation of rock volumes for hazard scenarios (see below).

The structural setting of the bedrock described above would seem to disagree with the interpretation of geophysical data reported in Figure 4.26, where the presence of four different layers characterised by progressively higher seismic velocity with depth is pointed out. Indeed, the inferred surfaces that have been traced in Figure 4.26 to separate layers characterised by different seismic velocities, are not to be understood as potential sliding surfaces, but simply as horizons (envelopes) that could separate rock layers showing different mechanical properties (here thought of as strictly dependent on fracture density).

Since the existence of a sliding surface has been postulated to justify the gravitational instability of the landslide (see scenarios), a critical analysis of the structural setting of the bedrock must be carried out to verify which among the pre-existent structural features could act as a potential sliding surface.

Furthermore, the physical and morphological characteristics of the structural system (roughness, morphology, opening, infillings, alteration etc.) are to be considered in order to evaluate qualitatively how much the slope is prone to instability.

Searching for kinematic relations between bedrock structures and landslide features is allowed only where the geometry of landslide body is strictly controlled by the tectonic features of the bedrock. At the Cassas site, this condition is verified, since the landslide has been subdivided into morphostructural sectors bounded by steep scarps parallel to regional fault systems and that correspond to local (on site) tectonic features.

Consequently, the role of each structural feature with respect to gravitational instability will be revised as follows:

- the schistosity is not prone to rock sliding, since it regularly dips to the SE (i.e. uphill).
- most of faults and shear zones of the Cassas bedrock are recemented by diffuse carbonate veins; furthermore, the fault surfaces are often rough, wavy or irregular.
- the bedrock of the landslide is strongly fractured, mainly by steep to medium-angle faults and shear zones whose dip direction is opposite or oblique to the dip direction of the slope.

Nevertheless, the intersection of the main structural systems points out that sliding movements toward the WNW and NW can be suggested (Fig. 4.29). These sliding directions that result from the intersection of F2sz shear zones with F1a discrete faults and by F2sz with F2 conjugate faults, are, respectively, oblique or quite parallel with respect to the dip direction of the slope (NNW). The WNW sliding direction is roughly perpendicular to the F3 system which corresponds to the main scarp which splits the landslide body into two parts longitudinally (Zones B and C). This sliding direction coincides with the displacement recorded by the monitoring system (C.T.M., 2003). In this kinematic framework, rock blocks bounded at the base by F2sz shear zones, laterally by F2 faults and at the rear by the F3 system could slide toward the WNW; in this case, the left lateral boundary of the landslide could hinder the sliding movement and could induce partial toppling of blocks.

The NW-dipping intersection line is, instead, more favourable to rock sliding, but could involve in fact only small-sized blocks, since the F2 faults are usually closely (dm) spaced.

In conclusion, fast sliding of huge rock masses is unlikely on the Cassas slope since the theoretical sliding direction that can be inferred from the intersection of the main fracture systems is oblique to the dip direction of the slope, or where sub-parallel, only small-sized blocks are involved.

Furthermore, the physical and morphological characteristics of fractures are not generally favourable to rock sliding. Consequently, complex debris slide and flows, instead of rock sliding, have been imagined for the assessment of hazard scenarios, as described in the following paragraphs.

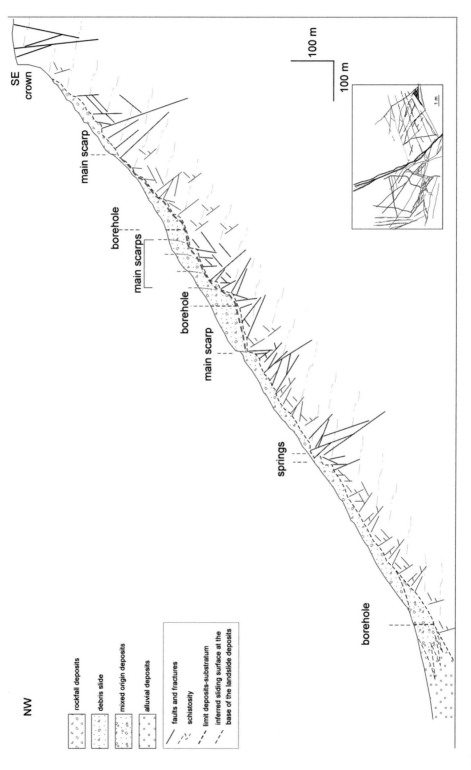

Figure 4.30. Geometry and cross-cutting relations of the faults and fractures in the bedrock refer to the actual configurations observed at mesoscale in the most intensively deformed zones, here shown in the right corner sketch.

4.3.4.4 *Scenario identifications*

Concerning the dangers recognised in this study, three main typologies of movement are described: rockfall, debris slide and flow. The rockfall involves only the main scarp sector, the debris slide and flow involve several zones of the Cassas accumulation.

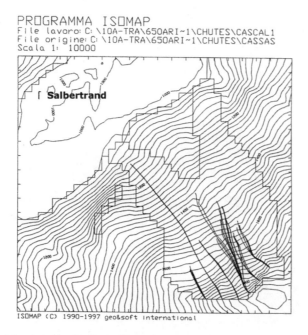

```
PROGRAMMA ISOMAP
File lavoro: C:\10A-TRA\650ARI~1\CHUTES\CASCAL1
File origine: C:\10A-TRA\650ARI~1\CHUTES\CASSAS
Scala 1: 10000
```

ISOMAP (C) 1990-1997 geo&soft International

Figure 4.31.   Rockfall trajectories (Epifani, 1991).

Rockfall accumulations are mainly located near the foot of the Cassas main scarp. The average volume of unstable blocks, as observed in the deposit, ranges from some m^3 to 10–15 m^3. The only rockfall accumulation of about 50'000 m^3 took place between 1957 and 1963; it is located at the main scarp foot. Concerning the movement of great volumes in the crown zone, the instrumentation data related to the open joints do not indicate any displacement at the moment. Rockfall trajectories (Fig. 4.31) have already been described (Epifani, 1991).

More difficult to explain is an interpretation of a general Cassas landslide evolution, in particular for Sectors B and C, inside the Cassas landslide accumulation of 1957, where important instability elements are shown.

The hypothesis expressed in the present work is that the movements in the debris body that were originated during the event of June 1957 remain constant over time, with possible acceleration during heavy rainfall (for example during the event of October 2000). The greatest volume corresponds to area B (about 1×10^7 m^3) which was also the mobilised volume in June 1957.

The debris body C presents a complex evolution with debris slide in the top sector, next to Sector B, and slow debris flow in the middle–lower area. During the paroxystic phase of 1957–1965, this area was involved in rapid debris flows partly accumulated here. Actually, slow flow movements seem mainly concentrated in the upper level, as indicated by inclinometer I_2 located at the foot of the accumulation (movement of about 30 mm/y to a depth of about 13 m; C.T.M., 2003). Volume estimation of the debris body is, therefore, about 1.5×10^6 m^3.

The continued movements measured over the last ten years in the upper part of the slope emphasised an acceleration starting from June 2001 (approximately 20 to 80 mm/y), caused by the heavy rainfalls of October 2000 and by the winter/spring snow melting.

The evolutionary hypothesis for these sectors is summarised as follows (Fig. 4.32):

- slow and continuous movements in area C towards the accumulation body;
- the slides in the upper part of area C produce instability in area B;
- due to progressive deformation of the debris accumulation or to triggering events (ex. rainfall) that produce acceleration, it is possible that one or more sliding surfaces are in limit equilibrium until the instantaneous failure and development of a paroxystic phase; parts or the entire body B and likely part of the body C would slide towards the valley floor;
- the water quantity would condition the typology and the velocity of the debris movement in the middle-lower area and subsequently the ways of accumulation in the valley floor;
- the sliding area B could also condition the stability of area E, as already happened in June 1957; the total volume of area E is about $1.5 \times 10^6 \, m^3$.

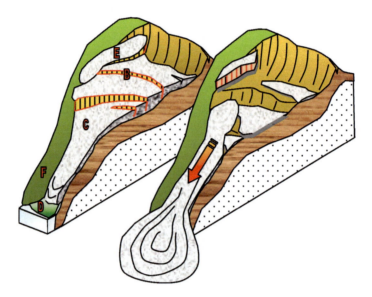

Figure 4.32.   Longitudinal sketch cross-sections of the Cassas landslide: present situation (the letters refer to Figure 4.28) and the evolutionary hypotheses.

The movement could include a total volume of about $10 \div 12 \times 10^6 \, m^3$ with accumulation both on the slope and on the valley floor. The accumulation would be lobe-shaped, considering the material typology and the shapes of the oldest accumulations on the same slope (Fig. 4.8). It is difficult to foresee the real extension of the accumulation area on the valley floor, as it will depend on the phenomenon velocity and on the real quantity of the material that reaches the floor. Besides, the "funnel-shaped" slope, tightening at the slope base, will strongly condition the extension of the accumulation.

In the risk analysis survey by S.I.T.A.F. (2000), a first attempt to determine the extension of the accumulation body was carried out using a numerical model (cellular automata modelling): it supplies the limits of the accumulation areas for different volumes, from 1×10^5 to $12 \times 10^6 \, m^3$. For volumes inferior to $0.15 \times 10^6 \, m^3$, the material accumulates entirely on the slope.

To complete the framework of the danger assessment, it is necessary to evaluate the areas external to the Cassas landslide. In the downhill sector (F in Fig. 4.28) some scarps and localised landslides are shown apparently as a continuum with the morphological elements of the Cassas landslide area. The morphological and structural-geological elements available do not permit the definition of a correct model of the whole slope; the lateral continuity of the morph-structures remains to be demonstrated. However, a generalised slope movement is improbable in the case of heavy rainfall, whereas, it is possible to observe local instability phenomena in the middle–lower area of the old accumulation at the eastern edge of the Cassas landslide. In the western area of the Cassas landslide (G in Fig. 4.28), characterised by a general stability, the only critical sector seems to be the rocky spur (T in Fig. 4.15), placed in the upper part of the slope, next to the main scarp of the Cassas landslide. In this area, there are fractured bedrock, trenches and small debris accumulations of gravitational origin. In this case, the hypothesis is that during a general slope movement this sector would also move and that part of the material would flow into the Cassas main accumulation and part would directly reach the valley floor. The volume of this unstable area is about $0.5 \times 10^6 \, \text{m}^3$.

In summary, the dangers identified in this study are:

– blocks or rock masses falling from the main scarp of the Cassas landslide (velocity of some m/s) with an accumulation zone at the scarp foot or at the slope. Each block has a maximum volume of some tens of m^3; the rockfall has a maximum volume of $10^5 \, \text{m}^3$;
– collapse of parts of the slope with debris slide and flow (variable velocity from some m/min to some tens of m/month) with accumulation partly on the slope, mostly on the valley floor. Maximum volume inferior to $10 \times 10^6 \, \text{m}^3$ (indicative volume useful for the delimitation of the area affected). It is excluded, at present, that the phenomenon could evolve as a debris avalanche;
– collapse of the whole debris body, also involving some parts external to the "1957 Cassas". Debris slide and flow (variable velocity from some m/min to some tens of m/month), rockfalls of the rock spur to the West of the Cassas landslide with trajectories that partly converge in the accumulation body of the Cassas landslide and partly reach the valley floor after passing over the wall above the service station; involvement of the E area (Fig. 4.28) with complex movements of debris slide type. Volumes of about $10 \div 15 \times 10^6 \, \text{m}^3$.

The affected areas have been defined according to considerations based on the available documentation and the geomorphologic analysis:

– for the rockfall phenomena: by rockfall trajectories (Fig. 4.31);
– for the complex debris slide and flow phenomena ($<10^7 \, \text{m}^3$): morphologically based hypothesis for volumes of about $1 \div 2 \times 10^6 \, \text{m}^3$;
– for the complex debris slide and flow phenomena ($>10^7 \, \text{m}^3$): morphologically based hypothesis for volumes of about $1.5 \times 10^7 \, \text{m}^3$.

4.3.5 *Occurrence probability*

In spite of the available information on the Cassas landslide, among which those related to the monitoring system data and to some weather data, the interpretations of the event probability of the different dangers previously recognised remain an hypothesis.

On the basis of the considerations explained in Chapter 3, it is possible to present the following interpretation.

4.3.5.1 *Rockfall phenomenon*

Detachment of rock blocks of variable sizes from 1 to $20 \, \text{m}^3$ are assumed to have annual frequency. In fact, although no direct observation or historical documents of such a phenomenon are available, it is supposed that cryoclastic processes and bedrock degradation cause continuous detachments of intensely fractured and released rock elements from the main scarp of the Cassas landslide; the absence of vegetation on the debris at the scarp foot underlines the high frequency of the phenomenon. Falls of rock blocks of larger size (of about $10^5 \, \text{m}^3$) have a lower frequency, about 10^1 years (in fact a rockfall of about 50'000 m^3 took place during the 50s–60s of the past century).

The rockfall frequency was applied to the maximum extension of the area affected by rockfall, without distinguishing between contiguous areas (high probability of rockfall trajectories) and areas far from the main scarp (low probability of rockfall trajectories).

4.3.5.2 *Complex debris slide and flow phenomena*

It is difficult to establish an occurrence probability for phenomena which would involve the whole slope. Considering the sparse historical data available, the heavy meteorological events of May 1728 and of June 1957, an interval of about 230 years is obtained. Such a recurrence time span is representative for the whole slope around Cassas and not only for the specific Cassas landslide, as the documents of 1728 refer to a generic "landslide opposite Eclause" (Fig. 4.8), therefore, on the right slope, that "temporarily would have dammed the Dora river" (Bogge, 1975).

An attempt to obtain the occurrence probability can be made by linking the observed effects on the slope to the probable triggering causes. It is not easy to discover a link between rainfall and the movement velocity, as the inclinometer measurements are carried out periodically at irregular intervals over a time span of less than 10 years. An attempt to correlate rainfall with movement was carried out by the society that manages the monitoring system (C.T.M., 2003). This study shows a relation between the "accelerations" recorded by the inclinometer I_3 (Fig. 4.27) and the average precipitation over the last 90 days; in particular, a prominent acceleration of the movement is recorded in correspondence with the maximum of the average precipitation over the last 90 days in October 2000 (Fig. 4.33). Considering this information and the rainfall data of the Salbertrand weather station, the return time for such a maximum of the average precipitation over the last 90 days was calculated using the Gumbel method.

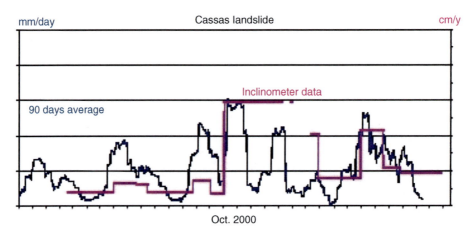

Figure 4.33.   The Cassas landslide: 90 days average precipitation and inclinometer data relationship (C.T.M., 2003, modified).

This elaboration gives a return time of **50 years** which can be considered the minimum threshold after which the Cassas landslide will have important accelerations induced by heavy rainfalls.

The recurrence time span is then assumed as ranging from **50 to about 250 years**. This recurrence time span is calculated for an event with movement of a debris volume less than that of the 1957 event ($<1 \times 10^7 \mathrm{m}^3$). During the 1957 event, a volume of about 10 million m^3 of bedrock, intensely fractured and released, slid along pre-existing surfaces (structural surfaces). After a first phase of acceleration, during which the landslide did not reach a velocity higher than that of a rock-debris avalanche, a period of progressive slowing of the movement followed which lasted about ten years. Gravitational phenomena that would involve volumes superior or equal to $1 \times 10^7 \mathrm{m}^3$ (until a max. of about $1.5 \times 10^7 \mathrm{m}^3$ for each event) have, on the basis of previous considerations, a **"return time" greater than 230–250 years**.

The return times cited above are indicative, considering the complexity and the large uncertainty of the system. For example, a hydrogeologic model of the Cassas landslide is not still available, permitting a relation among rainfall-infiltration-movements. The morphological features, the grain size distribution in the accumulation and the high fracture intensity suggest that the hydrogeologic system could be very complex. The available piezometric data confirm this interpretation, as the variation of the groundwater depth does not seem to be in direct relation with rainfall. Recent studies (Gianquinto, 2000), aiming to establish a hydrogeologic model for the Cassas slope, suggest the high permeability of both the landslide accumulation and the intensively fractured rock mass. The role of the latter in water circulation is unknown; it is not known if the bedrock has either a draining role or a feeding one or, probably, both functions, related to the overhanging debris accumulation.

For a quantitative risk assessment, the frequency of debris slides and flow have been applied only to the expansion areas of the gravitational phenomenon in the paroxystic phases (area affected on the valley floor). The high frequency values (= max frequency value recognised) have been applied to the whole accumulation body of the "1957 Cassas" landslide which, according to the instrumental data, is still continuously deforming.

Table 4.4.   Occurrence probability.

Scenario	Occurrence range (average)	Frequency
Rockfalls	1–50 (25)	1/25 = 0.04
Debris slide & flow ($<1 \times 10^7 m^3$)	50–250 (150)	1/150 = 0.007
Debris slide & flow ($1 \div 1.5 \times 10^7 m^3$)	250	1/250 = 0.004

4.4   QUANTITATIVE RISK ANALYSIS

Three scenarios for the Cassas landslide evolution have been assumed. Each scenario is characterised by a proper intensity, by a triggering and run-out areas (derived from the geomorphological model) and by an occurrence probability (as a result of the historical analysis of recorded events).

According to the proposed quantitative risk analysis, each element of the hazard scenario is used separately in the matrix calculation of risk evaluation. This process is applied to each recognised scenario; therefore, for the Cassas landslide three separate risk analyses have been carried out.

At the end of the matrix calculation, a unique value is obtained (total risk), divided into 4 categories: physical, economic, environmental and social, useful for the total risk assessment.

The three scenarios are:

A. **rockfall phenomenon**: volumes 10^0–$10^5 m^3$; area affected: obtained by rockfall trajectories;
B. **complex debris slide and flow phenomena**: volumes $<107 m^3$; area affected: morphologic hypothesis for volumes of about 1–$2 \times 106 m^3$;
C. **complex debris slide and flow phenomena**: volumes 1–$1.5 \times 107 m^3$; area affected: morphologic hypothesis for volumes of about $1.5 \times 107 m^3$;

Risk analyses are based on land use defined by Salbertrand town planning and by the areas affected as previously indicated.

4.4.1   *Elements at risk*

The elements at risk are recognised through the definition of the area affected by each danger in the scenarios previously specified. These elements at risk are shown in the following tables and in Figure 4.34.

Figure 4.34.   Elements at risk next to the Cassas landslide.

Table 4.5.   Rockfall scenario: element at risk identification.

Element at risk	Name	Notes
Forests (private and public properties)	Bosco Chapel (Piccolo bosco)	Area affected: 0.4 km^2 Persons affected: 2 (occasional presence of monitoring operators)
Special elements	Instruments of the Cassas monitoring system	Inclinometers Piezometers Geophones Extensometers Laser distance measuring

Table 4.6.   Debris slide & flow scenario ($<10^7$ m^3): element at risk identification.

Element at risk	Name	Notes
Forests (private and public properties)	Bosco Chapel (Piccolo bosco)	Area affected: 0.23 km^2 Persons affected: 2 (occasional presence of monitoring operators)
Rural areas	–	Area affected: 0.06 km^2 Persons affected: <5 (occasional farmer)
Transportation facilities (motorway and international railways with corresponding services)	Torino-Frejus motorway Motorway service station	Area affected: 0.05 km^2 Motorway affected: 400 m Persons affected: $50 \div 100$ (service station operators, motorway and service station customers)
Valuable environmental areas	Dora Riparia river bed	Area affected: 0.1 km^2
Special elements	Instruments of the Cassas monitoring system	Inclinometers Piezometers Geophones Extensometers Laser distance measuring

Table 4.7. Debris slide & flow scenario ($1 \div 1.5 \times 10^7 \, \text{m}^3$): element at risk identification.

Element at risk	Name	Notes
Forests (private and public properties)	Bosco Chapel (Piccolo bosco)	Area affected: 0.6 km² Persons affected: <5 (occasional presence of monitoring operators, people going to the Sapè of Salbertrand and Exilles)
Rural areas		Area affected: 0.08 km² Persons affected: <10 (occasional farmer)
Transportation facilities (motorway and international railways with corresponding services)	Torino-Frejus motorway Motorway service station Torino-Modane international railway	Area affected: 0.082 km² Motorway affected: 800 m Persons affected: 50 ÷ 150 (service station operators, motorway and service station customers) Railway affected: 900 m
Productive areas		Area affected: 0.09 km² Persons affected: 10
Residential areas	Gravere (Salbertrand)	Area affected: 0.005 km² Persons affected: <10
Valuable environmental areas Local services and plants	Dora Riparia river bed	Area affected: 0.13 km² Area affected: 0.03 km² Persons affected: <5
Special elements	Instruments of the Cassas monitoring system	Inclinometers Piezometers Geophones Extensometers Laser distance measuring

4.4.2 *Vulnerability of the elements at risk*

To define the vulnerability of the elements at risk, an analysis about involved assets typology and processes' intensity has been carried out.

In the rockfall scenario, the processes' intensity is not very high, as trees and debris would slacken the rock blocks rolling along the slope. The monitoring instruments do not resist rock block impacts, even those with low energy. In the technical reports related to the monitoring, difficulties in obtaining measurements due to damages to the instruments are recorded. Therefore, it has been decided to assign the elements at risk in the area affected a vulnerability value of 1 (100%).

In the debris slide and flow scenarios, a maximum intensity has been given to the recognised processes, as the involved volumes are huge (millions or tens of millions m³). Considering the elements at risk involved in the gravitational phenomenon (buildings, infrastructures, roads and railway), a vulnerability value of 1 (100%) was considered: in the area affected (failure, depletion and accumulation areas) the destruction is considered total. Also, at the valley floor, when the velocity tends to zero, the accumulation can be so big that all the assets can be completely destroyed, heavily damaged or buried and, therefore, permanently useless. The velocity and thickness distribution in the accumulation area depend on the initial velocity and on viscosity (water quantity in the material in movement).

In the quantitative risk analysis, it is important to consider the system uncertainty and most of all the delimitation of the areas affected, as it is the result of hypotheses and assessments. Therefore, the elements at risk next to affected areas cannot be considered to have zero vulnerability and, consequently, zero risk. In particular, concerning the debris slide and flow scenario $>10^7 \, \text{m}^3$, the area affected reaches the railway embankment, but it is possible that the whole railway would be impacted with heavy consequences for traffic in both directions.

4.4.3 *Expected impact*

Multiplying the values of the elements at risk and the vulnerability percentage (obtained as described in the previous section), the expected impact is obtained.

The affected areas recognised, even if hypothetical and approximate, underline the problem related to the presence of a motorway at the foot of the Cassas landslide. The A32 motorway Turin-Frejus tunnel, the only tunnel between Italy and France during the Mont Blanc tunnel closing due to an accident, is of international economic importance.

Concerning social consequences, being a heavy traffic road, the values are particularly elevated. If a landslide paroxystic phase occurs, probably many people would be affected.

This probability rises in the sectors involved in the event of 1957, when the service station "Gran Bosco" was built. Furthermore, the accumulation area at the valley floor is next to the tunnels on the left slope of the Susa Valley, being a problem for those who arrive from Turin if the traffic would not be stopped.

The same consideration must be made concerning the Turin-Modane railway, which runs in tunnels before arriving at the Salbertrand plain next to the Cassas landslide.

In order to state a precise quantitative damage evaluation for economic activities and assets, costs derived from economic analyses are essential (e.g. asset costs, reconstruction costs, turnover, profits loss, etc.). To assess environmental and social expected impacts (where an economic value is difficult to obtain), relative values indices are used.

4.4.4 *Risk assessment*

The last phase of the quantitative risk analysis concerns risk evaluation. This evaluation is obtained by multiplying the expected impact values, obtained previously, and the numeric values of the occurrence probability related to each scenario.

In the case of the Cassas landslide, as previously reported (§ 3.5), the values of the occurrence probability were obtained on the basis of an analysis of historical data (Table 4.4).

Risk assessment results are summarised in the images from G.I.S. and described in the related captions.

As shown in Figures 4.35, 4.36, 4.37 and 4.38, for the debris flow and slide scenarios, the major risks are connected to the landslide body on the slope: this situation is possible because the landslide on the slope is active and, therefore, with high frequency (≥ 1). *For the correct considerations of the real risk, only that which exists in the accumulation area on the valley floor must be evaluated.*

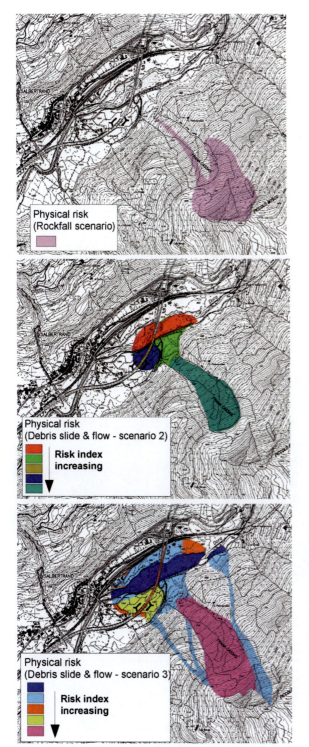

Figure 4.35. Map obtained by G.I.S. techniques concerning the physical risk for the 3 scenarios analysed.

For the rockfall scenario, no distinction exists, as it concerns a homogenous area. Risk values are not zero, as a very expensive monitoring system exists which is very important for the motorway and service station operation. A detailed analysis can be made considering only the instruments correctly located.

In the two debris slide and flow scenarios, the major values of physical risk are, first, those of the service station "Gran Bosco" and, second, those of the motorway and the areas classified as productive areas in the land planning.

81

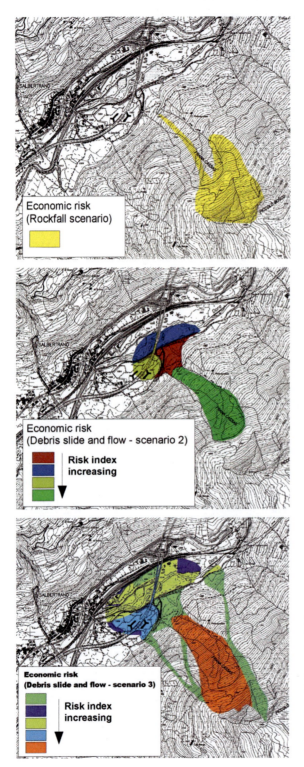

Figure 4.36. Map obtained by G.I.S. techniques concerning the economic risk for the 3 scenarios analysed.

As it can be observed in the images, the areas with major risk values are the A32 motorway (Turin-Frejus tunnel) and the service station. Particularly important is the Dora Riparia river, as well as the intrinsic value of the river, its waters are captured downstream of the Cassas landslide by a power plant.

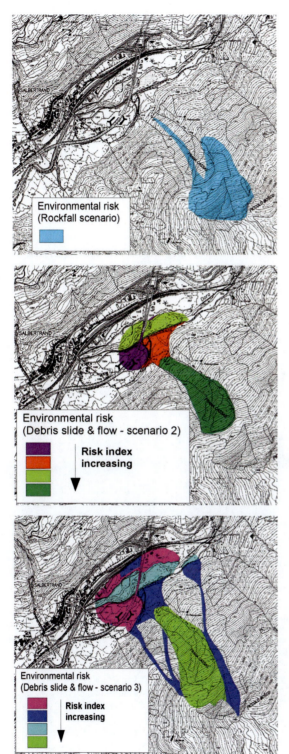

Figure 4.37. Map obtained by G.I.S. techniques concerning the environmental risk for the 3 scenarios analysed.

The main areas with major risk values are, obviously, the slope areas (in the Natural Park "Gran Bosco di Salbertrand") and the valley floor areas where the Dora R. river runs.

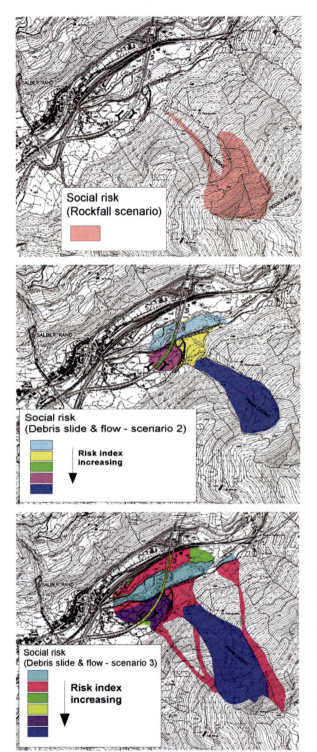

Figure 4.38. Map obtained by G.I.S. techniques concerning the social risk for the 3 scenarios analysed.

Concerning the rockfall scenario, the risk values are not zero as periodically 2 monitoring system operators are at the landslide.

For the debris slide and flow scenarios, the areas with elevated social risk are the Turin-Frejus motorway tunnel and the related service station. A detailed analysis shows how traffic variation due to time and seasons leads to different considerations about risk evaluation.

4.4.5 *Total risk*

In the images (Fig. 4.39) the results of the quantitative total risk analysis for each category (physical, economic, environmental and social risk) are reported. As explained in the chapter concerning quantitative risk analysis methodology, the total risk can be obtained adding the risk value for each category of each scenario.

As already detailed for the risk analysis, the high risk index obtained for the slope sectors involved in the active phenomenon must not be taken into account. This derives from the fact that for the Cassas landslide, a high occurrence probability value has been considered and this leads to a high risk value, even if linked to a low value of the affected assets. Therefore, for a correct risk analysis, it is necessary to consider the affected areas at the valley floor whose occurrence probability values are 10 times inferior to those on the slope and that reach high risk values for the important consequences evaluated.

Concerning the assets evaluation, the analyses are not detailed, but the relative values obtained by the methodologies have been applied. Probably, to obtain a correct framework of the risk situations related specifically to the Cassas area, careful social and economic evaluation should be done on the basis of the considerations of the next paragraph (§ "Indirect risk and regional impacts").

In any case, it is important to underline that the motorway and the service station affected by the landslide evolution present high risk indices regarding the physical, economic and social total risk. Particularly important in the analysis of the environmental total risk is the Dora R. river bed.

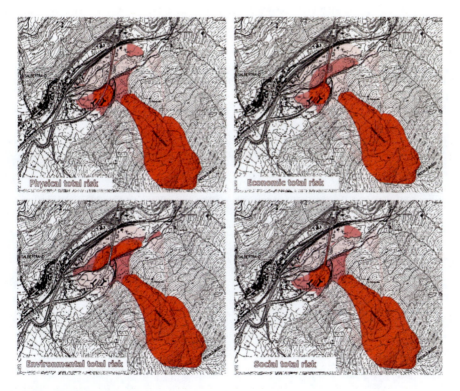

Figure 4.39.   Total risk maps.

4.5   INDIRECT RISK AND REGIONAL IMPACTS

It is particularly difficult to understand the indirect risks associated with the processes related to the recognised scenarios. This difficulty, in particular for debris slide and flow processes, is due to several uncertainties about the evolution of the phenomena, for example:

– What volume will be exactly involved in the movement?
– How much material will deposit on the slope and how much will reach the valley floor?
– Which is the maximum velocity of the material in movement?

This will depend on the water quantity and on the related viscosity and velocity of the mass in movement.

In the worst case, in which the catastrophic scenario shows the landslide blocking the valley, a dammed lake would be formed. It is presumed that the landslide would occur during a critical meteorological event, creating important flows along the Dora Riparia river (some hundreds of m³/s).

The combination of these processes (valley floor dam with tens of meters of landslide accumulation and river flood) would create a basin of some hectares in few hours. Some villages in the valley would be then overflowed with some damages to human assets and the evacuation of several people, rising the social risk value.

Moreover, due to lake formation, downstream villages could potentially be involved. Instantaneous lake depletion can, in fact, be caused by a rapid overtopping dam erosion, or a landslide dam collapse caused by piping and internal erosion. The processes that lead to a sudden lake depletion are mostly:

– threshold erosion with instantaneous dam break when the lake reaches the maximum level;
– piping erosion causing dam break.

In both cases, the propagation of a flash-flood wave in the valley would create important problems which are difficult to evaluate. In this specific case, no analysis of sudden collapse has been carried out yet.

Furthermore, due to a valley dam, the power plant of Chiomonte, which catches the water of the Dora Riparia river in Serre la Voute, would be closed with important economic consequences for the entire valley.

The presence of the motorway and of the railway in the landslide area has already been described. They have great importance from a military and economic point of view. The Frejus tunnel was the only passage to France (open all year) during the closure of the Mont Blanc tunnel following a bad accident. The economic damages caused by the motorway and railway closing would be of about 10^6 € with relapses on the economy of Italy and France (Fig. 4.40).

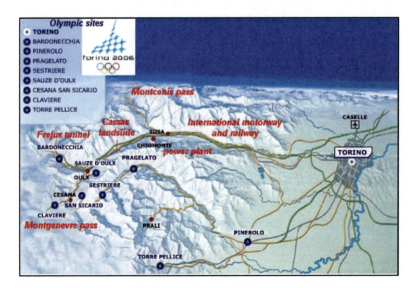

Figure 4.40.   Synthesis of the meaningful elements at risk.

The motorway and the railway play an important role in the economic development of these mountain areas, as they are linked to tourist resorts. The Susa Valley and the Chisone Valley are two important ski resorts connected to other ski stations in France: international ski competitions take place in this area every year and at this time new infrastructures are developing for the winter Olympic games of 2006.

4.6 REGULATION AND RISK MITIGATION MEASURES ALREADY AVAILABLE

The motorway and the service station "Gran Bosco" at the foot of the Cassas landslide raise some problems related to risk mitigation, problems never taken into account either in relation to the Salbertrand village or in relation to other infrastructures in this area. The S.I.T.A.F. (the society that manages the motorway) has been studying the best protection both for the motorway and the service station since the end of the '80s of the last century.

At the beginning of the '90s in the last century, S.I.T.A.F. built a reinforced earth wall, 120 m long and 5–6 m high, to protect the motorway service station from both debris flows and rockfalls. Rockfall fences (Fig. 4.41) were also installed at the accumulation foot and at the rock wall base over the service station.

Over the same period, a monitoring system was installed in order to measure the phenomenon with the following aims:

- to improve the knowledge of the phenomenon in order to evaluate the risk statistically and numerically;
- to create relations among instrumental data for landslide forecasting.

The monitoring system is composed of: inclinometers, piezometers, geophones, extensometers; recently, a distance measuring laser was installed to evaluate the surface movements of the secondary scarp crossing the "1957 Cassas". The rock wall over the service station is also monitored with automatic measurements in real time.

The monitoring system, managed by the motorway company, acts as a warning system for both the service station and the motorway. In future, it will be connected to a larger warning system, according to the general civil protection plans of the area, but at the moment the thresholds are not officially recognised.

Figure 4.41.   Rockfall fence at the Cassas landslide foot.

5
The Rosone landslide

G. Amatruda[1], S. Campus[2], M. Castelli[1], L. Delle Piane[3], F. Forlati[2], M. Morelli[4], L. Paro[2], F. Piana[4], M. Pirulli[1], R. Polino[4], M. Ramasco[2], C. Scavia[1]

[1]Politecnico di Torino; [2]Arpa Piemonte; [3]Consulting geologist; [4]CNR-IGG sezione Torino

5.1 INTRODUCTION

The entire southern side of the ridge bounded by the Orco and Piantonetto Rivers, in the Province of Turin, has been undergoing a slow process of deep-seated gravitational slope deformation (DSGSD). This deformation process, a vast landslide movement, is historically known as the "Rosone landslide" because it frequently involved the small village of Rosone located at the toe of the slope. At present, this phenomenon also involves the 99 MWh hydroelectric power plant of the Electricity Agency (AEM) of the city of Turin (Fig. 5.1). Coming from the Ceresole Reale dam, the water reaches the AEM facilities through a 17 km long pipe after spanning the entire length of the gravitational deformation, where it falls towards the power plant via a penstock with a drop of 813 m. At the toe of the slope, the only National Road (No. 460) from Turin to Ceresole Reale is located.

5.2 GEOLOGICAL BACKGROUND OF THE LANDSLIDE

The so-called Rosone landslide involves an area of about 5.5 km^2 and reaches a depth of over several decametres (Ramasco et al., 1989), and affects a 1300 m high slope, from an altitude of 2000 m at the ridge crest down to 700 m at the valley bottom (Fig. 5.2 and 5.3).

The morphological and structural characteristics of the area suggest subdividing it into three adjacent sectors, roughly corresponding to the villages of Ronchi Perebella, and Bertodasco (Fig. 5.2 and 5.3). These sectors reflect, respectively, final, early, and intermediate stages of the evolution of a site subject to a deep-seated gravitational process. On account of the presence of such widely diversified conditions in a single area, this site turns out to be particularly suited for stability studies and their comparative evaluation (Forlati et al., 1993).

Examining the area from west to east, the Ronchi sector is encountered first, characterised by a highly advanced stage of evolution of the deformation that has caused the disruption of the original rock formation. This sector may now be assumed to be substantially stable.

The central sector, around Perebella, reflects an early stage in the deformation process. It is clearly separated from the Ronchi sector by a N-S striking scarp a few hundred meters long and 50–90 m high. The bedrock in this sector is more damaged in the middle and lower parts.

The eastern sector (Bertodasco), which is inhabited and hosts the AEM electric power plant, shows a deformation stage intermediate between those of the Ronchi and Perebella sectors. Previous studies (Forlati et al., 1993) identify the Bertodasco sector as the one most likely to undergo a catastrophic evolution. A morphostructural analysis revealed the presence of three zones with different degrees of mobility, from the top to the floor of the valley, referred to as A, B and C (see below).

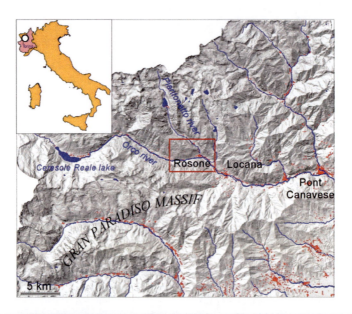

Figure 5.1. a) Digital Terrain Model of the north-western Piedmont with the location of the Rosone land-slide area (red square), main rivers and lakes (in blue), and villages and town (in red). b) General overview of the Orco and Piantonetto watershed (air photographs overlapped Digital Terrain Model) with instability phenomena surveyed in the surroundings of the Rosone landslide area (IFFI Project – Gravitational Phenomena Inventory in Italy, ARPA-Piemonte, 2004). The coloured areas show: DSGSD (orange), complex landslides (violet), rockfall (yellow), rock slide (light green), fall and topple areas (red).

90

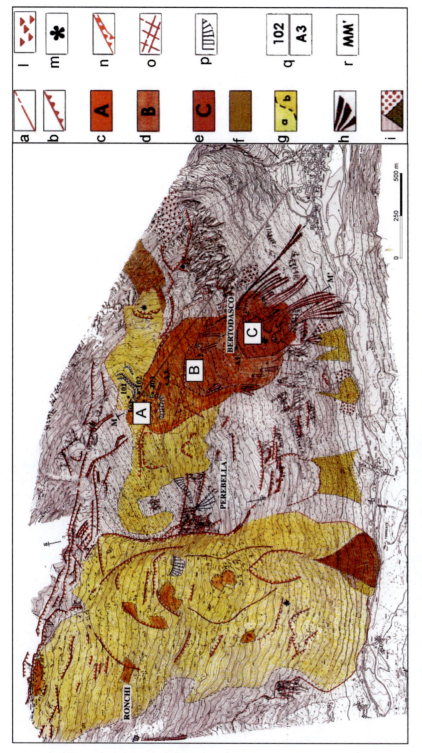

Figure 5.2.   Morphostructural map of the Rosone deep seated gravitational slope deformation showing the subdivision into distinct morphological sectors: Ronchi, Perebella and Rosone. Legend: a: boundary of main rocks subjected to mass movements; b: trace of the main slip surfaces or scarps; c: Bertodasco sector – Zone A; d: Bertodasco sector – Zone B; e: Bertodasco sector – Zone C; f: Ancient landslide deposits; g: (a) debris due to disruption of rock mass (b) fragmented rock mass where the original structural features are preserved; h: debris flows of the 1953 event (from 1954 aerial photos); i: debris cones; l: rockfalls; m: boulders; n: trenches and opened fractures; o: traces of the main joint systems; p: zone of extensive waste materials; q: boreholes location; r: M-M': trace of section of Figure 5.18 (Modified after Brovero & al., 1996b).

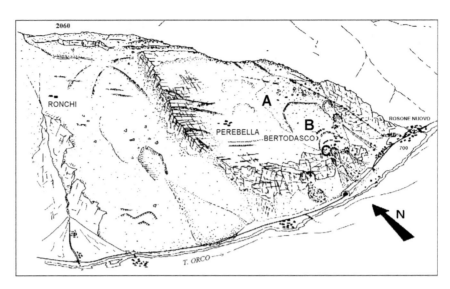

Figure 5.3.   Sketch of the Rosone landslide (Brovero et al., 1996b).

5.3   HISTORICAL BACKGROUND

The Rosone landslide underwent two major paroxysmal stages, one in the early 18th century (1705–1706 from an inspection report "*Atto di visita*" in Luino et al., 1993) and another in the fall-winter of 1953; they caused severe damage to many buildings and the disruption of cultivated fields.

From historical reports, in 1933–1934 the inhabitants were evacuated for about 7 months. In the early 1940's, a landslide occurred at an altitude of 1300 m: huge boulders toppled downhill and threatened Rosone and the eastern part of Bertodasco. In 1940, the Bertodasco sector was affected by increasing slide movements with the triggering of flows.

The landslide showed signs of re-activation in 1948 and 1951. In the fall-winter of 1953, following abundant precipitations, a portion of the slope collapsed. The movement of the ground in the Bertodasco area damaged or even destroyed some houses. The fall of debris and boulders from the slope overhanging Rosone caused the evacuation of the 250 inhabitants and their cattle. Due to the consequent damages, a new Rosone village (named Rosone *nuovo*) was built in 1956. This village is close to the AEM power plant.

Between 1953 and 1957, these movements gradually slowed down, as confirmed by topographic measurements carried out by AEM. In 1957, the two villages Rosone and Bertodasco were evacuated.

Between 1957 and the early 1960's, the movements accelerated and phenomena similar to those recorded in 1953 were observed, though less severe. Further phenomena were recorded in the fall of 1963 and in the spring of 1964 and 1969, 1988, 1993.

In the 1993 and 2000 important displacements were also recorded (several centimetres) immediately after heavy meteorological events.

At present, significant movements continue to take place in the upper part of the Bertodasco sector, as borne out by the topographic measurements performed on the anchoring blocks of the penstock.

A summary of the historical events is shown in Figure 5.4.

5.4   REGIONAL FRAMEWORK

5.4.1   *Climatic and water conditions*

The climate is characterised by a pre-alpine regime. The thermometric regime is characterised by 4–6 frost months per year. The Ceresole Reale station, located at an altitude of 1600 m,

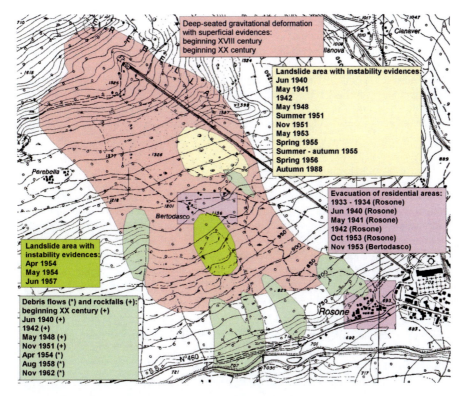

Figure 5.4. Historical data synthesis concerning slope instability (Rosone landslide).

registered 184 frost days (1950–1986) with a probability of 80–100% of frost days from December to March.

On the Graie Alps, the pluviometric rate is of a pre-alpine type and the average annual number of rainy days varies from 90 to 110. On the Orco basin, average annual precipitations are about 1224 mm/year and average annual rainy days are 96. The average daily intensity is about 12.8 mm/day. Average annual rainfall has been determined over a period of 42 years, from 1938 to 1980. The above-mentioned results refer to a station placed at an altitude of 700 m, at the bottom of the slope in question.

Since 1990, a new pluviometric and thermometric station has been activated at Bertodasco (1120 m) corresponding to the average altitude of the slope.

The seasonal distribution of rainfall can be summarised as follows:

May: the highest monthly rainfall value (160 mm), the greatest average number of rainy days (12 days) and the highest cumulative rainfall values are reached in three or four days (up to nearly 195 mm).

September: heavy downpours of relatively short duration.

October: the monthly average is similar to that of May (155 mm–160 mm), but the average number of rainy days is modest.

On a regional scale, snow rates mainly vary according to altitude. On the Graie Alps, snowfalls are frequent at an altitude between 2000 m and 2300 m and they have their maximum in April and a secondary maximum in February. At an altitude above 2300 m, snow rates have a single maximum in April. At an altitude between 1200 m and 1700 m, snow rates are unimodal, showing a maximum in February (Ceresole Reale at 1573 m). The maximum snow level recorded at Rosone is about 2 m (1994–1995), the minimum level is 0.1 m (1966) and the average is 0.35 m. In the Orco Valley, the maximum snow level was recorded during the 1989–1990 winter, preceded by the

93

minimum recorded during the 1986–1987 winter. The absolute monthly maximum in the last thirty years was measured in January at the Rosone station, located at and altitude of 700 m.

Temperature increases (10°–15°) have been recorded in the last decade for February, testified by rapid snow melting.

It is of interest to point out that two perpetual springs are found near Perebella and Bertodasco and some temporary springs at the toe of the slope and near the right flank of the Bertodasco sector. The flow of water in the slope occurs through the main discontinuities, which are generally open and persistent. Considering that the upper part of the slope acts as a reservoir, the largest part of the groundwater is cumulated when rainfalls coincide with the melting of the snow cover.

5.4.2 Regional geology and structural settings

The landslide site is located near the confluence of two glacial valleys, namely the Orco and Piantonetto valleys. In fact, the geomorphology features of the area are due to fluvial-glacial morphogenesis although modified by gravity deformations. Slopes are usually very steep and the valleys are narrow.

The Orco Valley is located in the central part of the Gran Paradiso Massif (Fig. 5.5). The geological unit belongs to the Pennine domain and it consists of a composite crystalline basement and a Permo-Liassic cover, locally preserved in the peripheral areas. The Gran Paradiso Massif consists of three different complexes (Compagnoni et al., 1974):

– Augen Gneiss Complex: augen gneisses and fine-grained gneisses with inter-bedded metabasites.
– Monometamorphic Complex of the "Money" area: albite micaschists and graphite-bearing, gneisses inter-bedded with quartzose meta-conglomerates.
– Erfaulet Ortogneiss Complex: leucocratic ortogneiss.

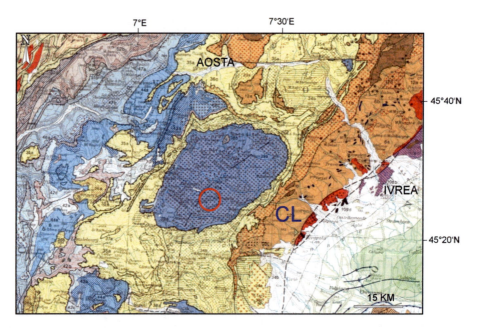

Figure 5.5. Geological map of a sector of the Western Alps (CNR, Structural Model of Italy 1:500000 scale). In the red circle the landslide site is placed within one of the main Alpine tectono-metamorphic units (Gran Paradiso Unit GPU) in the axial part of the chain, bordered on both sides by the Piemonte Zone (Penninic domain, in yellow). East of GPU are the Austroalpine and South Alpine domains (dark red and brown colour), the Canavese Line (CL), a major fault of the Alps and Po Plain, while west of GPU, Briançonnais (light blue) and Mont Blanc domains (red) are represented.

All of these units are locally covered by glacial deposits and they are cut by two main joint sets, mutually orthogonal, corresponding to E-W and N-S striking sub-vertical faults.

The metamorphic deformations are mainly recorded by a regional schistosity (Sr in Fig. 5.11) coeval with the development of regional folds that have been later refolded with structural styles (open folds).

The post-metamorphic deformations are characterised by major strike-slip tectonics and minor thrusting (Perello et al., 2004). Later extensional brittle faulting is also recorded.

5.5 HAZARD ANALYSIS

5.5.1 *Geomorphological analysis*

Following systematic geomorphological and structural analyses carried out in recent years, the Ronchi, Perebella and Bertodasco sectors are described here in detail:

– *The Ronchi sector*

In this area, the deep seated slope deformations have reached the maximum evolution level. The gravitational processes, having lasted here for a long time, have erased any evidence of glacial morphology and strongly modified the profile of the slope. In ancient times, the slope failed and rockslides occurred.

The slope shows, at present, large elongated depressions and double-crested ridges in its upper part, while in the intermediate and lower parts a prominent bulge formed and partially obstructed the Orco valley.

This morphology is mainly due to a settlement in the upper zone (immediately below the double-crested ridges), which occurred along several hundred meters of quite continuous scarps of circular shape and with sub-vertical slip surfaces; this settlement has also been partitioned along the discontinuities of the highly fractured rock mass and gave origin to many large undulations of the slope. It also induces the formation of a prominent bulge in the lower part of the slope, in response to dilatation.

Displaced material actually appears in the form of big rock blocks of several cubic meters (up to 2000–3000 m^3 meters for a single block) distributed, together with heterogeneous debris, over the whole area (Fig. 5.6). In the lower part of the Ronchi sector, some minor ancient landslides are present.

Figure 5.6.   Aerial view of the high part of the Ronchi sector. The boundary main scarp, on the right, as well as the intense disruption of rock mass, is depicted.

– *The Perebella sector*

This sector shows a preliminary stage of evolution (minor frequency of released joints and minor disruption of rock mass). Ridge-top trenches with graben-like depressions and double-crested ridges, also surveyed in the Rochi sector, characterise the upper part of the slope, where along the main discontinuities dislocative behaviour (shear and tension stress) is observed (Fig. 5.7). In the intermediate and lower parts of the slope, the gravitational deformation gave origin to many uphill facing scarps and in some places very unstable piles of boulders (Fig. 5.8).

The control of the structural setting on the instability phenomena appears quite clear in the middle part of the slope. The cross-cutting relationships between the discontinuity system sub-parallel to the schistosity, corresponding to the dip direction of the slope, and the two main sets of sub-vertical discontinuities develops a toppling mechanism of instability. The rock mass dilatancy due to severe confining stresses determines a particular "arch" slope configuration.

Finally, since the original (tectonic-related) structural features are still locally preserved in the Perebella sector, the cross-cutting relations among the main joint systems have been extrapolated over the whole substratum of the Rosone area (Fig. 5.9).

Figure 5.7.   Double-crested ridges along the Chisone-Piantonetto watershed of the Perebella sector.

Figure 5.8.   Up-hill facing scarps in the upper part of the Perebella sector.

Figure 5.9. Aerial view of the medium part of the Perebella sector: cross-cutting relations of the main steeply dipping joint systems are well exposed.

– *The Bertodasco sector*

This sector shows an intermediate degree of evolution with respect to that of the Ronchi and Perebella sectors. While the upper part of the slope is characterised by trenches perpendicular to the direction of the major movement, the middle part is affected by a rotational slide movement superimposed on a general toppling and planar slide. The disarticulation degree of the rock mass increases from the top to some hundred meters over the slope foot.

The morphostructural evidences and the different kinematic behaviour allowed the identification of three minor zones (A, B, C in Fig. 5.2 and 5.3).

– Zone A includes the upper part of the Bertodasco sector, where the movements of the slope (as a whole) and related morphological evidences, are poorly defined. Moreover, although disrupted, this zone still preserves its original structural features. This suggests the presence of minor translational movements, as also confirmed by the AEM monitoring system (Forlati et al., 1993).

The eastern lower part of Zone A shows more prominent movements, as indicated by opened fractures, and is affected by undulations and local bulges. The rocks here are split into blocks, although still aligned along the main joint systems.

– Zone B is bounded by a prominent curved scarps and by rectilinear scarps sub-parallel to the main joint systems. These scarps often act (at least since the 1953 crisis) as slip surfaces, with movements of 3–4 meters that occurred along quasi-planar surfaces, sub-parallel to the schistosity. These kinematics are strictly related to the shape and orientation of the rock blocks resulting from the intersection of the K1 and K2 joint systems. The gravitational instability of Zone B also induced a partial translation of the eastern lower part of Zone A, and caused the displacement of the pipes of the AEM power plant. At the eastern boundary of Zones A and B, toppling and rolling of rock blocks have been observed. These rock blocks frequently moved towards the ancient Rosone village and as a consequence the new Rosone village (Rosone *Nuovo*) was established.

– Zone C consists of very heterogeneous mobilised mass, as it has been affected by major gravitational movements since the 1930's. It consists of disengaged rock blocks of different size; a mixture of chaotic coarse gravel and fine material. The movements mainly affect the upper part of the zone along many concave and sub-parallel scarps characterised by an offset of 1 to 10 meters. These movements have also determined an abrupt change in slope morphology.

The movements in the upper part of this zone have caused, during the last fifty years, the break-down of part of the Bertodasco village, which is now completely abandoned.

The lower part of Zone C is instead affected by rock-debris and debris flows that have destroyed the National Road (SS. 460) that runs along the Orco valley (Fig. 5.10).

97

Figure 5.10.   View of the Bertodasco sector where the origin and deposit zones of debris flows are well exposed.

5.5.2 *Local geology and structural analysis*

5.5.2.1 *Lithological features*

The Rosone area is made up of intensively laminated orthogneiss, within minor levels of white micaschists and rare chloritoschists. The micaceous levels in the orthogneisses gave origin to anisotropy that causes subdivision of the rock masses into differently thick (cm to metre) massive slices, bounded by downstream dipping slip plains. These slices are sub-parallel to the main schistosity, which here is highly pervasive, as it represents the axial surface of isoclinal folds. At the slope scale, the anisotropy of the rock masses gave origin to plain-parallel surfaces that show high lateral continuity (Geoengineering, 1984; Brovero et al., 1996b; Perello et al., 2004). In some places, the white micaschists correspond to mylonitic levels.

5.5.2.2 *Mesostructural analysis*

➤ On site available data

(a) Regional schistosity

The main schistosity (Sr) dips to N150 with an inclination of 35° and is characterised by a regional metamorphic foliation corresponding to the axial planes of metric to hectometre-scale tight to isoclinal folds that refold an older pervasive foliation.

The schistosity is locally developed parallel to silvery micaschist levels (see below for further details) that represent high strain shear zones, due to metasomatic reactions along high pressure mylonitic shear bands (Dal Piaz & Lombardo, 1986). Though discontinuous, these silvery micaschists are to be carefully considered in the stability analysis of the landslide, since they are characterised by lower shear strength with respect to the average values of the Gran Paradiso orthogneiss.

A joint system is often developed parallel to the schistosity (Studio Geologico Italiano, 1984; Geoengineering, 1984).

(b) Fractures and fault systems

Two main joint systems have been recognised (Fig. 5.11). They are the K1 and K2 steeply dipping (75°–85°) systems, which strike N10 and N90–N100 respectively (Fig. 5.9).

Two other minor joint systems are also present. They are the K3 system, that dips 70° to N240 and the K4 system, that is steeply dipping and strikes N40 (Geoengineering, 1984).

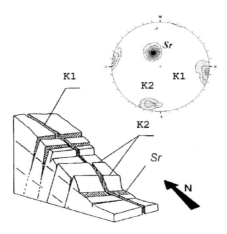

Figure 5.11.   3D sketch of the main tectonic discontinuities of the Rosone area. The poles of joint systems and schistosity are shown on a Schmidt diagram.

➢ Around site data

As mentioned previously, the Rosone landslide affects a pre-quaternary basement made of coarse-grained orthogneisses, deriving from alpine metamorphic evolution of hercynian granites. These granites were pervasively deformed along mylonitic shear bands, where metasomatic processes were prevalent, and caused a total rock volume reduction as great as 50%. The products of such transformations are known as white schists (Schreyer, 1977) or silvery micaschists (Compagnoni & Lombardo, 1974; Chopin, 1981; Dal Piaz & Lombardo, 1986).

Geological surveys on the Rosone landslide and the adjoining areas (SEA consulting, 2001) allowed the observation, within the orthogneisses, of the presence of several decimeter to decameter levels of *silvery* micaschists, locally folded with an axial plane corresponding to the main schistosity, but generally transposed and sub-parallel to it. The distinctive paragenesis is represented by quartz, phengite, chloritoid, ±talc, ±Mg-chlorite; in the Rosone area, the Mg-chlorite may predominate, originating chlorite-schists with poor mechanical characteristics (very low shear and compressive strength), with a general down-slope dip. Thick chlorite schists and silvery micaschist horizons locally border the Rosone landslide, cropping out in the Perebella area and in the adjoining upper Piantonetto Valley, and can be held responsible for deep instability processes within the pre-quaternary basement. The Rosone landslide is not the only example of such relationships between metamorphic structures and recent instability in the Orco valley: the Noasca landslide, about 10 Km west of Rosone, is another huge landslide whose sliding plane partially coincides with a metric level of down-slope dipping silvery micaschists.

The unfavourable outcrop conditions make very difficult to follow individual faults in the field, but correlations between faulting and development of deep instability in the Rosone area are pointed out by the parallelism between trenches and mesoscale faults and fractures. Nevertheless, the mapping of the main fault sets of the upper Orco valley (SEA consulting, 2001) highlighted the presence of several regional fault zones affecting the pre-quaternary basement and generating a related network of minor discontinuities (Fig. 5.12 and 5.13). The main fault zones of the area are the two E-W trending fault systems of the *Colle della Porta* (left side of the Orco valley) and *Lago Fertà* (right side) (Fig. 5.12). Between them, several kilometric WNW-ESE striking faults are present. The Rosone landslide is located in the step-over between two main WNW-ESE faults, where the relatively strong frictional deformation may affect a wide portion of the basement, generating several minor discontinuities sub-parallel to the major ones.

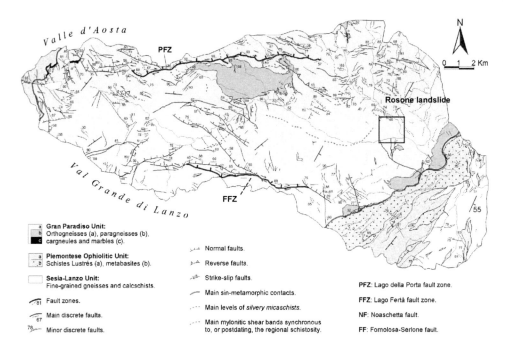

Figure 5.12. Structural sketch map of the upper Orco valley from 1:10.000 field mapping and 3D digital photo-interpretation.

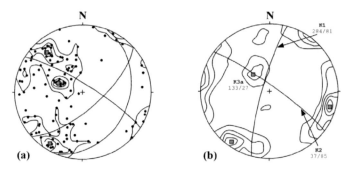

Figure 5.13. Schmidt equal-area diagrams – Poles of faults. a: 126 data and joints b: 1127 data of the Rosone-Noasca area. Data correspond to faults and joints observed at the mesoscopic (metric to decametric) scale. Great circles represent the average attitude of joint and fault sets evidenced by the contour lines. Joints K3a represent low-angle discontinuities sub-parallel to the regional schistosity; they may not correspond to levels of silvery micaschists.

5.5.3 Investigation and monitoring

In the 1960–1980 twenty year period, attention had been almost exclusively devoted to the sectors immediately adjacent to the AEM systems (penstock and reservoirs): percussion and rotary drillings with continuous coring were performed (maximum depth: 120 m) and both surface and deep movements were investigated (topographic and inclinometric measurements, respectively). The boreholes were instrumented with inclinometers and piezometers.

Since 1985, attention has been focused on the whole slope, with special emphasis on the Bertodasco sector: a systematic campaign has been conducted on the entire rock mass involved in the sliding activities.

The methods employed included:

- surface investigations for the characterisation of the rock mass and the main discontinuity systems;
- depth investigations consisting of: a) boreholes drilled by continuous coring. Inside the boreholes, the piezometric pressures and the deep-seated displacements by means of inclinometric readings have been recorded. Moreover, at the top of some boreholes the surface displacements have been measured by means of a triangulation net; b) two triaxial accelerometers, located on the main sliding body, useful to detect noises emitted by movements of rock masses within the landslide;

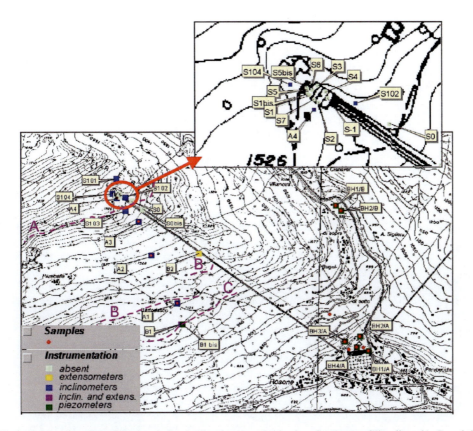

Figure 5.14. Location of the boreholes, inclinometers and seismic reflection profiling lines (A, B and C pink dashed lines) in the Bertodasco sector.

- laboratory tests and geomechanical characterisation of the samples obtained from the boreholes with the aim of evaluating the shear strength of both the discontinuity surfaces and the altered material obtained by drilling.

In order to survey the displacements around the Bertodasco sector, during the year 2000, a new integrated monitoring system was installed, with automatic data recording. It can be subdivided in:

- the geomechanical network, including inclinometers, piezometers and extensometers. In particular, 7 wire extensometers were placed along the main scarp which divides Zone B from Zone C and along some fissures inside the latter one; three thermometers allow compensation for thermal drift;

101

- the topographical network, consisting of a topographic total station and a GPS (Global Positioning System) net that uses 5 satellite receivers and 19 bench marks controlled with manual GPS measurements;
- the geophysical network, including a vertical deep geophone located in borehole A3, two 3D deep geophones located in boreholes B1 and B2 (see Fig. 5.14 for the location), 3D surface geophones and a vertical surface micro-seismic detector.

The results of the laboratory tests carried out on samples and of in situ tests on bedrock are summarised in Table 5.1.

Table 5.1.   Average values of some mechanical parameters related to the Rosone landslide.

Unit volume weight γ	26.2	kN/m^3
Uniaxial compressive strength C_0	74.9	MPa
Tensile strength T	7.1	MPa
Wave velocity p	3794	m/s
Base friction angle φ_b	35.1 ± 2.9	°
Joint compressive strength JCS	110	MPa
Barton joint roughness coefficient JRC	11	–

The surveys allowed the highlighting of the sliding surfaces' depth in the Bertodasco sector and in those immediately adjacent to the penstock and reservoirs (Studio Geologico Italiano, 1984; Forlati et al., 1993; Brovero et al., 1996b). Drilling campaigns were performed in 1959, 1960, 1970, 1982, 1984, 1991 and 1999; the results of those boreholes, later instrumented with inclinometers, are summarised in Table 5.2. Figure 5.15 shows the core drilled from borehole A2 in the depth range including the failure zone made of destroyed rock.

On the basis of the displacement versus depth plot registered by the inclinometers, it is possible to recognise the presence of a "knee" that can be interpreted as a failure zone. In Figure 5.16, the displacement versus depth plot from the A1 inclinometer is shown as an example. In Table 5.3, the measurements related to the sliding movements of the unstable rock mass, obtained by means of the inclinometers, are summarised.

Furthermore, seismic reflection profiling and a downhole log, performed along three lines and represented in Figure 5.17 (see Fig. 5.14 for the location), allowed the collection of data about the rock mass physical properties and landslide body thickness. The results, in terms of wave propagation velocity, underline two layers with different behaviour: the superficial one is characterised by a greater wave velocity, typical of a loose rock mass; the deeper layer reveals a lower wave velocity value that can be linked to a structured rock body. The surface between the two layers of rock can be linked with the previously mentioned failure zone, as obtained from the observation of the inclinometer measures and the drilling cores.

The recorded data of the monitoring system and the evaluations derived from the investigations permit the confirmation of the geomorphological hypotheses.

5.5.4   Danger characterization

5.5.4.1   Geomorphologic constraints
The danger characterisation was focused only on the Bertodasco sector due to the peculiar hazard conditions and availability of monitoring data recorded on AEM power plant and related pipes.
The Bertodasco sector

Two zones with major movements have been identified: Zone B mainly characterised by planar sliding surfaces, and Zone C, where rotational sliding is inferred (Fig. 5.2 and 5.3).

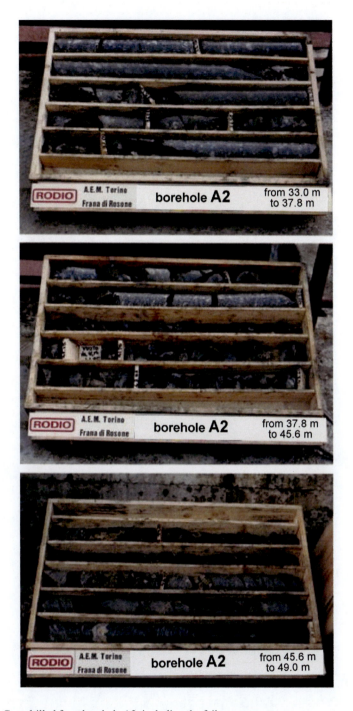

Figure 5.15.   Core drilled from borehole A2, including the failure zone.

The lateral boundaries of these sectors correspond to deep and opened fractures. The sliding surface (traced with a dashed line in Figure 5.18 at the base of Zones B and C) is marked by a sharp vertical change in the fracturing conditions of the rock mass. This change is taken as a failure band below which the rock mass is unaffected with respect to that involved in the landslide.

Table 5.2. Features and measures of the boreholes carried out immediately adjacent to the penstock and reservoirs. Here are shown only those instrumented with inclinometers in 1991 and 1999. The location of the mentioned boreholes is represented in Figure 5.14.

	Sampling type	Drilling date	Altitude	Depth	Meaningful Values
A1	Rotary drilling with continuous coring. Bored again in 1999 (solid drilling) until 51 m depth	1991	1143	102.2	0–48.7 m: alternating soil/rock 48.7–102.2 m: rock Failure surface intercepted at 38.98 m depth
A2	Rotary drilling with continuous coring	1991	1336	120	0–120 m: rock Failure surface intercepted at 45.7 m depth
A3bis	Rotary drilling with continuous coring. Bored again in 1999 (solid drilling) until 100 m depth	1991	14460	100.5	Failure surface intercepted at 71.3 m depth
A4	Rotary drilling with continuous coring	1991	1514	120	Failure surfaces intercepted at 30.0 m and 47.5 m depth
101	Rotary drilling with continuous coring	1984	1540	31.3	0–3.55 m: alternating soil/rock 3.55–31.3 m: rock with sampling >80% Failure surfaces intercepted at 3.1 m depth Water: 26.2 m
102	Rotary drilling with continuous coring	1984	1506	39.2	0–22.9 m: alternating soil/rock 22.9–39.2 m: rock with sampling >80% Sand and silt cohesionless level: 16–22.9 m Failure surfaces intercepted at 22.6 m depth
103	Rotary drilling with continuous coring	1984	1495	81	0–57.7 m: alternating soil/rock 57.7–81 m: rock with sampling >80% Sand and silt cohesionless level: 40–57.7 m Failure surfaces intercepted at 52.4 m depth
104	Rotary drilling with continuous coring	1984	1517	30.4	0–21.5 m: alternating soil/rock 21.5–30.4 m: rock with sampling >80% Sand and silt cohesionless level: 13–21.5 m Failure surfaces intercepted at 20.1 m depth Water: 25.4 m
B1	Rotary drilling with continuous coring. Television probe	1999	1105	108.3	0–43.5 m: alternating soil/rock 43.5–108.3 m: rock with sampling >80% Sand and silt cohesionless level: 37.3–43.5 m
B2	Rotary drilling with continuous coring. Television probe	1999	1290	99.7	0–40.3 m: alternating soil/rock 40.3–99.7 m: rock with sampling >80% Lugeon permeability test: depth [m]/UL extrapolated to 10 bars 62.95–64.45/68–78 72.70–74.70/10–15

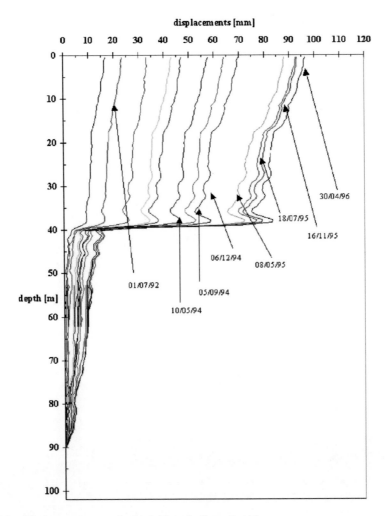

Figure 5.16.   Displacement versus depth plot from inclinometer A1.

Table 5.3.   Measurements related to the sliding movements obtained by means of the inclinometers.

| | Measurement period | | Surface displacements | | | |
	Initial	Last	Failure surface depth [m]	[mm]	Annual average [mm/year]	Dip direction [°]
A1	12/12/1991	3/6/1999	38.98	157.1	21.00	187.9
A2	12/12/1991	9/11/2000	45.72	131	14.69	170.1
A3	12/12/1991	5/10/1993	71.31	Closed	–	–
A4	12/12/1991	6/11/2002	30.00	73.4	6.73	158.7
101	6/12/1984	27/11/1993	3.05	57.7	6.42	147.9
102	6/12/1984	6/11/2002	22.55	116.3	6.49	157.1
103	6/12/1984	7/11/2000	52.42	117.1	7.35	165.5
104	6/12/1984	7/11/2000	20.12	89.5	5.62	149.3
B1	17/11/1999	25/10/2000	40.23	131.5	–	144.3
B2	17/11/1999	6/11/2002	37.78	56.2	18.91	162.7

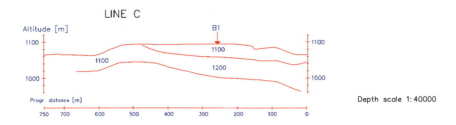

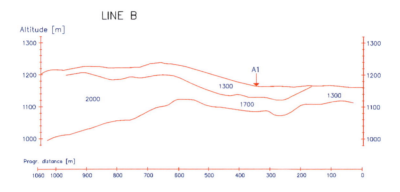

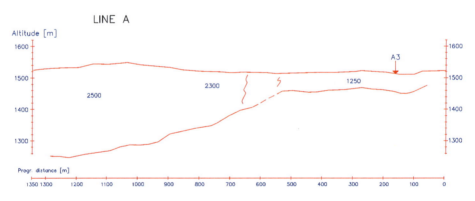

Figure 5.17.   Seismic reflection profiling performed along three lines in the Bertodasco sector.

The geometry of this surface is similar to the one derived on the basis of the geophysical data (see below).

In the lower part of Zone C, the main feature is the presence of heterogeneous debris made up of boulders of different sizes, due to rock-debris and debris flows.

5.5.4.2   Structural constraints

The main structural elements to be considered for modelling and hazard analysis (sketched in Fig. 5.18 and represented in Fig. 5.11) are:

- the schistosity and/or the joints that are sub-parallel to it;
- the mylonitic silvery micaschists interbedded in the orthogneisses that lie sub-parallel to the slope in the Rosone area;
- the sub-vertical joint systems (K1, K2 and other minor systems);

106

– the presence of an inferred deep failure zone at the base of the Bertodasco sector (Zones B and C) that has been thought here to be sub-parallel to the schistosity and related structures.

It is remarked here that the sliding surfaces of the Bertodasco sector seem strictly related to the regional schistosity and associated joints, dip parallel to the maximum inclination of the slope that in the Bertodasco sector. In fact, the disjointing of the rocks results mainly from the intersections of the schistosity and the perpendicular K1 and K2 joint systems. In this sense, the schistosity attitude (and those of related features such as the silvery mylonitic schists) is considered the triggering factor for the instability phenomena, which could strongly control the kinematic evolution of the sliding mechanism.

5.5.4.3 *Geometrical characterization*
The sliding surface, located at about 40–75 m below the ground surface, has been thought to be sub-parallel to the schistosity and the silvery micaschist levels. The geometry of this inferred surface is very near the one obtained considering the results of the inclinometers and the geophysical investigations performed in the landslide area (see below also).

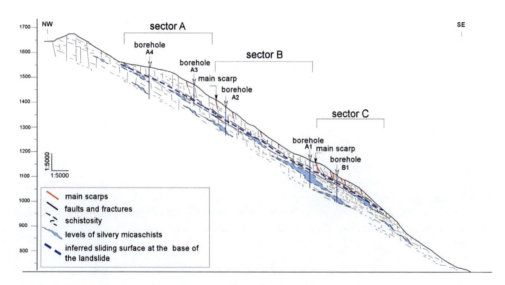

Figure 5.18.   Schematic cross section of the Rosone landslide (Bertodasco sector).

5.5.4.4 *Scenarios*
On the basis of geomorphological observation and field investigation, the rock slope foot, over deepened by glacial erosion and weakened by severe stress concentration, does not seem of particularly efficient significance in contrasting a global movement of the slope. As the deformation affecting the middle part of the slope will spread througt the basal part, a quite defined failure surface might be formed, leading eventually to a huge landslide and possibly severe rockfall and/or rock avalanche phenomena.

In detail, three different scenarios of evolution will be taken into account, with decreasing occurrence probability and increasing impact on land planning:

Scenario 1: collapse of Zone C (Fig. 5.2). The rock mass of Zone C is heavily fractured, therefore, continuous rockfalls can weaken the rock mass located in this sector. An avalanche involving a rock mass of about 2'200.000 m^3 could then occur.

Scenario 2: collapse of Zones C and B (Fig. 5.2). The collapse of Zone C may bring about the avalanche of Zone B at the same time. The total volume involved would be about 9'300'000 m^3.

107

Scenario 3: collapse of the whole landslide area (Fig. 5.2). The collapse of Zones C and B may induce the avalanche of the whole rock body involving a volume of about 20'500'000 m³. In Zone A, the quality of the rock body, which is much less fractured with respect to the other zones, and the recorded displacement data, smaller than those recorded in Zones B and C, suggest that this scenario must be considered less probable than the previous ones.

The volumes of rock mass involved in the scenarios have been roughly estimated. A precise estimation will be given on the basis of the volume computation made through the numerical triggering model.

5.5.5   Geomechanical modelling

5.5.5.1   Mechanical "triggering" model

The Rosone site was studied by means of a 3D equivalent continuous model with discontinuity planes that cut off the unstable mass from the whole rock slope.

The 3D analyses were carried out by means of the code Map3D (described in Chapter 3) based on an indirect Boundary Element technique, the Displacement Discontinuity Method. The numerical simulations carried out to investigate the triggering of the rockslide were focused on Zones B and C.

As previously seen, a complex description both from the geomechanical and the kinematical point of view arose from the geological investigations. At the present phase of the work, however, a simplified geomechanical model is assumed and simple numerical triggering simulations have been performed. Remarkable characteristics of these models are the 3D analysis conditions and the possibility of taking into account the measured and the simulated displacements. In order to be able to focus attention on these aspects, some simplifications had to be introduced from the geometrical and the mechanical point of view:

-   as the rock mass of Zones B and C is very fractured and disjointed, in the 3D geometrical model it was represented as an equivalent continuum and the discontinuity planes that belong to the principal joint sets were introduced as boundaries of the unstable volume;
-   the global behaviour of the unstable rock volume was analysed: this implies that it was possible only to reproduce the mean value of the measured displacements due to sliding. The numerically obtained value had to be referred to the whole investigated rock mass. It must be underlined that the considered mean value was evaluated with respect to time (i.e. during one year) and space (i.e. over Zones B and C);
-   the average displacements were supposed to be due to the overall creep behaviour of the rock mass, which is assumed to be concentrated only on the sliding surface. No influence on the displacements of water pressure variations with respect to time was considered, as the piezometric data were not sufficient.

On the basis of these considerations, the aim of the analyses was to reproduce the displacements measured in situ by the monitoring system and by means of the Bingham flow rule implemented in the code Map3D. This can be carried out in two steps: in the first one, the research of the mechanical parameters related to conditions of unstable equilibrium of the rock mass is performed by means of numerical simulations; on the basis of the results obtained, in the second step, the analysis of the behaviour of the unstable rock mass with respect to time is developed. In order to do this, a realistic value of the creep coefficient C was checked by comparing the evaluated displacements with those measured in situ.

In order to assess the 3D numerical simulation of Rosone, the first effort was devoted to building the geometrical model of the landslide. To attain this goal, the Digital Elevation Model (DEM) of the Orco Valley and the available data derived from the geophysical analyses and geotechnical site investigations were taken into account. The recorded data of the monitoring system and of the continuous corings reveal that the principal displacements occur 40–75 m below the ground surface (Brovero et al., 1996b). This suggests the existence of a deep-seated failure surface inside the slope. These conclusions are strengthened by the results obtained from the geophysical investigation. The interpretation of the position and shape of the failure surface has been performed by means of an interpolation methodology. Different interpolators were tested to estimate the failure surface (IDW, Kriging, Spline, TIN), and finally Spline was chosen because the in demand surface needs to be smooth and with no local anomalies, characteristics that are necessary for the foreseen numerical simulation.

The Spline interpolator is a general purpose interpolation method that fits a minimum curvature surface through the input points.

The dataset utilised for contouring is composed of: 33 points from seismic reflection profiling tests; 15 drillings; 5 inclinometers. Moreover, 28 points were added where the outcrop of the failure surface, according to the geomorphological studies, intersects the topography. In all, the final dataset consists of 81 points for a 400'000 m^2 surface. Each point gives the depth (meters) of the failure surface from the top surface. Unfortunately, the distribution of depth points is not homogeneous (Fig. 5.19) so, in those areas where no information is available, the derived surface position should not be very precise. The aspect of the derived failure surface is depicted in Figure 5.20. The mechanical properties of the sliding surface can be defined by a combination of the mechanical characteristics of the silvery micaschists and of the discontinuities sub-parallel to the schistosity.

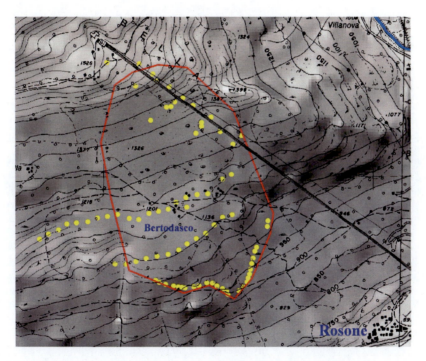

Figure 5.19.   Distribution of points utilised for contouring the inferred failure surface.

The rock volume involved by the inferred failure surface coupling its position with the topographic surface, obtained from DEM, has been also calculated. The side boundaries of the moving mass have been obtained from the geomorphological evidence (§ 5.5.2.2): hence, vertical joints have been introduced to reproduce the boundaries of Zones C and B (Figure 2). The estimated volume is about 11'000'000 m^3, in good agreement with the volume evaluated on the basis of the geomorphological studies (§ 5.5.4.4). In Figure 5.21, the 3D representation of the unstable volume obtained as described above is shown.

109

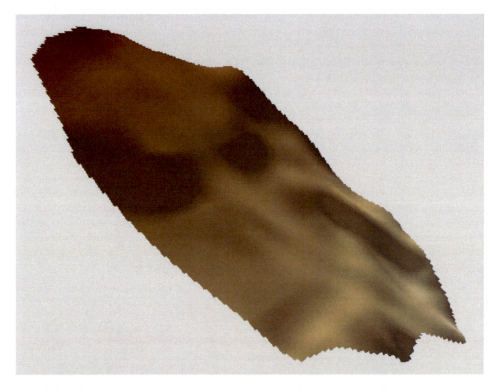

Figure 5.20.   Sketch of the failure surface as derived by means of the Spline interpolator.

The first aim of the 3D simulations was to obtain a realistic value of the in situ friction angle. This value is related to a heterogeneous material made of boulders, soil, water and voids considered as a whole; therefore, it is impossible to evaluate the friction angle by means of laboratory tests. A set of sensitivity analyses was performed in order to calculate the value of the friction angle that corresponds to the instability of the rock mass. These analyses were conducted without creep simulation.

The following step was the assessment of the value of the creep coefficient C. Activating the Map3D creep option, a simulation was carried out with an estimated value of C and the results were compared with the measured displacements. If these values were not comparable, more simulations were carried out in order to find a suitable C. The final set of mechanical parameters used for the numerical analyses is shown in Table 5.4.

The numerically computed displacements can be compared with the average annual displacement obtained by means of inclinometer measurements. Taking into account inclinometers A1, A2 and B1 (Fig. 5.14), the only ones inside Zones B and C, the measured mean value is about 18 mm/year with a dip direction of 167°; these values are very close to that obtained numerically, which are equal to 20 mm/year with 156° of dip direction.

Under the hypotheses made in the framework of the simplified geomechanical model, the obtained results allow the confirmation of the consistency among measured displacements and kinematical considerations, as carried out through geological studies, assuming reasonable values of the mechanical parameters.

The obtained results are very far from giving the probability of failure needed for the risk analysis, which is to be obtained through a probabilistic procedure. For this purpose however the variation of pore pressure in time and the decay of rock strength parameters in time should be known. These data are seldom known and are not known in the case of the Rosone landslide.

110

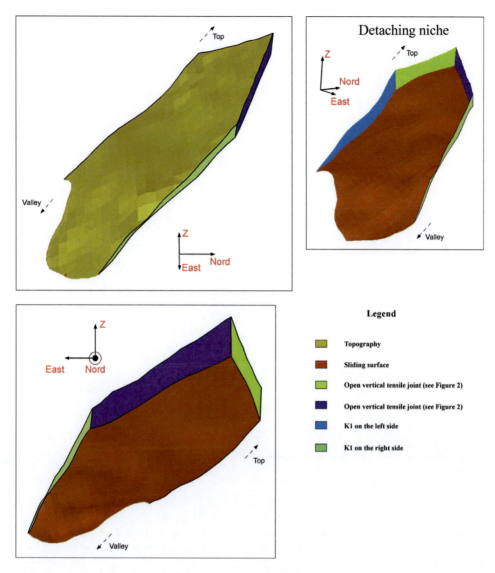

Figure 5.21. The unstable volume considered in the 3D numerical analyses.

Table 5.4. Mechanical parameters used for the Rosone land-slide analysis.

Equivalent continuous		
Unit weight, γ	0.027	MN/m^3
Young's modulus, E	7000	MPa
Poisson's ratio, ν	0.25	–
Failure surface		
Friction angle, φ	25	°
Normal modulus, K_n	5000	MPa
Shear modulus, K_s	1500	MPa
Creep coefficient, G	170	MPa × year

5.5.5.2 *Rosone run-out*

The three scenarios described in § 5.5.4.4 have been taken into consideration (Fig. 5.22), that is:
– Scenario C.
– Scenario C + B.
– Scenario C + B + A.

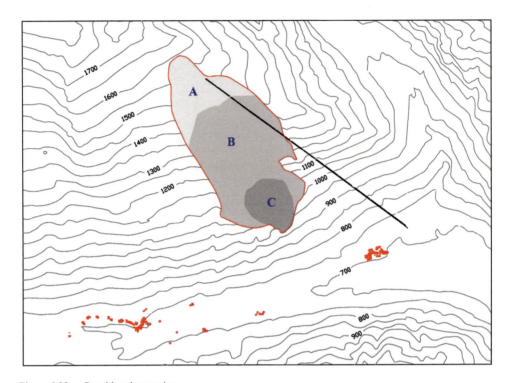

Figure 5.22. Considered scenarios.

Each scenario was analysed by Enel.Hydro (2001) through the application of a methodology elaborated by ISMES (Friz & Pinelli, 1993) that is based on the Perla & al. model (1980), the Li Tianchi model (1983) and on well-documented historical cases contained in a database drawn up by Dutto & Friz (1989).

In order to integrate the Enel.Hydro results, Scenario C was also analysed with the Hungr model (1995).

The results described here were obtained considering a volume estimates defined through the reconstruction of the possible shape of the failure surface, the topography configuration and with the hypothesis that, in correspondence of the boundary of the unstable area, the depth of the unstable mass is zero. Furthermore, due to the fact that the ISMES methodology considers the run-out volume (V), the initial volume (V_0) is multiplied by 1.3 (Table 5.5).

Table 5.5. Involved volume.

Scenario	V_0 [m³]	$V = V_0 \times 1,3$ [m³]
C	1'864'000	2'423'000
C + B	8'987'000	11'683'000
C + B + A	17'429'000	22'658'000

➢ ISMES methodology
To carry out an analysis using the ISMES methodology, the input data indicated in Figure 5.23 as F (front width), A and B are required. The output data obtained are the maximum axial travel distance (Ra) and the maximum lateral expansion (Sl).

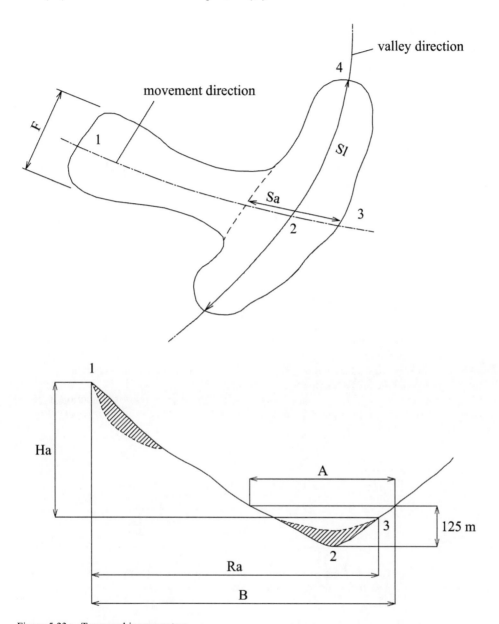

Figure 5.23.   Topographic parameters.

(a) Maximum axial travel distance estimate (Ra)
To define the maximum axial travel distance, Ra (Fig. 5.23):

– the unstable mass is considered as an adimensional block that slides on an assigned topography and that is subject to gravity and basal resisting forces (Perla & al., 1980).

113

– the dynamic friction coefficient (μ) and the ratio of involved mass (M) to aerodynamic strength coefficient (D) are considered constants and their value is defined through the back-analyses of historical cases.

The Ra minimum and maximum values were obtained by assuming a normal probability distribution, with the hypothesis that at least 95% of the estimated values result inside the considered range ($\pm 2\sigma$, with σ standard error) (Table 5.6).

Table 5.6.   Axial travel distance values.

Scenario	Ra_{min} [m]	$Ra_{average}$ [m]	Ra_{max} [m]
C	867	923	949
C + B	1297	1361	1438
C + B + A	1590	1861	1910

The Perla model was also applied to estimate the maximum lateral travel distance (Rl), this aspect supposes that, once the bottom of the valley is reached, the river direction is immediately followed by the moving mass (Table 5.7).

Table 5.7.   Lateral travel distance values.

Scenario	$Rl_{average}$ [m]	Rl_{max} [m]
C	1776	2245
C + B	2280	3475
C + B + A	2932	3526

(b) Maximum lateral expansion estimate (Sl)
To define the maximum lateral expansion, Sl (Fig. 5.23) a correlation between Sl and the front width (F) is introduced, and the conclusions of Li Tianchi (1983) about the relation between the run-out surface and the involved volume is confirmed.

The Sl variability range was obtained assuming a normal probability distribution, with the hypothesis that at least 95% of the estimated values result inside the considered range ($\pm 2\sigma$, with σ standard error) (Table 5.8).

Table 5.8.   Lateral expansion values.

Scenario	Sl_{min} [m]	$Sl_{average}$ [m]	Sl_{max} [m]
C	273	450	671
C + B	536	882	1315
C + B + A	660	1087	1619

Since it is not possible to foresee asymmetric behaviour of the mass in the run-out area, Sl is considered symmetric with respect to the axial direction of the movement.

➢ Hungr method
In order to integrate the ISMES results, the rock avalanche run-out in the case of Scenario C has been simulated by applying the DAN code (Hungr, 1995).

The geometry of the considered unstable mass is approximately the same as that assumed by Enel.Hydro but to allow a correct representation in the Hungr model a geometrical simplification is required (Fig. 5.24 and 5.25). In any case, it is important to underline that the volume used by Enel.Hydro, and consequently by DAN, is partially different from that estimated in the triggering model.

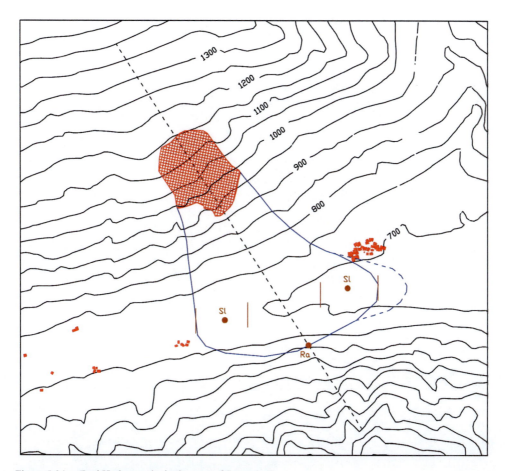

Figure 5.24.   Enel.Hydro results in the case of Scenario C.

To run an analysis, a frictional rheology has been chosen for the whole path and, since the ISMES method supposes an entrainment of material during the run-out phase, the code has been calibrated by modifying not only the friction angle but also the volume change rate.

The best results are obtained by considering a friction angle of 19.7° and a mass entrainment of about 730 m³/m (Table 5.9).

The output data obtained by using these parameters are shown in Table 5.10 and Table 5.11. In particular, a maximum front velocity of about 47 m/s and a maximum depth in the run-out area of about 30 m result.

➤ Final remarks concerning the run-out investigations

The methodology applied by Enel.Hydro yields values to assign to Ra and Sl. Since these are punctual parameters, Enel.Hydro indicates the shape of the possible run-out area through graphical interpretation of topography. In Figure 5.24 the continuous line represents Ra and Sl maximum values, under the hypothesis that the run-out area is symmetric with respect to the axial travel distance, instead the dashed line considers the hypothesis of a prevalent expansion of the mass in the direction of the valley.

The application of the Dan code allows an improvement of the knowledge of the mass behaviour during the propagation and the stop phases. Indeed, the obtained data are not only the run-out distance and the run-up on the opposite slope but also the landslide velocity, the expiring time and the depth of the displaced mass once it stops.

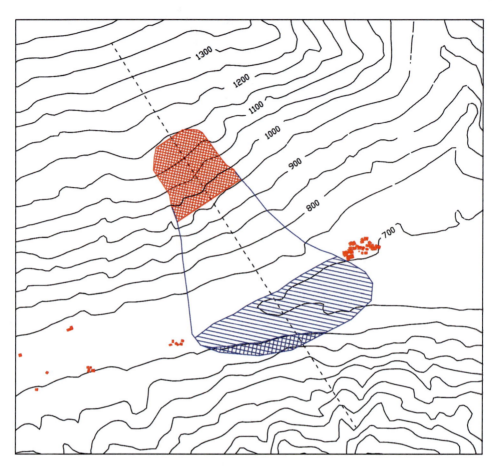

Figure 5.25.   DAN results in the case of Scenario C.

Table 5.9.   DAN input data.

Rheology	Frictional
Gamma [kN/m³]	
(*unit weight*)	26.50
Friction angle [°]	19.70
Pore pressure coefficient	
(*ratio of pore pressure to the total normal stress*)	0.00
Volume change rate [m³/m]	
(*entrainment or deposition of material*)	730.90

The analyses carried out underline that in the triggering case the displaced mass would deter-mine the river damming with disastrous consequences.

5.5.6   *Occurrence probability*

The estimation of the probability of sliding is one of the critical components of the assessment of landslide risk and hazard for natural and constructed slopes. The probability of sliding can be

116

Table 5.10. Run-out area depth and final front position.

$X_{profile}$ [m]	H [m]
698.0	4.70
723.2	12.50
743.4	19.71
759.4	25.27
775.0	28.27
788.9	30.69
801.2	31.82
812.4	31.77
823.0	30.65
833.0	28.49
842.6	25.58
851.4	22.38
859.5	18.20
867.4	12.86
874.5	6.70
880.9	1.68

Table 5.11. DAN output data.

Run-out time [s]	61.40
Front final position [m]	880.90
Rear final position [m]	697.99
Maximum velocity [m/s]	47.84 at X [m] = 640.63
Maximum front velocity [m/s]	46.57 at X [m] = 644.28
Slide volume [m^3]	Initial = 1856058.38
	Final = 2428449.50
Area in plan [m^2]	Initial = 65443.79
	Final = 103757.18

estimated using formal probabilistic analysis approaches which are inherently quantitative in nature, or using semi-quantitative methods based on historical records, geomorphology, rainfall, slope geometry, performance and other indications.

In the case of the Rosone landslide, quantitative methods cannot be applied because the uncertainty in the definition of the parameters involved is too large and it is not possible to define any reliable variability for them.

Since several historical records exist on the landslide activity in the area involving different volumes, the historical approach seems much more suitable to the case of Rosone: in this way it is possible to obtain some information related to periodic frequency.

From the historical data analysis (see Fig. 5.4), a frequency inferior to 10 years for both rockfall/debris flow phenomena and the Bertodasco sector instability is found. The latest signs were seen about 40 years ago. This indicates a relationship among these phenomena: the movements in Zone C cause the blockfall from the scarp over the National road 460. This situation is particularly apparent during heavy rainfalls.

The instability of Zone B is characterised by a recorded phenomena frequency between 0 and 30 years: the last event was recorded 15 years ago. The whole slope (Zones A + B + C) showed important movements during the 1953 event and probably at the beginning of the XVIII century, with an interval of about 250 years. It must be noted that these sectors are concerned by deformation movements which are almost continuous in time. Under particular conditions, they accelerate, causing morphological evidences on the slope and sometimes provoke damages. The recorded historical data and the temporal intervals of the occurrence probability refer to these paroxystic phases; the last ones could cause sudden collapses of the slope with the development of rock avalanche phenomena.

An occurrence time range was obtained for each scenario, considering the minimum and maximum time intervals between two consecutive recorded events. On the basis of their mean values, a frequency was then calculated in terms of events per year (Table 5.12).

Table 5.12.   Frequency calculated in terms of events per years.

Scenario	Occurrence range [years]	Average [years]	Frequency [event/year]
Rockfall & Debris flow	0 ÷ 50	25	1/25 = 0.04
Rock avalanche (Zone C)	0 ÷ 50	25	1/25 = 0.04
Rock avalanche (Zones B + C)	50 ÷ 250	150	1/150 = 0.007
Rock avalanche (Zones A + B + C)	>250	1000	1/1000 = 0.001

5.6   QUANTITATIVE RISK ANALYSIS

In the chapter dedicated to hazard analysis, various landslide scenarios were identified. Concerning quantitative risk analysis, only three rock avalanche scenarios were considered. These three scenarios refer to the evolution of three slope sectors with indications of instability or with movement data (recorded by monitoring systems). The slope sectors are: the upper zone (A), the intermediate zone (B) and the lower zone (C). The recognised dangers are linked to the following scenarios:

– collapse of Zone C and rock avalanche accumulation at the valley floor;
– collapse of Zones B and C and large rock avalanche accumulation at the valley floor;
– collapse of Zones A, B and C and huge rock avalanche accumulation at the valley floor.

The risk analyses of the scenarios are based on land planning of the Locana Municipality and on the landslide run-out areas obtained using the ISMES method (Friz & Pinelli, 1993).

5.6.1   Elements at risk

The elements at risk are recognised through the definition of the area affected by each danger in the scenarios previously specified. These elements at risk are shown in the following tables (Tables 5.13, 5.14 and 5.15).

Table 5.13.   Rock avalanche scenario (Zone C).

Element at risk	Denomination	Notes
Residential areas [valuable buildings or valuable rural centres (historical, architectural, artistic and/or cultural value)]	Rosone	Buildings affected: 2 buildings of the evacuated community (R.D.445/1908). Persons affected: 0
Forests (private and public properties)		Area affected: about 900,000 m^2
Heavy traffic or strategic roads	National road N° 450	Length affected: 650 m Annual traffic: about 170,000 vehicles
Lifelines	Power line "Valle Locana" 15 kV	Length affected: 700 m Pylons affected: 1

Table 5.14. Rock avalanche scenario (Zones B + C).

Element at risk	Denomination	Notes
Residential areas [valuable buildings or valuable rural centres (historical, architectural, artistic and/or cultural value)]	Rosone	Buildings affected: all the buildings of the evacuated community (R.D.445/1908). 32 buildings of new Rosone community Persons affected: 88
	Aghettini	Buildings affected: 8 Persons affected: 12
Forests (private and public properties)		Area affected: about 403,000 m^2
Heavy traffic or strategic roads	National road N° 450	Length affected: 1506 m Annual traffic: about 170,000 vehicles
Lifelines	Power line "Valle Locana" 15 kV	Length affected: 1000 m Pylons affected: 3

Table 5.15. Rock avalanche scenario (Zones A + B + C).

Element at risk	Denomination	Notes
Residential areas [valuable buildings or valuable rural centres (historical, architectural, artistic and/or cultural value)]	Rosone	Buildings affected: all the buildings of the evacuated community (R.D.445/1908). 62 buildings of new Rosone community. Persons affected: 186
	Aghettini	Buildings affected: 8 Persons affected: 12
	Casetti	Buildings affected: 7 Persons affected: 18
Forests (private and public properties)		Area affected: about 900,000 m^2
Heavy traffic or strategic roads	National road N° 450	Length affected: 2200 m Annual traffic: about 170,000 vehicles
Lifelines	Power line "Valle Locana" 15 kV	Length affected: 2500 m Pylons affected: 8
Secondary lifelines	Power line for the funicular service	Length affected: 600 m Pylons affected: 2

5.6.2 Vulnerability of the elements at risk

To define the vulnerability of the elements at risk, an analysis of the typology of the assets affected and the processes' intensity was carried out.

In the rock avalanche scenarios, a distinction between various degrees of energy based on land-slide velocity is useless in defining different degrees of vulnerability of the elements at risk. In the entire area affected (failure, depletion and accumulation areas) the destruction would be total. Also, at the valley floor, where the velocity tends to zero, the accumulation (some tens of m) would cause the complete destruction of the assets (heavily damaged or buried). Therefore, the degree of vulnerability is equal to 1 (for all vulnerability categories).

Considering areas next to the run-out zone, various hypotheses can be made about the wind and dust effects. The models known to the Authors, however, do not supply indications about the geometry of the area affected and the process intensity of wind and dust effects. Therefore, in this quantitative risk analysis, the effects associated with a rock avalanche have not been taken into account; qualitative considerations about these processes will be described in succeeding sections, deduced from case histories.

5.6.3 *Expected impact*

By crossing values for the elements at risk (assets, economic, environmental and social value) with vulnerability percentage, expected impact is obtained. It is possible to use a monetary index for assets and economic evaluation, derived from detailed economic analyses (e.g. costs of assets, reconstruction costs, turnover, profits loss, etc.). To assess expected environmental and social impacts where a monetary value is difficult to obtain, relative value indices are used.

In the quantitative risk analysis of the Rosone landslide, a detailed economic evaluation has not been considered at the moment. Arbitrary relative value indices have been temporarily applied to each category.

5.6.4 *Risk assessment*

The last phase of the quantitative risk analysis concerns the risk evaluation. Such an evaluation is obtained by multiplying the expected impact by the occurrence probability value related to each scenario. In the case of the Rosone landslide, the occurrence probability values have been calculated on the basis of the historical data.

The results of the risk assessment are shown in the following figures (Fig. 5.26, 5.27, 5.28 and 5.29) obtained by GIS layout and described in the captions.

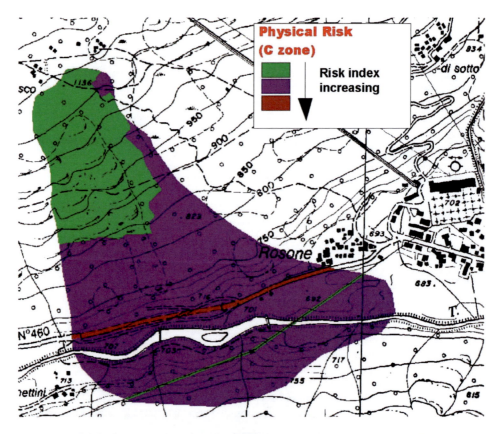

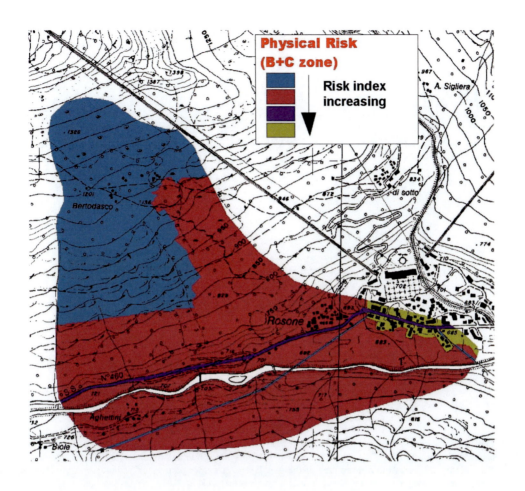

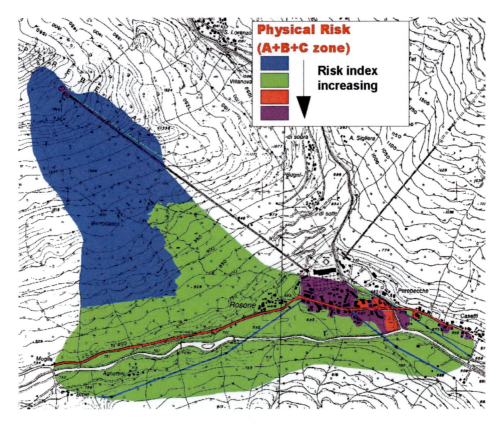

Figure 5.26.   Maps by GIS of physical risk for the three rock avalanche scenarios. The maps underline the areas with highest risk value, i.e. the National road no. 450 and part of the new Rosone community. Particularly relevant is the run-out area of the A + B + C scenario: part of the accumulation involves the hydroelectric power plant, part of the failure area involves the penstocks.

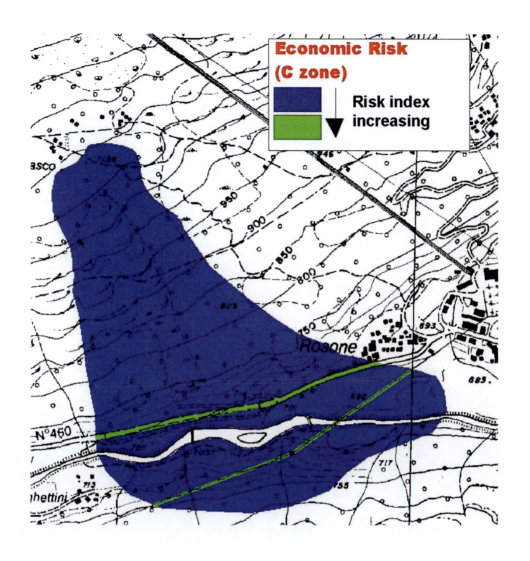

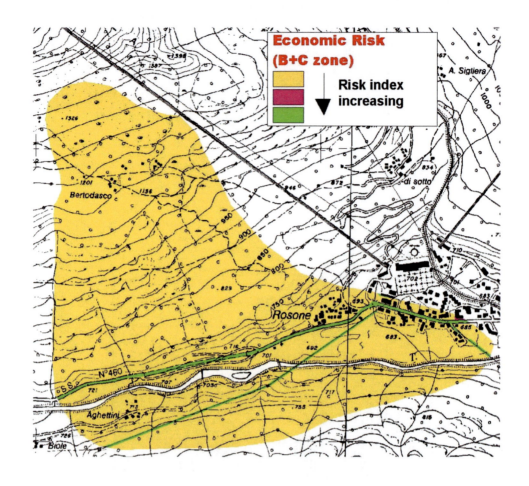

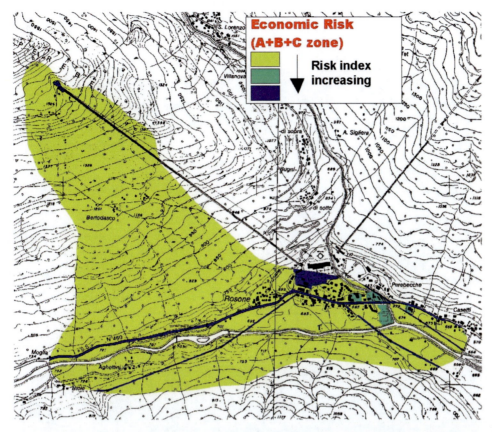

Figure 5.27.   Maps by GIS of economic risk for the three rock avalanche scenarios. The maps underline the areas with highest risk value, i.e. the National road no. 450 and the lifelines (electric power line "Valle Locana" and secondary electric power line for the funicular service). From an economic point of view the electric power plant of Rosone and the related infrastructures are very important, as they supply energy to a large part of Turin.

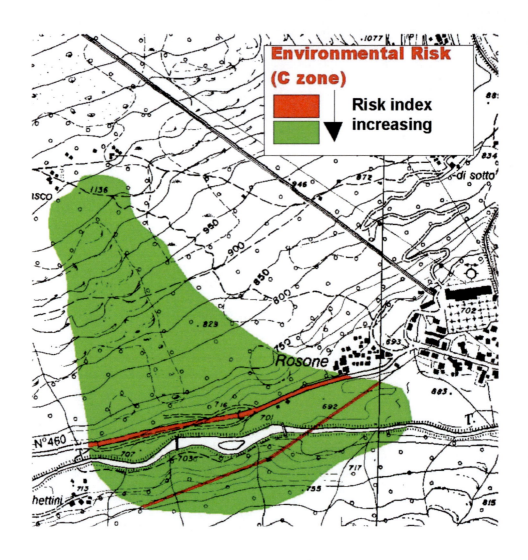

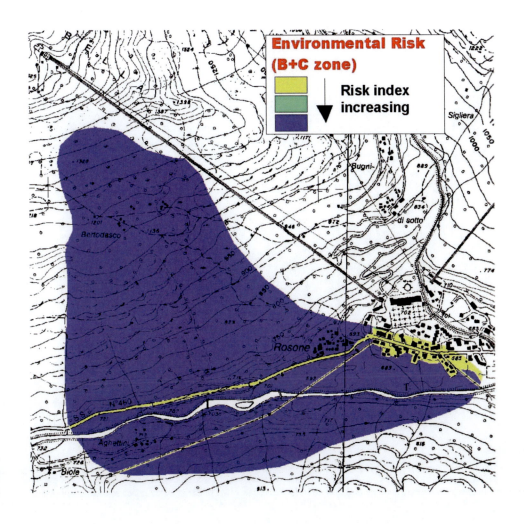

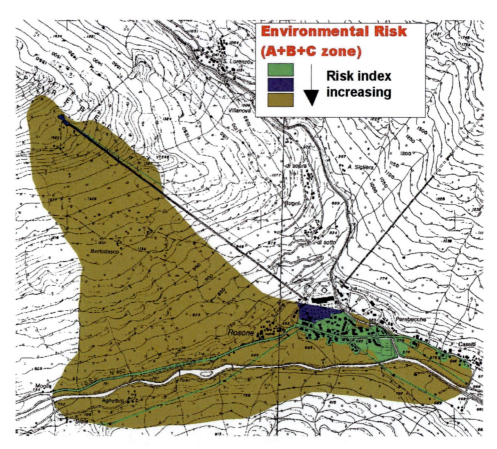

Figure 5.28.   Maps by GIS of environmental risk for the three rock avalanche scenarios. The maps underline the areas with highest risk value, i.e. the wooded areas and the valley floor. The Orco river is completely buried by rock avalanche accumulation, therefore, the risk value of the affected zones has no meaning.

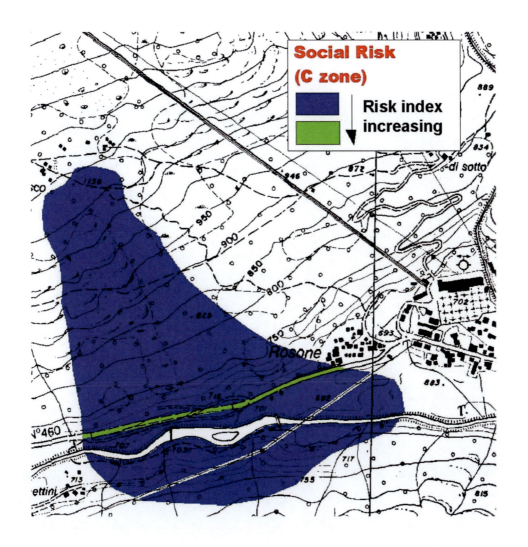

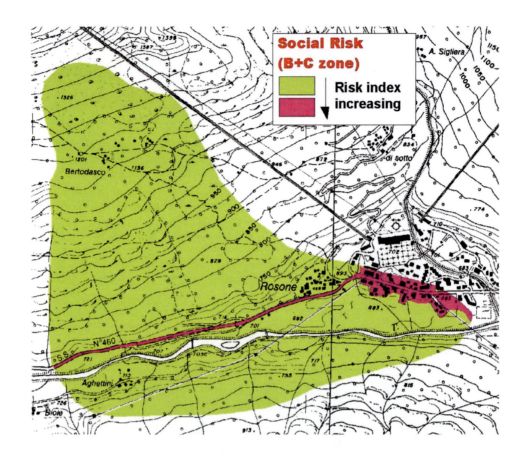

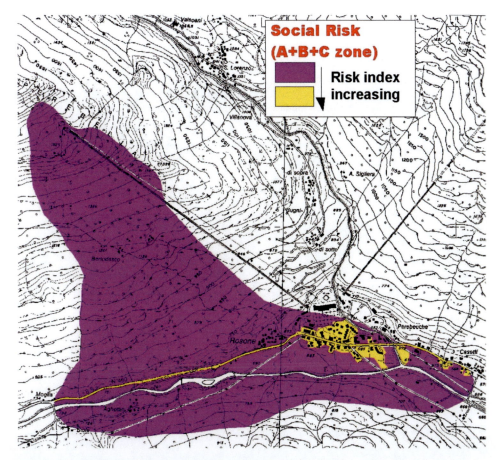

Figure 5.29. Maps by GIS of social risk for the three rock avalanche scenarios. The maps underline the areas with highest risk value, i.e. the urban areas of the new Rosone and Casetti communities. Also the National road no. 450 involved in the rock avalanche phenomenon has high risk values, being characterised by intense traffic during the tourist season.

5.6.5   Indirect risk and regional impacts

In this paragraph, some considerations concerning the indirect effects linked to rock avalanche are described (Fig. 5.30). The evolutionary scenarios of the Rosone site, besides rockfall, involve huge rock volumes (millions of m^3) at high velocity (tens of m/s), whose effects are not easily evaluated. The subsequent considerations are based on descriptive and qualitative evaluations derived from case histories (i.e. Hsü, 1975).

The large mass movement (tens of millions of m^3) that could occur in Rosone would produce a huge wind effect. The processes' intensity cannot be easily calculated but it is probable that they would be light laterally to the mass movement and very strong frontally. The same would occur for the dust that follows the wind effect.

At the moment, wind and dust effects cannot be obtained by a numerical model with confidence. Nevertheless, it can be stated that the damages produced by these indirect effects would be very severe, as the area next to the accumulation is densely populated.

Another effect to be considered in risk assessment is the formation of a landslide dam lake. This phenomenon can occur during a meteorological event with important flows along the Orco river (some hundreds of m^3/s). Instant damming of the thalweg by tens of meters of landslide accumulation and extraordinary flooding of the Orco River would form a basin of some hectares in a few hours. Many buildings in the Fornolosa village would be inundated and many people would have to be evacuated, increasing social risk.

Due to such a lake formation, downstream villages could potentially be involved. Instantaneous lake depletion could be caused by a rapid overtopping dam erosion, or a landslide dam collapse caused by piping and internal erosion.

In both cases, flash-flooding along the valley would cause enormous problems. At the Rosone site, a survey of the effects of flash-flood events has not yet been carried out.

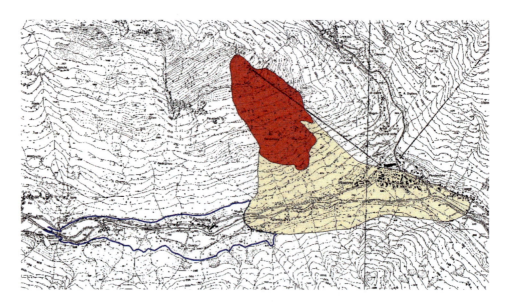

Figure 5.30.   Maximum extension of landslide dam lake (A + B + C rock avalanche scenario, Enel-Hydro, 2001).

5.6.6   *Total risk*

In the following maps (Fig. 5.31), the results of a quantitative analysis of total risk for each category (physical, economic, environmental and social risk) are shown. As clarified in Chapter 3, total risk is calculated by adding the risk values obtained in each category of each scenario.

The maps emphasise that the risk is strongly linked to the occurrence probability. The occurrence probability of a Zone C rock avalanche is 16 times superior to the occurrence probability of a Zone A + B + C rock avalanche and 4 times superior to the occurrence probability value of a Zone B + C scenario. Therefore, the major risk values are recognised in the areas of Scenario C.

In the case of these three rock avalanche scenarios, the processes' intensity is useless for the differentiation of the different risk degrees, as it is very high in all scenarios and destructive in all of the areas affected.

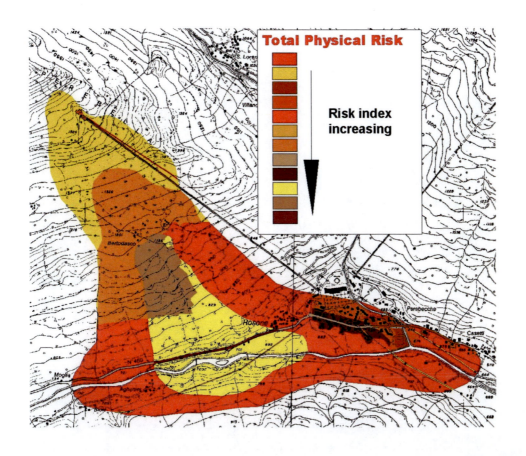

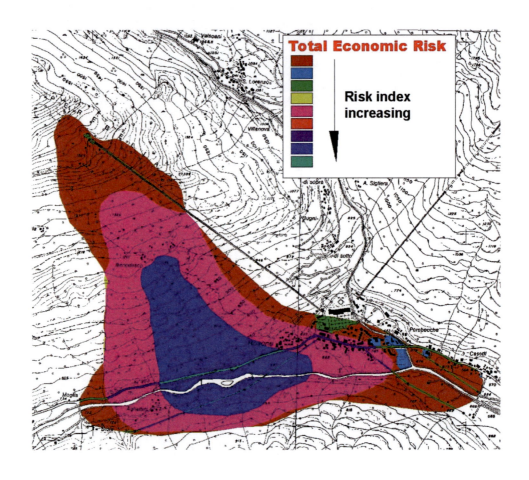

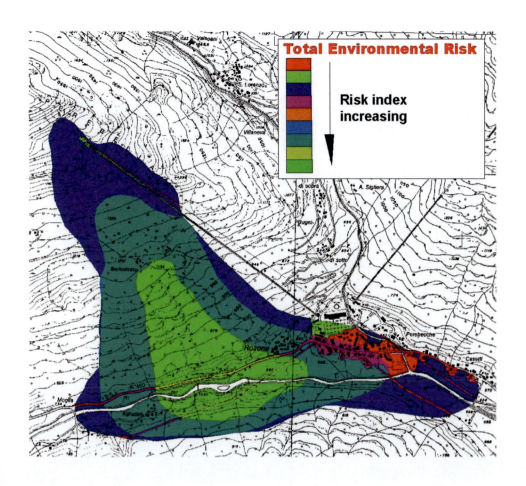

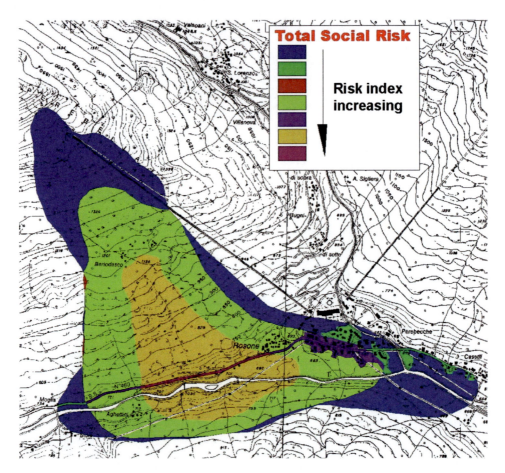

Figure 5.31.   Maps by GIS of total risk. The maps underline the areas with highest risk value for each category physical, economic, environmental and social.

6
The Oselitzenbach landslide

Gaetano Amatruda[1], Marta Castelli[1], Marcel Hurlimann[2], Alberto Ledesma[2], Michele Morelli[3], Fabrizio Piana[3], Marina Pirulli[1], Rainer Poisel[4], Riccardo Polino[3], Pere Prat[2], Alexander Preh[4], Werner Roth[4], Claudio Scavia[1], Ewald Tentschert[4]

[1]Politecnico di Torino; [2]Universitat Politècnica de Catalunya; [3]CNR-IGG sezione Torino; [4]Technische Universität Wien

6.1 INTRODUCTION

The area on the northern slopes of the so-called "Carnic Alps" in the southern part of Austria's Carinthia country is affected by numerous mass movements.

The Carnic Alps, as a member of the (tectonically defined) "Southern Alps" are divided from the "Central Alps" in the north of the area by a continental lineament called the "Periadriatic Lineament" leading from southern Switzerland to Slovenia.

Thus, the Naßfeld area, dewatered by a torrent called Oselitzenbach, belongs to a different tectonic unit and contains rock composition different from that of the Central Alps, i.e. mainly sedimentary beds of paleozoic age containing clayey schists, sandstones, marls and limestones. Due to the adjacent lineament (distance 1–3 km) the degree of tectonic influence is rather high.

The Reppwand-Oselitzenbach sagging zone consists of interlayers of competent and incompetent rocks which are highly deformed and disturbed both by tectonics and by long-term sagging and sliding processes.

The movements (measurements available since 1983) are very inhomogeneous, reaching from some cm/y up to more than 1 m/y in some areas.

The risks concern the destruction of the Naßfeld Road (National road B 90, crossing the Alps to Italy and connecting to a large ski area), as well as the possible spreading out of landslide material by the Oselitzenbach torrent (which occurred in part after heavy rainfalls and a consecutive flood in 1983).

Several measures have been taken to stabilise the landslide (drains, deviation of the torrent, debris dams in the river, anchoring of the road cuts and slopes), which led to some decrease in movement; however, they are still continuing, but with reduced amounts.

6.1.1 *Landslide description*

The catchment area of the Oselitzenbach torrent is considerably affected by two sagging slopes which were landslides in prehistoric times (Reppwand landslide, Schlanitzenalm landslide). The main sliding process of these slopes took place between the Rissian and the Würmian Ice ages (i.e. between 70'000 y–150'000 y BP); this is documented by Würmian age moraines partly overlaying the landslide area with no signs of displacement.

The main scarp of the whole Reppwand landslide (Fig. 6.2 and 6.3) shows a typical profile of the Southern Alpine sedimentary sequences from Devonian sandy – marly calcareous sediments up to lower Triassic carbonates.

The Reppwand sliding mass consists mainly of the marly elements and covers the Silurian-Devonian Hochwipfel schist basement (sandy–silty).

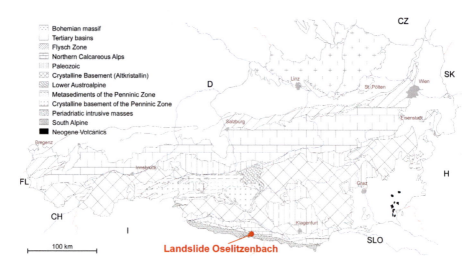

Figure 6.1. Geological sketch of Austria with the location of the site.

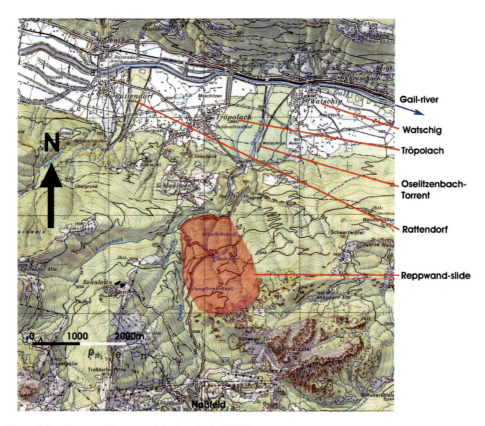

Figure 6.2. Topographic map; original scale: 1:50'000.

138

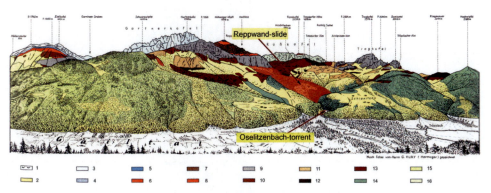

Figure 6.3.   Panoramic view of the area (Kahler & Prey, 1963).
Legend: 1 – landslide movements; 2 – moraine; 3 – dolomite; 4 – shelly limestone; 5 – shelly limestone, conglomerate; 6 – Werfener layers; 7 – Bellerophon dolomite; 8 – Grödener layers; 9 – Trogkofel limestone;
10,12 – Pseudoschwagerinen limestone; 13 – Auernig layers; 14 – Hochwipel layers; 15 – layered limestone;
16 – reef limestone.

Especially the toe zone of the Reppwand landslide is still quite active and is responsible for
debris-generating slope failures and the destruction of the Naßfeld Road. After sustained regional
rainstorms in Sept. 1983 with numerous embankment failures, an extensive project of construction work and research was initiated along the Oselitzenbach torrent and the sagging slope above.

Thus, the area of the eastern toe of the Reppwand landslide has been the object of numerical investigations as the displacements of this area are the largest and are, therefore, intensively monitored.

6.1.2   Historical background

The geological investigation of the area began in about 1850 by the first determinations of paleozoic
fossils, but the investigation of the landslide area started only 100 years later (Kahler & Prey, 1963).

There are no reports available about the landslide before 1915. During the Middle Ages up to
the beginning of the 20th century, the area of Naßfeld was only a summer pasture for the animals
of the valley villages. Only a small footpath led into the valley of Oselitzenbach.

During World War I, a military road was constructed. Since then, the movements along the road
have been known, but have not been investigated in detail.

In the seventies, the skiing area at the Naßfeld Pass was opened and then enlarged and the road
was upgraded to a primary road, connecting Austria and Italy.

In September 1983, heavy rainfall and flooding in the catchment area led to massive gravel
deposits at the alluvial cone at Oselitzenbach. As a consequence, erosion of the right abutment
took place and the Naßfeld Road became cut off (Fig. 6.4). Additionally, the old torrent control
dams along the upper and middle courses of the tributary torrent "Rudingbach" were destroyed.

In 1985, planning was started for an extensive construction programme carried out by the
Austrian Service for Torrent, Erosion and Avalanche Control as a reaction to the events in
September 1983.

As a consequence of repeated flooding, an advancing undercutting of the eastern toe of the
landslide arose. A run-out of 60'000 m^3 in volume took place (August 1987) and as a result, large
settlements up to several metres occurred. Additionally, the formation of new cracks up to 30 m
above the Naßfeld Road occurred.

The implementation of the extensive construction programme started in 1988. As a part of the
construction measures, a 400 m long new river channel was excavated in comparatively massive

Figure 6.4.   Destruction of Nassfeld Road Sept. 1983 by landslide movements.

Hochwipfel formations (Fig. 6.5), as well as a placement of landfill at the toe of the sagging mass using the excavated material (about 170'000 m^3).

Additional measures have been: drainage of the Quellenbach landslide (lower part of the ancient Reppwand landslide) between elevations 830–920 m a.s.l. and the draining off of surface watersabove 950 m a.s.l. as well to the area of the "Bodensee" Lake. These measures were necessary

Figure 6.5.   New rock channel of the Oselitzenbach torrent, fresh landfill at the toe of the sagging zone (brown and yellow) filling in the former ravine; in the background the Nassfeld Road is seen. Date of photo: 1989.

140

to prevent serious debris flow into the Oselitzenbach Torrent and in the receiving stream "Gail" and to prevent debris flow reaching and, possibly, destroying the villages "Tröpolach" and "Watsching". For economic reasons, the correction of all the debris sources from the very top downwards was not considered possible or necessary (Moser & Glawe, 1994).

6.2   REGIONAL FRAMEWORK

6.2.1   *Climatic and water conditions*

The Oselitzenbach torrent belongs to the Danube catchment system (1st order: Danube River, 2nd order: Drau River, 3rd order: Gail River, 4th order: Oselitzenbach Torrent).

The climatic conditions are influenced by an Alpine climate (westward winds), influenced additionally by Adriatic lows.

The principal rainfall takes place in summer (>50% of annual precipitation), but even heavy rainfalls can occur in October–November, especially when westward autumn weather occurs and meets Adriatic areas of low depression. The annual precipitation, therefore, shows a high value of 2963 mm/y (the highest value measured in Austria; the weather measurement station is situated at 1530 m a.s.l.)

The average annual temperature is: +7°C, the minima occur in January with a value of −18°C, the maxima in summer can reach +30°C.

The snow covering period depends on the altitude: at 1000 m a.s.l. the area is covered from November until March; at 2000 m from October to May. Snowfall may occur all the year, but mainly from October to April.

6.2.2   *Groundwater conditions*

The groundwater conditions, as understood up until now, differ in the various parts of the area.

The upper part shows a shallow infiltration, with groundwater coming to daylight as springs in the middle part. In the lower part, water losses are documented along deep rotational slide planes. The groundwater velocities in the upper part are about 0,2–5,0 m/h (evidenced by a tracer test, Hötzl & al., 1994).

There are three levels of springs: The highest level lies between an elevation of 1150 m and 1170 m. At this level, there are two lakes, the "Great Bodensee" and the "Small Bodensee", also there are springs west of the lake "Great Bodensee" and water outlets in the water logging zone east of the lake "Small Bodensee".

The area of the second level lies 20 m deeper (1130–1150 m), in the north of the lake "Great Bodensee".

At the third level, there are many selective water outlets below the Naßfeld Road (at an elevation of 700–800 m).

The "Bodensee" lakes lie in a small depression, made up of secondary zones of movement. The greater lake is fed by 2 small tributaries and is dewatered by a small runoff river – the smaller lake has no tributary at the surface, but has a small runoff (3–5 l/s).

6.2.3   *Regional morphology*

The Carnic Alps, forming the boundary between Austria (Carinthia county) and Northern Italy (Province of Udine), are an East-West striking mountain chain up to 2700 m a.s.l. The main valley (Gailtal, 600 m a.s.l.) in the north is parallel to these chains, whereas the smaller tributaries cut transversally through the structures by steep and narrow valleys, mainly being cut after the last ice age.

The morphology of the mountain chain is dominated by Triassic and Devonian limestones in the highest peaks, and at the foothills and on some less inclined slopes by (partly sandy) shales and slates. During the Ice ages, the valleys were covered by ice up to 2000 m a.s.l., leaving only the

141

highest tops uncovered. The carving of the valleys from U-shaped glacier valleys to V-shaped erosion valleys took place after the last period of the Ice Age (over the last 20'000 years). By these processes, some slopes became partly eroded at their toe area.

6.2.4   Regional geology and structural setting

As a part of the Carnic Alps, there are evidences of a complicated tectonometamorphic history since the first Variscian deformations and by the Alpine orogeny. The Variscian record is partly preserved because the Alpine overprint was generally weak.

Figure 6.6.   Geology (after Schönlaub & al., 1987).
Legend: 1: young river deflection, torrent debris; 2: mudflow debris; 3: alluvial cone; 5: blocky material, rock fall material; 20: ground moraine; 43, 44: Schlern dolomite; 45–47: shelly limestone; 49, 50: Bellerophon dolomite; 51: Grödner sandstones; 53: Trogkofel breccia: dolomite to conglomerate; 55–57: upper Pseudoschwagerinen formation; 61: lower Pseudoschwagerinen formation; 62–65: Auernig formations; 68–71: Hochwipfel formation; 81: Eder limestone (grey, banded limestone).

The Variscian basement is represented by an epizonal tectonic unit (laminated limestones, marbles and phyllites) and the anchizonal Hochwipfel nappe (Fig. 6.6 and 6.7). The stratigraphic order begins with light metamorphic rocks (lower paleozoic), overlain by intensely folded and fractured schists (mainly sandstones and marlstones of Devonian-Permian Age). The post-Variscian Permo-Carboniferous to Triassic Sequence rests on top of the Hochwipfel nappe above a major unconformity (Schwarzwipfel fault, Fig. 6.7).

In the main valley of the Gailtal, near the site, the continental tectonic divide of the "Periadriatic lineament" passes through (dividing the Northern from the Southern Alps from the Piemont to Slovenia). This leads to parallel faults of the periadriatic lineament (North West-South East), to some Riedel shears (diagonal) and extension joints (A-C- Joints, North-south)(see Fig. 6.7).

The investigated area mainly consists of two different geological formations (see Fig. 6.6), the Hochwipfel schists (Silurian upper Carbonian) and the Naßfeld schists (upper Carbonian to lower Permian).

The Hochwipfel schists contain sandy siltstones; they have higher strength than the more fractured Naßfeld schists. The heavily fractured Naßfeld schists are a local term (a formation combining the Aueringg schists, some intercalations and the Rattendorfer schists). The rock menu mainly contains a melange of conglomerates, limestones and sandy siltstones.

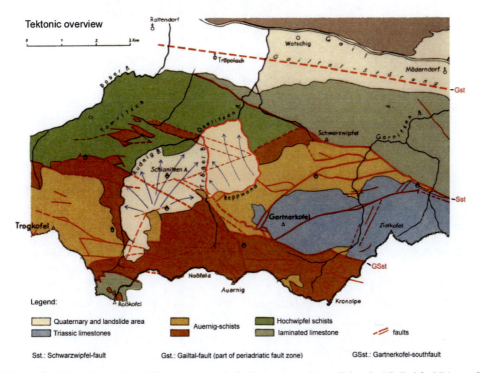

Figure 6.7. Tectonic overview of the area: the main fault zones are sub-parallel to the "Gailtal fault" (part of the periadriatic lineament).

6.3 HAZARD ANALYSIS

6.3.1 *Geo-morphological analysis*

The Oselitzenbach has four small tributaries within the sagging mass, forming the Ostelitzenbach torrent after being combined.

Through the sagging mass of the Reppwandgleitung (between the last two Ice Ages, Riss and Würm) the valley floor became very narrow and the torrent eroded the toe of the mass.

The lowest part of the Oselitzenbach is a steep inclined valley reaching the more deeply situated main Gail Valley, carved out after the last Ice Age over the last 20'000 years.

Above the main scarp of the mass, there are outcrops of massive limestones with steep slopes. Within the deformed mass, there are many small trenches (Fig. 6.8). Within the sagging mass, there are many secondary scarps, forming different elements of the deformed mass. As shown in § 6.3.3, there are different movement rates in the various sections.

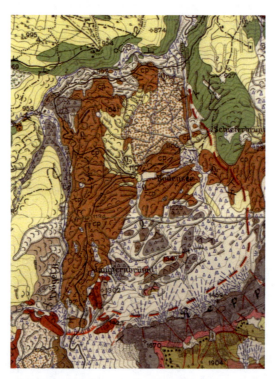

Figure 6.8.   (Kahler & Prey, 1963): geologic–geomorphologic map; many trenches are indicated, resulting from different secondary rotational slides;
Braun: Naßfeld layers; Yellow: moraine; Green: Hochwipfel layers; Dark grey: Trogkofel limestone; Light grey with triangles: blocky material; Pink: Grödner sandstones; Grey with points: detrital formation from the Rattendorfer layers.

6.3.2   *Local geology and structural analyses*

6.3.2.1   *"On site" available data*
The project area covers an alpine valley with elevations ranging from 590 m to 2000 m a.s.l. within the catchment area of the Oselitzenbach. The "rock menu" consists mainly of:

– Hochwipfel schists (see Fig. 6.9): dark greyish sandy-siltstones, scarcely calcareous. Anchizonal light metamorph, well bedded, with interlayers of clayey material; partly graphtolith schists. Mainly hard and brittle behaviour. Occurrence: mainly in the north of the mass movement, being the northern embankment of the river bed. Age: Silurian to upper Carbonian
– Auernigg schists (Fig. 6.10, 6.11 and 6.12): Conglomerates, partly transgressive formed limestone beds; lower group, less limestone: dark greyish fine graded sandstones, partly sandstone schists.
lower group, rich in limestone: sandy schists and sandstones, partly limestone beds.
middle group, less limestone: Sand & quartz conglomerates, scarcely schists.
upper group, rich in limestones: sandy schists with mica and sandstones & conglomerates
upper group, less limestone: schists & sandstones, shales dominating
Age: upper Carbonian – lower Permian

Figure 6.9.   Hochwipfel schists.

Figure 6.10.   Lower Auernigg schists, sandy.

– Pseudo-Schwagerinenkalk: Limestones, rich in the fossil "pseudoschwagerina": dark grey limestones, scarcely bedded. Age: Upper Carbonian
– Rattendorfer Schists: limestones, dark, grey. Sometimes thin interlayers of siltstone. Age: Lower Permian.

In the heavily disturbed, sagging areas, the Auernig and Rattendorfer schists cannot be precisely distinguished in the field. In this case, they are summarised by the term "Naßfeld schists".

Figure 6.11.   Upper Auernigg schists, right corner – centre: sandy variety, left above: marly beds.

Figure 6.12.   Auernigg (Nassfeld) schists, intensively fractured.

– Both stratigraphically and topographical above (outside the landslide area), there are also Triassic sediments: The main member is the Trogkofelkalk, consisting of limestones, partly dolomitic lesser bedded, massive beds, light grey, wide jointing.

6.3.2.2   *"Around site" structural analyses*
Within the slide area itself, the structural features are rotated and/or dislocated and, therefore, they do not represent the tectonic stress field.

In the Reppwand (main scarp), the general bedding has a dip of 30–50° with a dip direction of 200° to the south. The bedding planes are usually flat, partly polished and may contain mylonitic layers. The thickness of the bedding is in the range of cm–dm.

The main joint system dips 80° to the SW (dip direction 230°), is medium rough and has an average trace length of 5–10 m. Some clay seams may reach a thickness up to several cm.

A second joint system dips 75° to the NW (dip direction 310°), has usually rough joint surfaces and can be covered by mm thick seams of clay.

146

The rocks underlying the slide consist of the "Hochwipfelschichten" (grey schists and sandstones) and have good outcrops in the new river channel. Foliation there dips to the N (350/50–60), the primary joint system dips to the NW (300/45). A secondary system dips steeply to the NE (035/75).

The main fault zones are (sub-)parallel to the Periadriatic lineament (carving the main valley of the Gail River), striking WNW.

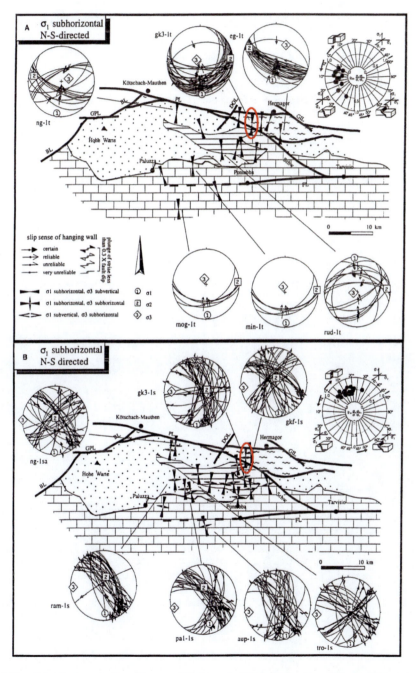

Figure 6.13.   Joint diagrams in the wider area (Läufer et al., 1997); the location of the landslide is indicated by a red ellipse.

The fault-slip data presented in Figure. 6.13 indicate a dextral-transpressional regime for the underlying units (Hochwipfel and Eder nappes). Faults which reveal stress tensors with σ1 orientated sub-horizontally (N-S) and σ3 switching between a sub-vertical (thrusting, Fig. 6.13 A) and sub-horizontal, E to W orientated position (strike-slip, Fig. 6.13 B) result from N to S directed contraction which is responsible for the formation of several positive flower structures in the vicinity of the Periadriatic lineament.

The geotechnical properties of the rocks in the front area of the Reppwand landslide can be defined as following:

– Naßfeld series: the rocks are heavily disturbed, therefore, they can be defined as a cohesive soil
– Rock structure is disintegrated to a block-talus with a fine-grained matrix
– Embankment (rock channel) with sandstones and schists, well-jointed
– Well-jointed, loosed limestones with wide open fissures at the main scarp

6.3.3   *Investigation and monitoring*

As a consequence of intense movements in August 1987, an extensive construction and monitoring program was started (see § 6.1.2). Since October 1988, the toe zone of the landslide has been intensively monitored by means of:

– 60 geodetic points (partly measured since 1985)
– 1 inclinometer-

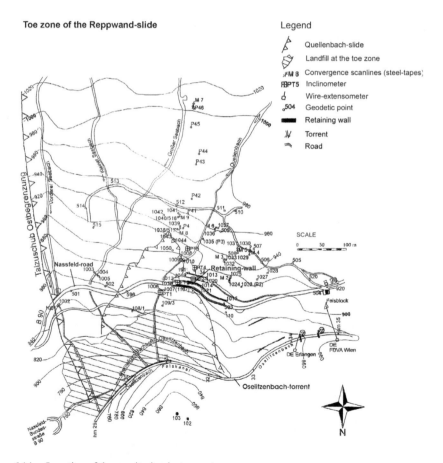

Figure 6.14.   Location of the monitoring instruments.

- 1 wire extensometer, l = 45 m, digital monitoring, measures every 2 min.
- 9 convergence scanlines (fixed endpoints), measurement with steel tape (Fa. Soil instruments MK II) , accuracy 1/100

Seven boreholes were drilled, one of them equipped with an inclinometer (BL depth: 52 m). Figure 6.14 shows the investigated area and the location of the monitoring instruments.

A comparison of the average values shows significant stabilisation after the finalisation of the construction measures in most areas. Periods A (1988–1991, before the construction measures) and B (1991–2000, after the construction measures) have, therefore, been compared showing the following:

- Areas of low velocity (<5 cm/y) significantly increased in Period B, whereas zones of high velocity (>10 cm/y) decreased.
- Above the fill, an extensive stabilisation was recorded in Period B.
- The most active zone in both periods is the upper area of the Quellenbach landslide, which, however, shows a stabilisation as well in Period B from 10–15 cm/y down to 7–15 cm/y. Figure. 15 shows the map of the movements, which were measured between 1991 and 2000.
- Relatively high movements can still be observed east of the point RP below the Naßfeld Road.

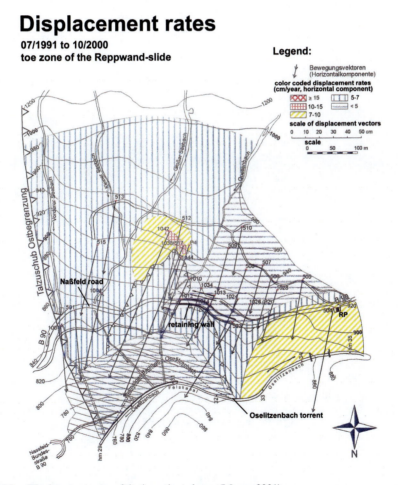

Figure 6.15. Displacement rates of the investigated area (Moser, 2001).

Within diverse homogeneous regions, there are different homogeneous regions, having different displacement rates. Since the construction measures were finished in 1991, the displacement rates decreased.

Table 6.1. Displacement rates.

Homogeneous region	1988–1991	1991–2000
"Seebach" above road	0,9 cm/month	0,5 cm/month
"Quellenbach"	1,3 cm/month	0,65 cm/month
Roadmaster hut	0,4 cm/month	0,35 cm/month

The refraction seismic section through the investigated area (Fig. 6.16) shows a three layer structure for the Naßfeld schists (Weidner, 2000):

– The lowest layer does not show any displacements at the moment although it is influenced by the Reppwand landslide.
– The middle layer (15–30 m thick) shows small displacements but is not as fractured as the upper layer.
– The upper layer (10–15 m thick) is heavily fractured and behaves, therefore, as soil with low cohesion. This fact is the main reason for sustained displacements.

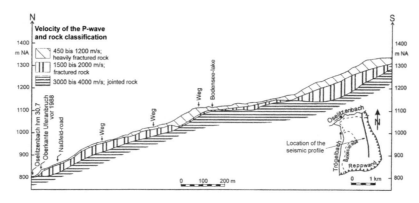

Figure 6.16. Seismic profile of the sagging slope (Weidner, 2000).

6.3.4 Danger identification

According to the results of the monitoring program and morphological studies (activity of landforms, cracks, etc.) three possible scenarios have been detected (Fig. 6.17 and 6.18):

1. Failure and detachment of the area displacing most at present
2. Bodensee landslide
3. Reactivation of the Reppwand landslide

Scenario 1: Failure and detachment of the area displacing most at present
Topographic monitoring, as well as wire extensometer readings and inclinometer readings, show that an area between 800 m and 1000 m a.s.l. is moving with displacement rates of 7 cm per year at present (Fig. 6.15: displacement rates MOSER) endangering the Naßfeld Road (see § 6.3.5).

Scenario 2: Bodensee landslide
Figure 6.17 shows that a sliding plane surfacing at the flattening of the Bodensee (1107 m a.s.l.; see § 6.3.1) was assumed to exist by various authors (e.g. Moser & al., 1988; Moser &

150

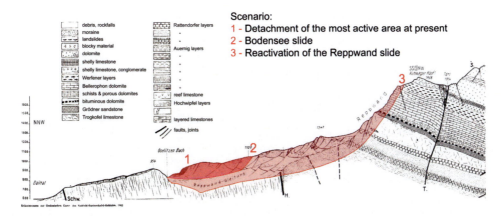

Figure 6.17.   Possible scenarios.

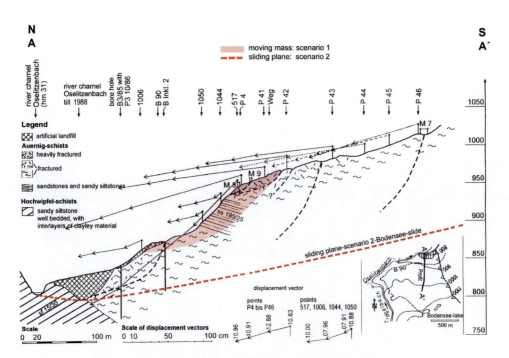

Figure 6.18.   Sliding planes of the possible scenarios at the toe zone of the landslide.

Windischmann, 1989; Kahler & Prey, 1963). The monitoring programme reveals that the area between Oselitzenbach and Bodensee is moving with displacement rates of 5 cm per year at present. Most probably, the Bodensee landslide will shear through the Hochwipfel schists' abutment at the toe of the slope for the Reppwand landslide mass.

Scenario 3: Reactivation of the Reppwand landslide
Morphological studies have indicated Graben-like structures parallel to the edge of the wall on the upper slope surface of the Reppwand displaying no signs of activity at present. However, a reactivation of the Reppwand landslide is seen as possible (Fig. 6.17; see § 6.1.1 and 6.3.2).

151

6.3.4.1 *Geometrical characterisation and geomorphologic constraints*

Scenario 1: Geophysical investigations (see § 6.3.3) as well as numerical modelling (see § 6.3.5) have shown that the sliding mass has a depth of some 25 m. The monitoring programme, as well as numerical modelling (comp. 3.5; Fig. 6.15: displacement rates MOSER), gave an unstable area of some 52'850 m^2 and a moving mass of some 450'000 m^3 (Fig. 6.19).

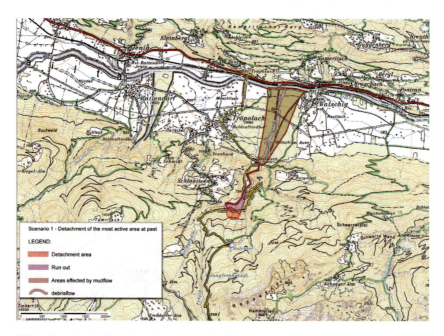

Figure 6.19.   Scenario 1 – Detachment and run-out of the most active area at present.

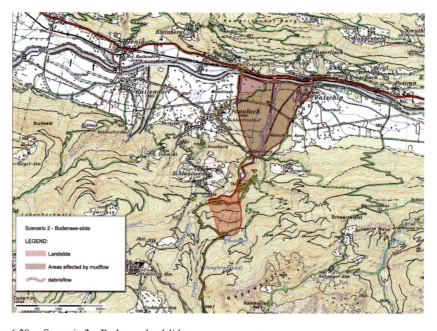

Figure 6.20.   Scenario 2 – Bodensee landslide.

152

Scenario 2: Due to the morphology (Vorderer Seebach), an unstable area of some 536'000 m^2 was assumed (Fig. 6.20).

Scenario 3: Geological investigations (Kahler & Prey, 1963) depicted an unstable area of some 2'911'000 m^2.

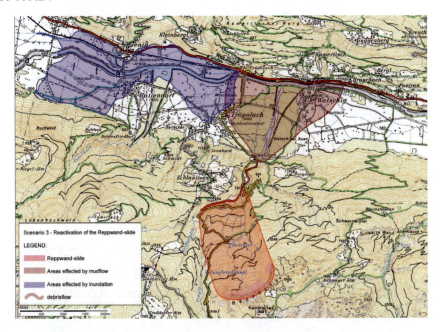

Figure 6.21.   Scenario 3 – Reactivation of the Reppwand landslide.

6.3.4.2   Structural constraints
Scenario 1: The whole sliding surface is running through the fractured Reppwand sliding mass. Thus, there are no structural constraints (Fig. 6.18).

Scenario 2: Almost the whole sliding surface is running through the fractured Reppwand sliding mass. Joints and faults in the Hochwipfel schists (see § 6.3.2) making an abutment at the toe of the slope and retaining the Bodensee landslide mass will form a sliding plane (Fig. 6.18).

Scenario 3: In the lower part, the assumed sliding plane follows the sliding plane of the ancient Reppwand landslide. In the upper part of the sliding plane, Graben-like structures parallel to the edge of the wall on the upper slope surface of the Reppwand coinciding with joint sets indicate sliding planes in the Bellerophon dolomite and the Trogkofel limestone (Fig. 6.17). Similar to the ancient Reppwand landslide, the Auernig schists will be sheared through during the development of the landslide (see Fig. 6.10).

6.3.4.3   Scenario identifications
Scenario 1: The run-out of the area displacing most at present was numerically modelled (see § 6.3.5). The calculations have shown that the Oselitzenbach ravine will be buried by 85'085 m^3 of debris flowing over the Oselitzenbach debris cone, burying 896'000 m^2 (Fig. 6.19).

Scenario 2: The run out of the Bodensee slide will bury the Oselitzenbach ravine by 3'375'000 m^3 of debris flowing over the Oselitzenbach debris cone, burying 200'000 m^2 (Fig. 6.20).

Scenario 3: The run out of the reactivated Reppwand landslide will bury the Oselitzenbach ravine by 16'500'000 m^3 of debris flowing over the Oselitzenbach debris cone, burying the villages of Tröpolach and Watschig and 2'678'000 m^2 of forest and rural areas and damming up the river Gail by 5 m (Fig. 6.21), thus causing an inundated area of some 4'852'000 m^2.

6.3.5 Geomechanical modelling

6.3.5.1 Mechanical "Triggering" model

The Oselitzenbach site has been studied by means of two 3D continuum models: the first model was made using the code FLAC3D (Itasca, 1997) based on the finite difference technique, the second one has been carried out using the finite element code DRAC (Prat et al., 1993).

Only Scenario 1 (Failure and detachment of the area displacing most at present) has been numerically investigated because the occurrence probability of this scenario is the highest. (see § 6.3.6).

Figure 6.22.   Investigated area of the Reppwand landslide.

The investigated area mainly consists of two homogeneous regions (see Fig. 6.24), the Naßfeld schists and the Hochwipfel schists. The heavily fractured Naßfeld schists consist of a melange of conglomerates, limestones and sandy siltstones. The Hochwipfel schists consist of sandy siltstones; they have higher strength than the more fractured Naßfeld schists.

Seismic surveys have revealed that the Naßfeld schists can be divided vertically into three layers corresponding to the degree of fracturation (Fig. 6.16). Since there are no main discontinuities at this site, only a continuum model has been analysed. The models, however, include four differ-ent material properties: the Naßfeld schists with three material properties and the Hochwipfel schists with one.

154

Figure 6.23.   View of the investigated area.

Figure 6.24.   Digital surface model.
Blue: Naßfeld schists – low strength, green: Hochwipfel schists – high strength.

➢ DRAC model

The Oselitzenbach test site has been analysed by a 3D continuum model (Fig. 6.25). The topography of the model was defined by the actual morphology of the slope utilising a digital elevation model (Fig. 6.24) with a cell size of 25 m. The final model consists of 15066 continuum elements.

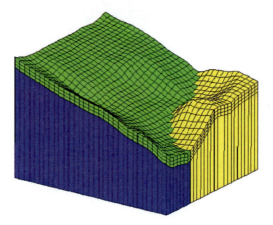

Figure 6.25.   Three-dimensional model of the Oselitzenbach slope including different material types.

A two-stage approach was used during the modelling. In the first stage, the upper limit of the glacier was set at 1700 m a.s.l., exceeding the surface of the entire study zone by several hundreds of meters, whereas in the second stage the glacier was removed. The boundary conditions were maintained constant during both stages and were set fully restrained at the base and restrained in the x and y directions at the lateral edges.

The values selected for the material properties are listed in Table 6.2. As mentioned above, the Naßfeld schists were divided into three layers due to their different degree of fracturation.

Table 6.2.   Material properties for the Oselitzenbach 3D model.

	γ[kN/m^3]	E[GPa]	ν
Naßfeld schists			
Upper layer	28	2,5	0,2
Middle layer	28	3,7	0,2
Lower layer	28	3,7	0,2
Hochwipfel schists	28	9	0,1
Glacier ice	9,2	9,6	0,33

The main objective of the analysis was the comparison of the different computer codes through back-analysis of the displacements measured during 1991 and 2000. Another aim was to check the sensitivity of the stability of the slope with respect to the material parameters. An elastic analysis will not present any failure mechanism, but it will provide some insight into the landslide behaviour, as areas with large displacements or tensile stresses may be potential zones of failure (Zettler & al., 1999).

Figure 6.26 shows a top view of the computed displacements in the northern (Y) direction (towards the Oselitzenbach). The results indicate that the location of the maximum displacements coincides rather well with the most unstable area detected by topographic measurements.

Figure 6.27 shows the displacement contours in the northern (Y) direction (towards the Oselitzenbach) on a section through the most unstable area. The results can be compared qualitatively with the results calculated by the FLAC3D model (Fig. 6.29 and 6.31) and show that both numerical models reveal maximum displacements in very similar zones of the landslide area.

Finally, Figure 6.28 indicates the computed deviatoric stresses (J2) along the same section as used in Figure 6.27, showing a measure of the shear stress distribution in the zone. The results show maximum shear in the channel bed of the Oselitzenbach Torrent and expose high values in the zone of the most unstable area.

➢ FLAC3D model
Two FLAC-models were investigated.

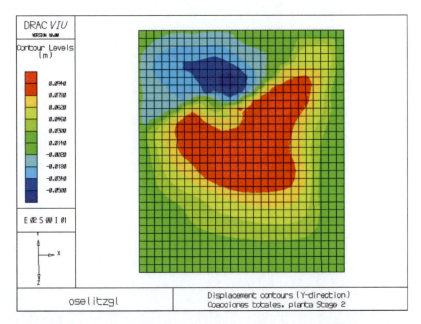

Figure 6.26. Displacement contours in the Y direction (towards the valley) – top view.

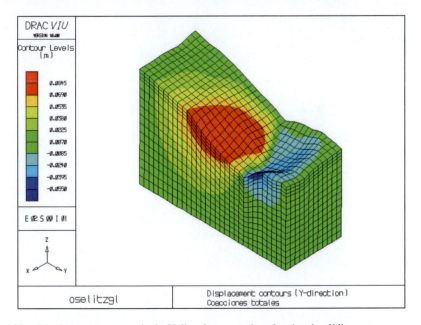

Figure 6.27. Displacement contours in the Y direction – cut view showing the sliding mass.

157

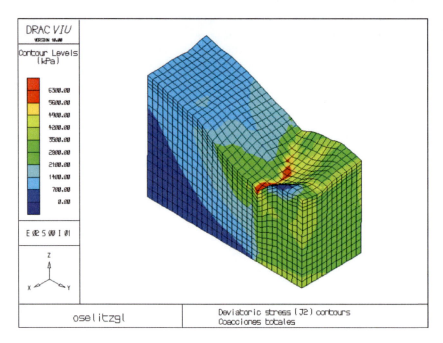

Figure 6.28. Deviatoric stress (J2) contours, showing maximum shear at the valley floor.

Model 1

The blocks of the Naßfeld schists and of the Hochwipfel schists are homogeneous. The mesh of the model (Fig. 6.29) is based on the digital surface model of the BEV (Fig. 6.24). The grid distance is 25 meters. A Mohr-Coulomb constitutive model was investigated using the following material properties:

Table 6.3. Material properties applied for the Naßfeld schists.

E [GPa]	v	$\varphi[°]$	c [kPa]
2,5	0,2	20	14

Table 6.4. Material properties applied for the Hochwipfel schists.

E [GPa]	v	$\varphi[°]$	c [kPa]
9	0,1	40	1e3

The density of the material is 2500 kg/m³. Figure 6.29 shows the distribution of the shear strain rate and the displacement vectors. In general, the displacements decrease continuously with depth resulting from creep (slope sagging). The distribution of the shear strain rate indicates a zone of maximum shear strain rate in a certain depth. Below this zone, displacements are zero, above they have a value increasing to the surface. Thus, the zone of maximum shear strain rate is the "sliding" zone.

Model 2

Based on the refraction seismic section, the area of the Naßfeld schist was built as a three layer model (Fig. 6.30). The upper layer was built with a continuous thickness of 15 m and the middle

158

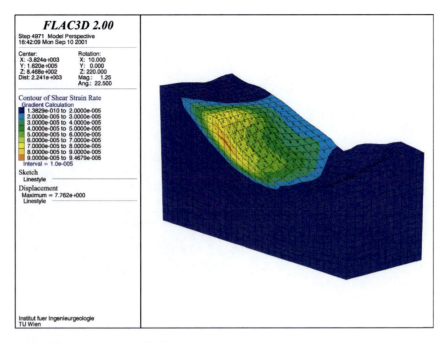

Institut fuer Ingenieurgeologie
TU Wien

Figure 6.29.   Shear strain rate and displacement vectors.

layer with a thickness of 30 m. The density of the material is 2500 kg/m³. A Mohr-Coulomb constitutive model was investigated using the following material properties:

Table 6.5.   Material properties applied for the Naßfeld schists.

	E [GPa]	ν	$\varphi[°]$	c [kPa]
Upper layer	2,5	0,2	18	14
Middle layer	3,7	0,2	25	20
Lowest layer	3,7	0,2	40	20

Table 6.6.   Material properties applied for the Hochwipfel schists.

E [GPa]	ν	$\varphi[°]$	c [kPa]
9	0,1	40	1e3

In general, Model 2 gave the same results as Model 1. The contour of the displacement magnitude, Figure 6.31, shows continuously decreasing displacements with depth.

As a result of the geological investigations and the monitoring program, a continuum mechanical approach was used. Thus, the analyses using FLAC gave an area with continuously decreasing displacements with depth down to a certain depth ("sliding" zone). The results are in good agreement with the results of the monitoring program. Figure 6.32 shows the map of the movements, which were measured between 1991 and 2000 and the area of movements from the FLAC3D analysis. From this a characteristic profile was deduced which was used for the two-dimensional run-out calculations.

The distribution of the shear strain rate indicates a zone of maximum shear strain rate at a certain depth. Below this zone, displacements and velocities are zero, above it they have a value increasing to the surface. Thus, the zone of maximum shear strain rate is the "sliding" zone.

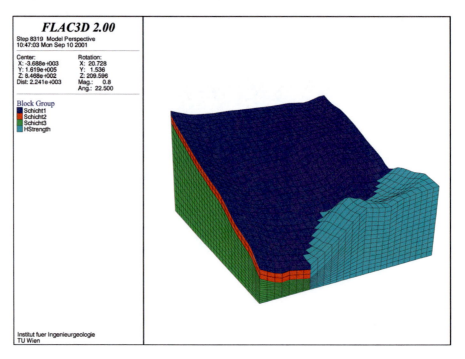

Figure 6.30.   Homogeneous regions of the model.

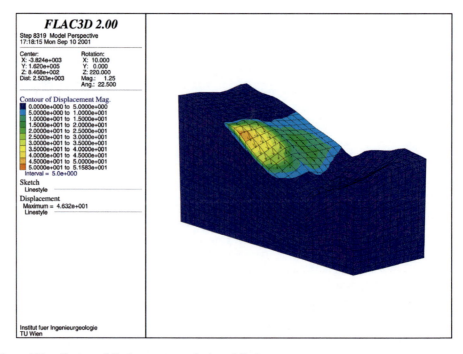

Figure 6.31.   Contour of displacement magnitude and displacement vectors.

160

This failure surface was used for the three-dimensional run-out analysis. Because of the more exact gradient, the contour of the velocity of the FLAC3D analysis was used to obtain the three-dimensional failure surface. In several cross-sections of the velocity contour, the position of the failure surface was defined and read out. In Figure 6.32 the path of the failure surface in the cross-section A-A is shown.

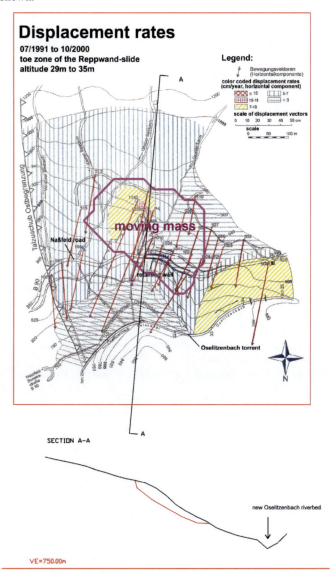

Figure 6.32.   Map of movements (Moser, 2001) and cross-section of the main movement directions.

➤ Comparison of the methods
The main objective of the analyses with DRAC and FLAC3D was the comparison of the different computer codes for the back-analysis of the displacements measured during 1991 and 2000. DRAC and FLAC are not directly comparable, as DRAC doesn't provide any plasticity models (e.g. Mohr-Coulomb). An elastic analysis will not present any failure mechanism, but it will provide some insight into the landslide behaviour, as areas with large displacements or tensile stresses may be potential zones of failure.

These results by DRAC (Fig. 6.26 and 6.27) can be compared qualitatively with the results calculated by the FLAC3D model (Fig. 6.29 and 6.31) and show that both numerical models reveal maximum displacements in very similar zones of the landslide area. The area of movements determined by FLAC as well as that by DRAC corresponds very well with the results of the monitoring program (Fig. 6.32).

6.3.5.2 *Mechanical "run-out" model*

The computer programs PFC2D and PFC3D from ITASCA (1999a, 1999b) were used as an All Ball and a Ball Wall model by TUW and the computer code DAN and the rock fall program ROTOMAP (Geo & Soft International, 1999, 2003) was used by POLITO in order to model the run-out behaviour of the Oselitzenbach landslide.

Generally PFC is not yet able to model the influence of water (e.g. pore pressure) on the run-out, whereas DAN needs an estimate of the detached rock mass and of the run-out direction (Hungr, 1995). An estimation of the detached rock mass is also needed for the PFC Ball Wall model, which was provided by a FLAC3D investigation (see § 6.3.5.1). Thus, the methods described should be used in combination, which makes a comprehensive assessment of the run-out as close to reality as possible.

➢ PFC Ball Wall Model

In the PFC Ball Wall model, the bedrock is simulated by linear (2D) and planar (3D) elements. In contrast to the All Ball model, where the relatively stationary bedrock is modelled by balls as well in order to model the failure mechanism of the slope and the detachment mechanism also, in the Ball Wall model only the detached rock mass is modelled by balls. Therefore, in the Ball Wall model, an estimate of the failure mechanism of the slope and of the detachment mechanism is needed as an input parameter. Consequently, in the Ball Wall model, the detached mass can be modelled with the help of more and smaller balls with the same computational effort. One goal of the investigations of WP4, therefore, was to compare the two different approaches.

The PFC Ball Wall model offers the possibility of making use of know-how related to run-out relevant resistances (factors of restitution, absorption, friction, etc.) applied in rock fall programs and, consequently, makes a realistic calculation of the run-out possible.

The combination with FLAC (see § 6.3.5.1) allows a realistic estimate of the detached rock mass on the basis of already existing experiences (see Fig. 6.32). An interaction of detachment and run-out, however, is not possible.

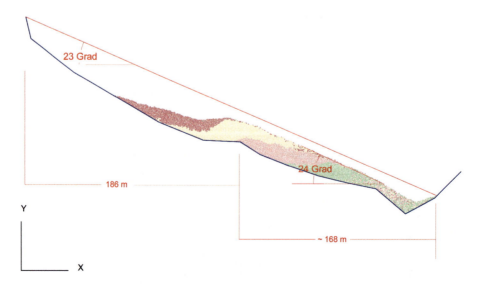

Figure 6.33.   Final state of the PFC2D calculation.

162

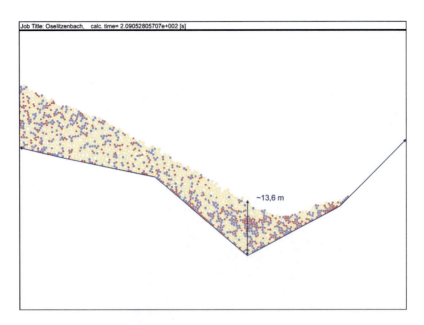

Figure 6.34.   Deposit in the valley after the rock mass fall.

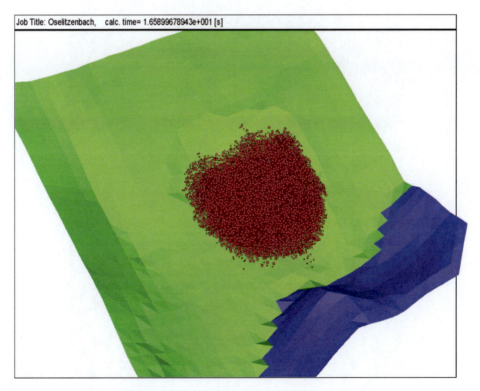

Figure 6.35.   3D run-out calculation after 16.5 sec.

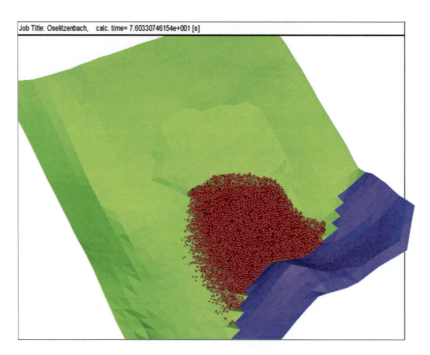

Figure 6.36.   3D run-out calculation after 76 sec.

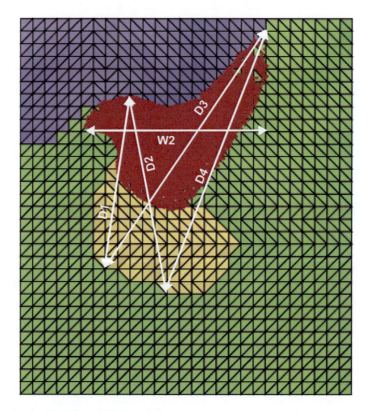

Figure 6.37.   Final state of the Ball Wall model.

164

Table 6.7.   Measurement lines and run-out distances.

Measurement line	Distance [m]
D1	346
D2	397
D3	573
D4	556
W2	377

➢   PFC All Ball Model

The PFC All Ball model allows the all in one calculation of failure mechanisms, detachment and run-out. For calculating the failure mechanism, however, quite a demanding calibration of materials is necessary. When comparing the volume of the detached rock with FLAC both methods correspond closely, which verifies the All Ball model. The surface of the model is rather rough due to the modelling of the bedrock by balls, which has to be considered at the calibration of the run-out parameters. This aspect needs further investigation.

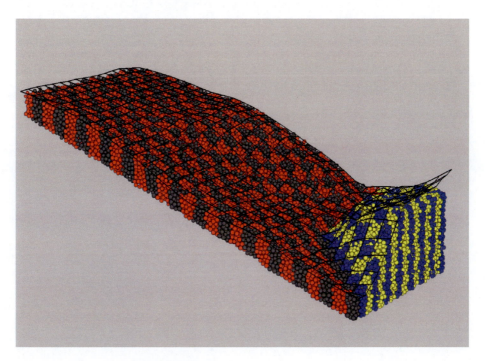

Figure 6.38.   Initial state of the 3D model.

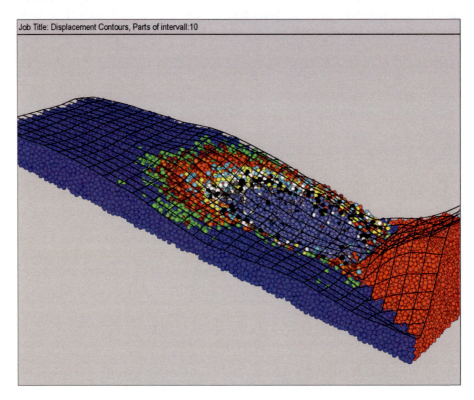

Figure 6.39. Final state of the All Ball model.

Figure 6.39 and Figure 6.40 show the final state of the All Ball model: the run-out is described with the help of four horizontal measurement lines (D1–D4), the path width by the lines W1 and W2 (see Table 6.8).

Table 6.8. Measurement lines and run-out distances.

Measurement line	Distance [m]
D1	330
D2	389
D3	373
D4	378
W1	204
W2	364

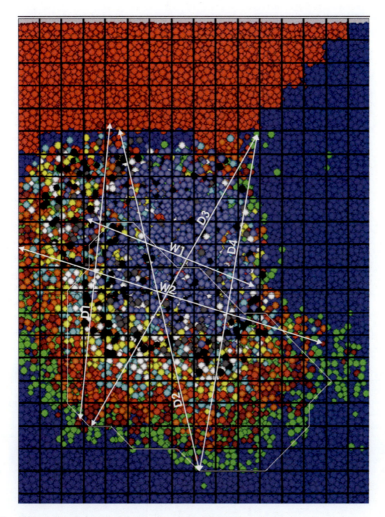

Figure 6.40.   Run-out distance (D1–D4) & path width (W1, W2).

➢   Comparison Ball Wall model – All Ball model
The area of the maximum displacement rates (light blue) in the All Ball model corresponds closely to the direction and width of the run-out in the Ball Wall model. The Ball Wall model, however, indicates a far greater travel distance (Fig. 6.37). This is due to the collision of moving particles with stationary ones (bedrock) in the All Ball model, with the moving particles losing energy additionally. This aspect also needs further investigation.

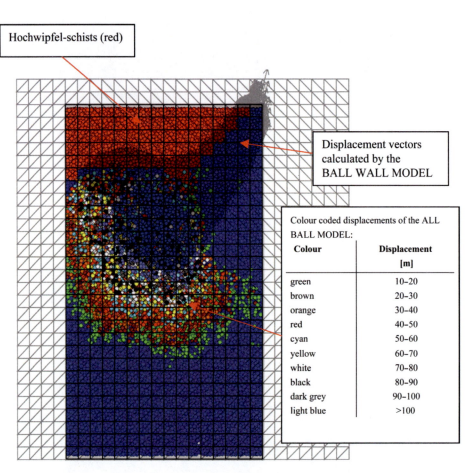

Figure 6.41. Comparison of the displacements calculated by the ALL BALL MODEL and the BALL WALL MODEL (plan view).

➢ Rotomap and Dan Code

Since in the DAN method the run-out direction and the path width are assigned a priori, in the investigations of WP4 they have been determined with the aid of a 3D rockfall program called ROTOMAP. The comparison of the results of the PFC2D Ball Wall model, which also needs an estimate of the run-out direction, and the results of the DAN model shows that two different run-out directions have been chosen and that these profile directions have an enormous influence on the results.

The run-out directions determined by Rotomap correspond very well to those of the 3D All Ball model in the West, whereas the run-out in the 3D All Ball model in the East indicates a smaller width (Fig. 6.44). The same applies to the 3D Ball Wall model (Fig. 6.43).

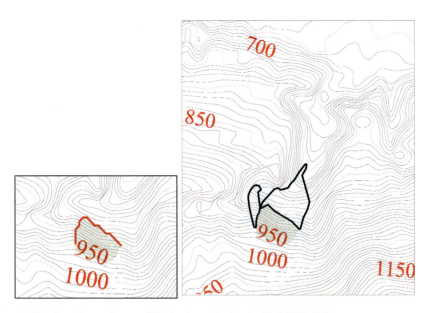

Figure 6.42. Detachment lines and block edges calculated with ROTOMAP.

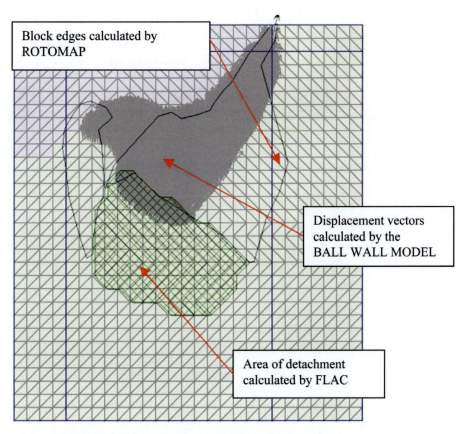

Block edges calculated by
ROTOMAP

Displacement vectors
calculated by the
BALL WALL MODEL

Area of detachment
calculated by FLAC

Figure 6.43. Comparison of the BALL WALL MODEL and ROTOMAP (plan view).

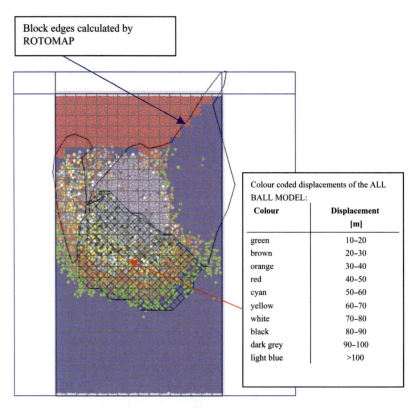

Block edges calculated by ROTOMAP

Colour coded displacements of the ALL BALL MODEL:

Colour	Displacement [m]
green	10–20
brown	20–30
orange	30–40
red	40–50
cyan	50–60
yellow	60–70
white	70–80
black	80–90
dark grey	90–100
light blue	>100

Figure 6.44.   Comparison of the ALL BALL MODEL and ROTOMAP (plan view).

In the case of the Oselitzenbach landslide, the comparison of the results of the methods (Table 6.9) shows that:

– the FLAC3D simulation as the basis of the DAN and of the 3D Ball Wall model give approximately the same detached rock volume as the All Ball model,
– the PFC – Ball Wall model and the DAN Code (depending on the pore pressure assumed) give the same travel distance of the run-out, whereas the PFC – All Ball model gives a much smaller value due to the rough surface of the bedrock caused by the bedrock built up by balls,

Table 6.9.   Comparison of the results of the methods.

Method	PFC – Ball Wall	PFC – All Ball	Rotomap + DAN Code
Detached rock volume [m³]	450'000 (input from WP3, FLAC)	377'000–495'000	450'000 (input from WP3, FLAC) (244'300–571'300 depending on the shape factor assumed)
Travel distance [m]	573	389	597 (for a pore pressure of 0.1) 470–833 (depending on the pore pressure assumed)
Run-out width [m]	377	364	340
Affected area [m²]	127'000	97'537	134.494 (Rotomap)
Maximum travel velocity [m/s]	29	4	21

170

- the run-out widths and the affected areas obtained by the models correspond more or less and,
- the maximum travel velocities of the run-out obtained by the models do not correspond. The slow travel velocity of the All Ball is caused by the rough surface of the bedrock built up by balls. These aspects should be considered during the calibration of the run-out parameters and has to be investigated further.

Thus, the combination of all three methods yielded:

- a detached rock volume of some 450'000 m³,
- a travel distance of the run-out of 600 m (due to the blocky nature of the Naßfeld schists a zero or low pore pressure is assumed to develop in the run-out mass),
- a run-out width of 360 m,
- an affected area of 130'000 m², and
- a maximum travel velocity of the run-out of 20 m/s.

These data mean that the Nassfeld road will be destroyed or buried with a maximum height of 7 m by the run-out in case of a slope failure over a length of 400 m. The Oselitzenbach torrent will be dammed up as well over a length of 460 m with a maximum height of 14 m, thus endangering the villages of Tröpolach and Watsching, including the adjoining agricultural and forest areas by debris flows.

➤ Occurrence probability

Scenario 1: Two events of debris flows have been recorded over an observation period of some 100 years. Thus, an occurrence period of 50 years has been assumed for Scenario 1.

Scenario 2: In historical times (about 1'000 years), no event as large as the possible Bodensee landslide has ever been reported. Thus, the worst case occurrence period for Scenario 2 is 1'000 years.

Scenario 3: The Würmian moraines show no dislocation at the edges of the Reppwand landslide. Thus, the Reppwand landslide is older than Würmian, meaning that it happened more than 70'000 years ago and that the worst case occurrence period for Scenario 3 is 70'000 years.

Table 6.10. Definition of the occurrence probability through the historical approach.

Scenario	Observation period (years)	Recorded events	Occurrence period (years)	Frequency (event/year)
1	100	1983 1987	50	1/50 = 0,02
2	Historic times	None	1.000	1/1'000 = 1,0e−3
3	70'000	None	70'000	1/70'000 = 1,43e−5

6.4   QUANTITATIVE RISK ANALYSIS

In Section 6.3.4, 3 scenarios of the Oselitzenbach landslide evolution were described. Each scenario is characterised by failure and run-out areas (derived from the geomorphological and geomemechanical models) and by an occurrence probability (as a result of an historical analysis of recorded events).

According to the proposed quantitative risk analysis, each element of the hazard scenario is used separately in the matrix calculus of risk evaluation. This process is applied on each recognised scenario; therefore, at the Oseltzenbach site three separate risk analyses were carried out.

6.4.1   *Elements at risk*

The elements at risk are recognised through the definition of the area affected for each danger in the scenarios previously specified (Fig. 6.19, 6.20 and 6.21). These elements at risk are shown in Tables 6.11, 6.12 and 6.13.

Table 6.11. Definition of the elements at risk: Scenario 1 (Detachment of the most active area at present).

Elements at risk	Name	Notes
Forests (private and public properties)		Area affected [km^2]: by landslide: 0,129 by mudflow: 0,896 Persons affected: 1
Heavy traffic or strategic roads	National road B90	Length affected [m]: 400 Persons affected: 3
Tourist accommodation		Persons affected: 1
Lifelines		Length affected [m]: 600 Pylons affected: 4

Table 6.12. Definition of the elements at risk: Scenario 2 (Bodensee landslide).

Elements at risk	Name	Notes
Residential areas [valuable buildings or valuable rural centres: historical, architectural, artistic and/or cultural value]	Watschig	Area affected [km^2]: by mudflow: 0,137 Persons affected: 15
Forests (private and public properties)		Area affected [km^2]: by landslide: 0,536 by mudflow: 1,652 Persons affected: 1
Rural area		Area affected [km^2]: by mudflow: 0,204
Heavy traffic or strategic roads	National road B90	Length affected [m]: 730 Persons affected: 3
Tourist accommodation		Persons affected: 5
Lifelines		Length affected [m]: 1370 Pylons affected: 9

Table 6.13. Definition of the elements at risk: Scenario 3 (Reactivation of the Reppwand landslide).

Elements at risk	Name	Notes
Residential areas [valuable buildings or valuable rural centres: historical, architectural, artistic and/or cultural value]	Watschig	Area affected [km^2]: by mudflow: 0,500 Persons affected: 129
	Tröpolach	Area affected [km^2]: by mudflow: 0,120 by inundation: 0,248 Persons affected: 180
	Rattendorf	Area affected [km^2]: by inundation: 0,271 Persons affected: 50
Forests (private and public properties)		Area affected [km^2]: by landslide: 2,911 by mudflow: 1,804 by inundation: 0,2 Persons affected: 1

(contd)

172

Table 6.13.   (*contd*)

Elements at risk	Name	Notes
Rural area		Area affected [km^2]: by mudflow: 0,204
Heavy traffic or strategic roads	National road B90	Length affected [m]: by landslide: 3450 Persons affected: 3
Secondary railway		Length affected [m]: by mudflow: 2422 by inundation: 3440
Tourist accommodation Lifelines		Persons affected: 5 Length affected [m]: by mudflow: 1935 by inundation: 2628 Pylons affected: by mudflow: 15 by inundation: 11

6.4.2   *Vulnerability of the elements at risk*

The vulnerability of the elements at risk was determined by an estimation of the effects on the elements at risk. Table 6.14 shows the vulnerability evaluation.

Table 6.14.   Vulnerability evaluation as a function of estimated effects on elements at risk.

Physical vulnerability			Social vulnerability	
Vulnerability description	**Loss range**	**Index**	**Vulnerability description**	**Index**
Intact structures	0	0	Non affected persons	0
Local damages	1÷25%	0.25	Non physical damages; evacuated persons	0.25
Seriously damages (possible to repair)	26÷50%	0.5	Physical damages (person continue theiractivities)	0.5
Mostly destroyed (difficult to repair)	51÷75%	0.75	Seriously wounded persons (50% disability)	0.75
Total destruction (out of use; e.g. >5% inclination)	76÷100%	1	Died, 51–100% disability	1

Economical vulnerability		Environmental vulnerability		
Vulnerability description	**Index**	**Vulnerability description**	**Loss range**	**Index**
Non interruption	0	Intact element	0	0
Short temporary interruption (hours to day)	0.25	Local loss	1÷25%	0.25
Average temporary interruption (days to week)	0.5	Seriously damages (possible to repair)	26÷50%	0.5
Longtemporary interruption (weeks to months)	0.75	Mostly destroyed (difficult to repair)	51÷75%	0.75
Permanent interruption	1	Total destruction	76÷100%	1

The following tables (6.15–6.20) show the considered values of the elements at risk and the evaluated vulnerability for each scenario.

Table 6.15. Scenario 1: Considered values of the elements at risk.

Considered values [VE]

Elements at risk	Physical	Economic	Environm.	Social
Forests/rural area (private and public properties)	1	1	3	1
Heavy traffic or strategic roads	3	4	1	2
Tourist accommodation – buildings	3	3	1	1
Lifeline	1	2	1	0

Table 6.16. Scenario 1: Vulnerability.

Vulnerability [V]

	Physical	Economic	Environm.	Social
Vulnerability	0,25	0,5	0,25	0,25

Table 6.17. Scenario 2: Considered values of the elements at risk.

Considered values [VE]

Elements at risk	Physical	Economic	Environm.	Social
Residential areas	4	1	1	3
Forests/rural area (private and public properties)	1	1	3	1
Heavy traffic or strategic roads	3	4	1	2
Tourist accommodation – buildings	3	3	1	1
Lifeline	1	2	1	0

Table 6.18. Scenario 2: Vulnerability.

Vulnerability [V]

	Physical	Economic	Environm.	Social
Vulnerability	0,5	0,5	0,5	0,5

Table 6.19. Scenario 3: Considered values of the elements at risk.

Considered values [VE]

Elements at risk	Physical	Economic	Environm.	Social
Residential areas	4	1	1	4
Forests/rural area (private and public properties)	1	1	3	1
Heavy traffic or strategic roads	3	4	1	2
Secondary railway	1	2	1	0
Tourist accommodation – buildings	3	3	1	2
Lifeline	1	2	1	0

Table 6.20. Scenario 3: Vulnerability.

Vulnerability [V]

	Physical	Economic	Environm.	Social
Vulnerability	0,75	0,75	0,75	0,75

6.4.3 *Expected impact*

Multiplying the values of the elements at risk (VE, see Tables 6.15, 6.17 and 6.19), and vulnerability percentage V (obtained as described in the previous section), the expected impact C is obtained as C = VE × V.

To state a precise quantitative damage evaluation for economic activities and assets, costs derived from economic analyses are essential (e.g. assets costs, reconstruction costs, turnover, profits loss, etc.). To assess expected environmental and social impacts (where an economic value is difficult to obtain), relative value indices are used.

In the quantitative risk analysis of the Oselitzenbach landslide, detailed economic evaluations were considered. Arbitrary relative values indices were applied for each category. In Tables 6.21–6.23 expected impact values of each scenario are shown.

Table 6.21. Scenario 1: expected impact.

Expected impact [C = VE × V]

Elements at risk	Physical consequ.	Economic consequ.	Environm. consequ.	Social consequ.
Forests/Rural area (private and public properties)	0,25	0,5	0,75	0,25
Heavy traffic or strategic roads	0,75	2	0,25	0,5
Tourist accommodation – buildings	0,75	1,5	0,25	0,25
Lifeline	0,25	1	0,25	0

Table 6.22. Scenario 2: expected impact.

Expected impact [C = VE × V]

Elements at risk	Physical consequ.	Economic consequ.	Environm. consequ.	Social consequ.
Residential areas	2	0,5	0,5	1,5
Forests/rural area (private and public properties)	0,5	0,5	1,5	0,5
Heavy traffic or strategic roads	1,5	2	0,5	1
Tourist accommodation – buildings	1,5	1,5	0,5	0,5
Lifeline	0,5	1	0,5	0

Table 6.23.   Scenario 3: expected impact.

Expected impact [C = VE × V]

Elements at risk	Physical consequ.	Economic consequ.	Environm. consequ.	Social consequ.
Residential areas	3	0,75	0,75	3
Forests/rural area (private and public properties)	0,75	0,75	2,25	0,75
Heavy traffic or strategic roads	2,25	3	0,75	1,5
Secondary railway	0,75	1,5	0,75	0
Tourist accommodation – buildings	2,25	2,25	0,75	1,5
Lifeline	0,75	1,5	0,75	0

6.4.4   Risk assessment

The last phase of the quantitative risk analysis concerns the risk evaluation. This evaluation is obtained by multiplying the expected impact values C (Tables 6.21–6.23) and the numeric values of the occurrence probability P related to each scenario (Table 6.10): R = C × P. Tables 24, 25 and 26 show the risk assessment results for each scenario. The same results are also represented by a zonation shown in Figures 6.45–6.48 for Scenario 1 and in Figures 49 and 50 for Scenario 2. Due to the extremely small risk values of Scenario 3, the physical, economical, environmental and social risk are not shown by a relative zoning (map of zones of equate risk) of the values.

Table 6.24.   Risk assessment for Scenario 1: R1 = C1 × 1/50.

Risk assessment [R = C × P]

Elements at risk	Physical risk	Economic risk	Environm. risk	Social risk
Forests/rural area (private and public properties)	0,005	0,01	0,015	0,005
Heavy traffic or strategic roads	0,015	0,04	0,005	0,01
Tourist accommodation – buildings	0,015	0,03	0,005	0,005
Lifeline	0,005	0,02	0,005	0

Table 6.25.   Risk assessment for Scenario 2: R2 = C2 × 1/1.000.

Risk assessment [R = C × P]

Elements at risk	Physical risk	Economic risk	Environm. risk	Social risk
Residential areas	0,0020	0,0005	0,0005	0,0015
Forests/rural area (private and public properties)	0,0005	0,0005	0,0015	0,0005
Heavy traffic or strategic roads	0,0015	0,0020	0,0005	0,0010
Tourist accommodation – buildings	0,0015	0,0015	0,0005	0,0005
Lifeline	0,0005	0,0010	0,0005	0,0000

Table 6.26.   Risk assessment for Scenario 3: R3 = C3 × 1/70.000.

Risk assessment [R = C × P]

Elements at risk	Physical risk	Economic risk	Environm. risk	Social risk
Residential areas	4,3e−05	1,1e−05	1,1e−05	4,3e−05
Forests/rural area (private and public properties)	1,1e−05	1,1e−05	3,2e−05	1,1e−05
Heavy traffic or strategic roads	3,2e−05	4,3e−05	1,1e−05	2,1e−05
Secondary railway	1,1e−05	2,1e−05	1,1e−05	0,0e+00
Tourist accommodation – buildings	3,2e−05	3,2e−05	1,1e−05	2,1e−05
Lifeline	1,1e−05	2,1e−05	1,1e−05	0,0e+00

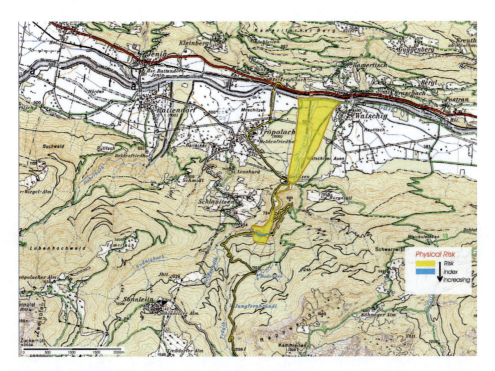

Figure 6.45.   Physical risk in Scenario 1.

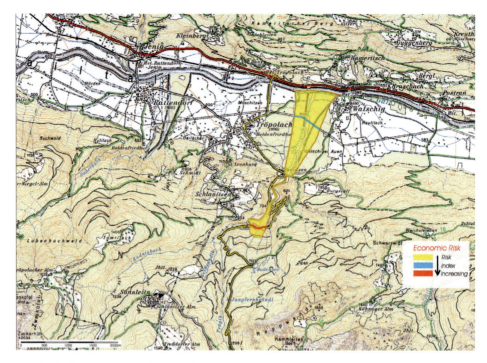

Figure 6.46.   Economic risk in Scenario 1.

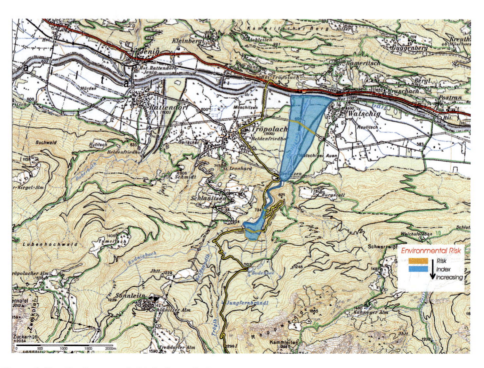

Figure 6.47.   Environmental risk in Scenario 1.

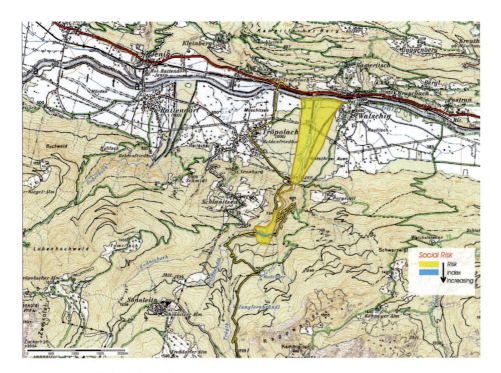

Figure 6.48. Social risk in Scenario 1.

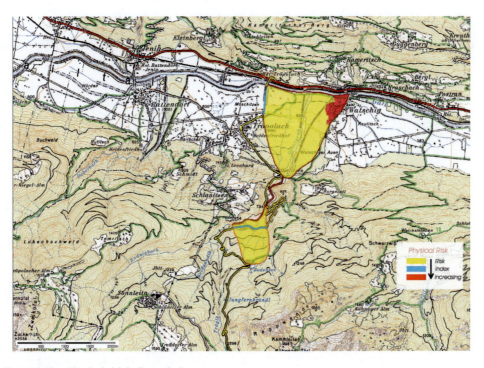

Figure 6.49. Physical risk in Scenario 2.

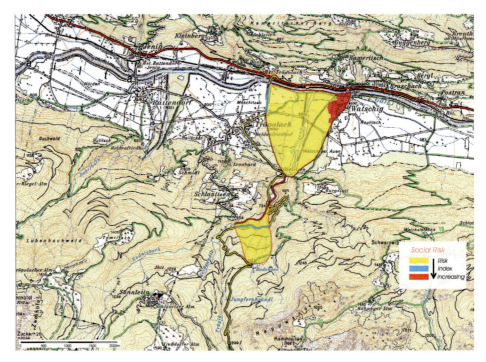

Figure 6.50.   Social risk in Scenario 2.

6.5   REGULATION AND RISK MITIGATION MEASURES ALREADY AVAILABLE

6.5.1   *Viability*

In order to prevent serious debris flow in the Oselitzenbach torrent and in the receiving stream "Gail", the following measures are possible:

– Linear measures: "Steps" on stream bed and debris retention in the flood outlet areas
– Forestry measures to improve the outlet conditions
– Measures such as draining and landfills to reduce the massive slope movements.

6.5.2   *Facilities*

An extensive project of construction work and research was started in 1988 to prevent debris-generating slope failures (especially at the toe zone of the Reppwand slide) and the destruction of the "Naßfeld Road". As a part of the construction measures, a 400 m long channel was excavated in massive Hochwipfel formations (comp. 1.2) and a landfill was made at the toe of the excavated material as well (about 170'000 m^3).

Additional measures were: Drainage of the Quellenbach landslide between 830 and 920 m a.s.l. and the draining-off of water, also above 950 m a.s.l. to the area of the "Bodensee" Lake. These measures were necessary to prevent serious debris flow in the Oselitzenbach Torrent and in the receiving stream "Gail" and to prevent debris flow from reaching the villages "Tröpolach" and "Watsching". For economic reasons, the correction of all the debris sources from the very top downwards was not deemed possible or necessary.

Additionally, a monitoring program was started in October 1988. The toe zone of the landslide has been intensively monitored as presented above in § 6.3.3. At present a risk management system according to the results of the IMIRILAND project is in preparation by local authorities (e.g. WLV – National Authority for torrent and avalanche control) and the project members.

7
The Ceppo Morelli rockslide

G. Amatruda[1], M. Castelli[1], F. Forlati[2], M. Hürlimann[3], A. Ledesma[3], M. Morelli[4],
L. Paro[2], F. Piana[4], M. Pirulli[1], R. Polino[4], P. Prat[3], M. Ramasco[2], C. Scavia[1], C. Troisi[2]

[1]Politecnico di Torino; [2]Arpa Piemonte; [3]Universitat Politècnica de Catalunya; [4]CNR-IGG sezione Torino

7.1 INTRODUCTION

The Ceppo Morelli rockslide extends on the left side of the middle Anzasca valley in the Pennine
Alps, northern Piedmont (Italy), a few kilometres from the border with Switzerland (Fig. 7.1). The
national road N° 549 runs on the valley bottom for about 30 km, from the Piedimulera village
(Toce River plain) to the small village of Pecetto di Macugnaga (at the foot of the Monte Rosa
Massif). The road is the only link to Macugnaga, a popular tourist resort close to Monte Rosa. The
Anza River drains the valley.

Attention has been focused on this well-known rockslide as a consequence of relevant slope
instability phenomena which occurred during the October 2000 heavy meteorological event.

Figure 7.1.   Digital Terrain Model of the north-western Piedmont with location of the Ceppo Morelli landslide
area (red star) main rivers and lakes (in blue), and villages (in yellow). General overview of the left slope of
Anzasca valley (air photographs overlapped by Digital Terrain Model) with Ceppo Morelli landslide (white arrow).

7.1.1 Landslide description

The direct effect of the catastrophic October 2000 heavy meteorological event was the reactivation
of a large ancient rockslide, which affects the middle-lower part of Monte Rubi's southern slope
(1850–1200 m a.s.l.). The slide involves a massive gneissic rock mass. The overall surface is
around 160'000 m^2 and the volume estimation is between 4 and 6 × 10^6 m^3.

Generally speaking, the whole unstable area can be divided in two main portions, a lower one, where the rock mass, although fractured and partially toppled and translated, is still ordered, and an upper one, constituted by coarse blocky debris. The upper part (delimited by tension cracks) moved due to the October 2000 event. Displacements reaching more than several centimetres (up to 400–500 cm) have been observed down a slope of about 30°–35°, inducing a lower part reaction, especially some hundred meters over the slope foot (1250 m). Numerous rockfalls reached the bottom of the valley (800 m a.s.l.) with boulders up to 300 m³. Several boulders reached the National Road No. 549. As the deformation affecting the lower part of the unstable area increases, a composed surface might be formed, leading to a huge rockslide and possibly severe rockfall and/or rock avalanche phenomena.

Major mass falls may reach two small villages dating back to the XVI century: Prequartera and Campioli located at the bottom of the valley.

7.1.2 *Historical background*

The site has been involved in repeated rockslides and falls since ancient times. A chronicle dating from the XV century reports failures in the years 312 and 843 A.D.; in the latter episode 25 knights were killed (allegedly carrying a sort of treasure). Remaining vivid in local history after several centuries (though unfortunately lacking written reports), the quoted events probably refer to mass failures rather than simple rockfalls.

"The year of the fall, 1816," is engraved on a 1'000 m³ boulder close to the Anza River. Local residents report that rockfalls are extremely common. In April 1977, severe rockfalls damaged the national road and some boulders up to 1'000 m³ reached the valley-road (Fig. 7.2).

Figure 7.2. April 1977: rock fall involving the national road N 549 (boulder of about 50 m³, prismatic shape).

Figure 7.3.   Rock fall triggered during the October 2000 heavy meteorological event. The boulder (right red arrow) damaged the national road (red line) and stopped near a house.

During the October 2000 severe event, several boulders up to 300 m³ reached the river and the National Road (Fig. 7.3). Some local rockfalls occurred in June 2002.

7.2   REGIONAL FRAMEWORK

7.2.1   *Climate and water condition*

Annual rainfall has been monitored over a period of 35 years, from 1951 to 1986. Its annual average value of about 1'594 mm, with 119 rainy days per year and average intensity of 13.4 mm/day, has been derived from the basis of data recorded in 14 stations belonging to the Idrografico e Mareografico National Service. Monthly rainfall peaks have been monitored at the Anzino station, located at the bottom of the valley, at an elevation of about 670 m and 6.5 kilometres away from the Ceppo Morelli rockslide. The months of May and October are characterised by the highest monthly rainfall value (198 mm and 190 mm) and by the largest number of rainy days (12) (Biancotti & Bovo, 1998a and b).

The high intensity event which occurred on 13–16 October 2000 produced a total precipitation of about 558 mm. The average daily rainfall value recorded from 13 to 15 October was almost equal to the highest monthly rainfall value recorded over a period of 35 years (Anzino).

The annual mean temperature is 9°C and there is an average of 93 frost days per year. Snow information is related to the Alpe Cavalli gauges (a station located at 1'510 m a.s.l.); data refer to 1966–1996 continuous measures: max. snow level 785 cm (1971–72), min. 137 cm (1985–86), average 438 cm. During the late spring days, the snow cover melts away quickly. As a result, a considerable quantity of water can seep down into the sliding body. The flow of water in the slope can then occur through the main discontinuities, which are generally open and persistent. Considering that the upper part of the slope acts as a reservoir, the largest amount of groundwater is cumulated when heavy meteorological events occur or when rainfalls combine with snow melting.

7.2.2 Regional morphology

The Anzasca valley is mainly characterised by fluvial and glacial morphogenesis and by landslide activity (rockfalls, large rotational and translational slides, deep seated gravitational deformations). The Anzasca valley is steep and impervious, characterised by steep rock walls seldom interrupted by scree deposits and dejection cones. The bottom valley altitude ranges from 247 m (Piedimulera village) to 1'378 m (small village of Pecetto di Macugnaga) along a 30 km length. Just beyond the village of Ceppo Morelli, the stream profile is interrupted by the Morghen rocky step that bounds the Macugnaga plain.

7.2.3 Regional geology and structural setting

The landslide site is found in the Monte Rosa basement nappe (Upper Pennine Units) represented by polymetamorphic augengneisses and paraschists. This basement is characterised by several Alpine ductile deformation phases (Fig. 7.4).

At the regional scale, the basement is deformed by the Vanzone antiform with a typical SW plunging axis and NW dipping axial plane.

The regional sin-metamorphic ductile deformation is characterised by the development of regional schistosity related to large-scale folding events that gave origin to local macrostructures such as the Vanzone antiform.

The regional brittle tectonic deformation is characterised by mesofaults and joint systems, partially parallel to the axial planes of the nVanzone.

Recent alluvial-colluvium deposits are represented by heterogeneous and incoherent masses of soil material and rock fragment deposits.

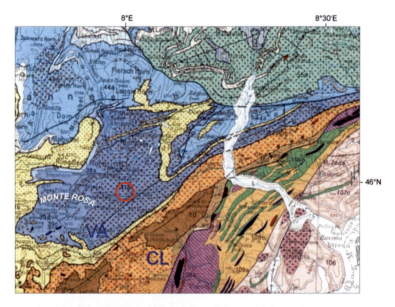

Figure 7.4. Geological map of a sector of the Western Alps (CNR, 1990 – Structural Model of Italy 1:500'000 scale). In the red circle is the landslide site placed within one of the main Alpine tectono-metamorphic unit (Monte Rosa Unit MRU), mainly made up of augengneisses and micaschists folded by Vanzone antiform (VA) and bordered on both sides by the Piemonte Zone (Penninic domain, in yellow). East of MRU are the Austroalpine domain (dotted orange and brown) while east of the Canavese Line (CL), a major fault of the Alps, are the southern Alps with the granulites of the Ivrea-Verbano Zone (Violet, light brown and green) and granitoids and micaschist of the Serie dei Laghi (pink and dotted pink).

184

7.3 HAZARD ANALYSIS

7.3.1 *Geomorphological analysis*

This high part of the Anzasca valley is characterised by a steep rock wall and a narrow, almost flat alluvial valley floor. These morphological features were caused by glacial erosion of highly fractured bedrock. The toe of the slopes is generally covered by very thick but discontinuous detrital fans and scree deposits, as in the sector below the Ceppo Morelli landslide, found at the base of the left slope, between the villages of Campioli and Prequartera.

This sector is studded with erosional scars separated by ridges of relatively safe rocks, both of which features represent the peculiar morphological aspects of the Ceppo Morelli landslide.

In order to understand the rockslide behaviour and to assess its probable impact on land planning, an intensive field survey and a systematic aerial photo analysis were conducted. The latest one was carried out on different flights (1970, 1978, 2000, and 2001), as a consequence of the 1977 and 2000 heavy meteorological events. Topographic and geometric attributes of the slope were derived from an available Digital Elevation Model (DEM). The entire studies and investigations have evidenced a morpho-dynamic situation, sufficiently clear in terms of recent and ancient slope evolution and in agreement with the geological and structural observations carried out during field surveys.

The landslide has been subdivided into three areas (see Fig. 7.5):

A. Detachment area
B. Rock mass translation area
C. Accumulation and landslide affected area

A. Detachment area (upper part)
The detachment area characterises the upper and the right boundaries of the landslide. It presents two main discontinuity systems with NE (K1) and NW (K2) directions (see Fig. 7.6).

The K1 system, together with minor N-S striking discontinuities, corresponds to the right lateral boundary of the Sector A and has hindered landslide movement towards the SW. A third, complex system consists of two families of SSW-dipping joints that developed on the schistosity planes. These joints are labelled here S1 and S2 since they are roughly coincident with the schistosity itself.

B. Rock mass translation area
This area, which corresponds to a dislocated rock mass, has been subdivided into three main sectors characterised by different morpho-structural features (Fig. 7.5):

- Sector B1 consists of an intensively fractured rock mass and it has been affected by significant movements (up to 5 m) towards the SSW (Fig. 7.7) that occur on a sliding surface roughly sub-parallel to the schistosity. The displacements are well evidenced by continuous open cracks in the upper part of this sector.
- Sector B2 consists of a less fractured rock mass; it is suggested here that the B2 sector has been affected only by minor movement toward the SW since it still preserves its original structural setting.
- Sector B3 consists of a dislocated rock mass and it is separated from the B2 sector by a pronounced NE-striking open fracture. In Sector B3, the inclination of the slope is greater than in Sector B2 and the rocks show bulging deformations and toppling. The lower boundary of Sector B3 does not correspond to any failure surface, but it represents the morphological toe of the landslide that, although partially masked by plant cover, can be recognised by an alignment of scars, source areas of debris flows and rockfalls.

C. Accumulation and landslide affected area
This fan-shaped area lies below the supposed landslide toe and consists of debris featuring big rock blocks. On the basis of the geomorphological features, it has been subdivided into two sectors:

- Sector C1 is characterised by fan-shaped debris deposits that probably fell down from scars placed in the SE part of Sector B3 and that should have reached the area of the Prequartera village. The

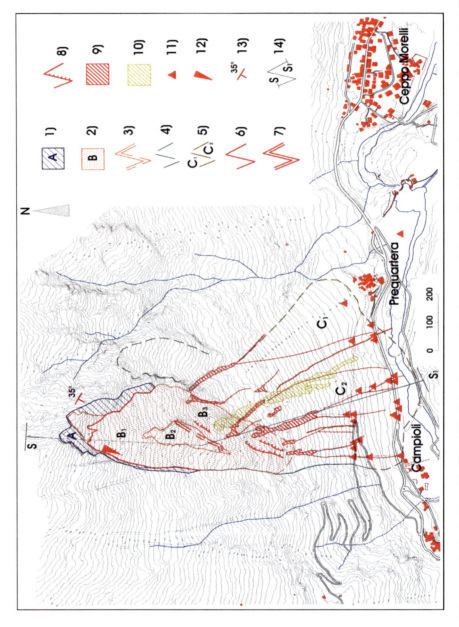

Figure 7.5. Main features characterising the Ceppo Morelli landslide: 1) detachment area; 2) rock mass translation area; 3) landslide boundary; 4) crown of an ancient rockfall; 5) accumulation and landslide affected area; 6) continuous open crack delimiting the landslide in the upper part; 7) main open cracks 8) main fissures observed in the disaggregated rock mass; 9) debris flows and rockfall trajectories; 10) debris flows and rockfall trajectories (1977 event) 11) main boulders fallen in October 2000; 12) sector where relevant movements occurred in October 2000; 13) main schistosity dip direction; 14) section S-S1 (see Fig. 7.18).

186

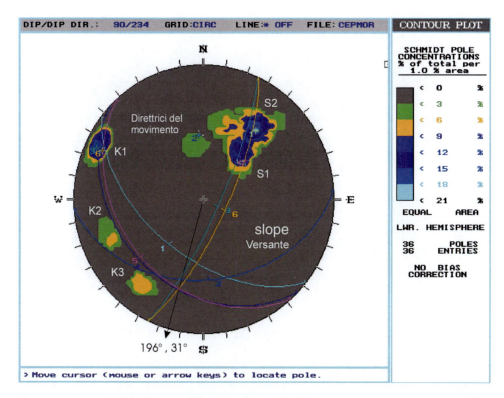

Figure 7.6. "On-site" available data; Schmidt diagram (lower hemisphere) counter of the poles distribution of joints. Legend: S1 and S2: schistosity; K1, K2 and K3: joint systems. The arrow indicates the sliding direction (Regione Piemonte, 2000).

geomorphological features of debris suggest that they are relatively old, although more rockfalls occurred as recently as during the 2000 meteorological crisis.
- Sector C2 is larger and more active than C1 and rockfalls and debris flows currently occur. This sector shows an advanced central lobe where larger blocks have reached and exceeded the valley road.

7.3.2   Local geology and structural analysis

As a consequence of the 2000 severe meteorological event, a rapid field reconnaissance was carried out (Regione Piemonte, 2000). New structural analyses were then performed with the principal aim of achieving improved knowledge of the physical and geometrical characteristics of the main structural elements.

This analysis was based on two data sets:

a) available geological data mainly collected at the landslide site during the October–November 2000 rapid field reconnaissance ("on site data"),
b) new data consisting of:
   – mesostructural "around site" data;
   – data from remotely sensed and aerial photo lineaments.

187

Figure 7.7.   The topographic control points allowed the identification of the direction and magnitude of the slip surface.

7.3.2.1   *"On site" available data*

The available data refer only to Sectors A and B1 of the landslide and underline the presence of three discontinuity systems and a composite metamorphic foliation, summarised in Figure 7.6. On the basis of these observations, interpretations of the landslide kinematics were attempted, since the orientations of the prominent discontinuity systems roughly correspond to the morphological boundaries of the rockslide.

7.3.2.2   *"Around site" structural analysis*

Structural analyses were performed around the unstable area, in order to:

- obtain a better geological characterisation of the structural systems described above;
- verify if the main morphological boundaries between the internal areas and the sectors of the landslide could be related to main tectonic structures;
- define the hierarchical and cross-cutting relations of the "around site" structures to compare them with those observed in the landslide area, and consequently derive some tentative "extrapolation rules";
- give information on the mechanical properties of the rocks and the structural elements to be used in the geomechanical model.

The results of these analyses will be described in the following sections.

➢ *Mesostructural analyses*

The analyses were performed at the boundary of the landslide area, near the boundaries between the different sectors and, in few cases, on isolated rock blocks that, although displaced, did not suffer major tilting or rotations.

These analyses pointed out that the available data (Fig. 7.6), although collected only in the upper part of the slope, are significant for the whole landslide body and that they can consequently be assumed as statistically representative.

Furthermore, a better geometrical and kinematic characterisation of all the structural elements was given in order to obtain conceptual tools for the correlation of the different structural elements.

➤ *Regional schistosity*

The schistosity regularly dips to the SSW, both within and outside the landslide area. Nevertheless, in the SW part of the landslide, it is characterised by a steeper inclination (50–60°); in this sector it has consequently been labelled as "S2". On the other hand, the more widely diffused low to medium angle schistosity has been indicated as S1. The schistosity also dips to the NE in some places within the landslide, probably due to the presence of open folds or the local tilting of isolated rock blocks.

Since S1 has a relatively constant orientation, it can easily be extrapolated along strike and at depth for at least several hundred meters.

Minor quasi-planar fracture sets are locally developed on the schistosity planes. They are mainly metric surfaces dipping to the SSW at a low angle with respect to the schistosity inclination. These surfaces, named here ZT (Fig. 7.8 and 7.9), are interpreted as boundary planes of minor brittle shear zones which have been locally reactivated by frictional processes having occurred on the schistosity surfaces. Minor NNE-dipping conjugate shear zones are also present.

Furthermore, subordinate, conjugate steep or medium angle joints (K) are also developed. They consist of two conjugate systems dipping to the SSW and the NNE, respectively. The crosscutting relations of ZT and K allow their interpretation as synthetic and antithetic Riedel shears within the ZT shear zones. This interpretation is corroborated by the scale invariance of the ZT-K crosscutting relations.

The schistosity is refolded by centimetre to decimetre-scale crenulation folds that gave origin to sub-horizontal crenulation cleavages, locally evolved into open joints (Ft, see below) due to unloading processes.

Finally, hydrothermal alteration bands have been locally observed on the schistosity planes. These are metre thick lenses or beds where decimetre-scale carbonate nodules can be observed, locally affected by karstic dissolution processes. These bands are characterised by poor, or very poor, rock quality index.

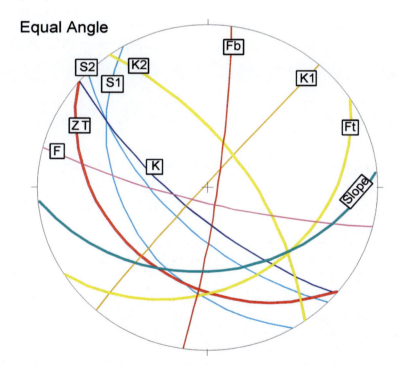

Figure 7.8. Around-site data. Schmidt diagram (lower hemisphere) average cyclograph trace of the main discontinuity systems. Legend: S1 and S2: schistosity; ZT: shear zone; K, K1 and K2 joint systems; F: regional faults and Fb: right-lateral boundary fault; Ft: release joint system.

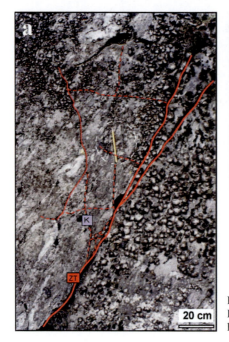

Figure 7.9a–b. Crosscutting relationship between ZT and K at different scales observed at the western boundary of the landslide.

➢ *Fractures and fault systems*

One of the most prominent is the K1 (Fig. 7.8 and 7.10a) system that is considered here as a comprehensive discontinuity system whose strike ranges from NNE to NE, and which has been synthetically represented in Figures 7.6 and 7.8 as a single average plane. K1 is indeed characterised by different sub-systems:

- roughly NE-striking highly persistent fractures (hundreds of meters) and largely spaced (from 10 to 100 m) joints;
- faults, well developed in the lower part of the slope, namely where the slope shows a greater inclination (at about 1250 m, see Fig. 7.5 Sector B3). These last surfaces strike from NE to ENE and consist of very steep planes, tens of meters long and dipping both SW and NE. It is assumed here that these faults belong to a fault zone; this interpretation is mainly based on the geometric array of F elements (scattered and anastomosed discontinuities) at both the regional and local scales.

The western boundary fault of the landslide (Fb) also strikes sub-parallel to the K1 system. This fault is characterised by a cataclastic belt tens of meters long; minor slip planes within this belt indicate S-ward right-lateral movements.

Other metric NNE reverse faults (Fi) have been also locally observed (Fig. 7.10a). The NNE striking steeply dipping Fb and Fi displace the less inclined ZT shear zones.

A second system (K2) (Fig. 7.8 and 7.10 a, b) consists of sub-vertical NW striking fractures, that tend to the NE in the upper sector of the landslide and that are more persistent and frequent on the left side on the landslide. These fractures consist of discontinuities tens of meters long that split the rock mass into blocks of different shape and dimension, due to some irregularities in the fracture spacing. In some cases, the K2 discontinuities correspond to reverse fault surfaces.

A third discontinuity system is mainly made up of joints (Ft) (Fig. 7.8 and 7.10b and c) that strike sub-parallel to the K1 system but show considerably lower inclination. The Ft mainly consists of release joints (Engelder, 1985) originated by unloading processes, since their attitude, generally less inclined than the slope, allows the sliding of rock wedges defined by the intersections of the fracture systems described above.

190

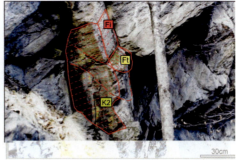

Figure 7.10. Crosscutting relationship and geometry of the main structural features. a) ZT, K1 and K2 observed in the western surroundings of the landslide. This relationship shows one of the peculiar rock block shape observed at the western of the landslide; b) Fi reverse fault, K2 fracture and Ft release joints observed at the western boundary of the landslide; c) major Ft release joints observed in the western of the landslide; d) Sigmoidal fracture systems associated to the F strike slip faults observed on the opposite side of landslide.

Finally, a fourth discontinuity system consists of regional EW steep strike-slip faults (F) (Fig. 7.8 and 7.10d), well exposed in the lower part of the slope and particularly on the opposite side of landslide, in the right slope of the Anzasca valley. Spacing and persistence of this system are particularly evident at the plurimetric scale. Sigmoidal fracture systems are usually associated with the F planes at different scales, where they delimitate rock slices interpreted as *strike-slip duplexes* (Woodcock & Fischer, 1986).

The hierarchical and geometrical relations of fault and fractures described above need to be compared to the geomorphological data and observations. Remotely-sensed satellite and aerial photograph data are very useful for this purpose, since they allow geometrical and morphological comparisons of the detected lineaments with respect to the effective structures well exposed on the field. Such comparisons allow the extrapolation of the fracture network that has been locally

191

observed in the field to relatively larger areas surrounding the studied site, and vice-versa, in order to better constrain the fracturing conditions in the numerical model.

➢ *Remotely sensed and aerial photo lineaments*

The geometric and hierarchic relations between the discontinuity systems observed on Ceppo Morelli landslide were also analysed at the regional scale by means of interpretation and statistical analysis of Landsat TM lineaments (Fig. 7.11) and at the scale of the slope by photo lineament detection on the DVP stereo viewer (digital video plotter, Geomatic Systems, Inc.) (Fig. 7.12). The photo lineament analysis was carried out on the NE boundary of the landslide and over its SW surroundings. Their combined use led to a clearer identification of the geometry and distribution density of the structural elements in the "around site" landslide area.

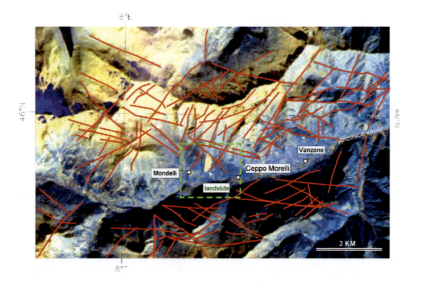

Figure 7.11.   Landsat TM image processed (colour combination of bands 7-5-4 RGB) with lineaments (in red colour). The green dashed line is the area represented in Figure 7.12.

Morpho-structural and spectral criteria (the last only for satellite Landsat images) were used for the lineament detection. The association and correlation of lineaments was carried out using the "Geometrical Lineament Identification" method (Morelli, 2000; Morelli & Piana, 2003).

The identified lineaments, represented here in Figures 7.11 and 7.12, were first analysed statistically and then compared with structural field data.

The statistical analysis was made to characterise the relationships between the azimuthal frequency distributions and lineament lengths for cumulative lengths and length classes. The statistical description is illustrated in the rose diagrams in Figure 7.13.

There are evident similarities between the aerial photos and the satellite lineament data. The former shows a distribution range from N20 to N60 with a maximum at N40. These lineaments are mainly concentrated in the NE. The greatest lineament length was found along the N80-90 direction.

The satellite data display the same distribution for both frequency and length diagrams with three average maxima at N290, N350-20 and N50. The lineaments can also be distinguished in terms of their length to determine the relations between frequency and length. Frequencies were divided into three length classes according to statistical distribution.

The aerial photo data show shorter and intermediate lineaments mainly at N20-60 and two relative maxima of shorter lineaments at N330-10 and on the intermediate length at N290-310. The greatest lengths (mostly >80 m) are along the N80-90 direction.

192

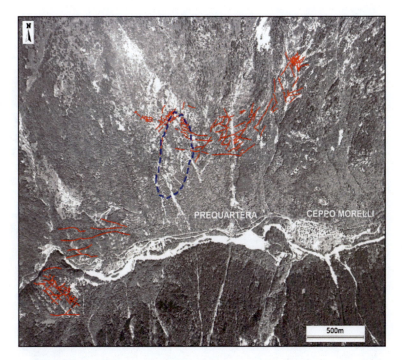

Figure 7.12. Aerial photograph with the identified photo lineaments (red lines) and the Ceppo Morelli landslide (blue dashed line).

The satellite data display a wider length distribution with short, intermediate and long lineaments with an average direction of N50-60 and mainly short and intermediate lengths along the N360-20, N280-290 and 290-310 directions.

In summary, in both the aerial photo and the satellite data sets, four main lineament systems can be recognised: the NE-SW (Ln1CM) and WNW-ESE (Ln2CM) striking systems display higher azimuthal frequency and longer lineaments; the NW-SE (Ln3CM) and N-S (Ln4CM) striking systems show, instead, lower azimuthal frequency and lengths.

The statistical and the geometrical pattern distribution of each system have been compared with the structural field data.

- The Ln1CM system consists of more widely distributed length and direction lineaments. In plan view, this distribution is described by linked lineaments that show an anastomosed geometric pattern, well evident throughout the whole slope. This geometry points to the presence of strongly fractured zones probably associated with steep fault systems consisting of many hierarchic orders of fault elements. This geometric pattern could be related among the F fault systems and the K1 fracture system.
- The Ln2CM system, well represented in the longer length classes, also consists of regularly and widely spaced lineaments, well exposed in the surroundings of the Ceppo Morelli landslide. This geometric pattern suggests that it could correspond to regional tectonic structures. In the field, the Ln2CM system corresponds locally to the F strike-slip fault system, not easily recognisable at the landslide site but well exposed in the surrounding of the landslide.
- The Ln3CM system is uniformly distributed over the Ceppo Morelli slope. It crosses the Ln1CM and Ln2CM lineaments or ends against them. This system partially includes the K, K1 and ZT systems, when the latter has greater inclinations.
- The Ln4CM system consists of fewer and shorter lineaments than the others. Its lineaments often connect with those of the LnM2 and the LnM3. This geometric relationship defines same

193

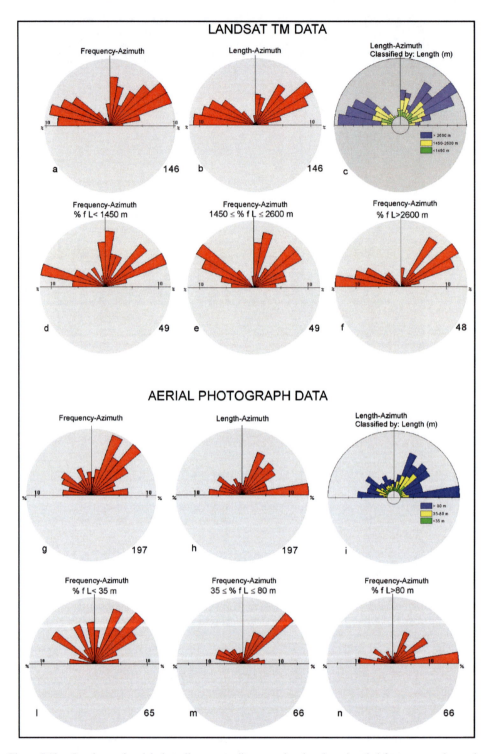

Figure 7.13. Landsat and aerial photo lineaments diagrams showing the azimuthal frequency and cumulative lengths (a, b and g, h) and azimuthal frequency divided into classes of lengths (c–f, and i–n). L: length of the lineaments; f: azimuthal frequency of the lineaments.

hectometre to decametre-scale sectors whose geometry is similar to that explained in the field by the crosscut relations among Fb, K2 and K1. This relationship is also in agreement with the shape of the many tilted rock blocks observed at the toe of the landslide.

The geometric and hierarchic relationships observed at the outcrop scale are also confirmed by the statistical distribution and geometric characteristics of the lineament systems, analysed at both the regional and the slope scales. This correspondence allows extrapolation of the geometry and hierarchy of the structural elements to different scales.

7.3.3 *Investigation and monitoring*

After the October 2000 intense rainfall event, the Ceppo Morelli rockslide has been monitored by means of (Fig. 7.14):

- 25 topographic benchmarks, measured by means of an automated total station located in a shelter on the facing slope;
- 9 wire extensometers, provided with 3 thermometers that allow compensation for thermal drift;
- 1 rain gauge.

All of the instruments are connected to automated data recording and transmission systems. All data are transmitted to a remote monitoring station by means of a radio modem and GSM modules.

At the moment, only the topographic monitoring data are available. These data are subdivided into three phases because of two maintenance interventions, which have caused different placements of the automated total station.

Figure 7.15 shows total displacement vectors from 5/8/2001 to 11/5/2003 (i.e. Phase 3); during this period the following displacements were measured:

- the topographic benchmarks in the upper part of the sliding body (Sector B1) show displacements on the order of 85–145 mm, with a velocity of about 4 to 12 mm/month; displacement vectors are mainly oriented southwards. Exceptions are represented by Point 21, which shows

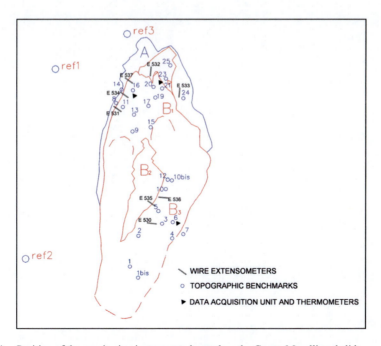

Figure 7.14.   Position of the monitoring instruments located on the Ceppo Morelli rockslide.

195

the highest total displacement (262 mm), and Points 8, 14, 21 and 23, located on the scarps of the detaching niche, which seem to be still;

- the topographic benchmarks in the lower part of the slide body (Sector B3) show displacements up to 400 mm (Point 3), with a velocity between 3 and 19 mm/month, displacement vectors are scattered between SSE and SSW.

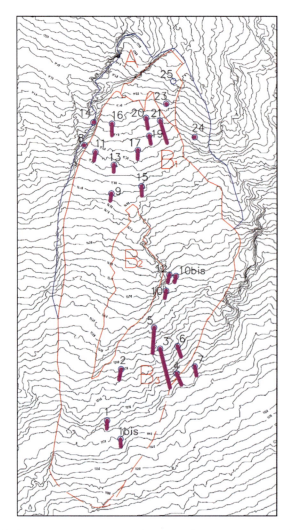

Figure 7.15.   Total displacement vectors measured by means of the topographic benchmarks from 5/8/2001 to 11/5/2003. Displacement vector scale: 1:35.

The recorded displacements confirm what was found out by means of the geomorphological analysis (see § 7.3.1): Sectors B1 and B3 show different behaviours, in particular with respect to velocity, which is greater in the lower part. This is explainable by thinking of the disengaging of the rock mass and of the greater inclination of the slope in Sector B3.

Attention has been focused on Points 3 and 5, located in Sector B3, and on Points 13 and 21, located in Sector B1 (Fig. 7.16) resulting in more detailed observation of the variation of distance

196

between the theodolite position and some topographic benchmarks: it is possible to notice the influence of daily rain on the displacement velocity. The highest observed rates of movement were reached during the heavy peaks of rain recorded in May, June and November 2002, all greater then 115 mm of water. Concerning this topic, for example, with regard to Point 3, the displacement rate measured in 6 days, from 4 June to 10 June 2003, is similar to that measured over 153 days, from 11 June to 10 November 2003, but the velocity is much different: 11.6 mm/day versus 0.43 mm/day. Thus, the rain seems to cause a velocity increase of more than 1 order of magnitude. With reference to the peak rain in November, it should be noted that measurements were interrupted for about 40 days. No indications can therefore be obtained concerning the velocity, although it is possible to recognise an increase in displacement rates in correspondence with this period.

In the upper part of the landslide, both displacements and direction of movements are quite homogeneous. The general direction of movement is southward, *i.e.* down the slope. This seems to indicate that the upper part of the landslide tends to move as a single body following the main foliation but, since the south-eastward movements are hampered by the cliff which limits the landslide on its right flank, the movements then evolve following the line obtained from the intersection of the foliation with the boundary fault, Fb.

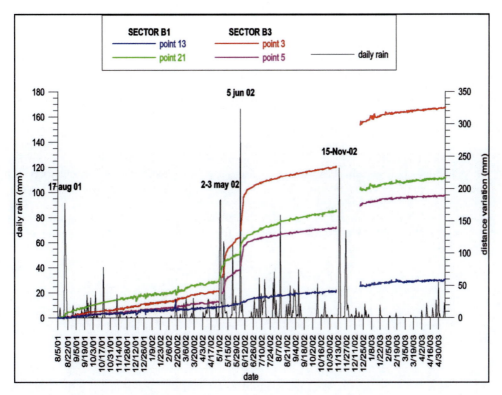

Figure 7.16. Variation of distance between the theodolite position and some topographic benchmarks versus daily rainfall from 5/8/2001 to 11/5/2003.

Here, it is remarked that the entity of the displacements measured through the monitoring system is an order of magnitude smaller than the displacement values estimated in situ as a consequence of the intense rainfall event of October 2000.

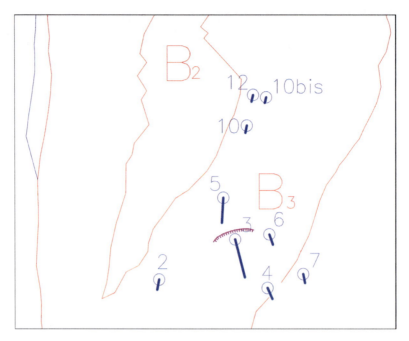

Figure 7.17.   The new sub-vertical joint developed inside Sector B3 after the intense rainfall event of October 2000. Displacement measured from 5/8/2001 to 11/5/2003. Displacement vector: 1:40.

The professionals responsible for the maintenance of the monitoring have noticed a new sub-vertical joint inside Sector B3 during their latest inspections (Fig. 7.17). Focusing attention on that sector, the displacement paths show two homogeneous areas with different behaviour patterns: points located downhill from the joint move with a dip direction that is close to that of the Ft system, whereas the others, near Sector B2, follow a southward direction, which results from the combination of the foliation and/or the ZT system with the boundary fault Fb. The different behaviours can be linked to different kinematics: sliding on the foliation in the upper part and toppling for the lower and more overhanging rock mass.

7.3.4   *Danger characterisation*

The knowledge of the landslide's morphological and structural features combined with the displacement analysis indicated useful directions for the modelling of the phenomenon which will be discussed in the following sections.

7.3.4.1   *Geometrical characterisation*

The geometrical and structural constraints allow the estimation of the thickness of the landslide body, with a maximum (40 m) near the crown and minimum (20 m) near the western landslide boundary (Fb fault). The rock volume affected has been estimated at $5 \times 10^6 \, m^3$.

7.3.4.2   *Geomorphological constraints*

The geomorphological analyses allow the subdivision of the landslide body into several sectors, bounded by morphostructural features (see also § 7.3.4.3). The geometry of the sectors is to be considered as a constraint for the modelling of the landslide body.

The following description of the sectors corresponds to Figure 7.5.

198

Sector A: defines the upper boundaries of the landslide; the eastern boundary of the landslide corresponds to stiff scarps, roughly coincident with the K2 discontinuity system.

Sector B1: shows evidences of SW-ward movements; these movements are hindered to the West by the rock wall corresponding to the Fb fault on the right side of the landslide. The N-S Fb can thus be considered as the western boundary of the landslide, and it can be prolonged at least down to the lower edge of Sector B3.

Sector B2: suffers minor translations and it is less damaged than the other sectors.

Sector B3: is more damaged than the others, since it is cut by the closely spaced K1 and F systems that subdivide it into small rock blocks. The greater inclination of the slope induces relatively strong gravity instabilities, as shown by toppling of blocks.

Sector C1: conveys most of the debris as indicated by its pronounced lobe at the base of the slope.

Sector C2: the western boundary corresponds to a NS elongated ridge that could constrain the falling trajectories of the boulders to the right.

7.3.4.3 *Structural constraints*

The discontinuities to be taken into account in a slope stability analysis of Sector B3 are (Fig. 7.8):

ZT (boundary planes of shear zones sub-parallel to the schistosity). These are characterised by a low internal friction angle (inferred to be about 16°, see for ref. Hobbs et al., 1976) and by low persistence. The role of rock bridges aligned parallel to the ZT surfaces can thus be taken into account in modelling.

Fb (boundary fault); it is characterised by a cataclastic belt tens of meters in length (breccias and polished surfaces with a low internal friction angle, inferred to be about 16°).

F (boundary fault); it is characterised by fault zones arranged by very steep anastomosed faults. Spacing and persistence of this system are particularly evident at the plurimetric scale.

Ft joints are closely spaced and intensively affect the rocks, mainly within the landslide. These joints, probably originated by unloading processes (release joints, Engelder, 1985) show inclinations ranging from few to 30 degrees.

K1 joints and maybe faults which are more closely spaced than the other systems in Sector B3. Their mutual array, at different scales, suggests that they could be segments of a NE striking fault zone.

K2 fractures and minor faults which are more persistent and pervasive on the left side of the Sector, namely in the upper part.

The combination of the surveyed discontinuity systems leads to the definition of several kinematic scenarios of sliding and toppling:

- the intersection of surfaces ZT and Fb can be used for the definition of the sliding direction of wedges isolated from the main rock body by means of tensile joints belonging to the F system;
- the intersection of surfaces Ft and ZT can be used for the definition of the sliding direction of wedges isolated from the main rock body by means of tensile joints belonging to the K1 system;
- toppling of blocks bounded by surfaces K1, K2 and Ft.

7.3.4.4 *Comments on the mechanism and on displacements*

Different kinematic behaviours have been observed in each sector of the landslide. In particular, the upper part of Sector B1 seems to move toward SSW, along sliding directions subparallel with the line resulting by intersection of ZT and Fb (Fig. 7.8). The lower part of Sector B1 is affected by SSW directed sliding on the western side and SSE directed sliding in the eastern one. This is interpreted as due to the different orientation of the slope, that respectively dips to SSW and SSE. In this interpretation, the gravitational instability should be extended only at shallow depth within the landslide.

In the lower part of the landslide (Sector B3), both sliding and toppling have been found. Sliding could occur toward SSW (Fig. 7.15), while toppling could be originated by the partial reactivation of pre-existing vertical fractures (K1 and K2) and cracking of the rock bridges along the Ft system.

In Figure 7.18, a schematic cross-section of the Ceppo Morelli landslide is shown, where the main structural elements and their cross cutting relations are outlined. The highlighted sliding surface in the upper part of the landslide (sliding zone in Figure 7.18) has been inferred on the basis of geomorphological data. The occurrence of an effective failure surface can be assumed in modelling the entire landslide as a whole. In this case, this surface would form by partial reactivation of pre-existing fractures or faults, and partially by rupturing of the rock bridges eventually placed between the different discontinuities. Orientation, shape and mechanical properties of rock blocks can be extrapolated by combining the main discontinuity systems described above.

It is here remarked that the kinematics analysis and modelling of the landslide could require a basal sliding surface that should not be thought as a continue sliding surface. It can be rather imagined as a composed surface that might be formed in reliability terms.

7.3.4.5 Scenarios

On the basis of the previous considerations, three different hazard scenarios are thus possible, with decreasing occurrence probability and increasing impact on land planning:

- Scenario 1: toppling and sliding of rock boulders and debris falls (features 9 and 10 in Figure 7.5). As pointed out by the historical research, the site has been involved in repeated falls since

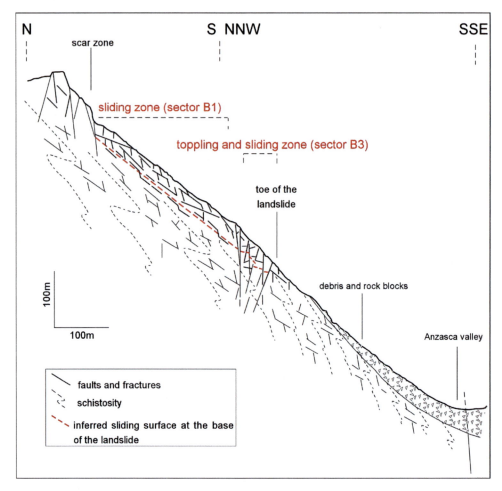

Figure 7.18. Schematic cross section of Ceppo Morelli landslide (overall trace in Figure 7.5). The thicker lines in the sliding zone indicate discontinuities which are generally open.

ancient times. The sites from which rock blocks are released are located all along the lower boundary of the unstable Sector B3. Rockfalls range from small cobbles to large boulders hundreds of cubic metres in size.

- Scenario 2: collapse of Sector B3 (Fig. 7.5). Continuous rockfalls can weaken the rock mass located in this sector. An estimated avalanche involving a rock mass of about 1'000'000 m³ could then occur.
- Scenario 3: collapse of the whole Sector B (Fig. 7.5). The collapse of Sector B3 could lead to a huge rock avalanche involving the entire Sector B, which corresponds to a volume of about 5'000'000 m³.

7.3.5 Geomechanical modelling

7.3.5.1 Mechanical "Triggering" model

The Ceppo Morelli site has been studied by means of two geomechanical models: the first is a 2D discontinuous model, the second a 3D equivalent continuous model with discontinuity planes that cut off the unstable mass from the whole rock slope.

The 2D analyses were carried out using the Finite Element Method (code DRAC, Prat et al., 1993). The approach applied during the numerical modelling consisted of two stages: a first stage during which the calculation of the initial stress conditions of the slope was carried out using topographical information (the actual morphology, where possible) and incorporating the effect of the glacier weight; a second stage in which the glacier was removed and the stress field computed. The objectives of the simulations were twofold. First, the general stress field of the slope was calculated in order to obtain information on the failure mechanism. The interaction of the upper part (B1) that consists of a rather well-structured rock mass and the lower part (B3) that is characterised by a disjointed rock body, was an additional aim. Second, the influence of different material properties and mechanical models for the cracks (interfaces in the discontinuous model) were analysed.

The 3D analyses were carried out by means of the code Map3D (described in Chapter 3) based on an indirect Boundary Element technique, the Displacement Discontinuity Method. Attention was focused on the lower part of the unstable area (Sector B3), indicated by the geological and geomorphological studies as the key volume, in the sense that the failure of B3 could trigger that of the whole landslide.

As previously seen, a complex description both from the geomechanical and the kinematical point of view was arisen from the geological investigations. At the present phase of the work, however, simplified geomechanical model and numerical triggering simulation are assumed. Remarkable characteristics of these models are the 3D analysis conditions and the possibility of taking into account the measured and the simulated displacements. In order to be able to focus the attention on these aspects, some simplifications had to be introduced from the geometrical and the mechanical point of view:

- as the rock mass of B3 is very fractured and disjointed, in the 3D geometrical model it was represented as an equivalent continuum and the discontinuity planes that belong to the principal joint sets were introduced as boundaries of the unstable volume;
- the global behaviour of the unstable rock volume was analysed: this implies that it was possible only to reproduce the mean value of the measured displacements due to sliding. The numerically obtained value had to be referred to the whole investigated rock mass. It must be underlined that the considered mean value was evaluated with respect to time (i.e. during one month) and space (i.e. over Sector B3);
- the average displacements were supposed to be due to the overall creep behaviour of the rock mass, which is assumed to be concentrated only on the sliding surface. No influence on the displacements of water pressure variations with respect to time was considered due to the lack of piezometric data.

On the basis of these considerations, the aim of the analyses was to reproduce the displacements measured in situ by the monitoring system: it is possible to evaluate the viscous plastic creep

201

behaviour of the rock mass by means of the Bingham flow rule implemented in the code Map3D. This can be carried out in two steps: in the first one, the research of the mechanical parameters related to conditions of unstable equilibrium of the rock mass is performed by means of numerical simulations; on the basis of the results obtained, in the second step, the analysis of the behaviour of the unstable rock mass with respect to time is developed. In order to do this, a realistic value of the viscous modulus G was checked by comparing the evaluated displacements with those measured in situ.

The 2D model was developed along a N-S cross-section of the entire slope. The topography of the model was defined using a DEM with a cell size of 10 m. A two-stage approach was applied during the modelling. In the first stage, the upper limit of the glacier was set at 1500 m a.s.l.; in the second stage, the glacier was removed. The mesh consisted of 1780 continuum elements and two discontinuity sets: the main schistosity, assumed as ZT, was introduced by southward dipping joints with a spacing of 10 m; the fracture system K2 was represented by sub-vertical joints with a spacing of 20 m. These discontinuity sets were modelled with zero-thickness interface elements. The constitutive law adopted was linear elastic for the continuum material as well as for the interfaces. The values selected for the material properties of the rock mass and the interfaces are given in Tables 7.1 and 7.2, respectively.

Table 7.1.   Material properties adopted for the 2D model continuum elements.

	Unit weight γ (kN/m^3)	Young's Modulus E (GPa)	Poisson's Ratio ν
Gneiss	28	50	0.3
Glacier ice	9.2	9.6	0.33

Table 7.2.   Material properties adopted for the 2D model interface elements.

Normal stiffness Kn (MPa/m)	Tangential stiffness Ks (MPa/m)	Cohesion c (kPa)	Friction angle φ
1000	100	0	30°

The opening of the interfaces due to the removal of the glacier is presented in Figure 7.19. During the simulations, it was noticed that the role of the continuous elements was rather limited. That is, deformation and displacement are controlled by the interfaces. Therefore, the joint constitutive parameters and the joint geometry became the key aspects of the analyses.

The 3D geometrical model of Sector B3 of the Ceppo Morelli landslide was carried out on the basis of a simplified DEM (cell size 170 m) of the topography in order to reduce the computational time. The boundary discontinuity planes were modelled as follows (Fig. 7.8): the upper tensile joint was simulated by a plane representing the F system; the left side boundary was given by two planes belonging to the K1 and Fb joint sets, respectively the discontinuity between B2 and B3 and the boundary fault; the lower boundary was represented by the outcrop of the sliding surface assumed as belonging to the ZT system. In Figure 7.20, the 3D representation of the unstable volume is shown.

The first aim of the 3D simulations was to obtain a realistic value of the in situ friction angle. This value is related to a heterogeneous material made of boulders, soil, water and voids considered as a whole; therefore, it is impossible to evaluate the friction angle by means of laboratory tests. A set of sensitivity analyses was performed in order to calculate the value of friction angle that corresponds to the instability of the rock mass. These analyses were carried out with no creep simulation.

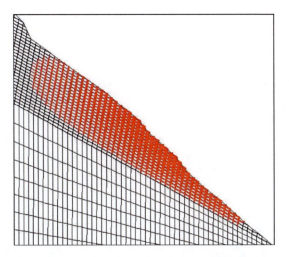

Figure 7.19.   Opening of the interfaces (red lines) at the final stage of the Ceppo Morelli model. Width of the red lines is proportional to the opening displacement of interfaces.

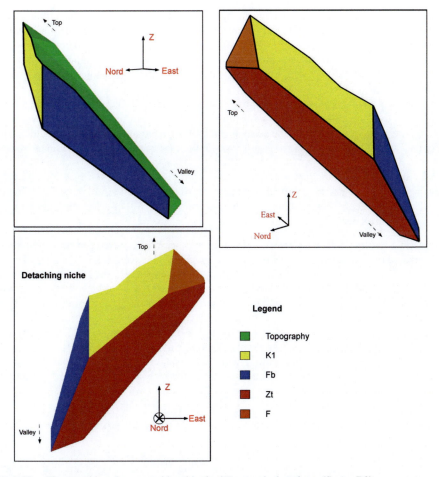

Figure 7.20.   The unstable volume considered in the 3D numerical analyses (Sector B3).

The following step was the assessment of the value of the creep coefficient C. Activating the Map3D creep option, a simulation was made with an assumed value of C and the results were compared with the measured displacements. If these values were not close to each other, more simulations were carried out in order to find a proper C. The final set of mechanical parameters used for the numerical analyses is shown in Table 7.3.

Table 7.3. Mechanical parameters used for the 3D model of the Ceppo Morelli landslide.

Equivalent continuum			
Unit weight of volume	γ	0.027	MN/m^3
Young's modulus	E	7000	MPa
Poisson's ratio	ν	0.25	–
Failure surface			
Friction angle	φ	25	°
Normal modulus	K_n	5000	MPa
Shear modulus	K_s	1500	MPa
Creep coefficient	C	100	MPa × Month

The computed displacement due to one month's creep can be compared in magnitude and direction with the mean value of the displacements obtained by means of topographic measurements. Taking into account points 1, 1bis, 2, 5, 10, 10bis and 12 (Fig. 7.15), the only ones in Sector B3 referred to a sliding kinematics, the measured mean value is about 4 mm/month with a dip direction of 182°; these values are very close to that obtained numerically, which are equal to 3.8 mm/month with 180° of dip direction.

The obtained results are very far from giving the probability of failure needed for the risk analysis, which is to be obtained through a probabilistic procedure. For this purpose however the variation of pore pressure in time and the decay of rock strength parameters in time should be known. These data are seldom known and are not known in the case of the Ceppo Morelli landslide.

7.3.5.2 *Mechanical "Run Out" model*
The morphostructural and kinematical characteristics of the Ceppo Morelli instability phenomena have allowed the recognition of three possible scenarios (see § 7.3.4.5). The run-out phase for the three scenarios has been simulated as follows:

- Scenario 1 (rock fall). The trajectories of the falling rock blocks have been simulated through a 3D lumped mass method (ROTOMAP code, Geo & Soft Int., see § 3.5.5.3). Values of the restitution coefficients in the normal and tangential direction equal to 0.32 and 0.80 respectively, and a friction coefficient of 0.65 have been assumed. The choice has been done on the basis of data obtained from the literature and through back analyses of the trajectories of already fallen rock blocks.
- Scenarios 2 (rock avalanche of about $1 \times 106\,\text{m}^3$) and 3 (rock avalanche of about $5 \times 106\,\text{m}^3$). The rock avalanche run-out has been simulated through the coupled use of ROTOMAP and DAN (Hungr, 1995) codes. A frictional reology, with friction angle equal to 30° and pore pressure coefficient equal to 0 (i.e. no water) has been assumed in the analyses carried out through DAN code.

➢ *Scenario 1*
The aim of the analysis is to reproduce the rock falls occurred in the past and to forecast new rock falls affecting a larger area.

To get realistic results the assumed range of velocity for the released blocks changes if they originate from Line 1 (Fig. 7.21) or from the Line 2 (Fig. 7.22) of the considered unstable area as indicated in Table 7.4.

Table 7.4.   Range velocity for the released blocks.

Line	v_{min} [m/s]	v_{max} [m/s]
1	1.0	1.8
2	0.6	1.0

These different ranges of velocity are justified by the possibility that some of the blocks crossing Line 1 can originate in the upper part of the unstable area and than gain a higher velocity.

In addition to rock fall trajectories (Fig. 7.21 and 7.22), the analyses allow the distribution of the kinetic energy of the blocks (Fig. 7.23 and 7.24): the slope can then be subdivided in homogeneous sectors, each associated to a specific value of the medium energy. The distribution obtained in terms of energy represents an important tool to be taken into account for risk and vulnerability evaluations.

Due to the fact that ROTOMAP is a "lumped mass" code, the energy values are formulated as $E = v^2/2$ [J/kg]. Hence, in order to link the obtained results to a possible real situation, an indicative mass should be hypothesised for the boulders. The maximum mass of the boulders observed at the bottom of the valley (around 2.7×10^6 kg, corresponding to a volume of 1000 m³) seems a reasonable value.

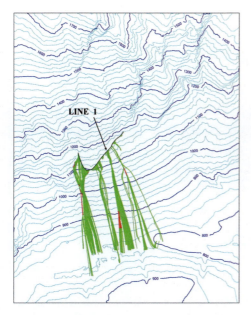

Figure 7.21.   Scenario 1: rock fall trajectories from Line 1 (in red: blocks in flight).

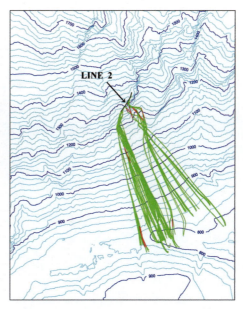

Figure 7.22.   Scenario 1: rock fall trajectories from Line 2 (in red: blocks in flight).

205

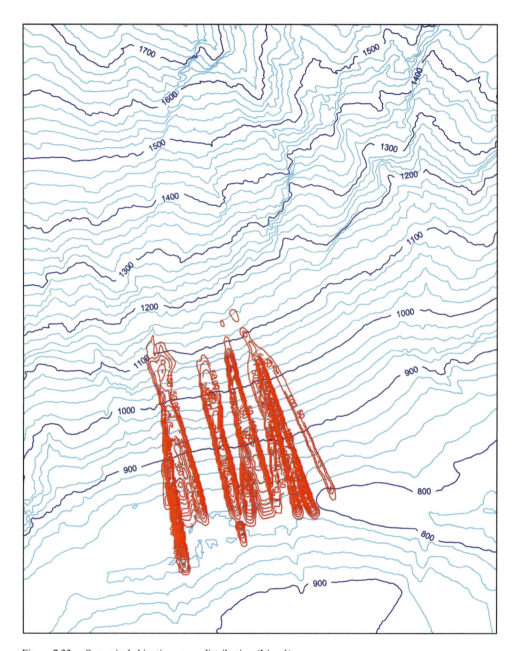

Figure 7.23.   Scenario 1: kinetic energy distribution (Line 1).

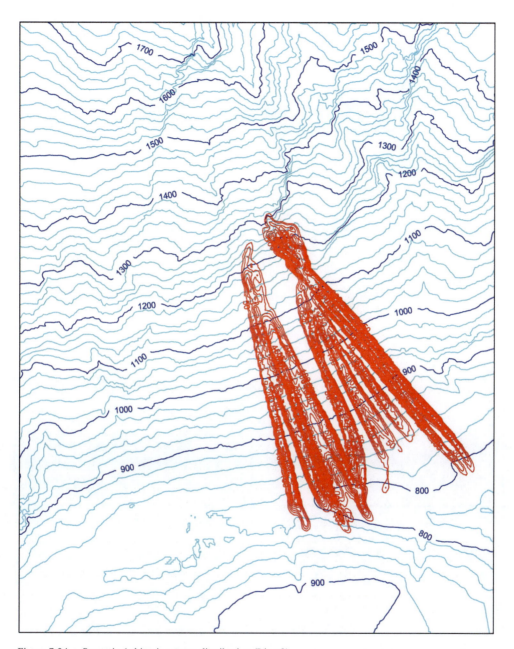

Figure 7.24. Scenario 1: kinetic energy distribution (Line 2).

➤ *Scenario 2*

On the basis of geomorphological and geostructural considerations (see § 7.3.4.5), Scenario 2 is considered the most prone to evolve as a catastrophic rock avalanche.

The direction of propagation chosen for the run-out analysis is identical to the dip direction of the joint system Ft (Fig. 7.25).

Together with the run-out area (Fig. 7.26), the value of the avalanche front velocity is defined along the section taken into account (Fig. 7.27); the front velocity distribution is then supposed to be the same for all the points located at the same height.

As the energy distribution for the rock fall case, the homogeneous areas obtained in terms of velocity distribution are an important aspect, that should contribute to the risk and vulnerability evaluation.

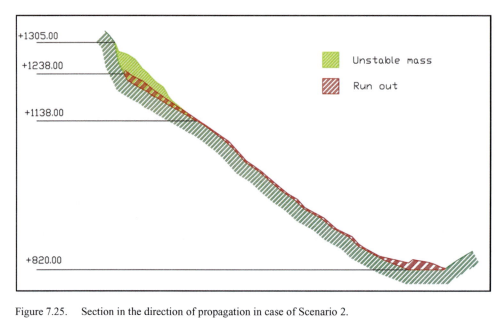

Figure 7.25.   Section in the direction of propagation in case of Scenario 2.

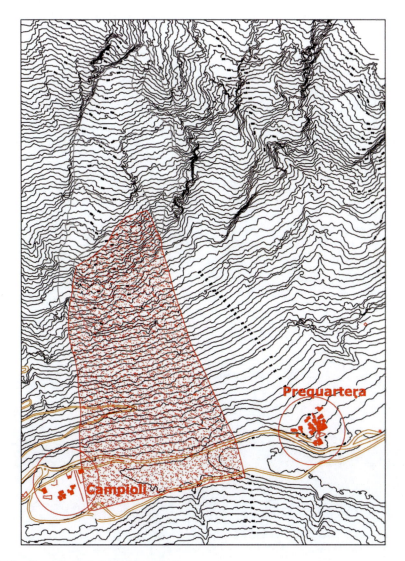

Figure 7.26.   Run out area in case of Scenario 2.

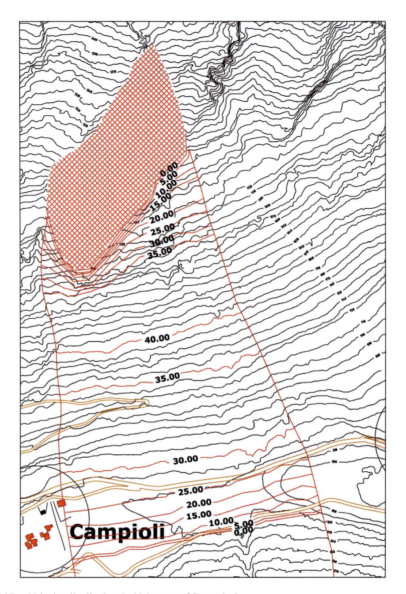

Figure 7.27.   Velocity distribution (m/s) in case of Scenario 2.

➢ *Scenario 3*
This is the most catastrophic hypothesis, considering the possibility that the whole volume is involved in a rock avalanche phenomenon.

Although extremely unlikely, this scenario has been considered since it would generate particularly catastrophic consequences for the hamlets, the river and the road that are located in the bottom of the valley. The run-out area has been obtained as described above for Scenario 2 (Fig. 7.28). In the same way as described above, the front velocities can be obtained.

➢ *Conclusions*
The results obtained applying this methodology represent a useful contribution to vulnerability and risk evaluations since the rock fall analyses give the possibility of calculating the energy

210

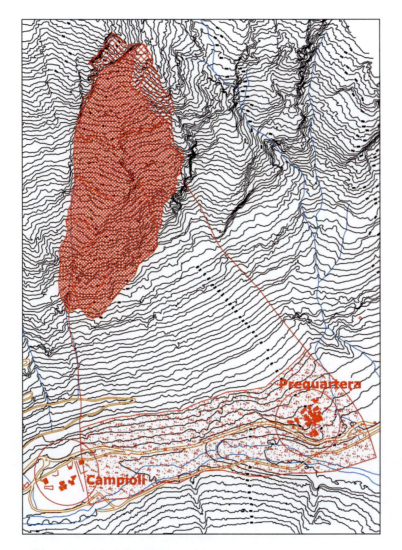

Figure 7.28.   Run out area in case of Scenario 3.

distribution per unit mass, and the avalanche analyses give both the definition of the run-out areas and the front velocity distribution.

In particular, the simulations allow the determination of different hazard situations for the hamlets located at the bottom of the valley. The Prequartera village can be involved in a rock fall phenomenon (scenario 1) and in the most catastrophic rock avalanche (scenario 3), whereas the expansion of the mobilized mass in both the rock avalanche cases (scenarios 2 and 3) could be the cause of a danger condition for the Campioli area.

Finally, an evolution in terms of rock avalanche (scenarios 2 and 3) would determine the road interruption and the river damming with disastrous consequences.

7.3.6   Occurrence probability

As previously discussed (§ 3.1.5.6), the probability of occurrence can be defined as the chance or probability that a landslide will occur. It can be expressed in relative (qualitative) or probabilistic (quantitative) terms.

In the case of large landslides, a rigorous, quantitative, procedure should include the application of formal mechanical-probabilistic methods, taking into account the uncertainty in all the geometrical and mechanical parameters (slope geometry, shear strength, piezometric pressure). In the case of the Ceppo Morelli landslide, however, a similar procedure cannot be applied because the uncertainty in the definition of the parameters involved is too large and it is not possible to define any reliable variability for them.

An alternative procedure takes into account the relationship between the intensity of triggering events such as rainfalls or earthquakes and the landslide occurrence through the use of empirical correlations or neural network methods. Again, the lack of data makes this approach inapplicable.

Since several historical records exist of the rockfall activity in the area involving different volumes, the historical approach seems much more suitable to the case of the Ceppo Morelli landslide: in this way it is possible to obtain some information related to periodic frequency.

An occurrence range of time (see also Corominas et al., 2002) was obtained for each of the three scenarios of evolution described in § 7.3.4.5, considering the minimum and maximum time interval between two consecutive recorded events. On the basis of their mean value, a frequency was then calculated in terms of events per year (Table 7.5).

Table 7.5. Definition of the occurrence probability using the historical approach.

Scenario	Recorded events	Occurrence range (years)	Average (years)	Frequency (event/year)
1	1940 Oct. 1971 Apr. 1977 Oct. 2000 Jun. 2002	0–30	15	1/15 = 0,066
2	(Intermediate)	30–200	115	1/115 = 0,009
3	312 843 1816 2000	200–1000	600	1/600 = 0,0016

It should be noted that the events recorded are related to two different types of danger: the fall of single blocks and catastrophic events involving a larger mass. The first has been associated to a rockfall "Scenario 1" (low energy phenomena and high frequency) and, conservatively, the second has been considered as representative of phenomena involving the whole mass (rock avalanche "Scenario 3": high energy phenomena and low frequency). The intermediate occurrence range has finally been associated to the rock avalanche "Scenario 2" (medium energy phenomena and frequency).

7.4   QUANTITATIVE RISK ANALYSIS

Each of the described scenarios is characterised by its own intensity, failure and run-out areas (derived from the geomorphological and geomechanical models) and by an occurrence probability (as a result of an historical analysis of recorded events).

According to the proposed quantitative risk analysis (see Chapter 3), each element of the hazard scenario is used separately in the matrix calculus of risk evaluation. This process is applied on each recognised scenario; therefore, for the Ceppo Morelli site, 3 separate risk analyses were carried out.

At the end of the matrix calculation, a unique value is obtained (total risk), divided into 4 categories: physical, economic, environmental and social, useful for the total risk assessment phase.

7.4.1 Elements at risk

The elements at risk (Fig. 7.29) are recognised through the definition of the area affected for each danger in the scenarios previously specified. These elements at risk are described in Tables 7.6, 7.7 and 7.8.

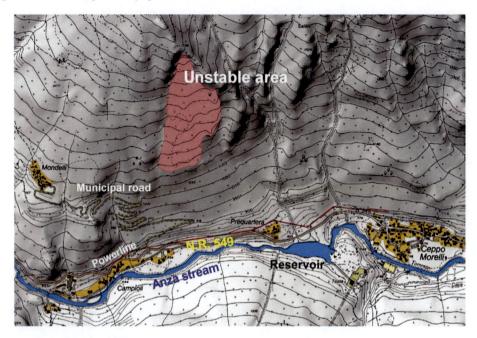

Figure 7.29.   Elements at risk.

Table 7.6.   Definition of the elements at risk: Scenario 1 (Rockfall).

Elements at risk	Name	Notes
Residential areas [valuable buildings or valuable rural centres: historical, architectural, artistic and/or cultural value]	Campioli	Area affected: 3 400 m^2 Number of buildings affected: 5 Persons affected: 12
	Prequartera	Area affected: 6 300 m^2 Number of buildings affected: 18 Persons affected: 42
Forests (private and public properties)		Area affected: 228 000 m^2
Heavy traffic or strategic roads	National road No. 549	Length affected: 800 m Annual traffic: 100 000 vehicles
Secondary roads	Mondelli municipal road	Length affected: 230 m Annual traffic: 1 000 vehicles
Lifelines	Power line Valle Anzasca 15 kV	Length affected: 1 025 m Pylons affected: 4

7.4.2 Vulnerability of the elements at risk

To define the vulnerability of the elements at risk, an analysis concerning typology and process intensity of the concerned assets was carried out. Geomechanical models give important data about the energy of impact of rock blocks, while structural and architectural analyses of the elements at risk give data about strength at impact.

213

Table 7.7. Definition of the elements at risk: Scenario 2 (Rock avalanche $1 \times 10^6 \, \mathrm{m}^3$).

Elements at risk	Name	Notes
Residential areas [valuable buildings or valuable rural centres: historical, architectural, artistic and/or cultural value]	Campioli	Area affected: $2\,300\,\mathrm{m}^2$ Number of buildings affected: 3 Persons affected: 3
Forests (private and public properties)		Area affected: $248\,500\,\mathrm{m}^2$
Heavy traffic or strategic roads	National road No. 549	Length affected: 543 m Annual traffic: 100 000 vehicles
Secondary roads	Mondelli municipal road	Length affected: 400 m Annual traffic: 1 000 vehicles
Lifelines	Power line Valle Anzasca 15 kV	Length affected: 527 m Pylons affected: 1

Table 7.8. Definition of the elements at risk: Scenario 3 (Rock avalanche $5 \times 10^6 \, \mathrm{m}^3$).

Elements at risk	Name	Notes
Residential areas [valuable buildings or valuable rural centres: historical, architectural, artistic and/or cultural value]	Campioli	Area affected: $2\,500\,\mathrm{m}^2$ Number of buildings affected: 3 Persons affected: 3
	Prequartera	Area affected: $6\,300\,\mathrm{m}^2$ Number of buildings affected: 18 Persons affected: 42
Forests (private and public properties)		Area affected: $406\,280\,\mathrm{m}^2$
Heavy traffic or strategic roads	National road No. 549	Length affected: 810 m Annual traffic: 100 000 vehicles
Secondary roads	Mondelli municipal road	Length affected: 400 m Annual traffic: 1 000 vehicles
Lifelines	Power line Valle Anzasca 15 kV	Length affected: 722 m Pylons affected: 3

Starting from energy data, different considerations concerning damages and losses of the elements at risk were obtained for the 3 scenarios for the Ceppo Morelli site, as impacts are strongly linked to rock volumes and distribution.

Concerning the rockfall scenario, the run-out model yields an energy zonation (J/kg) per unit mass, according to rockfall velocity. Considering a mass of $2.7 \times 10^6 \, \mathrm{kg}$ (maximum mass of the boulders observed at the bottom of the valley, see § 7.3.5.2), impact energy can be evaluated and a structural analysis supplies different values for physical, economic, environmental and social vulnerability of the elements at risk, through the indexes (0.5, 0.75, 1) described in Table 3.8 (§ 3.6.3.5).

For instance, buildings, mostly old, do not have a particular strength at impact, as they are made of stones and wood, without concrete. Blocks impacting on roads usually damage the wearing course as happened during the 1978 and 2000 events, interrupting road use (see Fig. 7.30).

Regarding human lives, vulnerability depends on where people are (inside or outside the buildings): in this case, the worst situation was considered, i.e. all persons without adequate protection.

Concerning rock avalanche scenarios (both 1 and 5 million m³) and the different energy degrees on the basis of the landslide velocity (measured in m/s in the run-out geomechanical model), it is

Figure 7.30. Few damages next to a house in Campioli village (Ceppo Morelli Municipality). This block of about 4 m³ rolled towards left, slightly damaging the road without reaching the house.

useless to define the different elements at risk by vulnerability degrees. In the rock avalanche area (failure, depletion and accumulation areas, see Fig. 7.26 and 7.28) the destruction would be total. Also along the road at the bottom of the valley, where the velocity tends to zero (see, for example, Fig. 7.27), the accumulation (some tens of meters) is so great that all the assets would be completely destroyed, heavily damaged or buried and, therefore, permanently useless. Following these considerations, a vulnerability degree equal to 1 is used for all the vulnerability categories.

7.4.3 *Expected impact*

Multiplying the values of the elements at risk (VE, see § 3.1.6.2), and of the vulnerability percentage (V), the expected impact (C) is obtained by C = VE × V.

In the quantitative risk analysis for the Ceppo Morelli rock slide, detailed economic evaluations were not considered for elements at risk and arbitrary relative value indices have been temporarily applied for each category (Table 7.9).

Concerning rockfall danger (Scenario 1), 3 different expected impact values were calculated, as 3 vulnerability degrees (0.5, 0.75, 1) had been previously differentiated on the basis of the energetic impact thresholds of the block falls.

All the considerations were made through the automatic application of a G.I.S. technique, whose graphics are not included in this section.

Table 7.9. Considered values of the elements at risk.

Element	Physical cons.	Economic cons.	Environm. cons.	Social cons.
Residential areas (valuable buildings or valuable rural centres)	4	1	1	4
Heavy traffic or strategic roads	3	4	1	3
Secondary roads	1	2	1	2
Secondary lifelines	1	2	1	0
Forests/Rural area (private and public properties)	1	1	3	1

7.4.4 *Risk assessment*

The last phase of the quantitative risk analysis concerns the risk evaluation. This evaluation is obtained multiplying the expected impact values (C) and the numeric values of the occurrence probability (P) related to each scenario (Table 7.5): $R = C \times P$.

Figures 7.31–7.33 show the physical risk assessment results as a G.I.S. layout for the 3 scenarios (rockfall: $P = 1/15$; rock avalanche, $1 \times 10^6 \, m^3$: $P = 1/115$; rock avalanche, $5 \times 10^6 \, m^3$: $P = 1/600$). In the same way, Figures 7.34–7.36 show the economic risk assessment results, Figures 7.37–7.39 show the environmental risk assessment results and Figures 7.40–7.42 show the social risk assessment results for the same 3 scenarios.

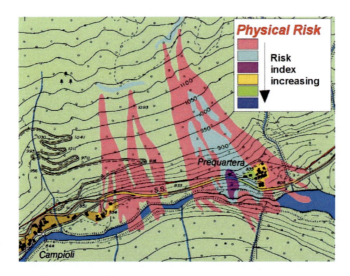

Figure 7.31. Map obtained by G.I.S. techniques concerning the physical risk for Scenario 1 (rockfall). The areas with elevated risk values represent the National Road No. 549 and the uphill part of the Prequartera village, where a church is located.

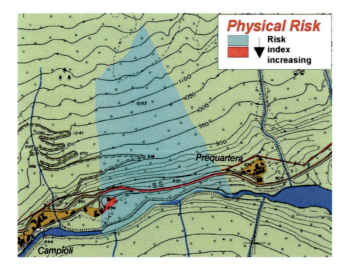

Figure 7.32. Map obtained by G.I.S. techniques concerning the physical risk for Scenario 2 (1 million m^3 rock avalanche). The areas with elevated risk values are the National Road No. 549 and part of the Campioli village involved in the rock avalanche process.

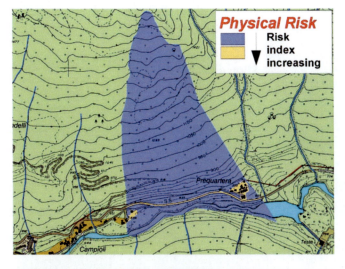

Figure 7.33. Map obtained by G.I.S. techniques concerning the physical risk for Scenario 3 (5 million m³ rock avalanche). The areas with elevated values are the National road No. 549, part of the Campioli village and the entire Prequartera village.

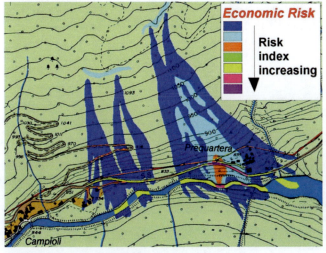

Figure 7.34. Map obtained by G.I.S. techniques concerning the economic risk for Scenario 1 (rockfall). The main areas with elevated risk values represent communications and services. The National Road of Valle Anzasca is highlighted, being the only important road of the valley from an economic point of view.

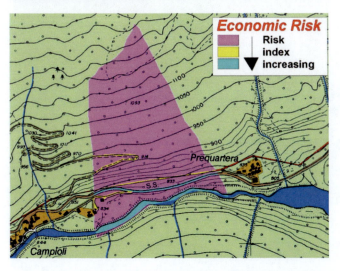

Figure 7.35. Map obtained by G.I.S. techniques concerning the economic risk for Scenario 2 (1 million m³ rock avalanche). The main areas with elevated risk values are communications, services and the main Valle Anzasca River (the Anza stream). The interruption of the river due to the rock avalanche accumulation produces important economic consequences, as the river feeds the Prequartera power plant reservoir.

217

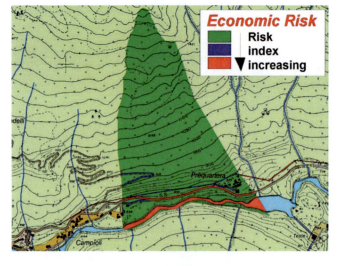

Figure 7.36. Map obtained by G.I.S. techniques concerning the economic risk for Scenario 3 (5 million m³ rock avalanche). The main areas with elevated risk values are communications (0.8 km. of the National Road No. 549), services (more than 700 m. of power lines) and the main river of the valley (the Anza stream). As previously stated, the Anza stream is economically essential to this valley, feeding a power plant reservoir also involved in the rock avalanche phenomenon.

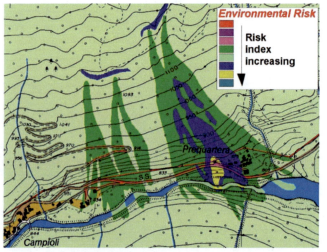

Figure 7.37. Map obtained by G.I.S. techniques concerning the environmental risk for Scenario 1 (rockfall). The main areas with elevated risk values represent the wooded areas near the Prequartera village and part of the village itself (it is a medieval village of great environmental and cultural value).

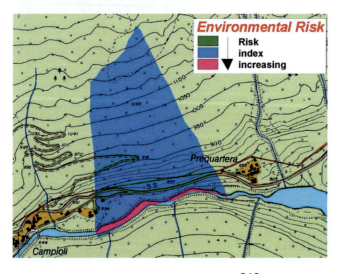

Figure 7.38. Map obtained by G.I.S. techniques concerning the environmental risk for Scenario 2 (1 million m³ rock avalanche). The main areas with elevated risk values are the wooded areas and parts of the Anza stream involved in the gravitative phenomenon evolution.

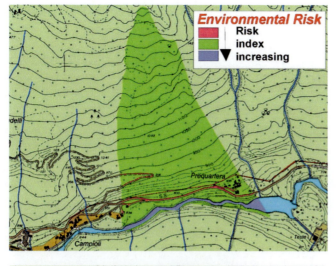

Figure 7.39. Map obtained by G.I.S. techniques concerning the environmental risk for Scenario 3 (5 million m^3 rock avalanche). The main areas with elevated risk values are the wooded areas and a part of the Anza stream involved in the rock slide evolution.

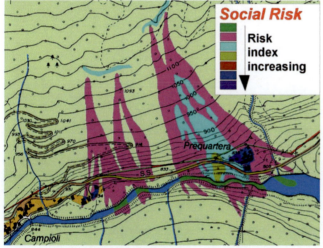

Figure 7.40. Map obtained by G.I.S. techniques concerning the social risk for Scenario 1 (rockfall). The areas with elevated values correspond to the villages of Campioli and Prequartera. In the last one, it is possible to distinguish the risk value according to the impact energy of different blocks.

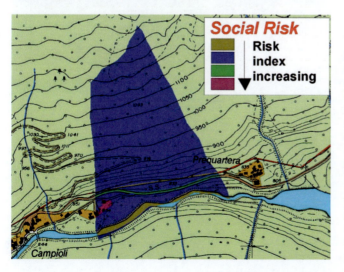

Figure 7.41. Map obtained by G.I.S. techniques concerning the social risk for Scenario 2 (1 million m^3 rock avalanche). The areas with elevated risk values are situated in the Campioli village. The National Road No. 549 of Valle Anzasca is also involved in the rock avalanche phenomenon and, therefore, presents elevated risk assessment values, due to the process type (characterised by a rapid evolution), to the large part of the involved road (0.55 km) and to the traffic of this road at certain periods of the year.

219

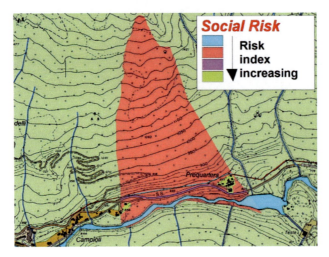

Figure 7.42. Map obtained by G.I.S. techniques concerning the social risk for Scenario 3 (5 million m³ rock avalanche). The areas with elevated values are the villages of Campioli and Prequartera; the last one would be completely involved and destroyed. About 1 km of the National Road No. 549 is also affected by the rock avalanche phenomenon, being a heavy traffic road in the tourist season.

7.4.4.1   Total risk

To obtain a risk assessment synthesis for the Ceppo Morelli site, total risk was calculated as described in § 3.1.7. The result is shown in Figure 7.43, showing the risk value sum for each

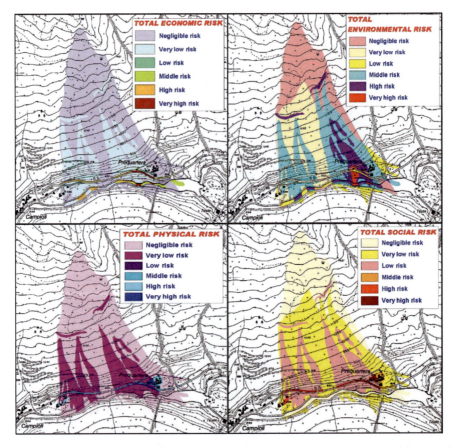

Figure 7.43.   Maps (obtained by GIS technique) about total risk assessment, divided into 4 risk categories.

scenario. Again, total risk is divided into 4 risk categories: physical, economic, environmental and social. In these maps, the comparison of the risk among the different areas is simplified.

The areas affected by high intensity and high occurrence probability processes present the most elevated risk values. At the Ceppo Morelli site, the area affected by rockfall shows the highest risk values, as the occurrence probability of this scenario is more than 40 times higher than the occurrence probability of Scenario 2 and about 7 times higher than the occurrence probability of Scenario 3.

7.4.5 Indirect risk and regional impacts

In order to complete the risk analysis, some considerations concerning the indirect effects associated with the rock slide evolution (not considered by the geomechanical models) are nonetheless necessary. The landslide scenarios of Ceppo Morelli show, as well as rockfalls, some phenomena involving huge rock masses (millions of m^3) at high velocity (40 m/s corresponding to about 150 km/h), whose effects are not easily estimated.

In particular, the great mass movement (5 millions m^3) that could occur at Ceppo Morelli according to the most catastrophic scenario would produce a huge wind effect. The process intensity cannot be easily estimated but a qualitative evaluation can be made on the basis of rare case analyses described in the literature (e.g. Hsü, 1978). It is probable that the phenomenon would be light laterally with respect to the mass movement and very strong frontally. The same would occur for the dust that follows the wind effect.

At the moment, wind effect and dust cannot be input in a numerical model without mistakes. Nevertheless, it can be stated that the damages produced by these indirect effects are of small extent, as a wooded slope without buildings stands in front of the mass movement. Some damages could occur to Campioli and Prequartera buildings not directly involved in the mass movement.

Another effect to be considered in risk assessment is the formation of a landslide dam lake (Fig. 7.44). This phenomenon can occur during a meteorological event with important flows along the Anza stream (flow of about 900 m^3/s, see Bossalini & Cattin, 2002). These processes together (instantaneous damming of the valley-road by tens of meters of landslide accumulation and extraordinary flood of the Anza stream) would form a basin of about 2 million m^3 in few hours. Many buildings in the Campioli village would be inundated and many people would have to be evacuated, increasing the social risk (see, for example, the case of Randa (CH), shown in Figure 7.45).

Moreover, due to this lake formation, downstream villages could potentially be involved. Instantaneous lake depletion can, in fact, be caused by a rapid overtopping dam erosion, or a landslide dam collapse caused by piping and internal erosion.

In both cases, flash-flood along the valley would cause enormous problems. In Ceppo Morelli, a study has been carried out to detail a civil protection plan in case of a flash-flood event (Bossalini & Cattin, 2002). Trying to delineate the areas affected by flash-flood, a simplified approach has been used in this study, similar to the rock slide risk assessment previously detailed. The elements at risk (land use, human lives, buildings and civil engineering structures) have been identified along the entire valley bottom (Fig. 7.46 and 7.47). In this risk analysis, physical,

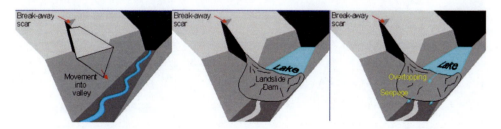

Figure 7.44. Mechanism and consequences of lake damming due to a huge landslide. (source http://www.kingston.ac.uk)

Figure 7.45. Flood effect in alpine area due to landslide dam (Randa, CH, 9th May 1991). Effects also on buildings 1 km far from landslide site. (source: www.crealp.ch).

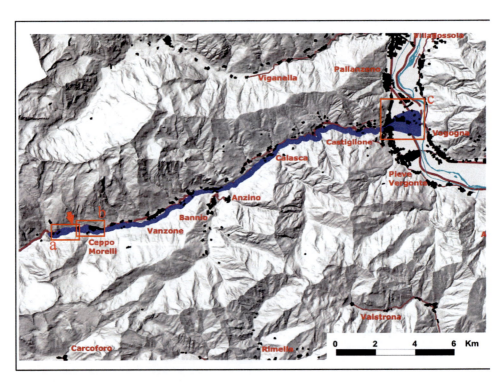

Figure 7.46. Areas that might be affected by a flash flood due to rockslide dam breaking (blue colored). The zones in orange squares are detailed in Figure 7.47.

economic, environmental and social vulnerability was not evaluated separately due to a very large potentially affected area.

To complete the description of indirect effects associated with the Ceppo Morelli rock slide, it is important to recall that a so-called "Vajont effect": waves produced by the rockfall in the reservoir can get over the dam located near Prequartera causing flooding along the Anza stream. Nevertheless, as the water volume in this reservoir is only some hundred of m^3, strong consequences at the bottom dam are not foreseen.

In Figure 7.48, a map of the main direct and indirect risks considering the worst catastrophic rock avalanche scenario is shown. This is obtained through the qualitative evaluation described in this section.

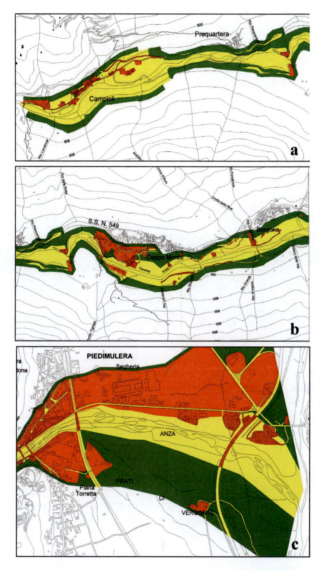

Figure 7.47.   Risk maps of zones in orange squares in Fig. 7.46. Green = moderate risk, yellow = medium risk, orange = high risk, red = very high risk (Bossalini & Cattin, 2002).

223

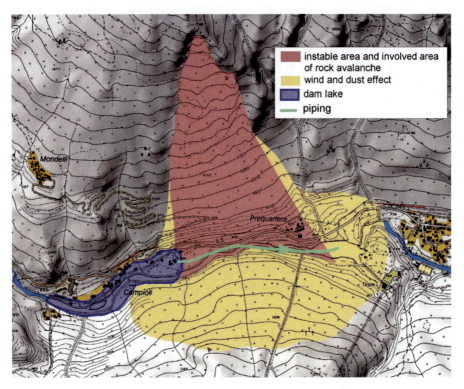

Figure 7.48. Map of direct and indirect effects associated to rock avalanche phenomenon. Borderlines are drawn by qualitative evaluation.

7.5   REGULATION AND RISK MITIGATION MEASURES ALREADY AVAILABLE

7.5.1   *Viability*

After the many rockfalls of October 2000, the risk conditions of the National Road were considered too high to be accepted. Since the road is the only possible connection for Macugnaga, a well-known ski resort, a provisional by-pass road was constructed along the right banks of the Anza River, in order to avoid the most dangerous zone. The road is about 1 km long and construction took about 30 days; the total cost was on the order of 1.25 M€. The river is crossed twice by means of Bayley-type bridges provided by army engineers.

Before construction, the Regional Geologists made a running evaluation of the risk condition of the provisional road. Results showed that the road is protected from rockfalls but may be affected by major mass falls (e.g. a rock avalanche). A rock avalanche would obviously destroy the provisional road.

Opening of the road was thus made conditional on the results of the monitoring system. If some defined threshold values (both for rainfall and displacements) are exceeded, the road is blocked by local authorities. Up until now this has happened only once, during the heavy rains of November 2002.

In order to provide a final solution to road practicability problems, a tunnel is currently being studied. The tunnel, about 800 m long, will totally avoid the rockslide and will allow traffic flow to Macugnaga even in the case of large failure scenarios. The total cost will be about 38 M€.

7.5.2   *Endangered buildings*

After the October 2000 flood, the mayor evacuated two small villages, Campioli and Prequartera, since a 70 m³ boulder stopped a few meters from a house of in Campioli and a 10 m³ boulder almost reached Prequartera. An analysis of possible remedial measures was carried out by the regional geologists and by the Politecnico of Turin, as follows.

Concerning Prequartera, which has about twenty houses, the analysis indicated that, due to its position and to the general features of the landslide, the small village may be affected by rockfalls but is out of the main area of influence of mass falls. The small village has been protected by means of a reinforced earth rockfall protection wall.

The wall is about 120 m and 5 m high; average base width is about 8 m. The wall was designed in order to withstand impact energies up to 8 MJ; the total cost was about 0.25 M€.

As for Campioli, the situation is much more critical, for it is just on the edge of the zone of possible invasion in the case of failure of the lower part of the landslide body. The evacuated part of the small village consists of seven houses. One of them, built in the seventies, may also be affected in the case of common rockfalls; in fact, a 70 m^3 boulder stopped a few meters from the building in October 2000. The house will be demolished and relocated elsewhere.

The remaining six buildings pose more problems, in that they form a historic nucleus built around 1650, a marvellous and perfectly preserved sample of typical Walser architecture, protected by the *Sovraintendendenza ai Beni Architettonici* (National agency for the protection of the Architectural heritage).

The conditions of risk related to this nucleus are too high to be accepted for residential use but acceptable for discontinuous use, so the more likely solution will be:

– purchase, by the local town authority, of the houses;
– construction of a reinforced earth rockfall protection wall, to offer partial protection.

7.5.3   Alarm system

The monitoring system described in § 7.3.3 is connected to an alarm procedure.

The data are examined on a regular basis by a group of professionals: the level of attention depends on the warning codes emitted daily by the *Sala situazione rischi naturali della Regione Piemonte* (*Regional natural risks monitoring office*). If any threshold values (relative to both rain and displacement rates) are exceeded, an alarm system is triggered in order to close the main road to traffic and to activate a proper civil protection plan. The plan also includes the possible effects of river damming.

8
The Sedrun landslide

Ch. Bonnard & X. Dewarrat
Ecole Polytechnique Fédérale de Lausanne

F. Noverraz
Consulting geologist, Lausanne, Switzerland

8.1 INTRODUCTION

8.1.1 *Landslide description*

The slope movement phenomenon named the Sedrun Landslide is located in eastern Switzerland, namely in the western part of the Canton of Graubünden, in the higher stretch of the Vorderrhein valley. It takes place in the rock slope called Cuolm da Vi overlooking the villages of Sedrun and Camischolas which form part of the community of Tujetsch. The unstable area extends between the Val Strem valley to the west, which deeply intersects the mountain range limiting the northern slope of the Rhine valley, and the temporary Drun Tobel creek to the east (Fig. 8.1).

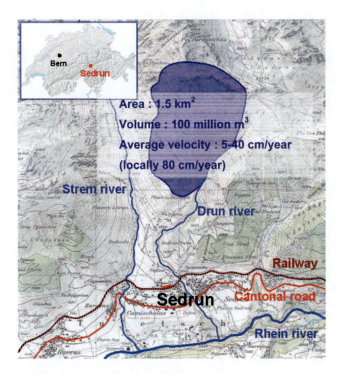

Figure 8.1. Layout of the Sedrun Landslide and its main characteristics (map published with the authorization of the Swiss Office of Topography, October 2001).

The instability in this mountainous and austere area, covered by very scarce vegetation and completely without buildings, is mainly the result of a toppling effect concerning the whole Cuolm da Vi zone. Due to sub-vertical schistosity and to several fault families, this phenomenon has produced a thick bending in the rock layers, quite visible from Val Strem, in particular in the upper part of this rock mass. This process could trigger rockslides, rockfall or local slides in some parts of the rock mass. Such sliding appears to be more likely in the top of the area, in particular at the highest point of the zone called Cuolm Parlets Dadens, but also on the right side of Drun Tobel creek, in the zone called Plauncas Tgamos.

The unstable area, covering 1.5 km^2, can be quite well determined on the basis of its morphology, the presence of major cracks more or less opened by the movements in the upper part and due to numerous monitoring data (see § 8.3.3). As no boreholes have been carried out, its volume can only be roughly assessed at some 100 million m^3. The important movements recorded cause the formation of typical uphill facing scarps in the upper part of Cuolm da Vi (see Fig. 1.3).

Some major infrastructures are located in the region of Sedrun, such as a construction site for the new Gotthard Tunnel (shaft for intermediate access), the Oberalp cantonal road, connecting the Cantons of Graubünden and Uri, and a regional railway line connecting Zermatt to St. Moritz (Glacier Express). Moreover, the village of Sedrun has been developed extensively over the last few decades, especially as a winter sport resort.

This landslide site has been studied particularly between 1993 and 1997 as a part of a large research project funded by the Swiss National Science Foundation, within the PNR 31 national project on climate change and natural disasters (VERSINCLIM project – Noverraz et al., 1998; Bonnard et al., 1996).

8.1.2 *Historical background*

The temporal origin of the movements observed in the Cuolm da Vi unstable zone is not easy to determine. Due to the late installation of the Swiss triangulation network in this area in 1942, and to the lack of a renewed determination of the position of the landmarks until 1973, past information concerning the evolution of movements is quite limited. No damage or special catastrophic event has taken place in Cuolm da Vi itself, as no building or infrastructure exists in that zone. The population was aware that some movements were occurring in this zone but did not pay much attention and was not overly concerned. The occasional snow avalanches originating from Val Strem sparked much greater concern.

However, a very remarkable evolution has recently been observed following several topographic surveys. The rock mass of Cuolm da Vi has shown some important movements in the last years, movements that were first visually perceived in 1957 (see Fig. 8.4; at that time the crack was only slightly open). The first geodetic control measurements which date from 1983 proved the existence of significant horizontal displacements, in particular 1.83 m for point No. 514 over a period of 41 years (see Fig. 8.5 below). A large number of movement measurements later confirmed this fast evolution over the last few years (see § 8.3.3). These movements seem to be much more important on top of the Cuolm da Vi zone, at an altitude of about 2'458 m a.s.l. (point No. 26), than in the lower part of the unstable slope. During the last 20 years, the elevation of this point has decreased from 2'458 m to 2'453 m a.s.l. However, no catastrophic landslide event has occurred yet, so this very active phenomenon only implies potential risk which is mostly ignored by the local population and even more so by the numerous tourists.

8.2 REGIONAL FRAMEWORK

8.2.1 *Climatic and water conditions*

At the top of the Cuolm da Vi area (altitude approximately 2'500 m a.s.l.), monthly average temperatures range from −9°C in January to +6°C in August, with important variations throughout the year, inducing a significant change in periods of snowmelt, especially as the main unstable slope faces south. The annual sunshine duration can be estimated to be 1'700 h, which is fairly high. The average snow cover lasts from six to seven months and may be quite thick (600 cm for a return period of 100 years at 2'500 m a.s.l.). Therefore, this zone is located at the lower level of the permafrost and might be affected by the progressive rise of this limit due to climate change. As far as the annual evapotranspiration is concerned, it ranges between 200 and 300 mm [data from the Swiss Hydrological Atlas-SHGN, 1992].

The average annual rainfall measured in the area of the Sedrun Landslide reaches a value of approximately 2'000 mm, dropping to 1'227 mm at the base of the valley, where the Sedrun rain gauge station is located. The high spatial variation is due to a marked topographic effect, which is typically encountered in Alpine valleys.

The extreme rainfall values for a return period of 100 years are 70 mm for a rain duration of 1 hour and 175 mm for a rain duration of 24 hours. The maximum annual precipitation amounts recorded at the Sedrun station were 1'844 mm in 1935, 1'809 mm in 1999 and 1'836 mm in 2002, this last value being mostly influenced by the extremely high precipitation in November (544 mm), but which fell mainly as snow in the area of Cuolm da Vi. Since the beginning of the measurements, no significant long-term tendency of variation has been observed, except a long dry period between 1959 and 1964 and a wet period between 1999 and 2002 (133% of the average long term rainfall over a period of four years). Therefore, it can be inferred that rainfall is not directly the cause of the increasing movements over the last 20 years.

Most of the extreme rainfall events in the area are produced by the same type of meteorological process: warm and wet air masses coming from the Mediterranean encountering the Alps are forced to reach high altitudes, leading to sudden condensation and discharge of their water content.

In the vicinity of the Cuolm da Vi area, there are two main streams determining the limiting groundwater conditions. The first one is the Val Strem stream, which runs along the western side of the unstable zone of Cuolm da Vi, but at a much lower level, producing a deep drainage effect for the landslide mass. To the east, the Drun Tobel creek, which is less important in terms of flow, but presents a very steep slope of 60% over 1'000 m, limits the unstable zone (Fig. 8.2). The Drun Tobel creek displays a temporary flow, being especially important in the snow melting season and

Figure 8.2.　Drun Tobel creek seen from the village of Sedrun. The unstable area extends up to the top of the mountain at the back.

collecting an important volume of surface water with a high bedload related to the mylonitic zone, which explains the presence of this creek (see § 8.2.2). It may induce important debris flows; its lower stretch requires regular maintenance.

On the right side of the Drun Tobel, as well as along the east slope of Val Strem, a series of temporary water outlets can be inferred by the presence of several gullies located between elevations 1'900 m and 2'000 m. In the Cuolm da Vi area itself, absolutely no drainage pattern is visible, which means that the net precipitation (rainfall and snow less evapotranspiration) infiltrates completely in the moving mass, in particular during snowmelt. However, no groundwater level data are available.

8.2.2   *Regional morphology*

The Sedrun Landslide is located in the basin of the Rhine river, namely on the slope forming the left bank of its tributary called Vorderrhein, near the Gotthard massif, just south of the Oberalpstock peak (3'327 m). The region consists of a deep glacier valley parallel to the main original geological structures of the Swiss Alps – Aar massif (N 70°E). The general slope presents a drop of nearly 2'000 m from the highest summit to the north (Oberalpstock) and of some 1'000 m from the top of the studied Cuolm da Vi landslide area (Cuolm Parlets Dadens, the peak of which is at an elevation of approx 2'450 m a.s.l.). Perpendicular to the Rhine Valley, a deep glacier valley (Val Strem) with steep slopes of some 40° intersects the main regional geological structures, displaying their conformation (see Fig. 8.3). The landslide area has been affected by the Würm glaciation which developed until some 12'000 years ago, as can also be seen on the gentle slope to the west of Val Strem.

The toe of the slope is incised by a deep erosion creek (Drun Tobel) resulting from a recent torrential erosion. It displays kakiritic and mylonitic rocks and favours the destabilization of the slope (Fig. 8.2). The villages of Sedrun and Camischolas are located on a vast alluvial fan, formed both by the Val Strem stream and the Drun Tobel creek. In most of the area, the forest cover is absent.

8.2.3   *Regional and structural setting*

As far as the stratigraphic units present in the zone are concerned, the metamorphic southern border of the Aar massif is in contact with the Urseren-Garvera-Furka zone and with the Tavetsch massif, which separate the Aar massif in the NW from the Gotthard massif to the SE.

The lithotypes encountered from north to south can be summarized as follows: granitic rocks of the Aar massif, gneiss and crystalline schists of the southern metamorphic border of the Aar massif, Verrucano and Permo-Trias units of the Urseren-Garvera-Furka zone, granitic and metamorphic rocks of the Tavetsch massif, granitic and gneissic rocks of the Gotthard massif.

The geological age of the region can be assumed as Pre-Hercynian and Permo-Trias.

In the area of Drun Tobel, chloritoschists and schistic gneisses are observed, as well as mylonites and kakirites resulting from an intense tectonic crushing without further crystallisation, so that this rock displays very weak mechanical characteristics.

By means of field mapping, it has been possible to detect in the region that the main schistosity is sub-vertical, and displaying an ENE-WSW direction, even nearly E-W at the scale of the massif, that is more on less parallel to the Vorderrhein Valley and perpendicular to the lateral small valley of Val Strem.

8.3   HAZARD ANALYSIS

8.3.1   *Geomorphological analysis*

The longitudinal profile of the Sedrun Landslide displays a fairly regular slope of around 30% in its central part, with a zone at an altitude of 2'200 m a.s.l. which is nearly horizontal over a length of 100 m. The global longitudinal profile between the summit of Cuolm da Vi (initial altitude

2'458 m a.s.l.) and the bottom of the Drun Tobel outlet, with its floor at 1'550 m a.s.l., leads to an average slope of 60% over a distance of 1'500 m. The exposure of the unstable zone is oriented towards the south (Fig. 8.3).

In the transversal direction, i.e. ESE-WNW, the profile of the unstable area displays lateral slopes of 78% in the direction of Drun Tobel to the east, and 68% in the direction of Val Strem to the west. Therefore, the most critical potential failure mechanisms are related to these lateral slopes rather than to the main NS slope, in whose direction the movements are observed.

The general instability corresponds to a quite particular type of mechanism, consisting of a deep toppling effect in the sub-vertical rock layers of gneiss. This toppling effect has its origin in a deep bending of rock layers under the effect of gravity. There is, thus, no failure surface in this kind of phenomenon, but it is possible to delineate an approximate limit of the zone affected by bending, as shown in Fig. 8.3.

The present instability process results from a more or less marked evolution of the toppling of rock layers, which will lead to rockslides or rockfall. The progressive inclination of the layers in the downstream direction induces the well-known "uphill facing scarp" type of morphology, with sometimes open grabens (see Fig. 1.3). This process has led some geologists to believe in an active alpine tectonic uplift, as this zone is located between the two most seismic areas in Switzerland, but this opinion can be contradicted on the basis of the displacement monitoring data.

In the described instability processes, some features have appeared or are likely to develop in the rock mass of Cuolm da Vi, being a direct consequence of this instability mechanism. These features are briefly described below and may be classified into four different groups: faults, rock-slides, uphill facing scarps and rockfalls.

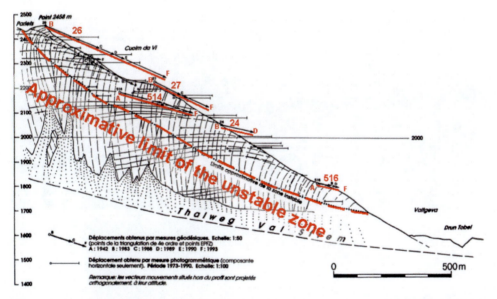

Figure 8.3.   Geological profile of the Sedrun/Cuolm da Vi Landslide with vectors of surface displacements and approximate limit of the unstable zone (red dotted line). The red vectors indicate the displacements measured during the geodetic campaigns : A in 1942; B in 1983; C in 1988; D in 1989; E in 1990; F in 1993. Black thin horizontal vectors indicate the displacements measured by photogrammetry between 1973 and 1990 (see § 8.3.3).

8.3.1.1   Faults in Cuolm da Vi

Apart from the three different systems of faults which cut into the rock mass of Cuolm da Vi (see § 8.3.2), there are other much larger faults which occur at particular places in the rock mass.

231

At an altitude of 2'404 m a.s.l., a very recent large fault appears, first seen in 1995, with a NS direction, transversal to the ancient fault already discovered in 1957 at the same location (NW-SE orientation). This old fault limits the unstable zone in movement to the east. It is due to the coexistence of two systems of fractures in the rock mass and has a length of 120 m and a width of 15–20 m (Fig. 8.4).

On the terrace, with a fairly horizontal inclination that extends at an altitude of 2'200 m a.s.l., there are some cracks filled with morainic material with some compression bulges. The orientation of these faults follows one of the families of fractures in the rock mass (S1) (see Fig. 8.5 and § 8.3.2). These presently active faults disappear towards the west in Val Strem, and to the east in Drun Tobel.

In 2002, a new major vertical crack, some 200 cm wide, oriented SW then turning W, has been observed uphill of Cuolm Parlets Dadens, in the small Parlets depression; its extension towards the W crosses the ridge along the Strem Valley leading to Piz Pardatschas. This crack was hardly visible 10 years before. However, it is not the upper scarp of the Sedrun Landslide as other recent cracks appear uphill of this major feature.

Figure 8.4.   Large open fault in the upper part of the Sedrun Landslide due to the joint effect of the intersection of faults and to gravity induced phenomena. Behind this crack, very reduced movements are observed (11 cm in 20 years for point No. 14).

8.3.1.2 Rockslides

To the west, along Val Strem, there are some instances of rock sliding, most of all at the top of the slope. Unfortunately there are no monitoring points in this area, but secondary scarps clearly appear and the rock mass is deeply cut by numerous faults and cracks. Limited phenomena have already occurred as the toe of the slope in Val Strem is covered with scree (Fig. 8.3).

To the east side of Cuolm da Vi, along the northern slopes of Drun Tobel, this phenomenon is even larger, in terms of spatial extension and activity, with several rockslides and rockfalls of limited depth being reactivated by seepage and rainfall, as this zone shows the presence of formations of mylonites and kakirites. A horizontal limit below which an activity is not perceptible is clearly visible in Drun Tobel (Fig. 8.2).

8.3.1.3   Uphill facing main scarps

Due to the mechanism of toppling in the upper rock layers, some uphill facing scarps appear near the top of the slope (see Fig. 1.3). They are the consequence of the bending of rock layers with initially sub-vertical schistosity on top of Cuolm da Vi. This toppling produces movements that are concentrated on the main tectonic joints, where the rock is less strong, due to the schistosity, faults and systems of faults. This effect produces a dislocation of the rock mass, leading to the apparition of some scarps or uphill facing scarps, which favour the potential development of mass movements. Most of these faults are widely opened inducing small grabens between 2'200 and 2'350 m. At a higher level, these phenomena disappear due to the more complete toppling and sliding process.

8.3.1.4   Rockfalls

Although no specific evidence of large rockfall is visible in the Sedrun Landslide area, except for some screes of larger blocks at the toe of the left bank of Val Strem and for minor events visible in Figure 8.5, in this zone, it is clear that main rockfall events are in a preparatory phase to the west of Cuolm Parlets Dadens at an elevation of approximately 2'400 m a.s.l., in the site called Gion Giachen. The very high drop down to the bottom of Val Strem (approx. 600 m), the very steep slope (some 80%) and the intensely cracked rock mass have to be interpreted as major features for a potential rockfall. However, it is difficult to forecast if these rockfalls could occur as a single main event or as a succession of smaller events.

8.3.2   Local geology and structural analysis

At the site of the Sedrun Landslide, the Aar massif forming the slope is mainly composed of granite and granodioritic gneiss with biotite or with two micas to the north, and the southern part is composed of chloritoschists and schistic porphyritic gneiss, mylonites and kakirites.

The degree of weathering of the rock of Cuolm da Vi (on the basis of examined rock exposures and by simple strength tests) could be classified as slight to moderate, and is related to the toppling effect. The rock presents a density of around 27 kN/m^3. The strength of the rock (estimated with the aid of a ball-pen or hammer) is extremely variable.

The rock in the northern part of the site displays a rough schistosity and a high strength (50–100 MPa). In the southern part, the rock displays a more developed schistosity and high tectonisation. This tectonisation leads to the formation of deep series of mylonites and kakirites (superficial tectonical crushing inducing the formation of a pulverulent rock without recrystallisation).

It is important to point out the presence of a schistosity in the rock layers, with a direction ENE-WSW. The dip of the schistosity is 75°–80° SSE at the end of Val Strem, then tends to be vertical below Cuolm da Vi, at a low altitude, and even dips NNW in the higher parts of the Cuolm da Vi slope that are visible from Val Strem, under the effect of a deep toppling mechanism (Fig. 8.3). This toppling progressively increases towards Drun Tobel, with a parallel increase of the kakiritisation of the rocks. This toppling effect even leads to almost horizontal rock layers near Cuolm Parlets Dadens.

Moreover, the main following structural systems have been detected as several families of main faults intersect the massif:

- S1: sub-vertical faults N 120°–130°
- S2: sub-vertical faults N 150°–165°
- S3: sub-horizontal faults dipping 15–25° E, sometimes SE

The S1 and S2 faults are present in the same zones and are apparently not related to a local evolution of the same system. Finally, some NS isolated faults can be observed.

These various fault families defining the local structural setting, detected by field survey in 1994 and on the basis of aerial photographs (Fig. 8.5), can be briefly described as follows:

Family S0: Main schistosity
Displays a direction between N 40° and N 90°, essentially N 70°, and a dip between 75° and 90° SSE (regional schistosity).

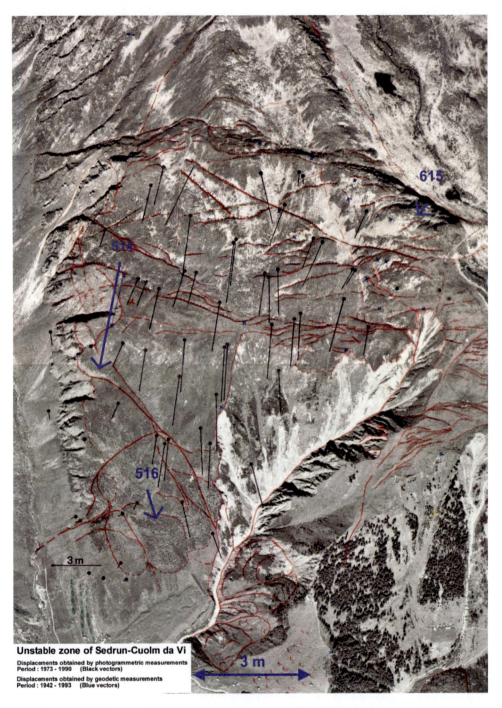

Figure 8.5.   Aerial photograph of the Sedrun Cuolm da Vi Landslide (1990) showing the main fault systems, as well as the displacements obtained by photogrammetry (black vectors – 1973/1990) and by geodetic measurement on 4th order topographic points Nos. 514, 516 and 615 (blue vectors – 1942/1993).

Family S1
It consists of sub-vertical faults with a direction N 120–130° and a dip of around 90°. These faults are not related to the schistosity of the massif, despite their quite similar orientation.

Their persistence varies from 100 m to several hundreds of meters; the opening of the faults varies from 0 to 1 m; the roughness can be classified as very rough near the surface. The infilling and water conditions in the massive and in the faults are unknown.

Family S2
It consists of sub-vertical faults with a direction N 150–165° and a dip of around 90°.

The persistence varies between 10 m and probably some dozens meters and the opening is some centimetres wide.

Family S3
It consists of sub-horizontal faults dipping 15–25° E, sometimes SE.

The persistence reaches several hundreds of meters, whereas the opening is not significant. They are visible in the eastern slope of Val Strem. Due to the toppling phenomena, these faults, which were originally dipping slightly opposite to the slope, display an inclination towards the slope, thus becoming able to induce a sliding phenomenon.

Family S4
It consists of isolated NS faults. One of them gave rise to a very large opening. They present a direction N 180° and a dip of around 90°. The persistence extends for some hundreds of meters.

The intersection of family S1 and S2 faults with S4 faults brought about the opening of a very large "fault" or hole due to gravity induced phenomena, below the crest at an altitude of 2'400 m (see Fig. 8.4).

8.3.3 *Investigation and monitoring*

The only monitoring systems that have been used in Cuolm da Vi consist of measuring the horizontal and vertical surface movements by means of different methods. Indeed no borehole has ever been carried out and no information on groundwater level is available, except for seepage horizons in Drun Tobel. Summing up the different geodetic measurement campaigns undertaken in Cuolm da Vi, the following actions and results can be mentioned.

8.3.3.1 *Measures carried out by the ETHZ (Swiss Federal Institute of Technology, Zurich)*
- Measurements organised to reconfirm the coordinates of the 4th order triangulation points around Cuolm da Vi.
- Local topographic measurements in 1988 and 1989 to determine the opening of some faults, by MM. Schwendimann, Pluss, Mark, Köchle and Lautenschlager.

The first geodetic measurements took place in 1983, when the ETHZ began to reconfirm the coordinates of the 4th order triangulation net in the area. The purpose of this campaign was to analyse the movements induced by probable slide processes and not those related to active faults already described in the zone for a research project for the National Science Foundation, by Eckardt, Funk and Labhart (1980 and 1983). This was the reason why this study focused on the zone of Cuolm da Vi, where a series of some 20 complementary points were installed. During these measurements, the movements of the triangulation points of the 4th order, namely Nos. 514, 516 and 615 were verified. Additional measurements were carried out again in 1988, then partially in 1989 and 1990 (only points 6 and 26 for this last year, by the survey office of the canton of Graubünden – see Fig. 8.6) [Bovay-Huguenin surveyors, 1994].

In 1997, new geodetic and GPS measurements were carried out. Point No. 26, which experiences the fastest movements, had to be replaced by point No. 70, located some meters away and which could still be measured in 2003 (its elevation decreased by 4.45 m over 20 years!).

Geodetic measurements were also extended to the zone of Alp Caschlé, to the west of Val Strem, which displays a similar glacier valley morphology. But no significant displacements were observed there.

In 1989, the survey office of the canton of Graubünden checked the triangulation network and signalled some points for a photogrammetric flight.

8.3.3.2 Measures carried out by the VERSINCLIM project of the EFPL (Swiss Federal Institute of Technology, Lausanne)

- Reconfirmation, by means of GPS measurements, of the position of the 4th order triangulation points and of the points installed by the ETHZ. This campaign was carried out by the office Bovay-Huguenin, official surveyors (Epalinges/Lausanne) in 1993, with the contribution of local surveyors (office Grünenfelder and Partners, in Domat-Ems).
- Determination of movements by photogrammetric measurements for the period 1973–1990 (Institute of photogrammetry, EPFL).
- Synthesis of movement measurements between 1943 and 1993 for the VERSINCLIM research project funded by the Swiss National Science Foundation (Noverrez et al., 1998).
- Field survey of major geological features and interpretation of the mechanism and causes of movements related to the geological context.

In 1993, within the VERSINCLIM project (Noverraz et al., 1998), a campaign including GPS (with Leica S200) and classic geodetic measurements was carried out by the Bureau Bovay-Huguenin, allowing a reconfirmation of 7 out of 12 points of the triangulation net (except some displaced or disappeared points and others with difficult access), and also of the majority of the points that had been laid out in 1983 by the ETHZ, namely 22 significant points. For all the surveyed points, a double independent determination has been carried out including either:

- two GPS determinations with 30 minute sessions
- one GPS session and one terrestrial geodetic measurement
- two terrestrial geodetic measurements.

At the same time, a photogrammetric determination of movements based on the comparison of photos taken in 1973 and 1990 was made by the Institute of Photogrammetry of the Swiss Federal Institute of Technology, Lausanne (EPFL) checking some 60 points on the slope of Cuolm da Vi.

Thus, considering the combination of the various measurements carried out, the periods of time for which a computation of the displacements was made are the following:

1942–1983; 1983–1988; 1988–1989; 1989–1990; 1990–1993, as well as 1973–1990, and, by deduction, for neighbouring points : 1942–1973 and 1973–1983.

The total significant displacements that were obtained by means of GPS measurements for the period 1942–1993 were 3.2 m, 0.9 m and 0.45 m, respectively, for points Nos. 514, 516 and 615 of the 4th order triangulation net, the global precision reaching less than 10 cm (including the uncertainty of the original 1942 determination). They vary between 0.94 m and 5.18 m for 10 out of the 22 ETHZ significant points between 1983 and 1993. The most important displacement (5.18 m in 10 years) is the one measured at the summit of Cuolm da Vi.

The displacements obtained by means of photogrammetry (Intergraph Image Station) between 1973 and 1990 give a more complete image of the spatial distribution of movements, showing a fairly homogeneous tendency within the unstable mass. The standard deviation on the measurements is ±20 cm, which is 10 to 15 times less than the observed movements.

Despite the particular care in carrying out all these measurements and in analysing them, it is quite possible that the short term observations (1988–1989, 1989–1990, 1990–1993), especially for the local points, may give too extreme velocity values, such as for point No. 26, which require further observations. This is due in particular to the fact that there may be a combination of global and local movements due to the intense cracking of the rock mass. The measurement conditions are also quite difficult and the number of really fixed points is reduced (often slowly moving pointed have been taken as reference points).

8.3.3.3 Measurements carried out by the consultants of the community of Sedrun

During the last five years, the geological office Bonanomi AG in Sedrun and the survey engineering office Donatsch in Landquart have been appointed by the Community of Tujetsch, including the village of Sedrun, to carry out a monitoring programme and follow the evolution of the landslide of Cuolm da Vi (Bonanomi & Donatsch, 1998).

The monitoring system first included the continuation of the survey measurements on the points located in the landslide area by the ETHZ and by the survey office of the canton of Graubünden (in 1989 and 1994), as well as on a few new points to the NW of the landslide, and then the operation of continuous distance measurement devices (since 2000) between a fixed point located in the valley near the village and two points near the top of Cuolm da Vi. This automatic system also includes a camera to visualize the evolution of the slope.

These data allowed the confirmation of the very important displacements recorded at point No. 26, which increased again, after a relatively quieter period from 1993 to 1998, to nearly 80 cm/year in 2000 and 75 cm/year in 2001. The other points on the slope below the peak are less active, as already observed, but several of them still reach annual velocities between 15 and 30 cm/year, which vary very little and, thus, do not show any clear acceleration trend.

Figure 8.6 below presents a synthetic view of the long-term tendency of variation of the velocities for several significant points, which clearly marks a considerable change in the development of the phenomenon since the 1970's, putting forward a potentially high hazard level.

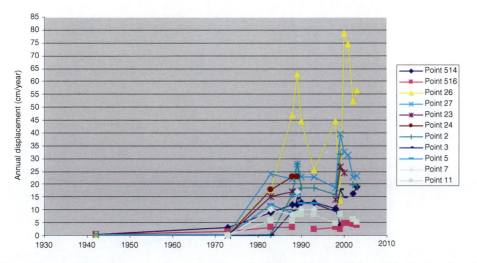

Figure 8.6.   Annual displacements of several points on the Sedrun Cuolm da Vi Landslide. The location of some of these points is visible in Figure 8.5.

The continuous measurements related to two specific points show a fairly constant tendency between June 2000 and the end of 2002 with a slight acceleration at the end of the spring. The average velocity recorded for these two points is included between 25 and 30 cm/year. Therefore the short term evolution seems less worrying than the long term one.

8.3.4   *Danger identification*

Because no boreholes are available at the Sedrun Cuolm da Vi Landslide, the volume of the unstable rock mass can only be assessed approximately to some 100 or $110 \cdot 10^6 \, m^3$, as a consequence of the small uncertainties related to the spatial limits of the landslide zone and above all to the presence or absence of a basal failure surface. But the existence of very important displacements, of major cracks extending over hundreds of meters and of severely dislocated rock zones visible at the surface, near the top of the unstable area, clearly proves the presence of a potential danger. In order to analyse the possible occurrence of hazardous phenomena, it is first necessary to understand the possible failure mechanisms that may occur, which leads to the description of several landslide scenarios that are likely to cause direct or indirect important damage to the village of Sedrun. Several realistic scenarios have been elaborated, based on the geology, morphology and survey data, and three of them have finally been selected as the most representative potential mechanisms, to be considered in the risk assessment procedure.

8.3.4.1 *First potential scenario with a high probability of occurrence*

This scenario implies the development of a limited slide, i.e. not a shallow phenomenon, but a massive slide (approx. 20'000 m³) on the right side of Drun Tobel, in the zone called Plauncas Tgamos where the granite gneiss formation to the north and the schistic gneiss formation to the south are separated by a zone of mylonites and kakirites prone to instability due to their low shear strength (Figs 8.2 and 8.7).

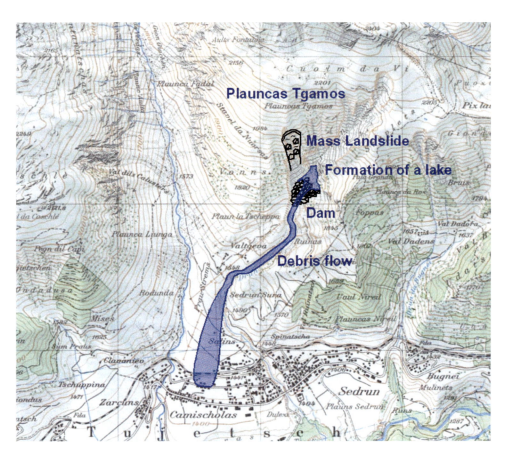

Figure 8.7.   Scenario No. 1 for the risk assessment studies.

This slide of limited volume, triggered on a very steep slope of the Drun Tobel, where similar events seem to have already occurred following intense rainfall, might suddenly dam this creek and induce the immediate formation of a small lake. As soon as the overflow begins, the dam will be eroded and cause a fast debris flow with an important mass of granular material, a part of which will be washed away from the creek bed itself.

Due to the presence of a sharp bend at the outlet of Drun Tobel Creek, at the approximate elevation of 1'500 m a.s.l., the high velocity debris flow will not stay in the thalweg and overflow on the alluvial fan of the Strem River, below the site called Valtgeva, upstream of the village of

238

Figure 8.8.   Bend of Drun Toble Creek where the debris flow considered in Scenario 1 will flow directly towards the village of Sedrun (the first visible houses are only barns).

Sedrun (Fig. 8.8). Therefore, the debris flow formed of coarse material would affect a part of the village downstream of the railway station. As the Val Strem alluvial fan is not very steep, the debris flow would reduce its speed and, therefore, only hit the basement of the first exposed houses, as well as damage the railway line.

This event has never occurred there in the past, according to historical information, but the presence of large blocks in the lower part of the Drun Tobel creek leads to the assumption that that it is a very reliable scenario, as the Drun Tobel streambed shows narrow stretches at some places. The very large fan deposits of the Drun Torrent indirectly prove that similar phenomena occurred in the past. Similar debris flows have seriously affected the Vorderrhein Valley downstream in November 2002, in particular in the village of Schlans, following extremely intense rainfall that had never been observed before, over a period of three days. But in the Cuolm da Vi area, the precipitation of this event only fell as snow. However this first scenario has not been modelled in a debris flow code, as the necessary assumptions related to the expected volume of implied materials and to the overflow process might induce a large range of flow velocities and very different potential consequences. The selected scenario has thus to be considered as a typical danger situation, but not implying a unique path, with a relatively high probability of occurrence (10 to 100 year return period).

8.3.4.2   *Second potential scenario with a medium probability of occurrence*
This scenario considers the occurrence of a large slide (more than 100'000 m^3) on the right side of Drun Tobel, namely in approximately the same zone as in Scenario 1, that is in a weak rock zone, but extending higher and wider in the lateral slope of the creek and implying a larger volume due to a regressive failure mechanism. This scenario takes the movement of point No. 27 into account, which reaches more than 5 m in 18 years (1983–2001) and is located in the upper part of the expected slide zone, as being a significant preparatory sign; but as no acceleration is perceived, the occurrence is not expected in the short term.

This potential event caused by intense or long lasting rainfall may induce a major debris flow which will not be contained by the natural thalweg, so that massive overflow will occur on the

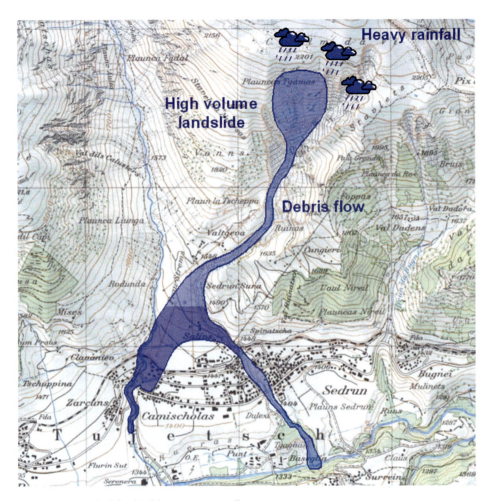

Figure 8.9.   Scenario 2 for the risk assessment studies.

alluvial fan of Val Strem, reaching the course of this river to the west of Sedrun; it will also extend along the Drun Tobel valley where it crosses the village of Sedrun down to the Rhine river. The narrow zone of impact in this last stretch may be widened by erosion phenomena which would destabilise the creek banks (Fig. 8.9).

The affected zone has been determined on the basis of the local topographical conditions of the slope in the lower part of Val Strem, referring to other similar situations which have occurred in Switzerland, in particular in August 1987 in Poschiavo and in September 1993 in Brig. For such a case, and considering the large uncertainties affecting the volume as well as the debris flow mechanism, mathematical modelling would not provide more reliable results to determine the expected zone of impact. Indeed these two first scenarios will be used in a comparison of quantitative risks with respect to phenomena of high and medium probabilities of occurrence (100 to 300 year return period for Scenario 2).

8.3.4.3   *Third potential scenario with a low probability of occurrence*
This quite different scenario considers a potential rockfall from the top of the Cuolm da Vi area, in particular from its peak called Cuolm Parlets Dadens at an initial elevation of 2'458 m a.s.l. It will occur in the direction of the Val Strem valley, as the slope is very steep to the south-west (about 80%), with an average height of fall of some 600 m. Due to the intense cracking and

240

de-structured aspect of the gneiss rock mass in this area of the Sedrun Landslide, as well as the recently appeared movements to the west of the peak following the deep toppling mechanism which weakens the rock strength properties, and despite of the fact that the present movements are not clearly oriented towards the SW, a total volume of approximately $5 \cdot 10^6\,m^3$ may be involved, either as a single event or as a succession of smaller rockfall events (Fig. 8.10).

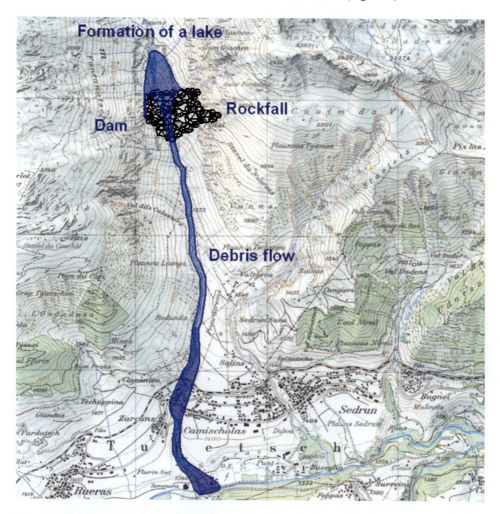

Figure 8.10.   Scenario 3 for the risk assessment studies.

Considering the induced energy of compaction related to the drop, such a phenomenon might certainly cause the formation of a large dam in Val Strem, which is very narrow at the expected point of impact of the rockfall (less than 300 m for a dam height of 100 m). It would thus form a large lake in the Val Strem valley, which would be filled rapidly, especially if this occurs in the spring during the snowmelt when the flow of the river is important (no discharge data exist, but according to the size of the drainage area and the amount of precipitation, it is expected to carry more then 1 m³/s in normal spring conditions – average annual discharge = 0.54 m³/s – and up to 15–20 m³/s during storm events).

Considering the original gneissic rock material, the dam would be formed of fairly coarse material, but the experience of Randa rockfall in 1991 with a similar rock formation shows that an intense compaction allows the dam to be fairly impervious except at the surface of the deposit. Even if the

241

seepage develops when the lake level reaches the lowest point of the deposited rockfall mass, it would probably not prevent an overflow which would cause a fast erosion and thus create a massive flood downstream. Considering the velocity of such flow induced by the steep inclination of the stream bed in Val Strem (approx. 10%), which would exceed 10 m/s, the flood would dash down the river bed and pass over its banks, devastating a part of Sedrun (Community of Camischolas) down to the Rhine river. This catastrophic scenario, assessed to have a return period of some 300 to 1'000 years, may have serious consequences on the Rhine valley downstream, in particular for a hydro-electric plant as well as one of the building sites of the Gotthard base tunnel and for its respective construction facilities. The present evolution of the affected zone near Cuolm Parlets Dadens may even lead to the assumption that the probability of occurrence might increase in the future.

8.3.5 *Geomechanical modelling*

In order to try to understand the development of deformations in the Cuolm da Vi area, geome-chanical modelling has been attempted at the Polytechnical University of Catalunya (UPC), aim-ing at the determination of the deformation field after the retreat of the glacier which occupied the Rhine Valley. However, this 2D-computation, taking the various crack and fault systems into account, has yielded minor displacements at the top of Cuolm da Vi and higher displacements at the toe of the slope, whereas the present monitoring data (extending over the last 60 years as well as over the last few years), indicate that the major displacements occur at the top of the slope. Such a situation can be explained as several mechanisms are affecting the slope (topple, eventually slide, as well as possible melting of permafrost) which are not directly related to the past history of the glacial retreat. Indeed, in such cases the classical "triggering" numerical modelling is not appropriate to supply a displacement pattern that can be directly used to express the hazard. The reference to the experience of potential landslide scenarios, based on historical events in other contexts, is more significant to serve as a basis, for the risk assessment, than the results of elasto-plastic mechanical modelling that is not sophisticated enough to consider a possible melting of the permafrost and creep effects.

8.3.6 *Occurrence probability*

As indicated in the three most reliable scenarios selected for the risk analysis described above, the occurrence probability considered here is not based on known past events at that site, but on experi-ence of similar events that are quite frequent when implying limited volumes (Scenario 1), less frequent when the slide displays a larger mass (Scenario 2) and rare in the case of a rockfall of sev-eral million m³ (Scenario 3).

For the risk computation, the following return periods have thus been assessed in a conservative way, especially for the last scenario.

- Scenario 1: 10 year return period.
- Scenario 2: 100 year return period.
- Scenario 3: 1'000 year return period.

It is clear that these investigation hypotheses should be confirmed at a later stage, for instance by more sophisticated numerical modelling or on the basis of new field observations, but above all when appropriate geological and geotechnical data will be obtained by boreholes, including underground temperature data.

8.4 QUANTITATIVE RISK ANALYSIS

8.4.1 *Elements at risk*

In the case of Sedrun, a large village with 1'100 inhabitants in 1950 and 1860 inhabitants in 2000, including the sector of Camischolas (figure for the whole community of Tujetsch), the potentially affected zones when considering the three selected scenarios are fairly limited, so that a detailed census of the elements directly at risk has been carried out. On the basis of the local management plan of the Community of Tujetsch (Fig. 8.11) and following investigations on site, a comprehensive list of the different exposed "objects" has been established, classified into 23 categories, as shown in Table 8.1. They first include the population, then the buildings and infrastructures, and finally the implied economic activities (transportation lines, for example). In winter, the population significantly increases, as the total number of holiday beds reaches 5'400.

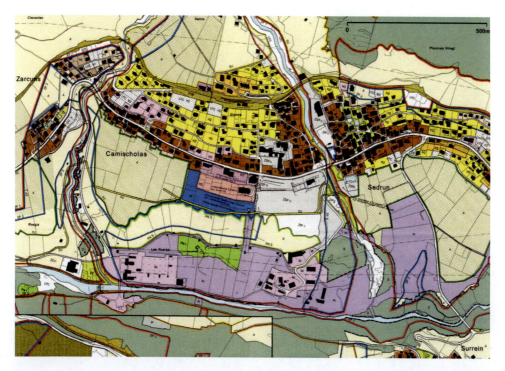

Figure 8.11. Local management plan showing the exposed objects. It can be noted that natural hazard red zones are shown on this map, but they are presently based on snow avalanche and debris flow hazard only. The purple zones along the Rhine River correspond to a hydro-electric plant, technical equipments and the construction facilities for the new Gotthard base Tunnel.

As far as the exposed persons are concerned, a fairly low average rate of occupancy is considered, due to the fact that most of the potentially affected buildings are holiday properties. Of course it cannot be excluded that many persons might be present outside of the buildings in the potentially affected are; but this assumption is extreme if the weather conditions causing at least Scenarios 1 and 2 are recorded. An appropriate alarm system may reduce such numbers significantly.

8.4.2 *Vulnerability of the elements at risk*

According to the type of structures built in the exposed areas and to the mechanism and intensity of the landslide events considered in the different scenarios, the vulnerability coefficients specifying the expected degree of damage have been assessed, based on the reference values given in the literature (Leone et al., 1996), as well as on the consideration of the type of building structure observed in the field. The selected values appear in Tables 8.2 to 8.4. It is clear that the

Table 8.1. Categories of exposed "objects" considered in the risk analysis.

People	Population at risk
Buildings	House 1st class
	House 2nd class
	House 3rd class
	Restaurant
	Railway annex building
	Garage
	Industrial building and equipments
	Barn, farm
	Hydro-electric plant
Networks	Road (value/m)
	Forest road (value/m)
	Bridge (value/m)
	Railway line (value/m)
	Electrical line (value/m)
Natural areas	Fields (value/m^2)
	Forest (value/m^2)
	River (value/m)
Vehicles, movables	
Socio-economic activities	Closed roads (value/day)
	Closed railway (value/day)
	Housing (value/day)
	Unemployment (value/day)
	Tourism (value/day)

considered scenarios described in § 8.3.6 which will not imply thick masses of debris in the village area justify fairly low vulnerability coefficients for most of the buildings. At this stage, no difference has been introduced between the buildings hit in a first line by the debris flows and those located downstream which will probably be less affected, due to the protection offered by these upstream buildings, especially in Scenarios 1 and 2. For Scenario 3, the considered vulnerability coefficients are higher due to the important impact of the flow.

8.4.3 Expected impact

The impact caused by each considered scenario results from the multiplication of the value of each "object" by the respective vulnerability coefficients; the sum of all expected impacts gives the global impact corresponding to each scenario (see Figs. 8.12, 8.13 and 8.15, as well as Tables 8.2 to 8.4). The monetary values included in the tables are in Swiss Francs and, therefore, must be multiplied by 0.64 to obtain amounts in €.

It is evident that Scenario 1 implies a reduced impact in terms of physical damage, mainly to buildings (1.82 mio CHF or 1.65 mio €) which are insured. The indirect impact on the economic and transportation activities will also be limited, as it will be possible to carry out the necessary repair works at very short notice (traffic interrupted during one day for the railway line). For the Drun Tobel creek, the damage to the right bank where the debris flow will leave the stream bed will be significant. But the suddenness of the event considered in Scenario 1 may imply human victims if the exposed persons are outside of their houses; however, this is fairly improbable considering the bad weather conditions expected. As far as the indirect impact on tourism is concerned, it is very difficult to assess, but the possibility of repairing the damage during the summer should not affect the next winter season.

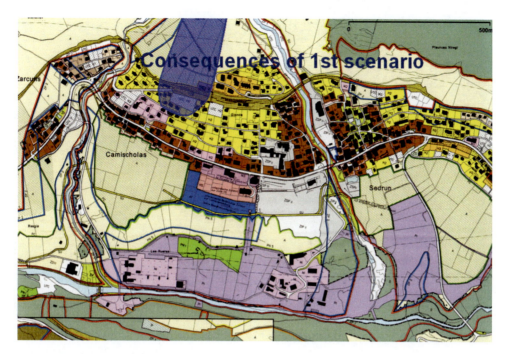

Figure 8.12. Impact of Scenario 1.

Table 8.2. Expected costs related to Scenario 1, excluding costs of directly affected persons. The number of persons at risk takes an average rate of occupancy into account.

		Number	Value	Vulnerability	Cost
People		25	?		
Buildings	House 1st class	4	1'000'000	0.2	800'000
	House 2nd class	2	750'000	0.25	375'000
	House 3rd class	2	500'000	0.2	200'000
	Restaurant	1	2'000'000	0.15	300'000
	Annex building	1	100'000	0.15	15'000
	Garage	2	100'000	0.15	30'000
	Barn	2	100'000	0.5	100'000
Networks	Road (value/m)	150	500	0.6	45'000
	Forest road (value/m)	100	200	0.8	16'000
	Railway line (value/m)	200	500	0.4	40'000
Natural areas	Fields (value/m^2)	75'000	1	1	75'000
	River (value/m)	200	2'000	0.5	200'000
Vehicles, movables					
Socio-economic activities	Closed roads (value/day)				
	Closed railway (value/day)		10'000	1	10'000
	Housing (value/day)				
	Unemployment (value/day)				
	Tourism (value/day)		750'000 (1'500 p/day)		
	Total costs (CHF)				**2'206'000**
	Total cost (€)				**1'412'000**

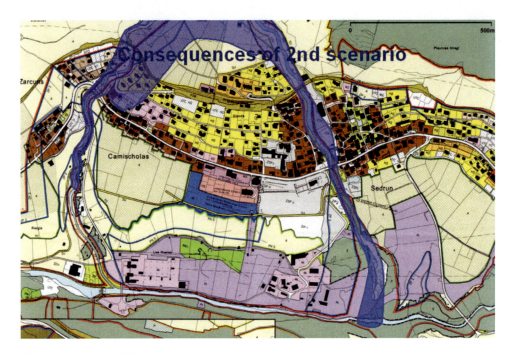

Figure 8.13.   Impact of Scenario 2.

Table 8.3.   Expected costs related to Scenario 2, excluding costs of directly affected persons. The number of persons at risk take an average rate of occupancy into account.

		Number	Value	Vulnerability	Cost
People		150	?		
Buildings	House 1st class	2	1'000'000	0.15	300'000
	House 1st class	1	1'000'000	0.2	200'000
	House 1st class	1	1'000'000	0.25	250'000
	House 2nd class	1	750'000	0.15	112'500
	House 2nd class	12	750'000	0.2	1'800'000
	House 3rd class	2	500'000	0.2	200'000
	Garage	1	100'000	0.15	15'000
	Barn	11	100'000	0.5	550'000
Networks	Road (value/m)	350	500	0.6	105'000
	Forest road (value/m)	600	200	0.8	96'000
	Bridge (value/m)	4/80	16'000	0.2	256'000
	Railway line (value/m)	250	500	0.4	50'000
Natural areas	Fields (value/m^2)	120'000	1	1	120'000
	River (value/m)	2'100	2'000	0.5	2'100'000
Vehicles, movables					
Socio-economic	Closed roads (value/day)	2	2'000	1	4'000
activities	Closed railway (value/day)	10	10'000	1	100'000
	Housing (value/day)				
	Unemployment (value/day)				
	Tourism (value/day)		750'000 (500 p/day)		
	Total costs (CHF)				**6'258'500**
	Total cost (€)				**4'005'000**

246

Figure 8.14.   Bridge for the cantonal road crossing the Val Strem creek that would be damaged by the debris flow considered in Scenario 2.

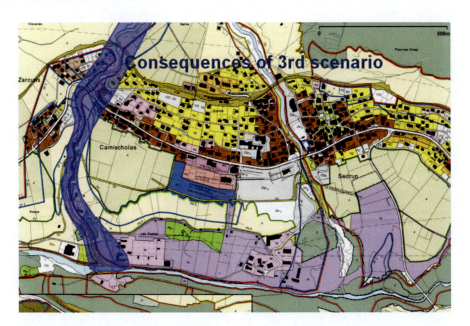

Figure 8.15.   Impact of Scenario 3.

In Scenario 2, the damage will affect several bridges, as the one shown in Figure 8.14, but it can be assumed that they will be only partially destroyed, due to their reinforced concrete structure and to the deep foundations of their piles. However, the damage to the river courses of the Strem River and the Drun Tobel Creek will be important in terms of environmental impact (2.1 mio CHF or

247

Table 8.4. Expected costs related to Scenario 3, excluding costs of directly affected persons. The total cost is given with and without the losses corresponding to a decrease in tourism activity.

		Number	Value	Vulnerability	Cost
People		120	?		
Buildings	House 1st class	2	1'000'000	0.3	600'000
	House 1st class	1	1'000'000	0.4	400'000
	House 1st class	1	1'000'000	0.5	500'000
	House 1st class	1	1'000'000	0.6	600'000
	House 2nd class	1	750'000	0.3	225'000
	House 2nd class	9	750'000	0.4	2'700'000
	House 3rd class	2	500'000	0.4	400'000
	Barn	5	100'000	0.8	400'000
	Barn	2	100'000	0.5	100'000
	Annex building	1	500'000	0.4	100'000
	Small annex building	3	10'000	0.3	9'000
	Small industrial cable lift	1	1'000'000	0.25	250'000
Electric plant		1	16'000'000	0.3	4'800'000
Networks	Road (value/m)	650	500	0.6	195'000
	Forest road (value/m)	100	200	0.8	16'000
	Bridge (value/m)	3/60	16'000	0.2	192'000
	Railway line (value/m)	150	500	0.6	45'000
	Electrical line (value/m)	200	150	0.5	15'000
Natural areas	Fields (value/m^2)	10'000	1	1	10'000
	River (value/m)	3'000	2'000	0.5	3'000'000
Vehicles, movables		3	100'000	1	300'000
Socio-economic activities	Closed roads (value/day)	10	4'000	1	40'000
	Closed railway (value/day)	20	10'000	1	200'000
	Housing (value/day)	2'000	150		300'000
	Unemployment (value/day)	4'000	120		480'000
	Tourism (value/day)		750'000 (1'500 p/day)		(22'500'000)
	Total costs (CHF) without loss related to tourism				**15'877'000**
	Total cost (€)				**10'161'000**
	Total cost (CHF) With expected loss related to tourism				**38'377'000**
	Total cost (€)				**24'561'000**

1.344 mio €). The damage to buildings will represent the higher part of the expected costs (approx. 3.43 mio CHF or 2.2 mio €). As far as socio-economic activities are concerned, a limited amount has been considered for closed roads, because the main access road to Sedrun along the Rhine valley to the east should not be affected and as it will be possible to re-establish a by-pass road through Camischolas for the traffic coming from the Oberalp Pass. No amount has been computed for the loss of tourist activity as the major facilities will not be concerned; but the reputation of the resort might be affected.

In Scenario 3, the direct damage to buildings is much higher than in Scenario 2 because of the higher expected vulnerability coefficients (5.93 mio CHF or 3.8 mio €), but one of the major indirect impacts will be the partial destruction of the hydro-electric plant near the Rhine River, as well as the damage to the construction facilities for the Gotthard base tunnel which includes losses of machinery (in the items of movables) and partial unemployment. Moreover, a delay in the achievement of this very important infrastructure for the whole European economy may have huge economic consequences that cannot be assessed in this research. The interruption of electricity production may also imply losses of millions of CHF. Finally if the expected cost without considering the loss related to tourism reaches nearly 16 mio CHF, the possible additional impact on tourism, assuming 30 days of total loss of tourist income, will be 22.5 mio CHF or 14.4 mio €. This value has to be taken into account as a rockfall may occur even in winter time. This last figure clearly puts forward the considerable importance of the indirect impacts on the local economy, which requires careful attention in management and protection plans.

8.4.4 *Risk assessment*

Table 8.5 below sums up the risks corresponding to the three selected scenarios in terms of physical, environmental and socio-economic damage. The number of potentially directly affected persons is also computed, but no monetary value is given. It clearly appears that Scenario 1 implies a relatively higher risk of assets and life because of its high assessed probability. In terms of social risk, Scenario 2 may appear more important as the impacted houses are numerous (19, dwelling houses) and will be more severely affected; but when the probability is considered, the risk to life is smaller than in Scenario 1. Nevertheless, the aversion factor for which a large number of victims in one event is less acceptable that the same number for several disasters will have to be taken into account as far as the risk perception is concerned. Finally in terms of indirect risks, not only at a local scale, but at a European scale, Scenario 3 will definitely imply very severe consequences in particular on regional tourism, electricity production and on the facilities for the future transportation system through the Alps.

Table 8.5. Computation of the risks involved by the three considered scenarios.

Scenario considered	1	2	3
Costs related to the corresponding scenario (approximate value)	1.4 million €	4.0 million €	24.6 million €
Number of persons directly at risk	25	150	120
Assessed probability	0.1	0.01	0.001
Corresponding monetary risk	140'000 €	40'000 €	25'000 €
Corresponding risk to life	2.5	1.5	0.12

As the three scenarios are independent of one another and will not necessarily occur within a short period during which damage will not yet have been repaired, the total risk implied corresponds to the sum of the three quoted figures, namely a monetary risk of some 205'000 € (yearly amount to spare in order to cover future damage) and a risk to life of 4.1 victims per year. Preventive measures are thus definitely required.

8.5   REGULATION AND RISK MITIGATION MEASURES

8.5.1 *Existing measures*

After a long period (1983–1993) during which the investigations carried out at the Sedrun Landslide in order to understand and determine the displacements were of interest only to scientists, and which led to the expression of various opinions concerning the possible mechanisms, the authorities of the community of Tujetsch became concerned about the marked evolution of displacements in the Cuolm da Vi area. Therefore, they appointed a consulting geologist (Bonanomi AG at Buguei-Sedrun) and a surveyor (Donatsch Ingenieur- und Vermessungsbüro SA, Landquart), in order to check the area and follow the evolution of the movements more closely. In accordance with this aim, several survey campaigns were organised annually (from 1998 to 2003) and from July 2000 on, a continuous distance measurement of two points located near the summit as well as one reference fixed point were installed and operated, with the possibility for access on-line to this information and also of visualising the slope by a picture taken at daily intervals from the site of the distance meter.

Moreover, it is foreseen by the community to carry out rockfall modelling in order to produce a hazard and risk map, according to information received in August 2002. But the community and its consultants are convinced that boreholes would neither be cost-effective nor contribute to the main aim, that is the safety of the population. This aspect is, however, still in discussion at the federal level.

8.5.2 *Proposed investigations to plan mitigation measures*

On the basis of all the existing information – as far as it is accessible – and considering the possible consequences of any hazard scenario as analysed in this chapter, four main recommendations in terms of need can be expressed at this stage to improve the safety conditions in the area of Sedrun:

1. Need for further investigation.
2. Need for preliminary protection measures.
3. Need for specific regulations.
4. Need for extended monitoring and warning system.

8.5.2.1 *Need for further investigation*
The present analysis, as shared by the IMIRILAND partners, does not conceal the fact that the real mechanism affecting the slope at Cuolm da Vi is not yet fully understood. This is due in particular to the lack of subsurface information concerning the geological conditions, the groundwater levels, the displacement pattern with depth and the temperature distribution. Surface displacement data are also lacking in some parts of the landslide, namely in the areas from which the expected scenarios presented in § 8.3.6 are likely to be triggered.

Before these investigations are carried out, any geomechanical modelling, such as the one which was attempted for the Sedrun Landslide within IMIRILAND project, will not reach a sufficient level of reliability and therefore will not help in producing significant hazard and risk maps. For the time being, an analysis based on possible feasible scenarios is still the best approach.

8.5.2.2 *Need for preliminary protection measures*
Although the potential landslide mechanisms and respective rock masses considered in the three selected scenarios are not yet completely defined, it is possible to plan preliminary protection measures along the expected trajectories, which will definitely contribute to slowing down the process or to limiting its extension. Therefore, there is a need to design and carry out such appropriate measures, according to the principles described in § 8.5.3. Indeed, even if the initial design is not the best, it will improve the safety conditions and should allow later modifications, when more reliable landslide scenarios are confirmed by geomechanical modelling.

8.5.2.3 *Need for specific regulations*
Presently the local management plan includes the consideration of a snow avalanche hazard, as it occurred in 1749, 1808 and 1817, uphill of the eastern part of Sedrun and of a debris flow hazard along the Val Strem and Drun Tobel river beds (see Fig. 8.11). The corresponding danger zones in the plan should also be extended to consider the hazards related to the presented scenarios, even if their delimitation cannot be expressed in a definitive way. Indeed, several examples exist in which, after a serious landslide event, provisional zoning has been adopted by the local authorities and building zones have been temporarily suspended until the real hazard could be properly determined.

These regulations do not concern the local management plan only, but also the preparedness plans, including the organisation of evacuation routes, the training of the population in presence of an alarm signal, the planning of emergency protection measures, etc.

8.5.2.4 *Need for extended monitoring and warning system*
According to the presently considered landslide scenarios, it is likely that the critical movements will not affect the whole zone of Cuolm da Vi, but only a small part of it. In these specific areas,

nearly no monitoring data are recorded due to the particularly difficult access; furthermore, no continuous measurements are available and the evolution of the respective slopes prior to failure will be difficult to detect visually. It is, therefore, necessary to install more monitoring systems; their cost-effectiveness has been proven on many occasions, when it is linked to the potential number of victims which may be evacuated in time on the basis of reliable monitoring data. On the other hand, these new devices with automatic transmission must include a two-level warning system allowing a preliminary scanning of the data by the authorities or their consultants before the public alarm is triggered.

8.5.3 *Proposed mitigation measures*

With respect to the three analysed scenarios, the following preliminary mitigation measures can be proposed for further design and implementation, according to the priorities set up for the most urgent protection plans.

8.5.3.1 *Measures corresponding to Scenario 1*
- Construction of check dams on the Drun Tobel creek so as to cause the deposition of sediments and to reduce the erosion mechanism upstream of the possible overflow zone (see Fig. 8.8); these works will also reduce the peak flow after the failure of the landslide dam, by ensuring the proper conditions for kinetic energy dissipation.
- Construction of a protection dam on the right bank of the possible overflow zone, in order to reduce or suppress the extension of the debris flow on the alluvial fan of Val Strem.
- An alternative proposal, or in conjunction with the former one, implies a riverbed improvement in the possible overflow zone so as to increase the flow velocity and avoid any sediment deposition. In such a case, all the downstream part of the Drun Tobel creek has to be protected so as to avoid scouring where the creek crosses the village.

8.5.3.2 *Measures corresponding to Scenario 2*
- Excavation of a 300 m long rectilinear channel between the Drun Tobel creek, at the beginning of its bend below the site of Valtgeva, and the Val Strem River, upstream of the railway bridge (see Fig. 8.9). This channel should be some 15 m deep; it will require the displacement of three ski-lifts on the alluvial fan of Val Strem. The normal flow of the Drun Tobel creek should be maintained in its original riverbed.
- River protection works along the lower part of Val Strem, in the area of Camischolas-Zarcuns, down to the hydro-electric plant near the Rhine.
- An alternative proposal may include the building of a high catch dam on the Val Strem River, upstream of Camischolas, to collect the debris flows and sediments coming from the channel connecting the Drun Tobel creek to the Val Strem River. This dam could be built with the excavation materials from the channel. It could also be useful for Scenario 3. Its height is difficult to assess, but will certainly exceed 20 m, which will cause a certain impact in the scenery.

8.5.3.3 *Measures corresponding to Scenario 3*
It is difficult to justify any major protection measures for such a process with a very low probability of occurrence. But the catch dam considered for Scenario 2 might prove to be efficient for this scenario also, even if it would not retain all the debris flow induced by the failure of the landslide dam. It would certainly reduce the peak discharge of this flood wave, block a large part of the transported sediments, and give some additional time to the population for its evacuation.

However, this dam will require a protected spillway to avoid its destruction in case of overflow and of obstruction of the torrent culvert by snow avalanches.

It is clear that all these mitigation measures have to be studied not only in the perspective of reducing the hazard, but also in a development prospective allowing a sustainable increase of tourist activities which are the major industry for the community of Tujetsch.

8.6 CONCLUSIONS

The Sedrun Landslide case study represents a typical situation of the difficult paradox that the populations of the alpine range have to face if they want to continue developing in their homeland. As the hazards implied do not seem obviously threatening, the population and the local authorities tend to ignore or understate the phenomena in order to avoid panic or fear and thus reduce the attractiveness of their resort. Meanwhile, they limit their investigations and do not favour a scientific debate of the risks involved. They rather favour new constructions which increase the value of the exposed objects. On the other hand, as time passes, the slope movements continue and their exceptional magnitude, especially at the top of the slope, necessarily implies an increased probability of occurrence of a significant event.

Therefore, the only way to face the landslide risks is to recognize their existence, to improve their knowledge by appropriate investigation and monitoring programs, to take provisional prevention measures in order to reduce the short term potential impacts, and once the potential mechanisms are better understood and modelled, to carry out protection works so as to limit the consequences of any possible landslide event.

The scientific investigations carried out for the Sedrun Landslide within the IMIRILAND project do not pretend to give final recommendations or advice as far as the mitigation actions are concerned. This case study was considered to show the applicability of landslide risk methodologies and demonstrate the possibility of expressing the relative magnitude of such risks according to the various scenarios analysed. It is clear that several parameters may change with time with respect to the evolution of the unstable slope, to the development of the tourist area and to the application of preliminary prevention measures. But it is vital to initiate a long-term process of risk analysis and management, so that the population and the tourists can be convinced that the situation is really "under control".

9
The Séchilienne landslide

Jean-Louis Durville, Laurent Effendiantz, Pierre Pothérat
Centre d'ÉtudesTechniques de l'Équipement de Lyon

Philippe Marchesini
Direction Départementale de l'Équipement de l'Isère

9.1 INTRODUCTION

The Séchilienne landslide has developed on the right side of the Romanche valley, 20 km to the south-east of Grenoble in the Isère Departement of the French Alps (Fig. 9.1). The national road RN 91 runs along the valley bottom and then to the Lautaret Pass (the link from Grenoble to Briançon and Italy), one of the higher alpine passes which is not closed in winter. Except for a narrow winding road, the RN 91 is the only access to numerous ski resorts as L'Alpe-d'Huez and Les Deux-Alpes.

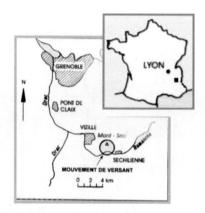

Figure 9.1.   Location map of Séchilienne.

9.1.1 *Landslide description*

The landslide is occurring on a slope extending from an elevation of 330 m a.s.l. at the bottom of the valley to 1150 m a.s.l. at Mont Sec (Fig. 9.2). The landslide itself extends from 600 m a.s.l. up to 1130 m a.s.l., over an area of approx. 70 ha. The limits are as follows:

– to the east: a major tectonic shear zone (N 20°), which is a clear-cut border,
– to the north (upwards): a scarp which has probably been formed by a large post-glacial sagging,
– to the south (downwards): the slope below 600 m a.s.l. is not moving,
– to the west-south-west: the movement is assumed to decrease continuously.

The landslide can be divided into three principal areas (Fig. 9.3): the frontal mass, the most active (several decimetres per year; source of many rockfalls) and disrupted part, the volume of which can be estimated at approximately 3 hm³; an intermediate zone with medium activity; the upper and north-western part, corresponding to an elliptic (probably) post-glacial sagging, with low velocities.

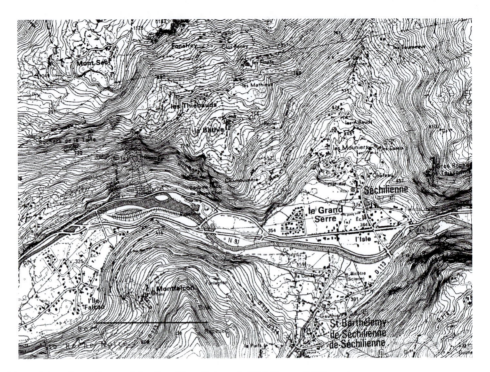

Figure 9.2. General map of the site of Les Ruines de Séchilienne (IGN map).

Figure 9.3. Aerial view of the frontal mass. The Séchilienne plain can be seen in the background.

9.1.2 *Historical background*

Many rockfalls have been reported during the last centuries resulting from the right side of the Romanche valley at the site currently called "Les Ruines". Some mining activity (metallic sulphurs) was carried out in the XIXth century and ended after the 1st World War. The oldest aerial photographs (1937 and 1948) show a slope morphology with elliptic sagging and NE-SW trenches and active screes. But it may be seen that a footpath remained passable through the frontal mass until after World War II.

An important reactivation was observed in the 1980's: rockfalls hit the national road RN 91 during the winter of 1985 and the existence of a large slope deformation was recognized (Antoine et al., 1987). Monitoring of the slope was set up and some protective measures were installed: a fence with an electrical wire alarm and traffic lights along the original portions of the RN 91, a dam with a storage volume capacity of about 2 hm^3 of debris and a diversion river bed for the Romanche River.

In 1986, a diversion road was opened to traffic; it lies in the middle of the valley on the other side of the river (the initial road ran along the foot of the slope) and it is, therefore, safe from rock-falls but remains exposed to rockslides of more than 1 or 2 hm^3 (Fig. 9.4). In 2003, a diversion gallery was driven in the opposite slope for the Romanche River.

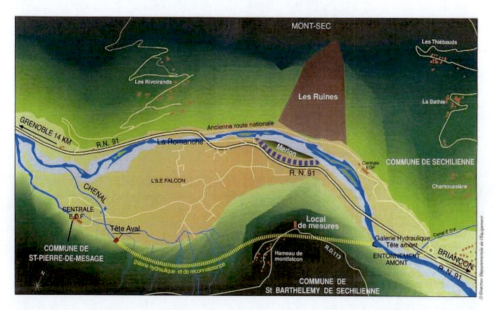

Figure 9.4.   Sketch of the main features of the site of Les Ruines: see particularly the original trace of the RN 91 (ancienne route nationale) and the dam (Merlon).

9.2   REGIONAL FRAMEWORK

9.2.1   *Climatic and water conditions*

The annual rainfall in Grenoble is nearly 1 m per year. Average rainfall (and snowfall) on Mont Sec can be estimated at about 1200 mm per year. High intensities occur mainly during the autumn rainfall, but water infiltration may be important during snow melting periods. No springs are present on the landslide slope.

9.2.2   *Regional morphology*

The higher peaks in the vicinity of Les Ruines are the Pic de l'Oeuilly (1500 m a.s.l.), north of the valley, and the Grand Serre (2140 m a.s.l.) to the south.

The Romanche valley has an average east-west orientation. It is a typical glacial valley originating near the Lautaret Pass. In front of Les Ruines, the valley is rather narrow (220 m) but enlarges
upstream (Séchilienne village) and downstream (L'Ile-Falcon, small village) (Fig. 9.5).

Figure 9.5.   General view of the Séchilienne landslide (from uphill). In the foreground, the small village of
Grand Serre in Séchilienne (see also Fig. 9.2).

9.2.3   Regional geology and structural setting

The landslide takes place in the external part of the crystalline Belledonne Massif in the French
Alps. It consists mainly of micaschists resulting from the metamorphism of old sedimentary
deposits (proterozoic or early paleozoic age). The main metamorphic episode is the hercynian one;
the alpine metamorphism is less pronounced (INTERREG 1996; Pothérat & Alfonsi, 2001).

Several tectonic episodes folded and faulted the massif during the hercynian and alpine tectonic
phases. The quaternary uplift of this part of the alpine chain is not yet complete.

The geological unit ("série satinée"), which includes the Séchilienne landslide, is bordered on the
western side by the Vizille fault and on the eastern side by the "median" syncline; both structures
have a N 20° azimuth, which is the elongation direction of the Belledonne Massif (Fig. 9.6).

9.3   HAZARD ANALYSIS

9.3.1   Geomorphological analysis

The Romanche River flows from east to west in the alluvial plain. Geophysical investigations performed in the valley in front of Les Ruines showed that the bedrock lies nearly 100 m below the
alluvial plain.

The south-facing slope of Les Ruines has an angle of 45° in its lower part and about 20° in its
upper part and near the crest. The main feature of the slope is the sagging of Mont Sec, bounded

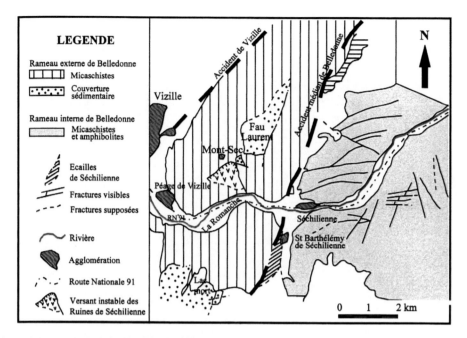

Figure 9.6. Geological sketch of the Séchilienne area.

by a 40–50 m high scarp with an elliptic shape. Several trenches N 70° cross the upper part of the landslide. The frontal zone (south-eastern part of the landslide) is presently wholly disrupted in its upper part, with deep fissures in the ground which are broadening and collapsing. Rockfalls and small debris flows run downhill from the frontal part (corridor of Les Ruines) and frequently reach the abandoned stretch of the RN 91.

Ancient quaternary glaciations (Riss and early Würm) covered, the Mont-Sec mountain and the present Romanche valley. During the Würm II period (−90 000 to −40 000 BP) the glacier covered the Mont-Sec mountain and filled the Romanche valley up to 1200 m a.s.l. During the Würm III period (−35 000 to −25 000 BP) the glacier was not higher than 600 m a.s.l., which is the level of the bottom of the moving zone.

9.3.2 Local geology and structural analysis

The metamorphic rocks (greenschist facies) that make up the slope are rather heterogeneous: micaschists (ancient pelitic rocks and sandstones) with variable quartz content producing variable resistance to weathering and erosion. To the north-east of Mont Sec, the metamorphic rocks are locally covered by carboniferous conglomerates or Mesozoic sediments.

Two main structural alpine directions are present at the site:

– strike-slip faults striking N 20–30° to the right, parallel to the "median" syncline,
– strike-slip faults striking N 120–140° to the left, conjugated to the preceding ones.

These faults divide the moving zone into four parts (Fig. 9.7). The geodetic measurements carried out over the past 20 years are fully coherent with these structural features.

The elliptic sagging of Mont Sec may be related to a cone-sheet structure due to a deep magmatic intrusion, possibly of Permian age (Fig. 9.8). This may be correlated to the radial and concentric filonian structures with sulphur mineralisation.

257

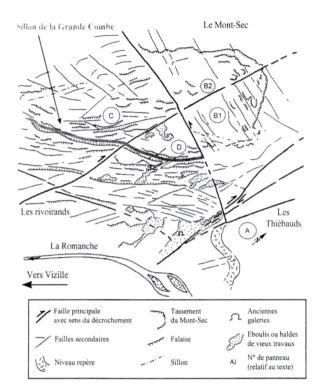

Figure 9.7. Structural sketch of the Séchilienne landslide. Zone A is stable. Zones B1, B2, C and D are formed by two conjugated faults. Thick lines: main faults. Thin lines: secondary faults.

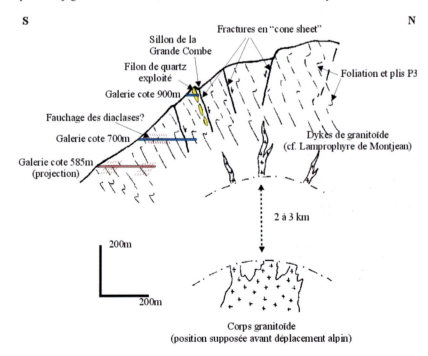

Figure 9.8. Geological section through the "sagging" of Mont Sec.

9.3.3 *Investigation and monitoring*

The geological survey included aerial photo interpretation, field observations, geological mapping, geophysical exploration and the boring of a 240 m long horizontal gallery (710 m a.s.l.). The cost of the gallery was about 380 k€.

Basic monitoring of the slope was set up by the Centre d'Etudes techniques de l'Équipement de Lyon (CETE) in 1984 (Evrard et al., 1990; Duranthon et al., 2003) and it was then progressively increased: wire extensometers, geodetic measurements of points on the slope and in the gallery, tiltmeters in the gallery, tacheometers and a new technique based on microwave radar (measures of distance from the opposite side of the valley: see Fig. 9.9), rain and snow gauges. Acoustic emission has been tested but did not give reliable results.

Some of these instruments are connected to an automated data recording and transmission system (33 extensometers, 54 distance measurements). Others are periodically surveyed, e.g. the geodetic measurements in the gallery.

The monitoring made it possible to define a zonation of the moving area (see § 9.3.4) and to make some inferences about the mechanism of deformation (see § 9.3.5.1): exclusion of simple

Figure 9.9.   Microwave radar located on the opposite slope.

259

sliding on a plane or circular surface for instance, proposal of a complex deformation including some kind of toppling.

The history of the velocities measured by extensometer A 13 is shown in Figure 9.10. One can see, apart from the dispersion of values due to the actual accuracy of the extensometer, the large seasonal variations and a trend to increasing mean annual velocities.

Geophysical prospecting (electrical and seismic) has recently been performed by Grenoble University. The results are not yet completely available but it seems that the electrical and seismic properties of the rock are quite changed by the deformation process, in comparison to those of the intact formations.

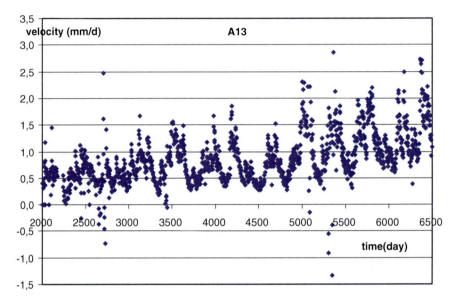

Figure 9.10. Rate of deformation of the A 13 extensometer (mm/day), located in the frontal zone. Day 2000 = 1990/12/05.

9.3.4 Danger identification

The hazard analysis is based on the existence of moving zones of various dimensions and velocities:

- the most active frontal zone (about $3 \, hm^3$ with velocity of 0.15 to 1 m/y) periodically releases rockfalls ($10–100 \, m^3$) through Les Ruines; larger rockfalls (10^4 to more than $10^6 \, m^3$) are expected to occur over the next few years,
- an active volume of $20–25 \, hm^3$ (0.05–0.15 m/y),
- a large slowly moving zone (the ancient sagging): 0.02 to 0.04 m/y.

An important question is to know the volume of the frontal mass which could be destabilised in the short term. Many estimates have been made, ranging from 2 to $5 \, hm^3$. In order to obtain the best estimate, it is assumed that the frontal zone is limited by a diamond-shaped polygon on the surface of the slope and by a basal surface dipping 30° towards the valley. This results in a volume of $3 \, hm^3$.

This volume is used as a basis in the short-term reference scenario for the local authorities.

9.3.5 Geomechanical modelling

9.3.5.1 Mechanical triggering model
Monitoring of the slope and of the survey gallery showed some toppling movement of large rock masses. Numerical modelling with the distinct element method (DEM) yielded some interesting

results despite the simplified 2D geometry and structure (only two families of discontinuities). It was shown that glacial melting induces slope deformation consisting of toppling in the middle of the slope, bulging of the lower part of the slope and sagging of the upper part, creating a major scarp at the top of the slope (Fig. 9.11).

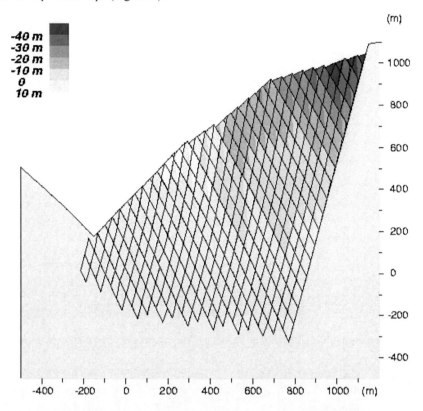

Figure 9.11.   Vertical displacement as given by a DEM numerical analysis (UDEC code).

There is no need of a basal failure surface to obtain such deformations, which reflect the post-glacial evolution of the slope fairly well. But the reactivation of the deformation observed since about 1950 is not explained by this type of model, which stabilises after an equilibrium deformed state has been reached; new processes (weathering and fracture growth) must be added to the model to obtain the recent evolution. The present deformation combines some toppling of the frontal zone (increasing from west to east), which induces the deformation of the sagged upper part. The geodetic measurements in the survey gallery confirm the toppling movement of large rock masses separated by weathered clayey joints (Fig. 9.12). One can notice that the end of the gallery is not in the stable rock mass (it is moving slowly).

An empirical hydrogeological model has been proposed which makes it possible to find a relationship between the velocities of the frontal zone of day n and the water input (net rainfall and snow melting) of days $n - 1, n - 2$, etc. including 40 days of data. Recently, however, the reaction of the moving mass to precipitation seems to be quicker, probably due to the deformation process and increase in permeability.

9.3.5.2   *Mechanical run-out models*
Various numerical analyses have been performed with different hypotheses for the unstable volume. Due to the narrowness of the valley at the foot of Les Ruines, the river may be dammed if a 3 hm^3 failure volume is assumed (it may be noticed that the estimated volume of the frontal mass is about

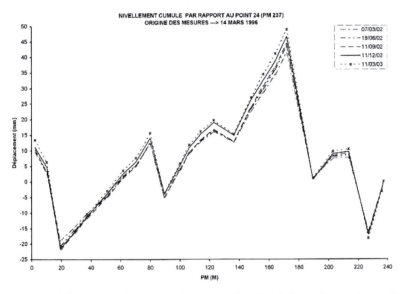

Figure 9.12. Vertical displacement (mm) along the survey gallery in relation to the end of the gallery (point 237 m on the right).

$3\,hm^3$, i.e. around the threshold between damming and no damming of the Romanche valley!). Figure 9.13 shows the result of one run-out model, supposing that $3.2\,hm^3$ ($4\,hm^3$ in the debris cone due to dilatancy) falls down in one step, a condition which strongly influences the shape of debris because of the increasing energy dissipation with volume (cf. the classical Heim or Scheidegger diagram of "Fahrböschung" versus volume); however, the model takes the progressive

Figure 9.13. Extension of the debris in the case of $3.2\,hm^3$ failure volume (from R&R consultants): the valley is assumed to be completely dammed. Topography: IGN map.

262

change of topography during the spreading process into account. For larger volumes, the steep opposite slope will induce the diversion of the debris upstream in the Romanche valley (towards the village of Séchilienne) and downstream (towards Ile-Falcon).

Figure 9.14 shows another result for the spreading of the debris using a digital elevation model and a mechanical process with N = 1000 pieces of rock sliding one after the other, given a law of energy dissipation.

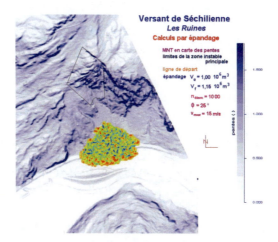

Figure 9.14. Debris cone resulting from a 1 hm^3 volume i.e., 1.15 hm^3 of debris simulation (courtesy of J.-F. Serratrice). It can be seen that the debris cone does not reach the earth protection dam in this case.

9.3.6   Occurrence probability

9.3.6.1   Scenarios of rupture of the slope

Different scenarios have been considered over the last years, including catastrophic failure of the whole landslide (nearly 100 hm^3). Two main groups of scenarios (in relation to the volume of rock involved) may be put forward at present, according to whether one considers the short term or the middle – long term (Panet et al., 2000). These groups have been defined in consideration of present scientific knowledge. They are based on physical considerations, such as measured velocities and opening of fractures.

— *Group 1* (short term i.e. <10 years): toppling and falls originating from the frontal zone; many rockfalls; failure of the whole frontal zone (about 3 hm^3), probably in several steps. There are two main possibilities: continuation of the current behaviour (rockfalls of a few cubic meters to hundreds or thousands of cubic meters) or a significant rockslide involving the whole fast-moving zone: the volume could be several million m^3 (2 to 3 in the more recent evaluation).
— *Group 2* (middle term: 10 to 50 years): the entire high or moderate velocity zone collapses. The volume could be 20 to 25 million cubic meters. Of course, intermediate scenarios could also occur.

9.3.6.2   Consequences according to the volume of the rockslide

In the same way as for the definition of the two groups of slope failure mechanisms, scenarios of the failure consequences may be defined according to the influence of the fall of debris in the bottom of the valley:

— *Scenario 1*: rockfalls of limited volume (e.g., less than 1 hm^3) do not dam the valley. The river will use the emergency riverbed. The debris cone does not reach or hardly reaches the RN 91.
— *Scenario 2*: a significant rockfall (3 hm^3) creates a small dam through the valley and a lake upstream. The dam is not very high (about 10 m) and the storage of water in the lake is about 200 000 m^3. A moderate volume of materials covers the RN 91.

263

— *Scenario 3*: a large rockslide occurs (about 5 hm³) and creates a large dam (approximately 15 to 20 m). The water storage is about 3 million cubic meters. The filling rate of the reservoir depends heavily on the discharge of the river, and may last between one or two days and a few hours; the upstream village of Séchilienne is partly flooded. The overtopping of the dam can cause its own destruction and then generate a dramatic flood which reaches the town of Vizille in about ten minutes.

— *Scenario 4*: a catastrophic rockslide occurs (10–25 hm³) and creates a very high dam (40 to 50 m). The reservoir (water storage between 10 hm³ and 20 hm³ of water) is filled within a few days (less than a day if it occurs during a rising period of the river). The village of Séchilienne is flooded. The overtopping of the dam and its destruction generate a dramatic flood, which reaches Vizille in about ten minutes and the industrial suburbs of Grenoble in 30–40 minutes.

Scenarios 1 and 2 are linked to the Group 1 slope behaviour. Scenarios 3 and 4 are related to Group 2. The occurrence of Scenarios 1 and 2 is considered to be probable in the next 10 years. Taking account of the current evolution of the slope, the occurrence of Scenario 3 is not considered to be realistic before about 10 years. It does not seem possible to define probabilities of failure for Scenario 4 because it is very difficult to get a precise idea of the morphology and of the hazard degree after a first rockslide involving around 3 hm³.

The possibility of windblast due to the rock avalanche is controversial; in any case it would only be significant in the case of Scenario 4.

9.3.6.3 *Scenarios related to the planning of the prevention measures*

The authorities in charge of the emergency planning have to make a distinction between the following: closure or not of the RN 91, flooding or not of the plain upstream, flooding or not of Vizille, etc.

Many studies have been devoted to this subject. In particular, evaluations of the risk of flood downstream have been carried out with three values of water storage (3 million, 9 million and 20 million m³).

From the knowledge of the slope and the results of these studies, various scenarios were defined. These scenarios make it possible to plan the actions of the authorities, either preventive ones (i.e., land use control) or curative ones (emergency plans). For each scenario, the authorities take into account a reasonably elevated evaluation of the risk, to guarantee a high level of safety, while not spending financial resources in a useless or premature way.

9.4   QUANTITATIVE RISK ANALYSIS

9.4.1   *Elements at risk*

The main elements at risk are, first, those affected by the run-out of the debris (hypothesis of 4 hm³ or more): the RN 91 (nearly 10'000 vehicles/day, more than 20'000 during the winter holidays), the small village of Ile-Falcon (initially 90 houses), a paper factory (initially 51 employees) and a small electric power plant. Considering the importance of the risk to the people, in 1997, the French State decided to purchase all the installations directly threatened by the rockfalls. This operation is about to be completed at present (2003).

Other zones are exposed to secondary (indirect) phenomena, in the case of Scenarios 3 or 4:

– In the case of high damming and rising water behind the natural dam formed by the debris: flooding of the major part of the village of Séchilienne (670 inhabitants),
– In the case of overtopping and rapid erosion of the dam: flooding downstream in the Romanche valley affecting the town of Vizille and its surroundings (10'000 inhabitants) and chemical industries near Grenoble, etc.

Economic consequences could be very important if the road traffic is blocked because no other route of reasonable capacity exists during the winter. About 10'000 people live in the upper valley. The winter tourist activity of the ski resorts in the upper valley (Les Deux-Alpes, l'Alpe-d'Huez, etc.: 80'000 beds) provides an essential income to the regional economy.

There are also large economic consequences according to Scenarios 3 and 4 for the industrial facilities located downstream (chemical industries, electric power plant, etc.).

The environmental consequences could also be very heavy: 4 km downstream of Les Ruines there is a drinking water supply zone serving 200'000 people. This pumping area would be flooded in the case of Scenarios 3 and 4. There are also chemical industries downstream: these Seveso type industries could be hit by the flood in Scenario 4 and this could have a huge environmental impact.

According to the given scenarios, the following consequences may be considered:

Scenario 1: SMALL ROCKFALL (1 000 to 1 000 000 m^3)
This scenario may actually be considered as a non-event: it would (probably) be a first step – though already significant – which would not result in real consequences for property, as described in Table 9.1.

Table 9.1.   Consequences of Scenario 1.

Romanche River	Cone scree reaches the river, likely locally diverting the flow towards the earth dam, inducing a beginning of diversion by the new bed. Pollution (suspended matter) of the Romanche River.
RN 91	Fall of blocks, temporary closure (several days); no significant damage.
Consequences downstream and upstream	Dust. Closure of short duration → economic impact on the ski resort of Oisans (80'000 tourist beds).

Scenario 2: ROCKFALL (3 million m^3) – SMALL DAM
Scenario resulting in a cone scree totally sealing the bed of the Romanche River and damming the valley at a moderate height. A lake of roughly 200'000 m^3 of water would form, as described in Table 9.2 (Fig. 9.15).

Table 9.2.   Consequences of Scenario 2.

Romanche River	Temporary cut, diversion and divagation in Ile-Falcon, pollution (suspended matter).
RN 91	Road covered and destroyed (along about 80 m) by the rockfall. Closure of long duration (several months): significant economic impact on the ski resort of Oisans.
Inhabited zones upstream	Dust, etc. Possible evacuation during the crisis for the closest areas. No significant consequence.
Nearby downstream (1–2 km)	Approximately 1 km of road destroyed and a bridge damaged. Divagation of the Romanche river flowing out of the failure point in the dam.
Water supply of Jouchy	Possible effects on drinking water for 200'000 inhabitants.
Side dams in Romanche	Possible localised degradations, without overflow.
Inhabited zones downstream (2–5 km)	Dust, etc. Temporary increase in the flow of Romanche River, without significant consequence.
Other consequences	Possible damage to a small hydroelectric power plant. Damage to local power lines.

Scenario 3: ROCKSLIDE (5 million m^3) – MEDIUM SIZED DAM
Scenario with significant hydraulic risk of flood downstream (city of Vizille). Formation of an already high dam, with the creation of a significant lake (approximately 3 million m^3), as described in Table 9.3.

Note: The diversion tunnel now makes it possible to be protected against a failure of the dam if the discharge of the Romanche River remains low.

265

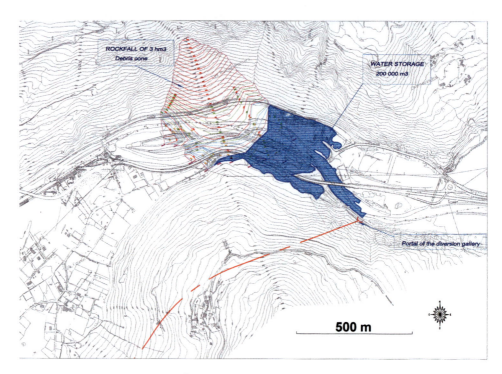

Figure 9.15. Lake formed in a 3 hm³ rockfall hypothesis. See also the debris cone across the valley (see § 9.3.5.2), overflowing the catch dam, and the location of the eastern portal of the diversion gallery.

Table 9.3. Consequences of Scenario 3 (in case of dam failure).

Romanche River	Temporary cut, diversion and divagation in Ile-Falcon, important pollution (suspended solids).
RN 91	Road covered by the debris: about 150–200 m long and 15 m high. Closure of long duration (several months): significant economic impact on the ski resorts of the Oisans area.
Inhabited zones upstream	Dust, etc. Evacuation during the crisis (several weeks) for the closest sectors. Possibility of flood for some of the houses located at the lowest level.
Immediate downstream (1–2 km)	Approximately 1 km of road destroyed and a bridge destroyed. Divagation of the Romanche River, damping out the peak flow following the dam failure.
Water supply area of Jouchy	Pollution of drinking water for 200'000 inhabitants. Partial destruction of the pumping wells and pipes.
Side dams in Romanche River	Overflowed, destruction (4–5 km).
Inhabited zones downstream (2–5 km)	Dust, etc. Evacuation during the crisis (several weeks) for the closest areas: 5'000 to 15'000 inhabitants affected. Damage to numerous buildings (several hundred). Damage to roads, bridges (destruction) within a radius of 20 km (local or national level).
Other consequences	Possible damage to several hydroelectric power plants. Damage to local and regional power lines and other electric installations (30'000 inhabitants affected). Flood of chemical factories, risk of chemical pollution, cessation of activity for a long duration.

Scenario 4: ROCK AVALANCHE (20 million m^3) – HIGH DAM

The consequences of the failure of the dam are assumed to be very significant, not only within the close downstream valley (areas of Vizille and Jarrie), but also in the urban area of Grenoble. This means that more than 100'000 people would undergo damage due to a significant flood. Buildings, chemical factories, electric installations would be damaged or destroyed. Roads and motorways of national importance and railways would also be destroyed or damaged. The importance of the zone in question makes it difficult to define an exhaustive list of the potential consequences of this scenario.

This scenario highlights very heavy consequences. However, it is currently not considered to be possible for a long time (several decades), because of the need of undergoing several stages of deformation in the slope before it could occur.

9.4.2 *Vulnerability of the elements at risk and expected consequences*

Table 9.4 roughly shows the vulnerability of the main exposed elements for the various scenarios, as far as direct and immediate effects are concerned. The scale that is used is a relative scale and ranges from 0 to 100 (conventionally the value of 100 has been attributed to the catastrophic consequences of Scenario 4 in the Grenoble area). Finally, the table presents the situation as it was five years ago: today some preventive actions have already been implemented (see § 9.5).

Table 9.4. Vulnerability of elements at risk.

Element	Scenario 1	Scenario 2	Scenario 3	Scenario 4
National road RN 91 and other roads, railway	0,05	0,2	1	5
Upstream economic consequences	0,1	0,8	10	20
Upstream local consequences (flood)	0,05	0,2	0,5	5
Small village of Ile-Falcon (94 houses, 1 school, 1 restaurant)	0	0,5	2	2
Area of Vizille and surroundings	0	0,1	1	5
Local factories (a paper factory and a small hydroelectric power plant)	0	0,05	1	5
Water supply	0	0,2	1	5
Grenoble and surroundings	0	0	1	100

9.4.3 *Risk assessment*

No precise evaluation of the total cost of the scenarios has been made. One can roughly evaluate the cost of Scenario 3 at several hundreds of millions of euros.

One should also add that the social and economic impact already exists: the development of the Séchilienne village has been stopped and it is also probable that the chemical industries of Claix and Jarrie downstream in the Romanche valley are no longer investing at their sites. Moreover, large expenses have been devoted over the past twenty years to investigation and monitoring of the landslide.

9.5 REGULATION AND RISK MITIGATION MEASURES ALREADY AVAILABLE

Taking account of:

– the volume of rock material involved,
– the jointed nature of the rock mass,
– the hydrogeological complexity,

267

no stabilisation technique seems to be either technically or economically feasible. We can mention that even a solution of massive blasting has been assessed.

The new alignment of the RN 90 (which was initially located on the right side of the Romanche River) was opened in 1986; it has been diverted to the left side of the river and two new bridges had to be built. A diversion channel for the Romanche River has been prepared and is protected by a catch dam. It is supposed that these protective measures would be efficient for a failure situation involving not more than 1 or 2 hm^3 of rock.

The houses in Ile-Falcon have been removed and people (200) relocated in safe areas (Fig. 9.16). A factory has been closed. The government has purchased all these buildings from the private owners (cost: roughly 20 M€).

Figure 9.16.   Ile-Falcon viewed from the frontal zone of Les Ruines.

A 2 km long exploration tunnel (Effendiantz et al., 2000) has recently been completed (10 M€); it was driven on the left side of the Romanche valley (see Figure 9.4). It could be used in case of catastrophic failure by the Romanche River but the discharge capacity (60 m^3/s) is only about a third of the maximum annual discharge. The need for a diversion suited to the floods of the Romanche River appears at the present time only for the scenarios likely to occur in the long run. One of the aims of monitoring is to determine the time when the implementing of this project would have to be decided.

The first emergency plan was worked out in 1993. A second version was prepared in 1999, corresponding to Scenario 3. Because of the recent analyses of the slope stability, and of the need for having an action plan adapted to the short term risks, a new version has recently been prepared. It corresponds to Scenarios 1 or 2.

Real time monitoring is a major component of this plan. Data from extensometers, distancemeters and rain and snow gauges are transmitted every 2 hours to the CETE in Lyon. In case of alert, safety measures would be progressively taken: closure of RN 91, call for the on site presence of experts, evacuation of people from the most dangerous zones, information for people living downstream, etc. The cost of monitoring is evaluated at 400 k€ per year (equipment value: 300–400 k€).

In the case of the occurrence of Scenario 1 or 2, the RN 91 would have to be reopened, probably after some clearing of debris and protective works. The main question would be to provide safe conditions for the workers and then to decide if the residual danger of rockfalls is over.

9.6 CONCLUSIONS

The Séchilienne and the La Clapière (Alpes-Maritimes) landslides are the two major active land-slides in France. At both sites, the landslide risk has been efficiently managed for more than s25 years, in spite of all imaginable types of difficulties. This has represented a scientific challenge as much as a political and financial headache...

Our knowledge of these large slope deformations is rather poor, therefore, there is no certain prediction of the future evolution of the system: the best approach is the construction of scenarios, the assessment and evaluation of these scenarios and their continuous updating in order that the local and regional authorities may take appropriate preventive measures and organise the pre-paredness in the manner most suited to the situation.

The history of Les Ruines de Séchilienne has not yet come to its end. The scientific experts have sometimes been mistaken, but the observations and data collected from the site allow continual understanding of the phenomenon and our ability to define the most probable scenarios, and therefore to prepare the mitigation decisions which are the responsibility of the political authorities.

10
Specific situations in other contexts

10.1   ROCKSLIDE HAZARD MANAGEMENT: CANADIAN EXPERIENCE

Oldrich Hungr
University of British Columbia, Vancouver, Canada

ABSTRACT: Over the last 30 years a number of hazard and risk studies involving large rock-slides have been completed in Western Canada. Several examples are summarised. Probably the most detailed hazard studies have been completed by BC Hydro, the publicly-owned electricity producer in the Province of British Columbia. Some hazard studies have also been commissioned by government regulatory agencies. No standard methodology can as yet be recommended for such work. Each case must be investigated on its own merits. Detailed geological work and moni-toring is an important part of hazard studies. Recent developments in run-out assessment tech-niques will be a useful additional tool.

10.1.1   *Introduction*

Some 1.2 million km^2 of Canadian territory, an area equal to more than half of Western Europe, contains mountainous terrain with sustained local relief in excess of 1000 m. Glacial erosion during the Holocene and Early Pleistocene left deep valleys and oversteepened mountain slopes. Large rockslides are relatively common in this area. Surprisingly, damage due to large rockslides is not a common occurrence. Evans (2003) – see references at the end of the book – reports that 238 lives have been lost to rockslides and rockfall in all of Canadian history, more than half of which occurred in only two events (the 1903 Frank Slide in Alberta and the 1915 slide at Jane camp, near Vancouver). The probable reason for this rather limited mortality is the remoteness and low population density of the mountain lands. The Pandemonium Creek rock avalanche (Evans et al., 1989) is an example that illustrates the point. The rock avalanche probably occurred in 1958 in a mountainous area approximately 300 km north of Vancouver. Several million of m^3 of rock failed from a cirque head-wall, passed over a glacier and traveled 11 km along a mountain valley, devastating some 8 km^2 of area (Fig. 10.1.1). Not only were there no fatalities, but the landslide was not even noticed by anyone! The extensive damage to the natural landscape was first identified on air photos only in the 1970's.

Nevertheless, the potential for damage is now increasing with growing population and increas-ing development of the area. Future damage reports could be considerably more serious than those in the past. Government and private agencies responsible for development regulation are aware of this and, more and more frequently, call on engineers and geologists to assess the hazards and recommend suitable remedial measures. This article summarises some examples of this process.

10.1.2   *Examples*

10.1.2.1   *Large rockslides on hydroelectric dam reservoirs*
B.C. Hydro, a crown corporation of the British Columbia Provincial Government, owns 60 hydro-electric dams. The reservoirs behind these dams form thousands of km of bank slopes in mountainous

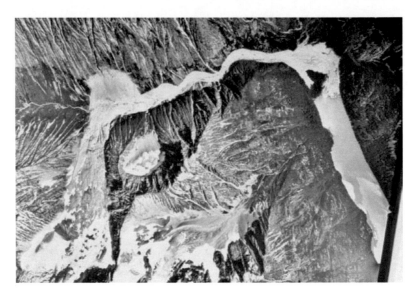

Figure 10.1.1.   Aerial view of the Pandemonium Creek rock avalanche in British Columbia. The landslide path is 11 km long.

terrain. Ever since the Vaiont disaster of 1963, there has been awareness of the danger posed to dams by large landslides displacing reservoir water. In the case of the dams on the Columbia River, for example, a landslide-generated displacement wave could potentially destroy a large earth or rockfill dam, generating a cascade of dam failures along a river corridor 1000 km long, reaching into the territory of the United States. Consequently, B.C. Hydro operates an intensive program of investigation and monitoring of reservoir slopes. Over the last 30 years, more than 2000 km of slopes have been subjected to systematic landslide hazard evaluations by B.C. Hydro and its consultants (Imrie and Moore, 1997).

Some very large specific rockslides have been studied in depth and subjected to remedial construction. The first of these, the Downie Slide, situated on the Columbia River 70 km upstream of Revelstoke Dam, is one of the largest recognized landslides in Canada. It involves a mass of $1.4 \times 10^9 \, \text{m}^3$ of metamorphic rock, situated above a fault dipping at about 20° towards the reservoir and more than 200 m beneath the slope surface (Imrie et al., 1992). The rockslide underwent a displacement of about 100 m in prehistoric time, but was in a dormant state until disturbed by water impoundment. During filling of the reservoir, a part of the slide mass began to move at rates of up to 100 mm/year. Although such slow movement posed no direct danger to the dam, there was concern about a Vaiont-like sudden displacement. B.C. Hydro carried out a detailed investigation with drill holes, testing and detailed suface mapping and then designed a dewatering system with a tunnel and fans of horizontal and inclined drainage holes. The total cost of the drainage system was 30 million $CDN. The drainage decreased the slope movement rates to about 10 mm/year. Design changes were implemented at the dam, in order to accommodate a displacement wave, predicted by physical modeling. Full time remote monitoring of displacement rates and piezometric levels is continuing.

Two somewhat smaller rockslides were identified by systematic mapping a short distance above the Mica Dam on the same river. Both were subjected to detailed investigations during the 1990's. In one case, the drilling information showed no adversely-oriented structure in the slope and the slide is believed to be benign. In the second case, called the Dutchman's Ridge landslide, a shear zone dipping 29° downslope was found (Moore and Imrie, 1992). The slope was drained using a drainage system similar to that of the Downie Slide, at a similar cost.

The small Wahleach power plant near Chilliwack, British Columbia, was in operation for 40 years, when cracks and buckling of an underground penstock were discovered during a routine

272

inspection. Subsequent detailed investigations showed that massive toppling of the granitic slope, separated by closely-spaced anaclinal joints, was occurring (Moore et al., 1991). An overall failure of the slope would endanger the powerhouse and enter the floodplain of the Fraser River, destroying a major highway, railway and pipeline corridor. The slope movement rates were reduced by improving drainage, by means of relocation of the power tunnel and penstock.

In each of the above cases, the slow slope movements observed were not of significant direct consequence. The main concern was about the possibility of a brittle catastrophic failure, endangering the area at the foot of the slope. The responsible authorities and their review boards concluded that there was insufficient proof of the absence of potential for such catastrophic detachment. Consequently, remedial measures were implemented at very considerable costs.

10.1.2.2 *Cheekye River valley, Mt. Garibaldi*

Mt. Garibaldi is a dormant volcano located approximately 100 km north of Vancouver. Following the last period of volcanism at the end of the Pleistocene glaciation, the stratovolcanic edifice was deeply dissected by erosion. The most prominent of the erosion features is the valley of the Cheekye River, whose headwall cuts nearly to the summit of the 2600 m high mountain (Fig. 10.1.2). At the outlet of the valley, only some 50 m a.s.l., there is a large recent fan built of sediment transported by the Cheekye River. Both ridges of the Cheekye Valley exhibit extensive sets of tension features, consisting of series of normal scarps and tension cracks (Fig. 10.1.3). Both the ridges and the high headwall are composed partly of uncemented volcanic breccias of Pleistocene age and partly of foliated and deeply altered metamorphic basement rock. Small and medium-scale rock instabilities are very common on the upper slopes of the valley and several small debris flows have occurred on the Cheekye River during the last 100 years.

The Cheekye Fan, with a surface area of about 10 km^2, is a very desirable area for urban development, being close to the rapidly growing community of Squamish. During the 1970's, several major developments, including a hospital, were being planned here. However, excavations revealed that the fan contains deposits of large debris flows, originating in the Cheekye Valley. A study of

Figure 10.1.2.   The headwall of Cheekye valley, showing a cross-section of Quaternary volcanics in the central part of the Mt. Garibaldi stratovolcano.

Figure 10.1.3.   A zone of cracks and scarps on the south ridge of the Cheekye valley. The valley slope is visible in the upper left corner of the nearly-vertical photograph.

landslide fan hazards on the fan was commissioned by the British Columbia Ministry of Environment in the early 1990's (Hungr and Rawlings, 1995). This was probably the most detailed quantitative study of landslide risk carried out in Canada. It was based on the assumption that the large prehistoric debris flows were generated by landslides originating from the slopes of the Cheekye Valley.

The study comprised two parts. The first was a detailed surface and geophysical investigation of the two ridges above the Cheekye Valley. Stability analyses based on the data collected were used to predict the likely maximum volume of detachment corresponding to the extent of the cracked zones. The second part was a detailed stratigraphic study of the Cheekye Fan deposits to establish the magnitude and frequency of past debris flows.

The stability investigation in the headwaters showed that the cracking was caused by an incipient translational failure of the volcanic breccia deposits over a zone of heavily altered basement rock, just below the gently sloping contact between the two units. It was concluded that a sufficiently large volume could detach during a future failure retrogression, with the possibility of sending several million m^3 of rock into the valley. Damming of the Cheekye stream and subsequent erosion of alluvial and colluvial deposits from the valley floor and sides could generate a large debris flow in the river. However, it was not possible to predict the probability of occurrence of such an event, especially because the tension features on the ridge were found to be currently inactive.

The stratigraphic study on the fan, using some 60 excavated test pits, revealed the presence of several large debris flow deposits, interbedded with fluvial gravels. Two of these could be dated using radiocarbon dating of organic remains buried in the debris. The largest deposit, called the Surface Diamicton, lies on the surface of the fan. It consists of an unsorted mixture of silty sand, gravel and boulders, similar in texture and mineralogy to the pyroclastic breccias of the ridges. Its total volume is about 7 million m^3 and its date of placement ranges between 1100 and 1300 years BP. A similar, although slightly smaller debris flow occurred about 5000 years ago. A compilation of the available data, combined with the estimate of total deposition volume during the last 5000 years, allowed the construction of a cumulative frequency-magnitude (CFM) curve (Sobkowicz et al., 1995).

Thus, the two parts of the study were found to be mutually compatible and the CFM curve was accepted as a basis for a quantitative hazard and risk assessment, assuming that the past behaviour was likely to extend into the future. Using the curve, a prediction of the probability of occurrence of three magnitude classes of debris flows was made. With the guidance of the identified deposits, a run-out analysis of debris flows between the fan apex and the hazard area was conducted. Based on this, a hazard intensity map was constructed as shown in Figure 10.1.4. Each of the hazard zones on the map is associated with estimated hazard intensity parameters (flow velocity, thickness of deposits) for each of the three classes of debris flow magnitude. This map in itself can be used by the planning authorities to determine whether a certain level of development is suitable or not.

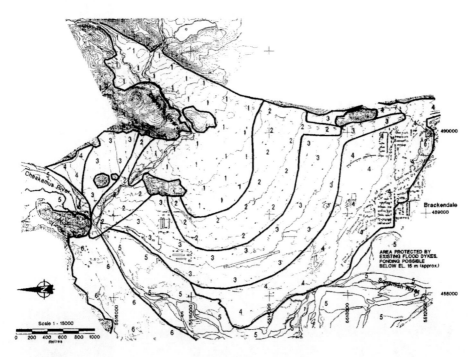

Figure 10.1.4.   Hazard zones as identified on the Cheekye Fan below Mt. Garibaldi (Sobkowicz, Hungr and Moyan, 1995).

Further analyses were carried out, assuming vulnerability of structures and inhabitants and calculating specific and total risks (Sobkowicz et al., 1995).

10.1.2.3   *Hope Slide*

The Hope Slide, which occurred in 1964 in the Coast Ranges, 200 km east of Vancouver, contained 60 million m^3 of metamorphic rock (Fig. 10.1.5). Cracks and linears existed on the crest of the slope prior to failure, but these were inactive and covered by forest (S.G. Evans, Geological Survey of Canada, pers. comm.). As a result, no warning of the catastrophic detachment was available and four travelers on a regional highway lost their lives in the slide.

The right flank of the slide contains a large mass of rock which moved several meters during the slide, opening large tension cracks, but did not accelerate and remained on the slope (Mathews and McTaggart, 1978). The British Columbia Ministry of Highways carried out a detailed investigation of this remnant mass. Run-out studies were conducted using a fragmental rockfall model. The highway was relocated to the opposite side of the valley based on the results.

10.1.2.4   *Frank Slide*

The Frank Slide of 1903 destroyed a part of the village of Frank in the Southern Canadian Rocky Mountains, causing the deaths of more than 70 people (McConnell and Brock, 1904) (Fig. 10.1.6).

Figure 10.1.5.   The source area of the 1964 Hope Slide. The disturbed rock mass is visible in the upper left corner of the image.

Figure 10.1.6.   An overall picture of the 1903 Frank Slide in Southern Alberta. A disturbed rock mass exists near the peak of the Turtle Mountain, in the upper left part of the picture (Photo courtesy Dr. D.M. Cruden, University of Alberta).

As in the case of Hope Slide, a portion of the limestone ridge east of the landslide source remained in a disturbed condition, with large open cracks. A study of the stability of this ridge was carried out recently and run-out estimates were made for a 5 million m^3 rock avalanche, using empirical techniques and a numerical Cellular Automata model (BGC, 2000). The latter model is a dry frictional model and is considered to represent a lower limit of possible run-out distance. Development constraints have been placed on the hazard area at the foot of the slope.

10.1.2.5   *Mt. Breakenridge*

Harrison Lake is an inland fjord located about 100 km north-east of Vancouver and occupying a deep glacial valley. The lake is about 40 km long. In 1992, an extensive area of slope disturbance was discovered on Mt. Breakenridge, above the northern part of the lake. (Fig. 10.1.7). Concern arose about the possibility of a catastrophic failure of the disturbed rock mass, causing a displacement wave and endangering a community at the south end of the lake, some 30 km away. A detailed surficial geological study was carried out. A parallel hydraulic study determined that danger to the community would arise only if the entire mass (200 million m^3) were to fail simultaneously. It was determined that the slide is a large flexural topple and that there is no large structural feature enabling an overall catastrophic failure of the unstable mass (Nichol and Hungr, 2002). Rapid failures are considered possible, but involving only parts of the slope. Such failures could create waves in the lake, but not of sufficient magnitude to endanger the distant community.

Figure 10.1.7.   The Mt. Breakenridge slope above Harrison Lake, British Columbia, disturbed by massive toppling.

10.1.3   *Conclusions*

Over the last 30 years, a body of experience with the management of large rockslide hazards has been gradually developing in Western Canada. It is, however, a difficult field and no general methodology can be recommended. Each case must be considered as a unique problem. In particular, the determination of probability of occurrence of a catastrophic failure remains exceedingly difficult. In a few cases, it was possible to discount the potential for a large-scale catastrophic detachment by means of identifying a non-brittle failure mechanism. Elsewhere, however, it was necessary to make the (sometimes conservative) assumption that a brittle failure could occur. Monitoring is an important part of risk reduction for large rockslides, although monitoring programs are not often adequately supported by public institutions. Corporations, such as B.C. Hydro, are an exception (Moore et al., 1991).

The estimation of possible failure consequences relies heavily on the prediction of run-out velocity and distance. None of the cases reviewed here could benefit from a proven technique of run-out prediction. Such techniques have only relatively recently become available and are currently being calibrated. (e.g. Hungr, 1995; Hungr and Evans, 1996; Hungr, 1997). Future hazard studies will benefit from these research advances.

10.2 LANDSLIDE RISK REDUCTION IN ECUADOR: FROM POLICY TO PRACTICE

P. Basabe

United Nations International Strategy for Disaster Reduction, Geneva, Switzerland

ABSTRACT: Within the framework of a multihazard project (PRECUPA) in the region of Cuenca City, in the south of Ecuador, geological and hydrometeorological hazards were studied, mapped and monitored between March 1994 and December 1998, including landslide, flood and earthquake assessments, real time network installation and the implementation of mitigation measures. One of the main components of the project included the detailed identification and mapping of landslides in inhabited or productive zones and the implementation of landslide risk reduction measures (non structural ones); such as the development of public awareness and community work, strengthening of local capacities, and encouragement of public commitment including the development and application of urban law and land-use planning.

10.2.1 *Introduction*

The location of Ecuador, a small country in South America located along a convergence zone of tectonic plates, has played a principal role in producing not only its beautiful geography, but also a high susceptibility to natural disasters such as earthquakes, tsunami and volcanic eruptions. Moreover, the influence of ocean currents exposes this country to intense hydrometeorological phenomena which cause landslides, debris flows and floods. These natural phenomena turn into hazards as population and activities increase without appropriate planning actions, causing a growing occupation of land in inappropriate zones, so that the vulnerability and risk factors continue to increase.

The economic impact of "natural disasters" in Ecuador is very important, as shown in Figure 10.2.1 for the period of 1980 to 2001. The gross domestic product (GDP) increase in percentage per year normally varies between 2.0 and 4.3; nevertheless, most of the GDP's decrease coincides with the occurrence of the main disasters having natural geneses in the same period. In 1982–83, the year of the El Niño phenomenon, the GDP decreased and reached a variation of 1.2% and −2.8%, respectively. In 1997–98 the impact of the "El Niño" phenomenon which caused floods, landslides and related damage was responsible for direct losses of 2'882 million dollars (CAF-CEPAL, 2000), i.e. 15% of the GNP, which is the highest value recorded in a South American country (Basabe, 1998). The GDP variation decreased to less than 2.0% in 1998 and even −7.0% in 1999.

In 1987 the decrease of the GDP (−6.0%) coincided with a mid-magnitude earthquake in the north-east of the country which severed the national oil pipeline into various parts. In 1988 the re-exportation of oil, including the 1987's quota, produced an abnormal growth of 10.5%. Finally, in 1993 the decrease in the GDP (from 4.3 to 2.0% in two years) coincided with the La Josefina large landslide which affected the third most productive region of the country, in the south of Ecuador.

This situation, and particularly the impact of La Josefina landslide, motivated the implementation of a pilot project in order to study and monitor natural hazards in a specific watershed as a geographic unit.

In the south of Ecuador, the main watersheds display loose rock formations which produce large landslides due to intense and long-lasting rainfall as well as to inappropriate human actions; thus, productive agricultural zones, inhabited areas and infrastructures are frequently affected. A serious event, called "La Josefina", occurred in March 1993, after a period of major rainfall, when a huge landslide blocked the Paute and Jadan valleys (Fig. 10.2.2), in the vicinity of Cuenca, the

279

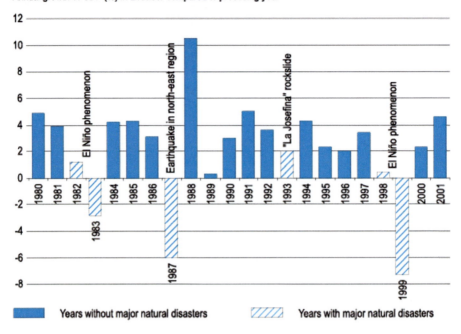

Annual growth of GDP (%) in Ecuador compared to preceding year

Years without major natural disasters   Years with major natural disasters

Source: PRECUPA/SDC project, Central Bank of Ecuador, 2002

Figure 10.2.1.   Comparison of "natural disaster" occurrence (floods related to El Niño phenomena in 1983 and 1998, earthquake in 1987 and La Josefina Landslide in 1993) and gross domestic product % change per year between 1980 and 2001.

Figure 10.2.2.   Major landslide "La Josefina" which took a heavy toll of some 80 persons and then caused intense destruction by flooding the valley upstream and downstream.

third major city of the country in the Southern Andes. Some 20 to 30 million m³ of rock which fell during the night produced the formation of a large lake causing extensive inundations upstream. After an emergency excavation of an outlet channel, some 33 days after the landslide, the huge natural dam was partially washed out by a regressive erosion mechanism, giving way to 170 million m³ of water in a few hours, which caused flooding and destruction down to 140 km downstream, in the Amazonian forest (Almeida et al., 1996).

As the United Nations had declared the 1990's the "International Decade for Natural Disaster Reduction", Switzerland answered the national and international plea by sponsoring the cited pilot project called "PRECUPA" which represents the abbreviation of "**Pre**vention-**E**cuador-**Cu**enca-**Pa**ute". This project gathered national, regional, local and university partners as Ecuadorian counterparts of Swiss Federal Institutes of Technology of Lausanne and Zürich, as well as Federal Institutes in a joint cooperation, through the support of the Swiss Humanitarian Aid and Disaster Relief Unit (now called Swiss Humanitarian Aid Unit – SHA), which is an entity of the Swiss Agency for Development and Cooperation (SDC).

10.2.2   PRECUPA Project aims and components

The objectives of the project included the assessment of the different natural hazards which could affect the high and central watershed of the Paute River, over an area of 3'700 km² (Fig. 10.2.3) in which some 700'000 inhabitants are living.

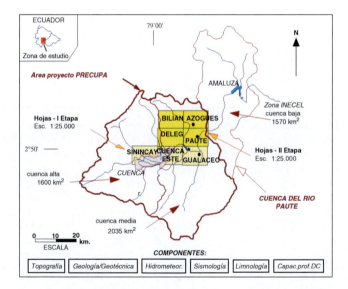

Figure 10.2.3.   Study zone of the PRECUPA project and location of areas where landslide and hazard maps at 1:25'000 scale were produced. The six components of the project are indicated below, as described in the text.

Moreover, this region is important because the lower part of the Paute watershed contains the major hydroelectric scheme of Ecuador which produced some 70% of the national electricity needs at the time of the La Josefina landslide (1993).

The PRECUPA project was developed in six components or fields of action (Fig. 10.2.3, lower part), following a systematic hazard assessment methodology (Basabe et al., 1998):

– Establishing complementary mapping and creating databases.
– Landslide identification and mapping, including rapid and slow onset as rockfall and debris flows.
– Landslide monitoring through geodetic networks for movement definition.

281

- Study and installation of real time hydrometeorological network, monitoring, data analysis and flood hazard identification and mapping.
- Study and installation of real time seismic network, monitoring, data analysis and seismic hazard identification and mapping.
- Monitoring of contamination of the residual lake which remained upstream of the slide after the natural dam failure.

These first five components required the planning, installation, operation and maintenance of several specific monitoring networks in order to obtain the data necessary for the identification and assessment of the corresponding hazards.

But the major aspect of the project included the development and application of non-structural disaster risk reduction measures. A large inter-institutional cooperation was developed with national, regional and local partners to create a multidisciplinary approach. Above all, primary importance was given to the professional training, in order to ensure the continuation of monitoring, analysis and implementation of disaster reduction measures at the end of the project. The measures also included Civil Defence strengthening, as well as the training of local and regional professionals operating in the application of hazard assessment and mapping within land use planning actions, as well as the promotion of regulations and laws as part of sustainable development and related programmes.

As the geodynamic phenomena represented the major hazard, the detection of landslides was extensive, including geological and geotechnical mapping, landslide and hazard maps as well as pilot studies on vulnerability aspects. This led to the production of seven complete maps at 1:25'000 scale which cover the most populated areas (Fig. 10.2.3).

Only the part of the project related to landslides, as well as the component of hazard management in land planning, will be developed in the following chapters.

10.2.3 *Assessment and mapping of landslide phenomena and hazards*

The stability of a slope is directly conditioned by the geological nature of the materials encountered, by their geomechanical behaviour as well as by the impact of external factors such as rainfall that causes the saturation of the ground, earthquakes and anthropic factors (Bonnard, 1994b). Thus, a slope will become unstable when the intrinsic preparatory causes combine with triggering causes, inducing a landslide phenomenon; even a small triggering cause may be sufficient to induce a marked acceleration of the instability phenomenon (Bonnard & al., 1995).

In the PRECUPA project, the most important preparatory causes were determined by the geological, morphological and geotechnical features of the slopes (presence of finely fractured marine siltstones, deep cuts in slopes due to rivers –canyons–, intense weathering and unfavourable dip of geological structures, leading to low geomechanical strength parameters).

The main triggering causes are the intense and long-lasting rainfall, as well as the anthropic actions such as deforestation, inappropriate cuts and fills for roads, quarry exploitation and undue use of soil.

Thus, 8 specific types of instability phenomena were identified, which can be grouped into three main categories (UNESCO, 1993):

- Slides, characterised by relatively slow movements, to which creep and superficial erosion can be associated.
- Rockfall, characterised by discontinuous fast movements developing mainly along planar discontinuities (dip, cracks), implying fresh or weathered rock.
- Flows, formed of mud or debris, which can develop at a high velocity along natural drainage channels.

10.2.3.1 *Methodology applied*
In order to follow a well-established methodology for the detection of landslide-prone areas, as well as to detect their level of hazard, the Swiss recommendations published by the Federal Offices

of Water Economy (OFEE), of Land Planning (OFAT) and of Environment, Forest and Landscape (OFEFP) were used (Lateltin, 1997; Loat & Petraschek, 1997). These recommendations were adapted to the local geological context for the PRECUPA project. Thus, some 28 maps at a 1:25'000 scale were produced, covering 900 km^2; in addition 20 specific landslide zones were studied and monitored, leading to maps at a 1:5'000 scale.

The methodology implied four phases:

1. Detection of landslide-prone areas and characterisation of rock outcrop and soil cover.
2. Improvement of available maps (especially 1:5'000 scale) and analysis of air and satellite photographs; comparison of photographs of different epochs in highly endangered zones; planning and installation of geodetic monitoring networks in most active zones (64 landmarks), including crack opening measurements (Fig. 10.2.4).

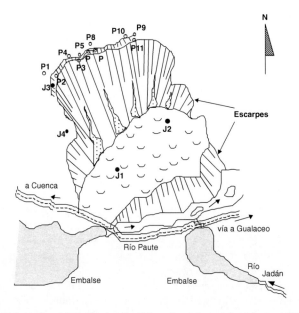

Figure 10.2.4. Location of geodetic landmarks [J] and crack opening devices [P] at the La Josefina Landslide, showing the residual lakes after the catastrophic outflow of May 1993.

3. Preparation of phenomena maps, including location of scarps, limits and accumulation zones, and determination of level of activity (active, dormant, relict) and of probable depth (Turner & Schuster, 1996). The main mechanisms are specified by a letter and the major phenomena studied in detail are identified by a code number.
4. Assessment of danger level depending on the preparatory and triggering factors, the observed level of activity, the intensity of the phenomenon, allowing the determination of the level of hazard in relation to the assessed probability and intensity (Fig. 10.2.5).

The reference values of probability and intensity obtained are introduced in standard diagrams which include both variables (Petraschek, 1995), and which were adapted for their use in the PRECUPA project.

10.2.3.2   Final product

The hazard map legend includes three colours as shown in Figure 10.2.5, as well as three letters for each zone. The first letter identifies the type of phenomenon; the second, as an index, indicates the probability of occurrence and the third, also as an index, states the intensity of the phenomenon.

283

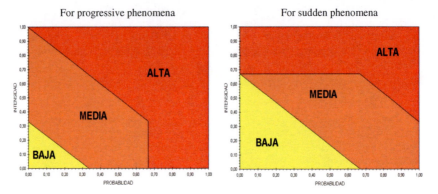

Figure 10.2.5. Assessment of level of hazard for landslides according to (Petraschek, 1995), adapted for the project, considering relative values of probability and intensity. "Baja" means "low". "Media" means "medium". "Alta" means "high".

Although this codification is qualitative, the values of probability and intensity are backed by characteristic values determined within the project which are based on quantitative conditioning or triggering parameters.

The index letters have the following meaning:

	Qualitative	Quantitative
A	High	$0.66 < x < 1.0$
M	Medium	$0.33 < x < 0.66$
B	Low	$0.00 < x < 0.33$

10.2.3.3   Vulnerability studies for geodynamic hazards

Looking at more significant parameters to introduce in natural disaster prevention studies, the PRECUPA project developed a methodology applying a vulnerability factor (Mora & Wahrson, 1994). It was used in particular in a case study in which the unstable zones were moving up to 84 m/year, affecting a rural zone of some 13 km^2 with a high hazard level, especially as it was in a process of urbanisation; this zone called Paccha is located 10 km east of the town of Cuenca.

For the computation of the vulnerability factor, a first assessment of the various hazard levels was carried out in the Paccha area (Fig. 10.2.7), dividing the landslide into six zones of potentially high vulnerability in which a population census was carried out, as well as a census of houses, economic activities and social conditions, with verifications in the field. Such an assessment allowed a better knowledge of the exposed zones which was expressed through four specific vulnerability factors, namely human, socio-economic, physical and preparation factors, the values of which vary from 0 (not vulnerable) to 1 (totally vulnerable). In the hazard map, these four factors, established by a standard characterisation, are shown as a dial (Fig. 10.2.6).

The human factor considers the level of knowledge of the population when facing an eventual hazard, considering that the landslide may be slow (as in the Paccha case) or violent (as in the La Josefina case), thus inducing a different level of impact. The socio-economic factor analyses the potential damage in the field, in terms of daily work carried out by the population, as well as by the impact on housing, health, education, employment, economic activities and standard of living. The physical factor considers potential damage to buildings and infrastructures, namely roads, electric lines, sewage and drinkable water systems, as well as schools, churches, health centres and houses. The preparation factor considers the level of knowledge of the population with respect to the Civil Defence programs and the preparatory activities that may save their lives, such as evacuation routes.

284

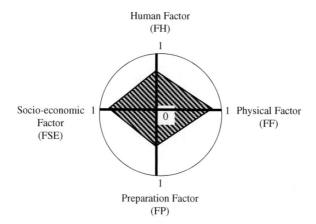

Figure 10.2.6. Graphical representation of the factors used in the vulnerability study.

Finally, it may sometimes be necessary to adopt a global vulnerability value with a weighted average, considering specific coefficients established according to the experience gained in the studied area. This global value was assessed as low when included between 0 and 0.4, as medium if it varied between 0.5 and 0.7 and as high when it reached a value between 0.8 and 1.0.

10.2.4   Landslide management and disaster risk reduction measures

10.2.4.1   Awareness raising and community work

The project took advantage of the traumatic situation and shock created by the La Josefina's impacts to raise awareness and contribute to the development of a culture of prevention. Every single result, main activity and training courses were accompanied with awareness campaigns through the media in the region. Bulletins of information, interviews in the radio and local television were common.

Every year the project delivered a detailed report to inform the community and authorities on the activities, results and progress made. The presentation of the report was organised by the municipality of Cuenca city by using the major meeting hall of the town. Authorities, local and regional governments and parliaments, national institutions, universities, scientific and social communities, as well as representatives of the neighbourhoods and of landslides prone areas were invited and informed in detail as appropriate. On the third year the project and its director were awarded with the "Honorific insignia VIRREY HURTADO DE MENDOZA for the high civil labor and technical assistance in the prevention of natural disaster project in Paute basin (PRECUPA), City of Cuenca Municipality and Parliament, Ecuador", 1997.

The first two years of the project were focussed in mapping and hazard assessment of landslides by covering 384 km^2 in 1994 and 512 km^2 in 1995 (Fig. 10.2.3). The communities located in main hazard prone areas were informed through a joint effort with Civil Defence and the Ministry of Education, regional branch. Specific booklets and comics (40'000 copies each) were created to inform the population.

The task of getting involved and not creating panic in the population is not easy. The project took advantage of a specific programme between Civil Defence (CD) and the Ministry of Education in which bachelor students need to develop a thesis or a social work on Civil Defence during one year. Every year some 800 hundred students were informed on landslide assessment and trained on CD and risk disaster reduction management. Every student had to apply her/his knowledge for the community. The work consisted to take care of 3 or 4 families living in the landslide prone areas to inform them about hazards and risk globally, as well as on landslide hazards, with emphasis on disaster risk reduction and CD measures. At the end of the social work the student delivered a report to the Ministry of Education and CD office who took care of the follow-up.

In the case of Paccha, the biggest landslide prone area $(13\,km^2)$ discovered (see § 10.2.3.3), the vulnerability assessment was based on the compilation of information through the census of population, housing, infrastructures, economic activities and social conditions (Fig. 10.2.7). The census was developed with volunteers of Civil Defence (CD) and the Ministry of Education, in conjunction with the social work, to inform the population on hazards, vulnerability, landslides and simple disaster risk reduction measures.

At the end of the project, the region of Paccha, which had been designated within the expansion urban zone of the city of Cuenca before the project, was redesignated as agricultural and recreational forest area. The same designation was done for flood prone areas along the streams; in this case the ordinance indicates longitudinal parks.

10.2.4.2 *Knowledge development*
One of the main goals of the project was to contribute to the development of local and national capacities, both by using national/local capacities in the execution of hazard mapping, assessment and monitoring and by training new professionals and future decision-makers, many of them were directly active in the project by secondments. The project also encouraged and used interns in all its components.

During the execution of the project, more than 80 technicians and professionals working in local or national institutions and interns participated and were trained in: techniques for the identification, mapping and monitoring of mass movements, use of satellite images, hazard and vulnerability analysis, community work, land-use planing and disaster risk reduction measures and management. Thus the gained know-how could lead to a practical application of the results at different levels (Basabe et al., 1998).

10.2.4.3 *Development of inter-institutional cooperation and application of results in disaster risk reduction measures*
From the beginning of the project, the development and implementation of disaster risk reduction measures were duly considered. Capacity building was therefore one of the main tasks; the project fostered both the participation of local and national institutions and professionals to encourage the inter-institutional co-operation and information sharing. This task in addition to knowledge development (see § 10.2.4.2) were the key factors to look after the implementation of the results in disaster risk reduction measures and the sustainability of the monitoring networks and related studies.

- National level

The studies carried out, as well as the maps, the GIS and the monitoring systems were handed over to the national institutions participating in the project, namely the National Direction of Civil Defence, the National Institute of Electrification and the National Meteorological and Hydrological Institute (INAMHI). These institutions have continued to use the results obtained and, in the case of INAMHI, to be part of the maintenance, data treatment and management of the hydrometeorological monitoring networks and forecast.

Furthermore, the information gathered was delivered to higher education centres and to the National Planning Council, now called the National Planning Secretary of the Presidency of the Republic (ODEPLAN). An agreement was signed with this institution so that the information and maps obtained could be used and the risk factor included in the development agenda and planning. Presently, ODEPLAN leads a ministerial and institutional working group in order to carry out a national plan for disaster prevention and risk reduction.

- Regional level

The studies carried out by PRECUPA project allowed the development of other projects in the region of Cuenca, in particular the one, financed by the European Union, which was set forth by the institution created for the rehabilitation of the zone affected by the La Josefina Landslide and which used the results obtained by the PRECUPA Project. Thus, the destroyed road was rebuilt,

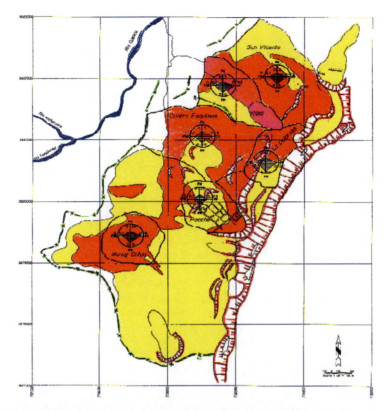

Figure 10.2.7.   Landslide-prone area hazard and vulnerability map in the Paccha zone.

considering the landslide phenomena and hazard maps and recommendations produced, so as to connect the cities of Cuenca and Azogues, upstream of the La Josefina site, with the towns of Gualaceo and Paute downstream of this point. In the same project, levees and dams were built to stabilise the riverbed. The obtained results were also useful for the Development and Redeployment Centre for Southern Ecuador (CREA) which used the installed monitoring systems and data.

• Provincial level
The provincial council of Azuay, namely the provincial government, received constant support by the PRECUPA project in order to assess the situation of landslide-prone areas in rural surroundings, even outside the zone of study. Such advice was directly useful to the population for opening new roads, building protection works or providing support to local communities. The results and maps of the project were also handed to the Provincial Council of Cañar, north of Azuay Province.

• University level
Several activities of the project were carried out with the participation of the regional universities, which also contributed to the training of students and assistants. For instance, the maintenance and operation of the seismic monitoring network, data base treatment, hazard and vulnerability assessment and further investigations are managed by the Cuenca University.

Up to now, three specialisation courses, in which the head of PRECUPA project intervened, were organised on disaster assessment, risk management, effect on environment and resisting structures, so that a deeper awareness was fostered concerning the sense of risk reduction.

- Local level

The main local counterparts of the PRECUPA Project were the Municipality of the Cuenca Canton and the Telecommunication, Water and Sewage Company of this canton. First, they participated in the studies, then they profited from the given courses, obtained useful results, satellite images, phenomena, hazard and vulnerability maps and studies, as well as some monitoring networks and equipment of the project in several fields. Both institutions mentioned assisted the Direction of Environment Management and with the help of the project, the Interinstitutional Commission of Environmental Management was created which also deals with risk aspects. This commission implies public and private institutions working in the canton.

Finally, the Cuenca Municipality used the results obtained by the PRECUPA Project in order to define ordinances for the use of the land at the level of the plot, in rural as well as in urban areas. These ordinances were considered in the process to obtain the decision of classification of Cuenca City as a Humanitarian Heritage by UNESCO in 1999, as well as cultural capital of America in 2001.

10.2.5 *Conclusions*

The participation of national and local institutions in the project, the knowledge development, the training of operational professionals and future decision makers, as well as the community and civil defence work towards a large dissemination of the results obtained, have guaranteed the continuity of activities and the implementation of specific risk reduction measures (Basabe et al., 1996). Thus much of the data and results could be incorporated in municipal ordinances and land-use planning. The conscience of the importance of natural hazard and risk management has clearly increased, allowing a better use of the land. This is expressed by a widespread popular slogan: "The knowledge and respect of Nature is fundamental to protect ourselves".

11
Suggestions, guidelines and perspectives of development

Ch. Bonnard[1], B. Coraglia[2], J. L. Durville[3], F. Forlati[2]
[1]Ecole Polytechnique de Lausanne; [2]Arpa Piemonte; [3]Centre d'Études Techniques de l'Équipement de Lyon

11.1   GENERAL OBSERVATIONS DRAWN FROM THE STUDIED CASES

11.1.1   *Assessment of obtained results in the studied cases with respect to hazard*

The presentation of the six most important case studies considered in the IMIRILAND Project shows that all sites imply a clearly hazardous situation which may induce significant damageable consequences in some of the studied scenarios. However, the assessment of the hazard intensity carried out in Chapters 4–9 cannot be considered as equally reliable in all cases. Indeed, several factors may impair the quality of the analyses developed:

– At several sites, a lack of basic data which could be technically obtained by field observation, long-term comprehensive monitoring and, if possible, boreholes including geotechnical labora-tory tests and groundwater measurements, but which are not available for several reasons, implies the necessity of formulating hypotheses in order to determine the characteristics of the landslide mechanisms and then to model landslide behaviour. Some of these hypotheses are soundly based on careful geological, morpho-structural and geomorphological investigations, but the degree of uncertainty of the resulting models is difficult to quantify.
– The major sources of uncertainty in the models considered lie in general in the pattern and evo-lution of groundwater conditions, especially for specific conditions which induce critical behaviour of the landslide; some of the models considered do not express groundwater pres-sures explicitly. Another main source of uncertainty lies in the dissipation of kinetic energy for rockfalls, either at the impact point of single blocks, or within the whole moving mass; conser-vative assumptions will lead to extensive hazard zones at the toe of the slope, whereas an opti-mistic selection of restitution coefficients or equivalent angle of friction may result in ignoring the residual level of hazard existing beyond the zone of impact.
– Finally, the selection of the models themselves aimed at describing the evolution of landslide masses may induce uncertain results which do not depend on the randomness of the material characteristics, as was shown by the Wall Ball and All Ball models used for the modelling of the Oselitzenbach landslide impact zone. Even though the extent of the zone was fairly similar in both computations, the velocity of the landslide process was quite different, which may induce radically different consequences in terms of infrastructure damage.

However, the difficulties in modelling natural dynamic processes such as large landslides will not be solved satisfactorily in the short term, especially as basic data will always be lacking due to financial reasons. It is thus more important to express the formulated hypotheses clearly and to build several scenarios of the landslide mechanism in order to qualify the meaning and reliability of the hazard zoning resulting from the modelling of such mechanisms. Considering the com-plexity of the case studies selected in the IMIRILAND project and the paucity of data related to the potential failure mechanisms, it can be stated that the determined hazard levels for the different scenarios are sufficiently significant to allow the development of a comprehensive risk analysis.

11.1.2 *Relative gradation of the scenarios in the risk assessment*

As it appears in all the case studies analysed, the potential development of large landslides into catastrophic phenomena cannot be expressed as a single, well-defined mechanism, as is the case, for example, when a minimum factor of safety nearly equal to one is reached for one specific failure circle in a homogeneous soil slope. Indeed, when the complexity of the geological and morpho-structural conditions is considered, as well as the possible triggering factors and failure sequences, it is evident that the landslide may evolve according to various mechanisms. The quantitative probability of each mechanism which may imply various intensities according to the volume mobilised, is difficult to establish, except when a series of past phenomena has been recorded.

It is, therefore, indispensable in a risk analysis to model several significant scenarios following a well-defined mechanism for which a correct hazard assessment can be performed, and then to attribute a relative probability to each of the scenarios. This procedure corresponds to real conditions as they may be observed by morphological studies in mountainous areas: phenomena of limited intensity (block falls, debris flows) are frequently observed at a given site (order of magnitude of return period: 10 years), medium-sized phenomena (rockfall, local slides) occur when ordinary triggering conditions are very severe (order of magnitude of return period: 100 years), and huge phenomena (rock avalanches, generalized catastrophic slides) only occur under extraordinary preparatory and triggering conditions or a combination, such as during glacial retreat or following a high magnitude earthquake (order of magnitude of return period: 200 to 1'000 years).

This type of analysis including the consideration of several scenarios is also essential to the risk analysis as it dictates not only the hazard level, but also the number and the vulnerability coefficients of the exposed objects. Even if an extraordinary catastrophic event has an extremely low probability of occurrence, which could seemingly justify that it not be taken into account (in the same way as the fall of a meteorite is not considered as a relevant criterion in most risk analyses), the importance of its expected direct and indirect consequences is such that the resulting risk may well be significant, with respect to the risk related to frequent phenomena.

It is not the responsibility of engineering geologists and scientists to fix the criteria for acceptable risks, but that of government officials and politicians. Nevertheless, all the possible risk scenarios have to be investigated so that progressive prevention and preparatory measures are taken, even if they are not all communicated to the population to avoid the development of an inappropriate feeling of panic.

11.1.3 *Consideration of the real value of exposed objects*

The notion of the monetary value of objects and beings (buildings, infrastructures, persons, animals, economic activities, the environment) exposed to a certain degree of hazard is very difficult to formulate in absolute terms.

As far as buildings are concerned, a large number of procedures aimed at assessing their intrinsic value exist based on the cost of construction and that of the plot of land, but they are often valid for short term evaluations only, as it is difficult to anticipate what will be the evolution of market prices, the attractiveness of the endangered zones and the consecutive possible annual rents as permanent or holiday housing in the long term; the value of the buildings also depends on their maintenance level. Therefore, a possible analysis grid consists in assessing the value of the building at the present cost of reconstruction, in case of total destruction, as is used by fire insurances; in this case the influence of the market is virtually neglected, except through the level of construction costs depending on the level of the local economy.

But this approach is open to criticism for zones exposed to the consequences of large landslides: indeed, if a chalet on a large slide has been seriously affected by surface movements, it is most probable that a new construction or even repair work will not be allowed in the same zone until it is properly stabilised. A chalet may even be tilted without any structural damage and completely lose its value (see Fig. 2.3). If the considered house proves to be located in a zone seriously exposed to rockfall, it will lose its intrinsic value even if it is not damaged, because most people are not ready to accept living with such risk (unless they have good reasons to do so, or do not believe in the threat).

In the case of the Sedrun landslide, for which the hazard level is not considered as severe by the inhabitants and, consequently, the risks do not affect the market prices in this tourist resort, the value considered for the buildings has been assessed on the basis of average construction prices in the area, which is quite appropriate as a large majority of the houses are recent and fairly standard (3 storied chalets or 5 storied buildings with 8 apartments). But, of course, the risk is assessed without considering the variation of prices over time. As far as the cost related to traffic interruption, it can be computed using the tables established for that purpose by the railway companies and road administrations.

In other situations where the value of exposed objects is much more difficult to assess (old buildings, complex infrastructures, industrial buildings), it is inappropriate to try to define a real value which might induce conflicts in the community. Therefore, relative values, as detailed in § 3.6, are more relevant for the comparison of the risks related to different scenarios. However, it is important to avoid practising an automatic analysis on the basis of incomplete data bases and to assess the relative importance of the exposed objects carefully not only based on their possible monetary value but also considering their patrimony value. This is particularly the case for forests, whose economic value is now very low, but which have a high environmental value because of their contribution to the landscape and to the safety of slopes against snow avalanches.

With the separate consideration of social vulnerability and the use of indicative coefficients, the value of life does not need to be expressed in monetary terms, but the social impacts can be analysed properly considering, in addition, the burden of an evacuation from an endangered zone as a social impact, even if no lives are directly threatened.

Finally, reference values for environmental vulnerability do not yet exist apart from forest maintenance costs, but the modern tendency to consider all ecological components as a patrimony of human ity encourages the identification and qualification of the value of the exposed environmental resources.

11.2 CONSEQUENCES OF THE RISK STUDIES ON LAND PLANNING PROCEDURES

11.2.1 *Possible actions to incorporate risk studies in land planning*

Even though some landslide zones may have been developed over several decades without any recognised evidence of potential dangers, the ignorance of such phenomena cannot be the rule and cannot serve only to cover the negligence of local authorities. So the major challenge, once a risk study related to a large landslide has been carried out, deals with the management of land in threatened areas.

The possible attitudes and actions when facing such a situation are very different when a master plan or a local plan is considered. At the national or regional level, landslide risk situations, once identified and recognised, are considered as local constraints which will be duly taken into account in the planning of new infrastructures (motorways, railway lines), as well as in specific development programs, for instance for flood protection. But such constraints do not substantially modify the development of projects, as they can be solved technically and, socially speaking, do not represent a major hindrance.

At the local level, on the contrary, large landslides may constitute a major preoccupation and induce risks which must be taken into account in land planning, for safety reasons as well as to comply with the legal prescriptions. The population of developed countries now demands better protection against risks, but at the same time tends to ignore the basic behaviours which could prevent the damageable consequences of specific natural hazards. Therefore, the local authorities have to develop an appropriate strategy to face this potential threat. Three distinct responses exist:

– **To accept** passively the occurrence of an unpredictable and very rare event which the population generally does not believe in, without assessing the potential consequences in an anticipated way. Such a policy must be considered as unacceptable in a perspective of sustainable development.

- **To resist** and oppose physical actions to the development of a dangerous potential event, so as to protect the primarily exposed elements. Such an attitude can lead to the construction of protection works, but in the case of large landslides it is often unrealistic and too costly except, for instance, when an important diversion road may be constructed as a tunnel below the landslide area; this is presently being carried out in the case of the Conters-Gotschnahang landslide for the bypass road avoiding the village of Klosters (Graubünden, Switzerland) and it is planned in the case of the Ceppo Morelli rockfall (see Chapter 7).
- **To take into account** the existence of the hazard and of the related risks and to adapt the land use and buildings to its possible occurrence. Such an attitude is, in most cases, the only reasonable one provided an adequate mitigation strategy is applied which considers the different protection objectives and analyses the residual risks in the case of an extremely rare scenario. Implicitly, this last option means that the population can live with certain risks in an exposed zone, even though, from a political point of view, the authorities state that they have taken all the required measures to avoid any risk. Explicitly, this means that the risk level has to be considered quantitatively in land planning policy, which is one of the goals of the IMIRILAND project.

This appropriate mitigation strategy thus requires different types of actions:

- Technical solutions, such as the construction of protection works (tunnels, protective dams) or the installation of warning systems and the implementation of an emergency plan.
- Political solutions, such as the adoption of an appropriate local development plan by the community, specifying different hazard and land use zones.
- Legal solutions, such as the adoption of laws and ordinances prohibiting construction in high hazard zones and the evacuation and expropriation of seriously endangered existing buildings.

11.2.2 Practical policy development

On the basis of the case studies examined within IMIRILAND project, as well as considering several other experiences related to large landslides, the local authorities in charge of the safe development of an exposed area have to analyse the following practical aspects:

- The construction of protection works, such as a tunnel for a diversion road or a protective dam to limit the extension of a zone exposed to rockfall, requires the definition of clear design criteria related to the landslide scenario selected, according to the needed degree of protection of the exposed objects. The determination of the protection objectives, depending on the value and political significance of the threatened elements, is a task for local authorities who must understand that protection works cannot be fully efficient in the case of extreme scenarios and that residual risks thus have to be considered as inevitable. Protection works may induce an exaggerated feeling of safety within the population if it is not duly informed of the limitations of these works. They may even be the indirect cause of an unforeseen disaster if these protection works are damaged by an unexpected scenario which then concentrates the destructive action of the landslide mass, particularly debris flows, on a limited portion of land.
- The installation of a monitoring and warning system requires not only the definition of an alarm strategy, which cannot be fully automatic due to the negative consequence of false alarms, but the preparation of appropriate evacuation plans and the elaboration of alternative routes in order to avoid congestion in exposed areas and to save lives, even if physical damage may occur. Such evacuation plans must be worked out in collaboration with the population and the scientists studying the landslide hazard, and tested during simulation exercises.
- The use of the endangered zones must be adapted according to the type of activities carried out or to the function of the existing or planned buildings. Indeed it is neither appropriate nor feasible to consider the exposed zones as banned areas in which no activity (e.g. farming, sports) and no constructions (e.g. secondary road, lifelines) should be carried out. But the users of these exposed zones, in particular when occasional sports or leisure activities are concerned, such as for camp grounds, must be aware of the possible dangers and of the conditions under which they might occur.

- In some cases for which an evacuation at short notice might prove to be inapplicable, such as for hospitals or retirement homes for elderly people, a preventive evacuation and a modification of building use must be carried out without waiting for a dangerous situation. These buildings may be still used for other functions, which do not induce the same risks.
- Easy access to appropriate information concerning landslide risks and the related consequences for persons and buildings must be organised by the local authorities, without dramatising the situation, but without hiding the reality of the danger and its potential development. All exposed persons, whether they reside in exposed areas or are tourists, must be duly informed and instructed on how they have to react in the case of a landslide event.
- Finally, in all of studied situations, it is not appropriate to apply one single solution but it is advisable to combine several measures within a comprehensive management strategy which has to be developed jointly by the local authorities and the scientists or experts in charge of the investigations on the landslide area. The efficiency and reliability of each solution should be properly assessed.

11.2.3 *Difficulties of application (population, authorities)*

It is necessary, however, to be aware that when facing landslide risk situations, particularly of very low probability of occurrence, many difficulties arise in the planning and implementation of safety measures. First of all, the population of the exposed area, which has lived there for centuries, might not believe in the seriousness of the hazard, in particular when no evident warning signs are perceptible for non-specialists or when the proneness of a serious event increases without a clear change of phenomenology (e.g. when cracks are opening exponentially in a cliff without releasing any blocks). Then, if the inhabitants are aware of a certain risk, they tend to understate it so as to preserve future possibilities of development on the land they own in exposed areas. Finally, in the clearly exposed areas, for which an evacuation is required, the owners will resist either because of a strong affective bond to their homeland, or because the compensation proposed by the state or the insurance companies does not reach their expectations. These difficulties should lead the authorities to develop communication actions involving the scientists which have to present what they know, but also their uncertainties.

As far as the authorities are concerned, the situation varies from one country to another, depending on the respective competencies of the local and regional authorities for the execution of safety measures (see § 11.4). It is clear that the municipal authorities, who have to be re-elected every four to six years, are not induced to take drastic safety measures limiting the use of the land, as this will displease the threatened owners and induce a feeling of injustice between the affected and non-affected citizens. Otherwise, they will take partial measures and then declare that the landslide risks are "under control", whereas the level of safety is only slightly or locally increased.

The fact that the adaptation of local plans in order to include new hazard zones or modify their requirements must be approved by municipal councils, makes it difficult to proceed to swift changes when the nature of the phenomenon evolves. Of course, the mayor generally has full power for the enacting of safety or evacuation measures when a critical situation occurs, but such actions are of limited duration. Therefore, it is not easy to reach an approved consensus or the rapid application of restrictive measures in order to reduce the landslide risks. But when the local authorities, and in particular the mayor, have not taken the required safety measures to avoid a disaster, they are legally considered to be responsible for the damage, in particular if human victims are reported.

11.3 CONSIDERATION OF RISKS IN THE DEVELOPMENT OF EXPOSED AREAS

11.3.1 *Possibilities of reducing risks by protection works or measures*

Risks induced by large landslides such as rockfalls or slides may theoretically be reduced by a modification of one of the five factors which intervene in the risk computation, namely:

- intensity of hazard assessed at the fall source zone or at the surface of the slide;
- intensity of hazard assessed in the foreseen fall impact zone;

– probability of occurrence of the hazard;
– vulnerability of the exposed elements (physical, social, economic, environmental);
– value of the exposed elements.

However, some of the corresponding actions, through landslide protection or stabilisation works, prevention, reinforcement or modification works of the exposed elements, as well as reassignment of the exposed structures and land, which induce a change in their value, are not always technically and financially feasible, mainly due to the considerable size of the landslides studied. Some induced aspects of the corresponding actions may also appear as unfeasible due to environmental protection constraints.

Practically, the economic feasibility of protection works depends essentially on the importance of the physical and economic risks involved, whereas the risk to life can be essentially limited by a comprehensive monitoring and warning system allowing the safe evacuation of the exposed inhabitants and the interruption of the threatened traffic. One significant case of reduction of hazard intensity was the stabilisation of the slope above the Tablachaca Dam on the Mantaro River in Peru, which represented a high hazard for the hydro-electric scheme as well as implying a potential loss of energy production. The slope was, therefore, stabilised, among other works, by a huge fill placed at the toe of the slope, in the reservoir, with a total volume of half a million m³, by a series of drainage galleries at three different levels, extending over a total length of 1'400 m, from which 3'000 m of radial drainage boreholes were drilled. In addition, more than 400 prestressed anchors were installed, some of which extended over a length of more then 100 m (Carrillo-Gil & Carrillo-Delgado, 1988).

In the case of the large slides considered in the IMIRILAND project, the reduction of hazard intensity related to the landslide mass itself may only be achieved by extensive underground drainage, as recently carried out at the Campo Vallemaggia landslide in Switzerland, but only if certain favourable conditions prevail, related to the permeability and groundwater pressure pattern (Bonzanigo & al., 2001). In fact the knowledge of water distribution and flow in large rock slopes is low and drainage may sometimes be an uncertain action. In the cases of rockfalls considered, no practical action can reasonably be developed in the source zone to reduce the hazard.

A possible action to reduce the intensity of the hazard in the potential impact zone is either the construction of a transversal earth dam to limit the extension of the rockfall, as carried out at the toe of the Séchilienne landslide (see Chapter 9), or the construction of a longitudinal earth dam to divert the falling mass from an area at risk, as built in several sites in Switzerland. For large rockslides, the common solution of transversal wire nets to retain blocks is not applicable due to the high energy of the falling rock mass.

If there is no possible direct action to reduce the probability of occurrence of the main landslide hazard cause, i.e. rainfall, the construction of deep drainage systems, even though they may not stop the continuous movement of a large slide, may significantly reduce the probability of experiencing exceptionally high groundwater pressure along the slip surface. So, this solution represents a significant contribution toward the reduction of the probability of occurrence of disaster, as shown for the Downie Slide by Patton (1984) (see § 10.1).

A major possibility of reducing risks exists by removing the exposed objects from the impact zone, thus bringing their vulnerability to nil, such as in the case of the construction of a tunnel for a road or railway below the landslide. In extreme scenarios, however, such a measure may not be sufficient and partial damage to the tunnel portals may occur. In the same way, the evacuation of the population in due time significantly reduces their vulnerability, but does not decrease it to nil as enduring long-term evacuation in temporary shelters may cause important psychological impacts. Economic vulnerability may also be reduced by organising alternate routes for traffic that might be blocked by the landslide mass or by avoiding the storage of goods in exposed zones, so that carefully planned management measures, based on the information provided by warning systems, may significantly reduce the economic risks.

Finally, the risk may be reduced, in the case of temporary threats, by reducing the value of the exposed elements, for instance the goods produced and stored on the landslide mass. A well-known extraordinary case was reported at the Cornillo landslide in Italy, when the acceleration of the moving mass induced severe damage to a ham factory; major efforts were done to evacuate all the stored meat before the building and access road were too seriously damaged as the value of the meat was much higher than that of the processing plant itself (Larini & al., 2001).

11.3.2 *Legal aspects related to protection measures*

In order to enforce a danger zone permanently in a local management plan, the procedures vary from one country to another, but the approval of the legislative municipal authority is often required, which means that the population has to be convinced of the pertinence or relevance of the proposed limits. Therefore, detailed information must be given by the scientists or the experts to the municipal authorities and the population in order to justify the proposed limits.

This is even more complicated to explain when different limits with different criteria are set up, as is recommended in Switzerland (see Fig. 8.11, showing the Sedrun present local management plan where two limits (red and blue lines) for avalanche, flood and debris flow danger zones appear). In this case, the results of the risk analysis will be useful to justify the proposed limits, economically and socially.

According to Swiss legislation, the local management plans must be revised every ten years so that, according to the evolution of the situation, it is possible to adapt the danger zones, as well as the rules which have to apply where prescriptions are only given for the exposed land use.

In the case of France, a specific law allows the expropriation of endangered areas by the State, which permits a satisfactory monetary settlement for the evacuated owners (see § 9.5 and 11.4). However, several open legal questions still arise in relation to the obligation of the French State to carry out preventive actions if any natural danger may possibly affect persons, whatever its probability of occurrence, according to a decision of the State Council. In particular, the application of the principle of precaution, which implies the application of preventive actions even if the threat cannot be demonstrated, may lead to an excess in terms of limitation of land use.

11.3.3 *Acceptable risks*

The use of several possible scenarios in hazard analysis stems implicitly from the fact that the maximum intensity of a given natural phenomenon, considered as a value that would never be exceeded, is difficult to assess; thus it is often judged as too extreme to consider as a reference value in land planning. Consequently, less critical scenarios are generally accepted to fix the limits of the protection zones in the local management plans.

This practice indeed means that, when facing an extreme scenario with a very low probability of occurrence, such as the failure of protection dams, damage will extend to zones which were supposed to be protected, implying residual risks. Such risks are, therefore, considered as acceptable and must imply limited consequences, such as material or environmental damage only, when the evacuation of the exposed population has been carried out.

It is clear that, despite the difficulty in terms of legal reference, the application of any methodology based on risk analysis necessarily means that some risks will be considered as acceptable by the authorities. Such risks can only be accepted by the population when they are duly informed of this possibility and have clearly expressed their will to assume the respective consequences. This situation often occurs in mountain villages, but tourists coming from safe areas and not aware of the potential consequences of natural phenomena may not be ready to tolerate a risk that is indeed unavoidable.

11.4   TENDENCIES IN RISK MANAGEMENT POLICY IN SEVERAL EUROPEAN COUNTRIES

11.4.1   *Introduction*

In the face of a natural hazard, risk management can be divided in several stages:

a) hazard assessment and vulnerability analysis,
b) risk evaluation and assessment,

c) risk prevention (protective works, land use regulation, monitoring, etc.),
d) crisis and post-crisis management,
e) feedback from experience.

The public authorities have to choose which kind of policy for risk prevention and mitigation they will apply. One can define three degrees of increasing involvement by the authorities:

- To produce and disseminate information about hazards and risks (items a) b) and e) above): in principle the citizens or the communities are therefore supposed to be aware of the dangers and they are supposed to take them into account in their behaviour ("liberal" policy);
- To make regulations (item c) – it is prohibited to…, it is compulsory to… – in order to oblige the citizens or the community to apply mitigation measures ("dictatorial" policy);
- To bring money for mitigation measures in order to incite and to help the citizens or the community to take preventive (item c) or rehabilitation (item d) actions ("État-providence" or Welfare State).

Generally, there exists a mixing of these attitudes in a given country, but each of them can be more or less stressed.

The comparison of risk management in different countries brings our attention to some particular points:

- Hazard mapping has developed in Europe since the 1970's. It is divided in most countries into regional mapping (typically 1/25 000 scale) and local mapping. The regional level is presented in Table 11.1.

Table 11.1.   General hazard maps produced by different European countries.

	Italy	Austria	Switzerland	France
Decision maker	Autorità di bacino/ Regione	Austrian geological survey	Canton	Préfet of the Département
Maps	Hazard map (1:25 000)	Hazard maps (1:25 000 to 1:50 000)	Danger (indicative) map (1:10 000 to 1:50 000)	Hazard map (1:10 000 to 1:25 000)

- Landslide inventories or databases are in progress in several countries. They are a useful tool to keep the information available and to prepare hazard maps.
- The vulnerability analysis is, in actual practice, most simplified, generally reduced to an identification of "zones of conflicts", where a high hazard level and some exposed properties coexist. Experience shows that it is only where very large landslides are threatening the lives and property of numerous people that an actual vulnerability and risk analysis is performed.
- The protective works are financed at different levels. Most frequently the cost is far too high for private persons and even for small communities. For instance:
  - in Switzerland: protective works funded by the federal, cantonal, municipal levels
  - in Italy: planning in Piano di Assetto di Bacino, funding by the Region
  - in France: recent possibility of funding by "fonds Barnier" (but the available budget in 2003 is rather small…).
- Land use regulations exist in all countries: it is probably the most effective means of regulating development but it is a long-term strategy which has been used in Austria and France since the 1980's and in Italy and Switzerland since the 1990's. The different kinds of hazard mapping should be used to update the municipal land use plans:
  - Italy: the "Piano Regolatore Generale" (from the "Piano di Assetto Idrogeologico"),
  - Switzerland: the "Plan d'affectation communal" (from the "Carte des dangers" generally established at the cantonal level),

- France: the « Plan Local d'Urbanisme » (from the « Plan de Prévention des Risques »),
- Austria: The « Flächenwidmungsplan » (from the « Gefahrenzonenplan »)
- In all countries, the different zones are typically:
  - zones where building is forbidden,
  - zones where building is permitted with conditions (preventive passive or active techniques),
  - zones without any particular problem.

In Switzerland, a fourth zone has been defined: zones of low probability of occurrence of a major landslide, where the preparedness for the event has to be developed (emergency plans).

- Insurance against natural hazard: in some countries the public authorities intervene strongly:
  - Switzerland: in 19 of the 26 cantons there is a cantonal insurance for buildings,
  - France: insurance against natural disasters is compulsory (law of 1982).

In other countries, only private insurance is available.

From a general point of view, some final remarks can be drawn from the comparison of different risk management policies in Europe:

- Legislation relating to hazard prevention has developed after the occurrence of some large disasters. In European countries, the legislative output corresponds to an increasing involvement of the governmental authorities in the face of the growing social demand for security.
- We may contrast France, which is a typical, centrally-run country (the government agencies decide where the hazardous zones are and what must be done), and countries like Italy and Switzerland, where the government defines the general framework, but the regional and municipal levels are the main players. The amount of negotiation among the different players (gov-ernment or region, municipality, citizens) is certainly different from one country to another.
- The local level authority (i.e. the mayor of the community) is generally primarily responsible for the crisis management.
- The legal framework is probably fitted to the most frequent events but it may not be suitable for very large landslides.
- The back-analysis of landslide disasters is not as well organised as for earthquakes, dam failures, etc.

In the next paragraphs, a brief description of risk prevention policy is given for Italy, France and Switzerland.

11.4.2 *Italy*

Before the 1990's, the policy in Italy was characterised by two principal behaviours:

- to take no action at all,
- to provide relief and rehabilitation assistance after disasters occurred.

The Government provided payment for damage suffered by private human activities or buildings also in known landslide prone areas.

At the beginning of 1970's, land use management was transferred to the regions. For example, in Piedmont the law 45/1989 regulates land use modification and transformation in areas subject to environmental protection. It does not define common lines on land use planning, as each case is considered separately and prescriptions are different.

As a result of technical and sociological advances, the concept of prevention and risk by appropriate land use is becoming increasingly important. The law 183/1989 introduces land use planning on a basin scale: the Government sets the standards and aims but it does not fix a methodology. The same law designates the Basin Authority, whose main goal is to draw up the Basin Plan, a tool for planning actions and rules for conservation and territorial protection. Concerning the Po basin, the last plan adopted is called PAI: it tries to verify the geological instability of the whole territory and to upgrade the land use planning. Moreover, this project classifies the territory into risk classes.

The catastrophic event of May 1998, which caused heavy damage and death in the municipalities of Sarno (Fig. 11.1) and Quindici (Campania), urged the government to provide answers for development regulation (to reduce or eliminate landslide losses). According to a decree named "the Sarno Decree", the Government detailed legislative measures at the national level, including the procedure to define landslide risk areas.

Figure 11.1.   Large debris flow which affected the town of Sarno.

Another important aspect of law 267/1998 promulgated after the Sarno Decree regards the development of "Extraordinary Plans" to manage the situation of higher risk where safety problems or functional damage are possible. In these areas, protection measures must be applied, executing projects for risk mitigation. In some regions, the actions have been extended to the whole area and in others they have been applied only to some significant cases; in the Piedmont Region, the Ceppo Morelli landslide has thus been classified as a very high risk area.

The national legislation defines general principles, functions, activities and the authorities involved. As a consequence, the regions apply restrictions on land use through different regional laws. In Piedmont, the local management plan includes danger/hazard zoning in order to identify landslide prone areas and to recognise danger/hazard zones on the basis of morphological features. For example, the Rosone and Cassas areas are classified as higher hazard zones.

In a state of emergency, a Piedmont regional law allows the stopping or the suspension of development in landslide prone areas. Consequently, new land planning must be implemented.

At present in Italy, a national inventory landslide map does not exist. Innovations in land use regulation could be introduced by a new project named IFFI (ARPA-Piemonte, 2004), an inventory of the landslides on the whole national territory. It represents a very important tool for the planners who finally have the detailed and most complete knowledge of the landslide occurrence on the whole territory. The recognition of different landslide types, the definition of their relationship with human interactions, the geomorphological and the geological structural setting, have allowed detailed territorial zoning. As a consequence, the Geological Services intend to use the inventory data to improve land use management and to evaluate the hazard and risk areas at the regional scale, as well as to support Government decisions.

11.4.3 *France*

Since the 1980's, several laws related to natural hazard prevention have been passed. Citizens are more and more concerned with their own safety and the idea of "fate" is no longer accepted. The prevention of natural or technological hazards has become a politically important topic.

In 1982, when H. Tazieff was the (first) minister in charge of hazard prevention, a law created and organised:

– the insurance against "natural disasters" (the word disaster has been broadly interpreted); property insured includes buildings and moveable property (including vehicles) insured against fire or any other type of damage (theft, etc.); the rates of additional premium are set by the Government (12%).
– the "Plans d'Exposition aux Risques" which regulate and plan the land use at the municipal level; they were changed into "Plans de Prévention des Risques" (PPR) in 1995; more than 10'000 communities are involved (5000 PPR have been produced up to now). The PPR are the main tool for the Government to encourage better land use in relation to natural hazard. In the most dangerous zones, building is forbidden; in zones of medium hazard, preventive measures have to be taken.

In 1995, a new law simplified the PPR procedure and created a possibility of expropriation by the Government when private buildings are exposed to a short-term danger.

In 2003, a law devoted to natural and technological hazard prevention was established. Under particular conditions, the Government could help citizens or communities by partially funding a local prevention policy. Moreover, the information for the citizens about natural hazards is strengthened: for instance the mayor has to publish information for the inhabitants every two years. In addition, a national database related to landslides is in progress; it deals with data about the landslides themselves, consequences and human response. Data are freely available on the Internet for every citizen.

Finally, in France there exist powerful legal tools in order to encourage a prevention policy. Nevertheless, land use regulations that were developed in the 1980's and 1990's have a long-term efficiency and are frequently not well accepted. Presently the public authorities are trying to set up a more positive policy based upon technical and financial help to favour the development of preventive measures by citizens or communities. The case of a very large landslide such as Séchilienne shows, however, that it is very difficult to organise the funding for monitoring and preventive works (investigation gallery, for instance) because of the lack of clarity related to the relevant responsibilities of the different partners (Ministry of Public Works and Ministry of the Environment, Département, Commune).

11.4.4 *Switzerland*

The 1979 Federal Law on Land Planning (LAT) requires that the areas seriously endangered by natural hazards be designated specifically in cantonal master plans. Their management, as well as the type of qualification and representation, then depends on the cantons. But a federal recommendation has been produced in 1997 specifying how to take landslides into account in land planning (OFAT, OFEE, OFEFP, 1997).

Several cantons have developed mapping projects in order to identify not only landslide-prone areas, but to establish danger maps (indeed, hazard maps) which will be considered in local management plans; the canton of Freiburg is one of the most advanced in this task. These maps distinguish red zones, in which a prohibition to build will be declared, blue zones, in which prescriptions will be enacted and only some buildings or constructions implying limited risks will be accepted, and yellow zones, in which the owners have to be informed of the existing limited danger.

If such zoning is already applied extensively for snow avalanche and flood hazards in local management plans, it is not yet common as far as landslide risks are concerned. In the two Swiss communities considered in the IMIRILAND project, the landslide risk was not explicitly considered in the local management plan. But in the planning of the diversion road to avoid the centre of Klosters, which was first designed at the toe of Conters-Gotschnahang landslide, the potential difficulties for such a construction were duly evaluated and finally a route passing partly on the other side of the valley and partly in a tunnel below the Gotschnahang landslide was adopted and is presently being carried out (Bonnard & Noverraz, 2001).

In other communities affected by slow movements due to large landslides, such as Grindelwald (Canton of Bern) or Lugnez (Canton of Graubünden, see Fig. 2.1), the zones reserved for construction are never located in red zones and only rarely in blue zones, so that no major conflict arises between the promoters and the authorities. But in the Canton of Freiburg, near the Black Lake (Schwarzsee), a development project was rejected after a long struggle as being too exposed to landslide hazard. Some years later, the slide suddenly accelerated...

11.4.5 Conclusions on risk management policies

A comparison between the main features of the risk management tools applicable to land planning allows the following conclusions that are summarised in Table 11.2 through their respective advantages and disadvantages:

Table 11.2. Comparison between positive and negative aspects in risk management.

	Italy	Austria	Switzerland	France
Positive aspects	Regions have technical know how, so the risk analysis can give a contribution to risk management	Classification by the "Brown zones of reservation": Areas of danger, such as rockfall, landslides... (statal bylaw – regulations 1976) Similar to the well known zones for avalanche protection	The hazard is classified in a homogeneous way (red zone, etc.)	The hazard mapping program is under way and should be completed in a few years. People and local authorities are more and more convinced that risk prevention is not only the government's duty but also their own's
Negative aspects	A common way to define risk levels does not yet exist	Only qualitative system, no vulnerability costs Not yet all zones mapped	No vulnerability and costs are taken into account	The PPR's are strongly heterogeneous. Nobody is really controlling their application for land use

11.5 SUGGESTIONS FOR FUTURE RISK MANAGEMENT STUDIES

11.5.1 Hazard studies

The definition and evaluation of a landslide hazard includes several parts:

- The precise nature and intensity of the expected phenomena: the landslide itself and the secondary or indirect effects (air-blast induced by the rapidly moving mass, damming of the valley by debris, etc.),
- An evaluation of the probability of occurrence for a given time period, generally a qualitative estimation of this probability.

With respect to cases like Rosone (Italy), Encampadana (Andorra) or Séchilienne (France), it appears that the deformation process of large rock slopes is not well understood yet. An effort should be made in order to elaborate more appropriate mechanical models and to estimate global (macroscopic) properties of the rock mass that could account for the observed deformation and morphology. This clearly relies upon a detailed structural field survey and should involve geologists in collaboration with mechanical and numerical modelling specialists, within multidisciplinary teams.

As for the time occurrence, a quantitative probability approach is not suited to the large (and rare) landslides of the alpine region because of the lack of statistical data. The prediction of a catastrophic failure at sites like Séchilienne or Rosone is based on extensive monitoring of the slope, which should give warning about failure a few months (or weeks, or days) before it occurs. It is also essential to obtain quantitative data related to the movement (direction, velocities, dependence on rainfall, etc.) for the understanding of the deformation process. New terrestrial, aerial or satellite imagery techniques will probably become an essential part of the monitoring systems of large landslides.

As for run-out modelling, several numerical tools are available deduced from solid, fluid or particulate mechanics, but not all give relevant results. The priority is to collect detailed information related to past or future events: comparison of pre- and post-topography, estimation of velocities, etc. These case histories will enable the calibration and validation of the mechanical models.

11.5.2 Consequence studies

One of the most significant consequences of large landslides in mountainous areas is the damming of the valley. It is, therefore, necessary to address the question of the stability and durability of the natural dam formed by the landslide debris. Hydrological simulations can estimate the flooding hazards, both upstream and downstream, in case of dam failure, but the results are directly dependant on the behaviour and failure process (permeability of the dam, overtopping consequences, internal erosion) of the dam. A careful analysis of past events should be carried out in order to be able to make reliable predictions about the dam stability.

The quantitative data related to the exposed objects may exist but are difficult to find and to collect. In the near future, data bases and geographical information systems will probably facilitate the obtaining of information related to the spatial distribution, the functionality and the cost of the main objects. As an example, the Swisstopo database includes a GIS layer containing all of the buildings in Switzerland. The evaluation of indirect costs, because of the closure of a road for instance, has yet to be formalised; economists should be involved in such analyses in conjunction with engineers.

In the case of slow movements, such as in Conters and Gotschnahang (Switzerland), it is difficult to evaluate the long-term effect of the moving slope and the potential consequences on infrastructures (tunnels, for instance). It should be possible to live and to develop buildings and infrastructures on such sites, but a better evaluation of the maintenance and protection costs due to the movement should be carried out.

11.5.3 Total risk and prevention

In order to develop risk prevention, some feedback is needed from the experience of past events, in the scientific, technical, social and economic fields. Lessons should be learned from risk management before, during and after the crisis. The appropriateness of the legal framework could be assessed and many aspects should be analysed; for instance, the changes in public opinion concerning the risk level, which is an essential "parameter" of risk management, should be analysed: these changes may be induced by some local (e.g. a small but visible rockfall) or external events (e.g. the landslide in a neighbouring country: see the influence of the 1987 Valtellina landslide in France), by a public statement coming from a well-known expert, by the realisation of some preventive works, etc. Cost estimates of past events, including all direct and indirect items, should also be made in order to assess the importance of landslide risks better.

In the face of a large landslide, involving several million m^3 of moving rock, most prevention engineering techniques have a limited efficiency. However, deep drainage may be considered. In order to optimise the drainage system, research should be pursued concerning the hydrogeology of rock slopes in mountainous areas (hydrogeology of discontinuous material).

As far as passive techniques are concerned, earth catch dams have been extensively used for rockfalls with energies less than 100 MJ or so. Data are lacking concerning the efficiency of large catch dams, particularly their resistance to an impacting mass of debris; a better understanding of the run-out process should give an estimate of the induced pressure on obstacles and allow an increase in the field of applicability of this technique.

11.6 PERSPECTIVES AND OPEN QUESTIONS

11.6.1 *Relevance of overall risk analysis in development policy*

The concept of vulnerability was indirectly referred to first by engineers, in considering construction value and building design criteria related to the level of resistance and deformation to physical forces exerted by the landslide moving mass. Over the last years there has been a significant and important development in the understanding about what makes people, as well as social, economic and environmental assets, susceptible to hazard of any nature (landslide, flood, avalanche). This has expressed a new range of socio-economic and environmental concerns which could be also expressed by the notion of vulnerability. So, the vulnerability meaning was extended to a reflection on the state of the individual and the collective physical, social, economic and environmental conditions. Governed by human activity, vulnerability cannot be isolated from ongoing development efforts and, therefore, it plays a critical role in the social, economic and environmental context of sustainable development.

There is a fundamental need in landslide risk assessment to recognise the relationships between population growth, the physical demands of human settlements, short and longer term economic trade-off, social and cultural questions, as well as the most appropriate use of available land. The determination and wide acceptance of the most suited use of land, whether it is privately or publicly held, is demanding enough. It becomes even more daunting if there are various points of view about the role that land use can or should play, in terms of reducing collective exposure to risk. Too often, the desire for short-term gains is prone to override anticipated benefits that stretch further into the future.

Anyway, land use planning that is carefully designed and rigorously implemented is the most useful approach to managing population growth and the physical demands of human settlements minimising associated risk. For this reason, land use management and the related aspects of regional or territorial planning have to be considered as a natural extension of conducting hazard assessment. A failure on the part of government to implement effective land use and planning practices could bring about unexpected effects. Consequently, inadequate, ill-informed or non-existent land use planning can contribute to increasing the vulnerability of the communities exposed to landslide hazard.

When studying land use planning measures, it is necessary to take into account the dangers and their effects, because of the high population density of the potentially involved area. The land planning must be adapted to the territorial and environmental reality. In order to increase land planning efficiency in minimising risk, the following important principles, generally valid for different kind of dangers (flood, flash-floods, avalanches, earthquakes, etc.), have to be taken into account:

- Land use management operates at different geographic scales which require different ranges of suitable management tools and operational mechanisms.
- Land use management involves legal, technical, and social dimensions. The legal and regulatory dimensions include laws, decrees, ordinances adopted by national and local administrations. The technical and instrumental dimension includes planning tools and instruments that regulate uses of land and balance the private and public interests. The social and institutional dimension includes people participation in land use management practices.

- The practice of land use management proceeds through three consequential and interrelated stages: strategic planning, administration and fiscal control and monitoring.
- Successful strategic land use management requires a clear legal, procedural and regulatory framework that defines the competencies of the various stakeholders and the "rule of the game", including the role of each actor in the different phases of the planning.

11.6.2 Gaps in landslide risk management studies

The IMIRILAND project develops a generalised method for risk assessment providing a basis in relation to large landslides, but also accounts for an insufficiently emphasised source of uncertainties and gaps to the global risk assessment process. Many of these gaps depend on the multidisciplinary and multisectoral nature of the approach and should be ascribed to the following themes.

a) Crucial quantification of the socio-economic and environmental impacts and related interconnected aspects

To take into account the large landslide risk means to change the proper assessment scale, i.e. to extend it from a local perspective due to the directly exposed elements at risk to a more extensive regional one, often not easy to identify, but usually of relevant impact. These risks are related to the indirect effects, e.g. chain effects, implying further impacts on social and economic assets etc… To assess the global risk thus means to take into account more vulnerability aspects, which often interact with one another, often implying different zones of influence both in terms of dimensions and spatial identity. These interconnected aspects also include the quality and availability of regional socio-economic, environmental and cultural data, as well as the capacity for understanding and managing this topic correctly.

While hazard analysis and physical aspects of vulnerability evaluation have been substantially facilitated and improved due to the use of GIS techniques, the inclusion of social, economic and environmental variables into GIS conceptual models remains a major methodological challenge. The need to assign a quantitative value to the variables analysed in the spatial model used by GIS is not always feasible for some social/economic dimensions of vulnerability. For instance, how can the social impact of a destroyed international railroad or the impact on cultural aspects be quantified?

Moreover, the diverse scales (individual, community, regional, national, international) at which different aspects of socio-economic and environmental vulnerability operate, make a representation through these techniques very difficult. The quality and detail of the information required by the analysis facilitated by GIS is in many cases non-existent. In general, the quality and availability of regional socio-economic and cultural aspects and respective data limit the accuracy of risk analysis and assessment. On the other hand, the possibility to measure and to quantify the socio-economic and environmental impacts correctly has been proved to be a very difficult task for risk analysis and assessment. Risk assessments need to reflect the dynamic and complex consequences involved, in order to incorporate them into a large landslide reduction strategy properly. Multi-hazards and comprehensive vulnerability/capacity assessments that take into account the changing patterns in landslide risk are departure points for raising risk awareness at all scales.

b) Real perspectives of risk management with respect to short and longer term

Prevention focuses mainly on repetitive events of low to medium impact. However, past experiences reveal that the safety of the community is threatened primarily by rare or extremely rare events, and in these cases prevention strategies and measures are quite insufficient. People, therefore, become unconsciously accustomed to the devastating effects of these extremely rare events. Scientific studies and research still need to be conducted to determine the conditions under which such events occur and to evaluate their effects.

Concerning risk management tools, one of the most effective methods of reducing the landslide losses is to delineate the danger/hazard zones in a management plan. In order to ensure short and long-term control, management plans have to be revised and updated (for example: every 5–10 years) so that, according to the evolution of the situation, it is possible to adapt the danger/hazard zones, as well as the rules which have to apply, where only prescriptions are given for exposed land use.

c) Lack of knowledge and of available tools to manage the temporal component of the studied landslides

The spatial scale of the examined instability problem and its related effects, the role exerted by climatic and meteorological triggering factors and the time-dependent rock mass behaviour influence the temporal scale of the analysis. What is the likely future behaviour of the slope (acceleration leading to a catastrophic failure, deceleration then stabilisation, change of speed with seasonal fluctuations, etc.)? Even though this question is raised systematically, at the present time it is rarely solved. Risk analysis requires the assessment of occurrence probability in time; however, the time element is of great uncertainty in landslide hazard/risk analysis. Occurrence probability which can be defined as the chance or probability that a landslide hazard will occur can be expressed in relative (qualitative) or probabilistic (quantitative) terms. A rigorous, quantitative, procedure could include the application of formal mechanical-probabilistic methods, taking into account the uncertainty in all the geometrical and mechanical parameters, including the variability of the boundary conditions (for instance pore pressure distribution in the slope, initial state of stresses etc.).

In the case of large landslides, however, a similar procedure cannot be applied because the uncertainty in the definition of the parameters involved is too large and it is not possible to define any reliable variability for them; finally, the number of events is not sufficient to obtain extreme values by extrapolation.

Moreover, the use of empirical correlations is often still difficult because of the lack and/or reliability of monitoring data. Otherwise, a periodic frequency of landslide occurrence can be estimated using semi-quantitative methods related to the assessment of historical data in conjunction with a detailed geological and geomorphological characterisation. In any event, occurrence probability assessment introduces severe uncertainties and, therefore, is almost unable to answer the following question raised by managers at this time: When will failure occur? Only when detailed displacement data are available for several days before failure is it possible to express a reasonable answer on the date of failure, as it was put forward during the second stage of the Randa rockfall (Noverraz & Bonnard, 1992).

d) Limits in the process of modelling

The term "modelling" may be used in many contexts with different meanings. On the one hand, for a field geologist, the model of a mountain area includes the geological origin of the massifs, the tectonic and morphologic process, the main structural units, etc. On the other hand, a model for a finite element expert involves a characterisation of the geometry of a body and its material properties as well as initial and boundary conditions. Both experts are required in order to produce a final comprehensive result, and the quality of that final result will depend of their inter-communication and exchange of experiences.

Moreover, when defining the modelling work, it is important to point out that the quality of the final result depends on the quality of geological and numerical analyses. First, a geological model should be built from field observation that requires some expertise in field of geology, since this process requires understanding of the landslide site at a very global level and implies a simplification of reality. Second, the results of the geological model have to be "converted" into a numerical model framework. This is not without uncertainties, because of the limitations and intrinsic drawbacks of numerical models simplifying the geological models (e.g. the too high number and the different levels of importance of geological parameters). Despite all limitations, to obtain a good characterisation of landslide behaviour, it is necessary to compare the results of numerical simulation with field information and to decide whether further analyses are required. This is done as feedback for the geological model, which may be redefined with this new information. Feedback for the geological model is an added value in large landslide analyses.

Nevertheless, it is important to underline that this type of analyses are full of difficulties despite the capabilities of the numerical techniques and the quality of geological models. Indeed, our knowledge is still far from being able to reproduce instability phenomena. For instance, it is impossible to predict, at the moment, the temporal occurrence of displacements leading to a failure condition, and thus the development and operation of alarm systems based on these analyses is still a matter of further research.

11.6.3 *Future potential developments*

Through the discussion of the results obtained (applying the risk assessment methodology pointed out) into different landslide risk situations in mountainous contexts, the IMIRILAND working group highlighted a need of improvement for the following principal aspects:

a) A global, integrated and multidisciplinary approach involving different disciplines

Sound methodology can only be derived from a multidisciplinary approach taking into account the results obtained by different disciplines.

In the light of this assumption, the risk management is based on the development of integrated phases. Moreover, each of these phases consists in a further multidisciplinary approach which must provide a synthesis of data and knowledge for the next step. These phases are described in the following points:

- analysis of instability phenomena which consists in the construction of a geological and numerical model of the landslide in order to understand the triggering and evolution of the instability phenomena. This phase is carried out by experts in geological, geotechnical and meteorological disciplines which define the hazard scenarios.
- risk analysis of the instability phenomena which consists of the evaluation of the direct and indirect consequences, in a wide sense, determined from the hazard scenarios and risk levels. This process should be carried out by experts in economic, ecological, political and social disciplines which define a quantitative risk analysis.
- Land planning devoted to mitigation of risk. This phase is carried out by local, regional and national authorities that provide legal, technical and administrative instruments to make decisions, in order to choose the most appropriate strategies.

The risk management quality depends not only on the quality of each phase but also on the communication among the experts of the consecutive phases. Practical difficulty exists since each phase is performed by experts of different disciplines and the integration of the works may suffer from communication problems.

From the experience gained in the IMIRILAND project and also from the previous experience of the groups involved, it is possible to improve the integration of knowledge in order to guarantee straightforward and efficient communication among the disciplines.

b) The role of the risk perception in the community

Hazard assessment uses specific procedures which are mostly restricted to a scientific community. On the other hand, vulnerability and consequence analysis makes use of more conventional methodologies and techniques, through which the community at risk may also play an active role. The distinction between risk assessment and risk perception has an important implication for disaster risk reduction. In some cases, risk perception may be formally included in the assessment process by incorporating people's own ideas and the perception of risk they are exposed to. In fact, the daily "living together" between the population and the environment has developed a high sensitivity in the community to search the safest areas and to perceive any incipient signs of slope instability. This sensitivity produces an increase of the acceptable risk level. The case of the Langhe area (southern Piedmont), for instance, underlines the importance of this sensitivity. In fact, despite the high number and the size of landslides (over 900 cases) as a consequence of heavy meteorological events of November 1994, no victims were recorded.

c) Maintenance and systematic updating of landslide impact data sets

Historical analysis of disaster data provides the information to deduce levels of risk based on past experience. Post-analysis of a catastrophic event can be essential for the data connected to the phenomenon, its impact and the way in which it is managed. In addition, historical databases are essential to identify the dynamic aspects involved in vulnerability. In this context, the refinement, maintenance and systematic updating of landslide data sets are vital for risk assessment as a whole. Data is the primary input for identifying trends in hazard and vulnerability, as well as driving the risk assessment and landslide impact analysis. There is a need to work towards the standardisation of methodologies and processes related to the collection, analysis, storage, maintenance of data.

11.6.4   *Contribution of IMIRILAND project to the management of large landslides*

The large number of still open questions related to the adequate management of large landslides might suggest that no final conclusions may be drawn from the research developed within IMIRILAND project. However the practice of multidisciplinary investigations and the interchange of methodologies have led to a much sounder assessment of involved risks which will be profitable to all land planners involved with such landslide problems. Moreover the complete analysis of several significant case studies with respect to hazard assessment and risk management shows that the scientists and the implied authorities can develop well-designed strategies together that will prevent increasing damage. It is thus possible to express confidence in the future improvement of large landslide management actions.

REFERENCES

Abele, G., 1974. Bergstürze in den Alpen. *Wissenschaftliche Alpenvereinshefte*, München. H 25, 230 S.

Almeida, E., Basabe, P., Jaramillo, H., Ramón, P. & Serrano, C., 1996. Desestabilización de laderas, por la crecida debido a la ruptura de "La Josefina". Libro: *"Sin plazo para la esperanza", reporte sobre el desastre. Escuela Politécnica Nacional (EPN)*. Quito-Ecuador, p. 213–223.

Antoine, P., Camporota, P., Giraud, A. & Rochet, L., 1987. La menace d'écroulement aux Ruines de Séchilienne (Isère). *Bull. liaison Labo. P. et Ch., n° 150–151*, pp. 55–64.

ARPA-Piemonte, 2004. *Il Progetto IFFI in Piemonte: Inventario dei fenomeni franosi in Italia.* Archives Settore Studi e Ricerche Geologiche – Sistema Informativo Prevenzione Rischi, ARPA-Piemonte [Agenzia Regionale per la Protezione dell'Ambiente], www.arpa.piemonte.it.

Baretti, M., 1881. *Relazione sulle condizioni geologiche del versante destro della valle della Dora Riparia tra Chiomonte e Salbertrand*. Tip. e Lit. Camilla e Bertolero, Torino.

Baretti, M., 1893. *Geologia della Provincia di Torino – con atlante fuori testo di 27 carte e 8 profili geologici.* Torino.

Basabe, P., 1998. *Amenazas por inundaciones y plan de contingencia ante el fenómeno "El Niño".* Comité de Emergencia Nacional del Paraguay y PNUD, Asunción–Paraguay, UNDHA, p.69+ anexos.

Basabe, P. & al., 1996. Prevención de desastres naturales en la cuenca del río Paute. Libro: *"Sin plazo para la esperanza", reporte sobre el desastre de "La Josefina"-Ecuador. EPN*. Quito-Ecuador, p. 271–287.

Basabe, P., Neumann, A., Almeida, E., Herrera, B., García, E. & Ontaneda, P., 1998. *Informe Final del proyecto PRECUPA* de cooperación entre el CSS, DNDC, INECEL, INAMHI, Munic. de Cuenca/ETAPA, U. Cuenca y Consejo de Programación por "La Josefina", para la prevención de desastres naturales en la cuenca del río Paute. Temas: Topografía/geodesia, Geología/geotecnia, Hidrometeorología, Sismología, Limnología, Capacitación/difusión y apoyo a Defensa Civil, Cuerpo Suizo de Socorro, 369 p. más anexos, Cuenca–Ecuador.

BGC Engineering Inc., Calgary, Alberta, Canada, 2000. Geotechnical hazard assessment, south flank of Frank Slide, Hillcrest, Alberta. Unpublished report to *Alberta Environment*.

Biancotti, A. & Bovo, S. (eds), 1998a. *Distribuzione regionale di piogge e temperature*. Studi climatologici in Piemonte, Vol. 1, 79 pp. Regione Piemonte, Università degli Studi di Torino.

Biancotti, A. & Bovo, S. (eds), 1998b. *Le precipitazioni nevose sulle Alpi piemontesi*. Studi climatologici in Piemonte, Vol. 2, 80 pp. Regione Piemonte, Università degli Studi di Torino.

Bieniawski, Z.T., 1974. Geomechanics Classification of Rock Masses and its Application in Tunnelling. *Proc. 3rd Int. Congr. Rock Mech.*, ISRM, Denver U.S.A. (II A): 27–32.

Bogge, A., 1975. *L'alluvione del 1728 in Val di Susa.* Centro Studi Piemontesi 4(2): 379–396.

Bonnard, Ch., 1984. Determination of slow landslide activity by multidisciplinary measurement techniques. In Kovari, K. (ed.), *Field Measurements in Geomechanics. Proc. Int. Symp, Zurich (CH), 5–8 September 1983* (1): 619–638. Rotterdam: Balkema.

Bonnard, Ch., 1994a. Recognition, Identification, and Control of Slope Instability Movements – General Report, *Proc. IVth Geoengrg. Int. Congress, Torino*, Vol. 3, pp. 685–692.

Bonnard, Ch., 1994b. Los deslizamientos de tierra: Fenómeno natural o fenómeno inducido por el hombre? *Memorias 1er simposio panamericano de deslizamientos de tierra*. Soc. ecuatoriana de Mecánica de Suelos y Rocas, Guayaquil, Ecuador, Vol. II, pp. 1–15.

307

Bonnard, Ch. & Noverraz, F., 2001. Influence of climate change on large landslides: assessment of long-term movements and trends. *Proc. Int. Conf. on Landslides: Causes, Impacts and Countermeasures, Davos (CH)*, 121–138. Ed. VGE.

Bonnard, Ch., Noverraz, F. & Dupraz, H., 1996. Long-term movements of substabilized versants and climatic changes in the Swiss Alps. *Proc. VIIth Int. Symp. on Landslides, Trondheim, Norway*, Vol. 3, pp. 1525–1530, Rotterdam: Balkema.

Bonnard, Ch., Noverraz, F., Lateltin, O. & Raetzo, H., 1995. Large Landslides and Possibilities of Sudden Reactivation. *Proc. 44th Geomechanics Colloquy, Salzburg*. Felsbau No 6/95, pp. 401–407.

Bonzanigo, L., Eberhardt, E. & Löw, S. 2001. Hydromechanical Factors Controlling the Creeping Campo Vallemaggia Landslide. *Proc. Int. Conf. on Landslides: Causes, Impacts and Countermeasures, Davos, Switzerland*, Ed. VGE, pp. 13–22.

Bossalini, G. & Cattin, M., 2002. *Studio dell'onda di piena conseguente ad una ipotetica frana in località Prequartera*. Internal report, Regione Piemonte, Comunità montana Valle Anzasca.

Brabb, E.E., 1984. Innovative approaches to landslide hazard mapping. *Proc. IVth Int. Symp. on Landslides, Toronto, Canada*. Can. Geot. Soc., Vol. 1, pp. 307–324.

Brand, E.W., 1988. Special lecture: Landslide risk assessment in Hong Kong. *Proc. Vth Int. Symp. on Landslides, Lausanne, Switzerland*. Vol. 2, pp. 1059–1074. Rotterdam: Balkema.

Broilli, L., 1974. Ein Felssturz im Großversuch. *Rock Mechanics, Suppl. 3, Wien*, pp. 69–78.

Brovero, M., Campus, S., Forlati, F., Ramasco, M. & Susella, G., 1996a. La Frana del "Cassas", Salbertrand, Val di Susa. In Regione Piemonte & Université J. Fourier (Eds.), *Rischi Generati da Grandi Movimenti Franosi*, Programma Interreg I Italia-Francia: 71–103.

Brovero, M., Campus, S., Forlati, F., Ramasco, M., Susella, G. Scavia, C., 1996b. La Frana di "Rosone", Valle Orco, In Regione Piemonte & Université J. Fourier (Eds.), *Rischi Generati da Grandi Movimenti Franosi*, Programma Interreg I Italia-Francia: 143–177.

Bunce, C.M., Cruden, D.M. & Morgenstern, N.R., 1997. Assessment of the hazard from rock fall on a highway. *Can. Geotech. J.*, (34), p. 344–356.

Bureau Bovay-Huguenin, géomètres officiels, 1994. *PNR 31 Versinclim – Etude du glissements de terrain de Sedrun (GR)* (unpublished report, commenting ETHZ reports).

Büros Bonanomi, A.G. & Donatsch, 1998. Gemeinde Tujetsch – Deformationsmessung Cuolm da Vi (unpublished report).

Büros Bonanomi, A.G. & Donatsch, 2003. Measurement data at Cuolm da Vi (unpublished CD).

CAF-CEPAL, 2000. *Las lecciones de El Niño, Volumen IV: Ecuador*. Memorias del Fenómeno El Niño 1997–1998 Restos y propuestas para la región andina. Corporación Andina de Fomento (CAF), Caracas, Venezuela.

Capello, C.F., 1941. Il lago quaternario della conca di Salabertano (Valle di Susa). *Boll. Com. Glac. It.* (21): 155–160.

Carol, I., Prat, P.C. & López, C.M., 1997. A normal/shear cracking model. Application to discrete crack analysis. *Journal of Engineering Mechanics* (123–8): 765–773.

Carrara, A., 1992. Landslide hazard assessment. *Proc. Ist Symp. Int. Sensores Remotos y Sistema de Inform. Geogr. para el Estudio de Riesgos Natur., March 10–12, 1992, Bogotá*, 329–355.

Carrara, A., Cardinali, M. & Guzzetti, F., 1992. Uncertainty in assessing landslide hazard and risk. *ITC Jour.*, v. 1992:2, 172–183.

Carraro, F., Dramis, F. & Pieruccini, U., 1979. Large-scale landslides connected with neotectonic activity in the Alpine and Apennine ranges. In *Proc. 15th Plenary Meeting of IGU-UNESCO Commission "Geomorphological Survey & Mapping", Modena, 7–15 September 1979*: 213–230.

Carrillo-Gil, A. & Carrillo-Degado, E., 1988. Landslide risk in the Peruvian Andes. *Proc. Vth Int. Symp. on Landslides, Lausanne, Switzerland*, Vol. 2, pp. 1137–1142. Rotterdam: Balkema.

Carta Geologica d'Italia in scala 1:50'000, 2002. *Foglio 132-152-153 Bardonecchia*. Servizio Geologico Nazionale APAT, Roma.

Castelli, M., Forlati, F. & Scavia, C., 2001. Landslides: the questions of the decision-maker and the answer of the modelisation, *Proc. 1st Int. Conf. Albert Caquot, Paris*, 3–5 October 2001.

Cherubini, C., Giasi, C.I. & Guadagno, F.M., 1993. Probabilistic approaches of slope stability in a typical geomorphological setting of Southern Italy. *Risk and reliability in ground engineering*, pp. 144–150, Thomas Telford, London.

Chopin, C., 1981. Talc-Phengite: a Widespread Assemblage in High-Grade Pelitic Blueschists of the Western Alps. *J. of Petrology*, 22(4): 628–650.

Compagnoni, R. & Lombardo, B., 1974. The Alpine age of the Gran Paradiso eclogites. *Rend. Soc. It. Min. Petr.*, 30 (1): 223–237.

Compagnoni, R., Elter, G. & Lombardo, B., 1974. Eterogeneità stratigrafica del complesso degli "gneiss minuti" nel massiccio cristallino del Gran Paradiso. *Mem. Soc. Geol. It.*, 13 (1): 227–239.

Consiglio Nazionale Delle Ricerche (CNR), 1990. Structural model of Italy, scale 1:500'000, sheet 1. Progetto Finalizzato Geodinamica, Firenze, Italy.

Consorzio Forestale di Oulx, 1955. *Capitolato di vendita di soprasuolo boschivo*. Not published, Archives Consorzio Forestale di Oulx, 13 febbraio 1955.

Corominas, J., Copons, R., Vilaplana, J.M., Altimir, J. & Amigó, J., 2002. Integrated landslide susceptibility analysis and hazard assessment in the principality of Andorra. *Natural Hazard*, Kluwer Academics Publisher: 1–15.

Corpo Forestale dello Stato, 1956. *Lettera inviata al Consorzio Forestale di Salbertrand del 26/1/1956*. Not published, Archives Consorzio Forestale di Oulx.

Crosta, G., Frattini, P. & Sterlacchini, S., 2001. *Valutazione e gestione del rischio da frana. Principi e metodi*. Volume 1. Milano: Regione Lombardia and Università di Milano Bicocca.

Cruden, D., 1991. A simple definition of a landslide. *Bull. Int. Ass. of Engrg. Geol.*, No. 43, pp. 27–29.

Cruden, D.M. & Fell, R. (Eds.), 1997. *Proc. Int. Workshop on Landslide Risk Assessment, Honolulu (Hawaii, USA), 19÷21 February 1997*. Rotterdam: Balkema.

Cruden, D.M. & Varnes, D.J., 1996. *Landslide types and processes*. In: "Landslides: Investigation and Mitigation". Transportation Research Board. National Academy of Sciences, pp. 36–75.

C.T.M. [Compagnia Torinese Monitoraggi], 2003. *Impianto di monitoraggio territoriale, geotecnico e sismico della Val Susa – Frana del Cassas. Impianto di monitoraggio inclinometrico – Misura strumentazione, Rapporto Interpretativo N° 14, giugno 2003* (not published). SITAF (Soc. It. Traforo Autostradale Frejus) s.p.a., Archives Settore Progettazione Interventi Geologico-Tecnici e Sismico, ARPA-Piemonte, Torino.

Cundall, P.A., 1987. Distinct Element Models of Rock and Soil Structure, *Analytical and Computational Methods in Engineering Rock Mechanics*, Ch. 4, pp. 129–163, E.T. Brown, Ed. London: Allen & Unwin.

Dal Piaz, G. & Lombardo, B., 1986. Ealy Alpine ecloite metamorphism in the Penninic Monte Rosa-Gran Paradiso basement nappes of the northwestern Alps. *Geol. Soc. Am.*, 164: 249–265.

Davis, G.H. & Reynolds, S.J., 1996. *Structural geology of rocks and regions*. 776 p. New York: Wiley & Sons.

Dela Pierre, F., Lozar, F. & Polino, R., 1997. L'utilizzo della tettono-stratigrafia per la rappresentazione cartografica delle successioni meta-sedimentarie nelle aree di catena. *Mem. Sci. Geol.* (49): 195–206.

Del Prete, M., Giaccardi, E. & Trisoro-Liuzzi, G., 1992. *Rischio da frane intermittenti a cinematica lenta nelle aree montuose e collinari urbanizzate della Basilicata*. Pubbl. n. 841 GNDCI (Gruppo Nazionale per la Difesa dalle Catastrofi Idrogeologiche). Potenza: Centro Nazionale delle Ricerche.

DRM (Délégation aux risques majeurs), 1990. *Les études préliminaires à la cartographie réglementaire des risques naturels majeurs*. Secrétariat d'Etat auprès du premier Ministre chargé de l'Environnement et de la Prévention des Risques technologiques et naturels majeurs. La Documentation Française.

Duranthon, J.-P., Effendiantz, L., Memier, M. & Previtali, I., 2003. Apport des méthodes topographiques et topométriques au suivi du versant rocheux instable des Ruines de Séchilienne. *Revue XYZ*, n° 94.

Dutto, F. & Friz, E., 1989. Ricerche sull'area di invasione di valanghe di roccia. *Rapporto interno CNR-IRPI/ISMES*, Febbraio, 16p.

309

Eckardt, P., Funk, H.P. & Labhart, T., 1983. Postglaziale Krustenbewegungen an der Rhein-Rhone-Linie. *Vermessung, Photogrammetrie und Kulturtechnik 2/83*, pp. 43–56.

Effendiantz, L., Marchesini, P. & Pera, J., 2000. La galerie de reconnaissance de Séchilienne. *Tunnels et ouvrages souterrains*, n° 161, pp. 303–306.

Einstein, H.H., 1988. Special lecture: Landslide risk assessment procedure. *Proc. Vth Int. Symp. on Landslides, Lausanne, Switzerland*, Vol. 2, pp. 1075–1090. Rotterdam: Balkema.

Eisbacher, G.H. & Clague, J.J., 1984. Destructive mass movements in high mountains: hazard and management. Ottawa: *Geological Survey of Canada, Paper 84*, 16, p. 230.

Enel.Hydro, 2001. *Attività di progetazione, fornitura ed installazione di un sistema di monitoraggio integrato del movimento franoso di Rosone (TO); Scenari di rischio*. Prog. ISMES 2338. Archives of Settore Studi e Ricerche Geologiche ARPA-Piemonte, Italy.

Engelder, T., 1985. Loading paths to joint propagation during a tectonic cycle: an example from the Appalachian Plateau, U.S.A. *J. Struct. Geol.* 7: 459–476.

Epifani, F.D., 1991. *Studio geologico e geomorfologico del versante a monte delle aree di servizio di Salbertrand* – Studio geologico Epifani – SITAF (not published), Archives Settore Progettazione Interventi Geologico-Tecnici e Sismico, ARPA-Piemonte, Torino.

Evans, S.G., 2003. Characterizing Landslide risks in Canada. *Proc. Geohazards 2003*, Edmonton Conference, Canadian Geotechnical Society, pp. 20.

Evans, S.G., Hungr, O. & Enegren, E.G., 1994. The Avalanche Lake rock avalanche, Mackenzie Mountains, Northwest Territories, Canada: description, dating and dynamics. *Canadian Geotechnical Journal* (31): 749–768.

Evans, S.G., Clague, J.J., Woodsworth, G.J. & Hungr, O., 1989. The Pandemonium Creek Rock Avalanche, British Columbia. *Canadian Geotechnical Journal*, 26: 427–446.

Evrard, H., Gouin, T., Benoit, A. & Duranthon, P., 1990. Séchilienne. Risques majeurs d'éboulement en masse. Point sur la surveillance du site. *Bull. liaison Labo. P. et Ch., n° 165*, pp. 7–16.

Fell, R., 1994. Landslide risk assessment and acceptable risk. *Can. Geot. J.*, (31): 261–272.

Fell, R., Hungr, O., Leroueil, S. & Riemer, W., 2000. Keynote Lecture – Geotechnical engineering of the stability of natural slopes, and cuts and fills in soil. *Proc. GeoEng 2000 Conf., Melbourne*, Vol. 1, pp. 21–120.

Finlay, P.J. & Fell, R., 1997. Landslides: risk perception and acceptance. *Can. Geot. J.*, (34): 169–188.

FLAC3D, 1997. *Itasca: Fast Lagrangian Analysis of Continua in 3 Dimensions (FLAC3D)*, Version 2.0, Manual, Itasca, Minnesota, Minneapolis.

Forlati, F. & Piana, F., 1998. Vincoli geologico-strutturali e caratterizzazione degli ammassi rocciosi. In: G. Barla Ed., MIR 98, *Conf. Naz. Mecc. Ing. Rocce*, VI ciclo, Torino, 1–11.

Forlati, F., Ramasco, M., Susella, G., Barla, G., Marino, M. & Mortara, G., 1993. La deformazione gravitativi di Rosone. Un approccio conoscitivo per la definizione di una metodologia di studio. *Studi Trentini di Scienze Naturali, Acta Geologica* 68: 71–108.

Fratianni, S. & Motta, L., 2002. *Andamento climatico in Alta Val Susa negli anni 1990–1999*. Regione Piemonte, Università degli Studi di Torino. Studi climatologici in Piemonte (4).

Friz, E. & Pinelli, P.F., 1993. Riceerche sull'area di invasione di valanghe di roccia. *Atti del convegno "Fenomeni Franosi", Riva del Garda*, 23–25 maggio 1990. Studi Trentini di Scienze Naturali, Acta Geologica, 68, 1, pp.71–108.

Geo & Soft International, 1999. *ROTOMAP for Windows*. Turin, Italy.

Geo & Soft International, 2003. *ISOMAP & ROTOMAP for Windows (3D Surface Modelling & Rockfall Analysis)*. User's Guide.

Geoengineering, 1984. Studio geologico-tecnico del settore di versante retrostante alla frazione di Rosone vecchio. Internal report Comune Locana.

Gianquinto, F., 2000. *Studio idrogeologico dei settori di versante interessati dalle frane del Cassas e di Serre la Voute*. ECOPLAN – SITAF, May 2000 (not published), Archives Settore Progettazione Interventi Geologico-Tecnici e Sismico, ARPA-Piemonte, Torino.

Giardino, M. & Polino, R., 1997. Le deformazioni di versante dell'alta valle di Susa: risposta pellicolare dell'evoluzione tettonica recente. *Il Quaternario, Italian Journal of Quaternary Sciences* (10–2): 293–298.

Hancock, P. L., 1985. Brittle microtectonics: principles and practice. *J. Struct. Geol.*, 7: 437–457.

Heim, A., 1932. *Bergsturz und Menschenleben*. Zurich: Fretz & Wasmuth Verlag, p. 227.

Hobbs, B.E., Means, W.D. & Williams, P.F., 1976. *An Outline of Structural Geology*. New York: Wiley & Sons.

Hötzl, H., Moser, M., Reichert, B. & Rentschler, K., 1994. Hydrologische Markierungsversuche in Massenbewegungen. Reppwandgleitung, Kärnten und Stubnerkogel, Salzburg. *Steir. Beiträge z. Hydrogeologie*, H. 45, S. 69–92.

Hsü, K.J., 1975. Catastrophic debris streams (Sturzstroms) generated by rockfalls. *Bull. Geol. Soc. of Am.* 86 (1): 129–140.

Hsü, K.J., 1978. Albert Heim: observations on landslides and relevance to modern interpretation. In Voight (Ed.), *Rockslides and avalanches* (1): 71–93.

Hungr, O., 1981. *Dynamics of rock avalanches and other types of mass movements*. Ph. D. thesis, University of Alberta, Canada.

Hungr, O., 1995. A model for the runout analysis of rapid flow slides, debris flows, and avalanches. *Canadian Geotechnical Journal*, 32: 610–623.

Hungr, O., 1997. Some methods of landslide hazard intensity mapping. *Proc. Int. Workshop on Landslide Risk Assessment*, D.M. Cruden & R. Fell (Eds.), pp. 215–226. Rotterdam: Balkema.

Hungr, O., 2001. A review of the classification of landslides of the flow type. *Environmental and Engineering Geoscience* (7): 221–238.

Hungr, O. & Evans, S.G., 1996. Rock avalanche runout prediction using a dynamic model. *Proc. 7th Int. Symp. on Landslides, Trondheim, Norway*, V. 1, p. 233–238. Rotterdam: Balkema.

Hungr, O. & Rawlings, G., 1995. Terrain hazards assessment for planning purposes: Cheekye Fan, B.C. *Proc. 48th. Canadian Geotechnical Conf. Vancouver, B.C.* 1: 509–517.

Hungr, O., Sobkowicz, J. & Morgan, G., 1993. How to economise on natural hazards. *Geotechnical news*, v. 7/1, pp. 54–57.

Hutchinson, J.N., 1992. Landslide hazard assessment. *Proc VIth Int. Symp. on Landslides, Christchurch, New Zealand*. Vol. 3, pp. 1805–1841. Rotterdam: Balkema.

Imrie, A.S. & Moore, D.P., 1997. BC Hydro's approach to evaluating reservoir slope stability from a risk perspective. *Proc. Int. Workshop on Landslide Risk Assessment*, D.M. Cruden & R. Fell Eds., pp. 197–205. Rotterdam: Balkema.

Imrie, A.S., Moore, D.P. & Enegren, E.G., 1992. Performance and maintenance of the drainage system at Downie Slide. *Proc. 6th. Int. Symp. on Landslides, Christchurch, New Zealand*, Vol. 1: 751–757. Rotterdam: Balkema.

I.S.R.M. [International Society for Rock Mechanics], 1978. Suggested Methods for the Quantitative Description of Discontinuity in the Rock Masses. *Int. J. Rock Mech. Min. & Geomech.* (Abstract, 15): 319–368.

ITASCA, 1997. *Fast Lagrangian Analysis of Continua in 3 Dimensions (FLAC³ᴰ)*, Version 2.0, Manual. Minneapolis: Itasca.

ITASCA, 1999a. *PFC²ᴰ, Particle Flow Code in Two Dimensions*, Version 2.0, Manual. Minneapolis: Itasca.

ITASCA, 1999b. *PFC³ᴰ, Particle Flow Code in Three Dimensions*, Version 2.0, Manual. Minneapolis: Itasca.

IUGS/WGL/CRA (International Union of Geological Societies/Working Group on Landslide/ Committee on Risk Assessment), 1997. Quantitative risk assessment for slopes and landslides – The state of the art. In Cruden, D.M. & Fell, R. (Eds.), *Proc. Int. Workshop on Landslide Risk Assessment, Honolulu (Hawaii, USA), 19÷21 February 1997*: 3–12. Rotterdam: Balkema.

Khaler, F. & Prey, S., 1963. *Erläuterungen zur geologischen Karte des Naßfeld-Gartnerkofel-Gebietes in den Karnischen Alpen*. Wien (GBA) (Erläuterungen geol. Karten).

Lapenta, M.C. & Trombetta, A., 1996. Caratterizzazione geomeccanica dei versanti rocciosi dell'alta Valle di Susa (Piemonte) in relazione ai problemi di stabilità. *GEAM (Associazione Georisorse & Ambiente)* (91): 31–39.

Larini, G., Malaguti, C., Pellegrini, M. & Tellini, C., 2001. La Lama di Corniglio (Appennino parmense), riattivata negli anni 1994–1999. *Quaderni di Geologia Applicata*, 8.2 (2001), pp. 59–114. Bologna: Pitagora Editrice.

Läufer, A.L., Frisch, W., Steinitz, G. & Loeschke, J., 1997. Exhumed fault-bounded Alpine blocks along the Periadriatic lineament: the Eder unit (Carnic Alps, Austria). In: *Geol. Rundschau*, Bd. 86, S. 612–616.

Leone, F., Asté, J.-P. & Leroi, E., 1996. Vulnerability assessment of elements exposed to mass-movement: working toward a better risk reception. *Proc. VIIth Int. Symp. on Landslides, Trondheim, Norway, Vol. 1*, p. 263–270. Rotterdam: Balkema.

Leroi, E., 1997. Landslide risk mapping: problem, limitations and developments. In Cruden, D.M. & Fell, R. (Eds.), *Proc. Int. Workshop on Landslide Risk Assessment, Honolulu (Hawaii, USA), 19÷21 February 1997*: 239–250. Rotterdam: Balkema.

Li Tianchi, 1983. A mathematical model for predicting the extent of a major rockfall. *Zeitschrift für Geomorfologie*, 27: 473–482.

Luino, F., Ramasco, M. & Susella, G., 1993. Atlante dei centri abitati instabili piemontesi. Gruppo per la difesa delle catastrofi idrogeologiche. *Prog. Spec. Studio centri abitati instabili, pubb. n° 964, CNR-IRPI, Regione Piemonte, Settore Prevenzione Rischio Geol. Meteorol. e Sismico*. p.245.

Mathews, W.H. & McTaggart, K.C., 1978. The Hope rock slides, British Columbia, Canada. In *Rockslides and Avalanches*, Voight, B. Ed., 1: 197–258. Amsterdam: Elsevier.

Mayoraz, F., Cornu, Th., Djukic, D. & Vulliet, L., 1997. Neural networks: A tool for prediction of slope movements. *Proc. 14th Int. Conf. on Soil Mechanics and Foundation Engineering, Hamburg*. Vol. 1, pp. 703–706.

McConnell, R.G. & Brock, R.W., 1904. The great landslide at Frank, Alberta. *Canadian Parliament Sessional Paper, No.25*, Department of the Interior, Government of Canada.

Moore, D.P. & Imrie, A.S., 1992. Stabilization of Dutchman's Ridge. *Proc. VIth. Int. Symp. on Landslides, Christchurch, New Zealand*, Vol. 3:1783–1788. Rotterdam: Balkema.

Moore, D.P., Imrie, A.S. & Baker, D.G., 1991. Rockslide risk reduction using monitoring. *Proc. Canadian Dam Safety Conference*. Vancouver: BiTech Publishers.

Mora, S. & Wahrson, W., 1994. Macrozonation methodology for landslide hazard determination. *Memorias 1er simposio panamericano de deslizamientos de tierra, Sociedad Ecuatoriana de Mecánica de Suelos y Rocas, Guayaquil*, Vol. I, pp. 406–431.

Morelli, M., 2000. *Analisi sulla possibilità di integrazione dei dati telerilevati in studi geologico-strutturali: applicazione nel dominio del Monferrato e delle Langhe.* Ph.D. Thesis, University of Torino, p. 168.

Morelli, M. & Piana, F. 2003. "Geometric filtering" of remote-sensed lineaments and search for geological rules of their distribution. Application to the Monferrato succession marly-arenaceous succession (NW-Italy). Abstract in *Proc. of Tectonic Studies Group AGM, Liverpool(UK), 8–10 January 2003*.

Mortara, G. & Sorzana, P.F., 1987. Fenomeni di deformazione gravitativa profonda nell'arco alpino occidentale italiano. Considerazioni lito-strutturali e morfologiche. *Boll. Soc. Geol. It.* (106): 303–314.

Moser, M., 2001. *Talzuschub Reppwand-Gleitung, Geologisch-geotechnischer Bericht für 2000 unter besonderer Berücksichtigung der Bewegunsmessungen (Periode 14–15, Juni 99 bis Oktober 2000).* Bericht WLV (Austrian Service for Torrent, Erosion and Avalanche Control), unpublished.

Moser, M. & Glawe, U., 1994. Das Naßfeld in Kärnten – geotechnisch betrachtet. In: *Abh. Geol. B.-A.*, Bd. 50, pp. 319–340.

Moser, M. & Windischmann, Th., 1989. Die Reppwandgleitung/Kärnten. Geologische und geotechnische Betrachtungen. In: *Oberrhein. geol. Abh.*, 35. pp. 157–176.

Moser, M., Angerer, J. & Seitz, S., 1988. Geotechnische Untersuchungsergebnisse im Rahmen des Verbauungsprojektes Oselitzenbach/Kärnten. *Interprävent (Ed): Interprävent 3, Graz.* pp. 77–102.

Mulder, H.F.H.M., 1991. *Assessment of landslide hazard*. Geograph. Sci. Fac., Elinkwijk Utrecht, p. 150.

Nichol, S. & Hungr, O., 2002. Brittle and ductile toppling of large rock slopes. *Canadian Geotechnical Journal*, 39:1–16.

Noverraz, F. & Bonnard, Ch., 1991. L'écroulement rocheux de Randa, près de Zermatt. *Proc. VIth Int. Symp. on Landslides, Christchurch, New Zealand*. Vol. 1, pp. 165–170. Rotterdam: Balkema.

Noverraz, F., Bonnard, Ch., Dupraz, H. & Huguenin, L., 1998. *Grands glissements de versants et climat*. Rapport final PNR 31 – Projet VERSINCLIM, 314 p. Zurich: v/d/f Verlag.

OFAT (Office Fédéral de l'Aménagement du Territoire), OFEE (Office Fédéral de l'Economie des Eaux) & OFEFP (Office Fédéral de l'Environnement, des Forêts et du Paysage), 1997. *Prise en compte des dangers dus aux mouvements de terrain dans le cadre des activités de l'aménagement du territoire, Recommandations Fédérales*. Série dangers naturels (O. Lateltin, Ed.), OCFIM (Office Central Fédéral des Imprimés et du Matériel), Bern, Switzerland, 42 p.

OFEE (Office Fédéral de l'Economie des Eaux), OFAT (Office Fédéral de l'Aménagement du Territoire) & OFEFP (Office Fédéral de l'Environnement, des Forêts et du Paysage), 1997. *Prise en compte des dangers dus aux crues dans le cadre des activités de l'aménagement du territoire, Recommandations Fédérales*. Série dangers naturels (R. Loat et A. Petraschek, Eds.), OCFIM (Office Central Fédéral des Imprimés et du Matériel), Bern, Switzerland, p. 32.

Panet, M., Bonnard, Ch., Lunardi, P. & Presbitero, M., 2000. *Expertise relative aux risques d'éboulement du versant des Ruines de Séchilienne*. Rapport pour le ministère de l'aménagement du territoire et de l'environnement. See the website: www.environnement.gouv.fr.

Paro, L., 1997. *Ricostruzione dell'evoluzione geologica quaternaria del versante destro della Valle di Susa nel tratto compreso tra Salbertrand ed Exilles*. Thesis in Geological Science, University of Torino; 4 maps published by Regione Piemonte.

Patton, F.D., 1984. Climate, Groundwater Pressures and Stability Analyses of Landslides. *Proc. IVth Int. Symp. on Landslides, Toronto, Canada*. Can. Geot. Soc., Vol. 3, pp. 43–59.

Perello, P., Delle Piane, L., Piana, F., Stella, F. & Damiano, A., 2004. Brittle post-metamorphic tectonics in the Gran Paradiso massif (north-eastern Italian Alps). *Geodinamica Acta*, in press.

Peretti, L., 1967. Collegamento autostradale del traforo del Frejus con Torino. *La Rivista della Strada* (36, 310).

Peretti, L., 1969. Premesse geoapplicative per la realizzazione coordinata dell'autostrada Torino – Oulx e della sistemazione idrogeologica della Valle di Susa. *Cronache da Palazzo Cisterna-Periodico della Provincia di Torino* (1).

Perla, R., Cheng, T. & McClung, D., 1980. A two-parameter model of snow-avalanche motion. *Journal of Glaciology*, Vol. 26, n.94, 197–207.

Petrascheck, A., 1995. *Mapas de peligros: Encuesta respecto de los niveles de peligrosidad*. Offices fédéraux suisses de l'économie des eaux (OFEE), de l'aménagement du territoire (OFAT) et de l'environnement, des forêts et du paysage (OFEFP), Berna. Inédito. 16 p. y anexos.

Polino, R., Dela Pierre, F., Borghi, A., Carraro, F., Fioraso, G. & Giardino, M., 2002. *Note Illustrative della Carta Geologica d'Italia alla scala 1:50'000 – foglio 132-152-153 Bardonecchia*. Servizio Geologico Nazionale APAT, Roma.

Pothérat, P. & Alfonsi, P., 2001. Les mouvements de versant de Séchilienne (Isère). *Revue française de géotechnique n° 95–96*, p. 117–131.

Potter, D., 1972. *Computational physics*. London: John Wiley & Sons.

PPR (Plans de Prévention des Risques Naturels Prévisibles), 1995. *Loi n° 95–101 du 2 février 1995*. Ministère de l'Ecologie et du Développement Durable, Direction de la Prévention, des Pollutions et des Risques (DPPR), France.

Prat, P.C., Gens, A., Carol, I., Ledesma, A. & Gili, J.A., 1993. DRAC: A computer software for the analysis of rock mechanics problems. H. Liu (Ed.), *Application of computer methods in rock mechanics*. Xian (China): Shaanxi Science and Technology Press. 1361–1368.

Programme INTERREG I Italie-France, 1996. *Risques générés par les grands mouvements de versants*. Regione Piemonte, Université Joseph Fourier. p. 207.

Puma, F., Ramasco, M., Stoppa, T. & Susella, G., 1984. Carta dei movimenti di massa nelle alte valli del Chisone e di Susa. – in Compagnoni et al., *Geotraversa della Zona Piemontese nella Valle di Susa, escursione pre-congresso (10÷11/9/1984) 72° Congresso Soc. Geol. It., Serv. Geol. Reg. Piemonte.*

Puma, F., Ramasco, M., Stoppa, T. & Susella, G., 1989. Movimenti di massa nelle alte valli di Susa e Chisone. *Boll. Soc. Geol. It.* (108): 391–399.

Ramasco, M. & Susella, G., 1978. *Studi geologici per il collegamento stradale tra il traforo del Frejus e Torino (tratto Bardonecchia-Susa).* Regione Piemonte, Dipartimento Organizzazione e Gestione del territorio.

Ramasco, M., Stoppa, T. & Susella, G., 1989. La deformazione gravitativa profonda di Rosone in Valle Orco. *Boll. Soc. Geol. It.* 108: 401–408.

Ramsay, J.G. & Huber, M.I., 1987. *The techniques of modern structural geology.* Volume 2: Folds and Fractures. London: Academic Press.

Regione Piemonte, 2000. *La frana di Prequartera, analisi del movimento franoso e valutazione delle potenziali fasi evolutive e schema del sistema di controllo del fenomeno franoso.* A cura del Gruppo Interdisciplinare di Studio Regione Piemonte e Politecnico di Torino e del settore Regionale Progettazione Interventi Geologico Tecnici e Sismici, Internal report (unpublished).

Rochet, L., 1987. Développement des modèles numériques dans l'analyse de la propagation des éboulements rocheux. *Proc. of the VIth Int. Congress on Rock Mechanics, Montreal,* Vol. 1, pp. 479–484.

Romana, M., 1990. Practice of SMR Classification for Slope Appraisal. *Proc. 5th Int. Symp. on Landslides, Lausanne, Switzerland,* Vol. 2: 1227–1231. Rotterdam: Balkema.

Sacco, F., 1898. *La geologia e le linee ferroviarie in Piemonte.* Torino.

Scavia, C., Barla, G. & Bernaudo, V., 1990. Probabilistic stability analysis of block toppling failure in rock slopes. *Int. Journal of Mechanics Mining Science & Geomechanical Abstracts* (27–6): 465–478. Oxford: Pergamon Press.

Schönlaub, H.P. & al., 1987. *Geologische Karte der Republik Österreich 1:50'000,* Blatt 198 (Weißbriach). Geol.BA. Wien.

Schreyer, W., 1977. Whiteschists: their composition and pressure-temperature regimes based on experimental, field and petrographic evidence. *Tectonophysics,* 43: 127–144.

Scioldo, G., 1991. ROTOMAP: analisi statistica del rotolamento dei massi. Ass. Min. Subalpina: *Atti Convegno "La meccanica delle rocce a piccola profondità", Torino,* 8184.

SEA consulting, 2001. Ductile structural setting of the Gran Paradiso unit and relationships with large scale landslides in the Orco valley (Italian western Alps). In Giulio Elter (Ed.), *Dalla Tetide alle Alpi; Proc. symp. Cogne, 21–22, Giugno 2001, 37–38.*

Segré, C., 1920. Considerazioni geognostiche sul tronco Bussoleno-Salbertrand (ferrovia Torino-Frejus) con riguardo speciale ai tratti franosi: provvedimenti. *Giornale del Genio Civile* (LVIII): 19–40.

Service Hydrologique et Géologique National, 1992. *Swiss Hydrological Atlas.*

S.I.T.A.F. [Società Italiana Traforo Autostradale del Frejus], 2000. *Frane di Serre la Voute e del Cassas, modellazione matematica e valutazione del rischio geologico.* Authors: C.T.M. – Polithema studio associato – Oboni associés, not published, Archives Settore Progettazione Interventi Geologico-Tecnici e Sismico, ARPA-Piemonte, Torino.

Sobkowicz, J., Hungr, O. & Morgan, G.C., 1995. Probabilistic mapping of a debris flow hazard area. *Proc. 48th. Canadian Geotechnical Conference, Vancouver, B.C.* 1: 519–529.

Spang, R.M. & Rautenstrauch, R.W., 1988. Empirical and mathematical approaches to rockfall protection and their practical applications. *Proc. Vth Int. Symp. on Landslides, Lausanne, Switzerland,* Vol. 2, pp. 1237–1243. Rotterdam: Balkema.

Studio Geologico Italiano (SGI), 1984. A.E.M. Torino Impianto S. Lorenzo (Locana). Relazione geologica, Allegato 1, Internal report Regione Piemonte.

Turner, A.K. & Schuster, R.L. (Eds.), 1996. *Landslides – Investigation and Mitigation.* Special Report 247. Transportation Research Board, Academy of Sciences, Washington D.C., p. 673.

UN (United Nations), 1991. *Mitigating Natural Disasters. Phenomena, Effects and Options. A Manual for Policy Makers and Planners.* UNDRO/MND/1990 Manual. New York: UN (1991-VIII).

UNESCO, 1993. *Multilingual landslide glossary.* The International Geotechnical Societes. UNESCO Working party for world landslide inventory. BiTech Publishers. ISBN 0-920 505-10-4, Canada.

Varnes, D. J. & The International Association of Engineering Geology Commission on Landslides and other Mass Movements, 1984. *Landslide Hazard Zonation: A Review of Principles and Practice.* Natural hazards (3), p. 63 Paris, France: UNESCO.

Varnes, D.J., 1978. Slope movement types and processes. In Schuster, R.L. and Krizek, R.J. (Eds.), *Landslides. Analysis and Control.* Transportation Research Board, Special Report 176: 12–33. Washington, D.C.: National Academy of Sciences.

Vialon, P., Ruhland, M. & Grolier, J., 1976. *Eléments de tectonique analytique.* p. 118 Paris: Masson.

Vulliet, L., & Bonnard, Ch., 1996. The Chlöwena Landslide: Prediction with a Viscous Model. *Proc. 7th Int. Symp. on Landslides, Trondheim, Norway,* Vol. 1, pp. 397–402. Rotterdam: Balkema.

Weidner, S., 2000. *Kinematik und Mechanismus alpiner Hangdeformationen unter bes. Berücksichtigung der hydrogeologischen Verhältnisse.* PhD Thesis. Uni Erlangen, unpublished.

Woodcok, N.H. & Fischer, M., 1986. Strike-slip duplexes. *J. Struct. Geol.* 8: 725–735.

Zettler, A.H., Poisel, R., Roth, W. & Preh, A., 1999. Slope stability analysis based on the shear reduction technique in 3D. *Proc. of the International Flac Symposium on Numerical Modeling in Geomechanics, Minnesota, Minneapolis,* pp. 11–16.

Author Index

MUSIC AND FAMILIARITY

SEMPRE Studies in the Psychology of Music

Series Editors

Graham Welch, *Institute of Education, University of London, UK*
Adam Ockelford, *Roehampton University, UK*
Ian Cross, *University of Cambridge, UK*

The theme for the series is the psychology of music, broadly defined. Topics will include: (i) musical development at different ages, (ii) exceptional musical development in the context of special educational needs, (iii) musical cognition and context, (iv) culture, mind and music, (v) micro to macro perspectives on the impact of music on the individual (such as from neurological studies through to social psychology), (vi) the development of advanced performance skills and (vii) affective perspectives on musical learning. The series will present the implications of research findings for a wide readership, including user-groups (music teachers, policy makers, parents), as well as the international academic and research communities. The distinguishing features of the series will be this broad focus (drawing on basic and applied research from across the globe) under the umbrella of SEMPRE's distinctive mission, which is to promote and ensure coherent and symbiotic links between education, music and psychology research.

Other titles in the series

Collaborative Learning in Higher Music Education
Edited by Helena Gaunt and Heidi Westerlund

I Drum, Therefore I Am
Being and Becoming a Drummer
Gareth Dylan Smith

The Act of Musical Composition
Studies in the Creative Process
Edited by Dave Collins

Studio-Based Instrumental Learning
Kim Burwell

Musical Creativity: Insights from Music Education Research
Oscar Odena

Music and Familiarity
Listening, Musicology and Performance

Edited by

ELAINE KING
University of Hull, UK

HELEN M. PRIOR
King's College, London, UK

ASHGATE

Published by
Ashgate Publishing Limited
Wey Court East
Union Road
Farnham
Surrey, GU9 7PT
England

Ashgate Publishing Company
110 Cherry Street
Suite 3-1
Burlington, VT 05401-3818
USA

www.ashgate.com

British Library Cataloguing in Publication Data
Music and familiarity : listening, musicology and
  performance. -- (SEMPRE studies in the psychology of music)
  1. Music--Psychological aspects. 2. Music appreciation.
  3. Musical analysis. 4. Musicology. 5. Music--
  Performance--Psychological aspects.
  I. Series II. King, Elaine, 1974- III. Prior, Helen M.
  IV. Society for Education, Music and Psychology Research.
  781.1'7-dc23

The Library of Congress has cataloged the printed edition as follows:
Music and familiarity : listening, musicology and performance / edited by Elaine King and Helen M. Prior.
    pages cm. -- (SEMPRE studies in the psychology of music)
  Includes bibliographical references and index.
  ISBN 978-1-4094-2075-0 (hardcover) -- ISBN 978-1-4094-2076-7 (ebook) -- ISBN 978-1-4724-0027-7 (epub) 1. Music--Psychological aspects. 2. Music appreciation--Psychological aspects. I. King, Elaine, 1974- II. Prior, Helen M.
  ML3830.M966 2013
  781'.11--dc23

                                                                            2012042526

ISBN 9781409420750 (hbk)
ISBN 9781409420767 (ebk – PDF)
ISBN 9781472400277 (ebk – ePUB)

Printed and bound in Great Britain
by MPG PRINTGROUP

Contents

PART III PERFORMANCE

List of Figures

List of Tables

List of Music Examples

Notes on Contributors

Artemis Apostolaki completed her PhD thesis at the University of Hull, investigating the role of solfège in music memorisation. Her background is in linguistics and music whereby she obtained a degree in Greek literature, specialising in linguistics, from the Aristotle University of Thessaloniki before completing a Masters degree in music performance at the University of Hull. She received a Gerry Farrell Travel Award from the Society for Education, Music and Psychology Research in support of her doctoral study. Her research interests include music perception and encoding, and music education in respect to aural skills and music reading. She is currently employed as a cello teacher.

Melissa C. Dobson completed her PhD thesis at the University of Sheffield, focusing on the factors that influence audience members' experiences and enjoyment of classical concert attendance. Her research interests lie in the social psychology of music; she is particularly interested in both performers' and audience members' responses to live music performance. Forthcoming articles are in press at *Psychology of Music*, *Music Performance Research* and the *Journal of New Music Research*.

Mine Doğantan-Dack is Research Fellow at Middlesex University in London where she is currently leading the music research programme. Born in Istanbul, she is a music theorist and pianist, performing regularly as a soloist and chamber musician. She has published articles on the history of music theory, expressivity in music performance, affective responses to music, chamber music performance, and phenomenology of pianism. Her books include *Mathis Lussy: A Pioneer in Studies of Expressive Performance* (2002) and the edited volume *Recorded Music: Philosophical and Critical Reflections* (2008). She was a finalist for the annual Excellence in Research Award given by the Association for Recorded Sound Collections. She is the founder of the Marmara Piano Trio (www.mdx.ac.uk/alchemy) and the recipient of an AHRC award for her research in chamber music performance.

Katherine A. Finlay is Lecturer in Psychology at the University of Buckingham. Her research interests are a fusion of clinical and music psychology research, with an emphasis on pain management. She read music as an undergraduate at Cambridge University and moved into psychology through an MSc in Music Psychology at Keele University. Her PhD at Edinburgh University investigated the role of audio-analgesia and psychological methods of pain control in working with acute and chronic pain sufferers. She has subsequently extended her research into the field of health psychology. She is a member of the Buckingham University Music Psychology Research Group and regularly performs as a flautist in recitals and orchestral concerts.

Jane Ginsborg is Associate Dean of Research, Director of the Centre for Music Performance Research and Programme Leader for Research Degrees at the Royal Northern College of Music, where she holds a Personal Chair. Formerly a professional singer, with a BA (hons) degree in Music from the University of York and an Advanced Diploma in Singing from the Guildhall School of Music and Drama, she subsequently gained a BA (hons) degree in Psychology at Keele University; she is a Chartered Psychologist and a Fellow of the Higher Education Academy. She was a Lecturer at the University of Manchester, a post-doctoral researcher at the University of Sheffield and Senior Lecturer in Psychology at Leeds Metropolitan University. She has published widely on expert musicians' preparation for performance, collaborative music making and musicians' health, is Managing Editor of the online peer-reviewed journal *Music Performance Research* and serves on the Editorial Boards of the *Journal of Interdisciplinary Music Studies* and *Musicae Scientiae*. In 2002 she was awarded the British Voice Association's Van Lawrence Award for her research on singers' memorising strategies.

Alinka Greasley is Lecturer in Psychology of Music at the University of Leeds. Her research lies within the field of social psychology of music and focuses on all aspects of people's engagement with music, including musical preferences, listening behaviour, understanding audiences and music festival research. She has a developing international profile of journal articles (for example, *Qualitative Research in Psychology*, *Musicae Scientiae*) and book chapters (for example, Oxford University Press, Ashgate) and presents regularly at national and international conferences. She is a Member of the British Psychological Society, the Society for Education, Music and Psychology Research, and a Fellow of the Higher Education Academy.

Susan Hallam is Professor of Education at the Institute of Education, University of London and currently Dean of the Faculty of Policy and Society. She pursued careers as both a professional musician and a music educator before completing her psychology studies and becoming an academic in 1991 in the department of Educational Psychology at the Institute. Her research interests include disaffection from school, ability grouping and homework and issues relating to learning in music, practising, performing, musical ability, musical understanding and the effects of music on behaviour and studying. She is the author of numerous books, including *Instrumental Teaching: A Practical Guide to Better Teaching and Learning* (1998), *The Power of Music* (2001), *Music Psychology in Education* (2005) and *Preparing for Success: A Practical Guide for Young Musicians* (in press); editor of *The Oxford Handbook of Psychology of Music* (2009) and *Music Education in the 21st Century in the United Kingdom: Achievements, Analysis and Aspirations* (2010); and has extensive other scholarly contributions. She is past editor of *Psychology of Music*, *Psychology of Education Review* and *Learning Matters*.

Jonathan James Hargreaves graduated from the University of York in 2009, and his PhD thesis, *Music as Communication: Networks of Composition*, investigated issues of communication between composers and listeners in twentieth-century music. In 2009 he worked as Research Associate for Heriot–Watt University, composing music for an experiment investigating the effects of music on dairy cattle, and in 2010–11 he was Research Consultant for the *Changing Key* project, a study of the potential for the role of music in school transition, funded by the Paul Hamlyn Foundation. He now lives and works in London, where he is a musician, arranger and educator. In March 2011, he co-founded The Octandre Ensemble, of which he is Co-Artistic Director and Conductor. Jonathan teaches in the Junior Department at Trinity Laban Conservatoire of Music and Dance.

Vanessa Hawes is Lecturer in Music at the University of Christ Church Canterbury. She completed her PhD at the University of East Anglia on *Music's Experiment with Information Theory* and is currently working on projects involving the combination of performance and analysis research. Her publications include articles and book chapters on music theory and applied musicology.

Elaine King is Senior Lecturer in Music at the University of Hull. She co-edited *Music and Gesture* (2006) and *New Perspectives on Music and Gesture* (2011), and has published book chapters and articles on aspects of ensemble rehearsal and performance, including practice techniques, gestures and team roles. She is a member of the Royal Musical Association (Council, 2009–12) and Society for Education, Music and Psychology Research (Conference Secretary, 2006–12). She is an active cellist, pianist and conductor.

Alexandra Lamont is Senior Lecturer in Psychology of Music at the University of Keele. She comes from a multidisciplinary background, having studied, taught and researched in the fields of music, education and psychology. Her research focuses on the question of why people approach music in different ways and this has involved studies with infants, children and adults from perceptual, cognitive, social and educational perspectives. She has published output in book chapters and journals including *Psychology of Music*, *Musicae Scientiae* and the *Journal of Early Childhood Research*. She is currently Editor of the journal *Psychology of Music*.

Rowan Oliver is Lecturer in Popular Music at the University of Hull. His research interests include African diasporic popular music, the nature of groove and the use of technology in performance. As a professional musician he has worked internationally with a number of artists, including six years as the drummer with Goldfrapp, and he continues to perform and produce in a range of genres. In 2007 his score for *Mouth to Mouth* won Best Film Music at the Annonay International First Film Festival and he is currently composing for several feature films in production. He is book reviews editor for the *Journal of Music, Technology and Education.*

Helen M. Prior (née Daynes) is Research Assistant at King's College, London. She is working with Professor Daniel Leech-Wilkinson on a project investigating *Shaping Music in Performance*, which will contribute to the AHRC Research Centre for Musical Performance as Creative Practice. She studied as an undergraduate at the University of Hull before achieving an MSc in Music Psychology at Keele University. She completed her PhD thesis, *Listeners' Perceptual and Emotional Responses to Tonal and Atonal Music*, at the University of Hull. She has lectured in Music and Music Psychology at the Universities of Hull and Sheffield. She is a member of the Royal Musical Association and Society for Education, Music and Psychology Research.

Henry Stobart is Reader in Music/Ethnomusicology at the Department of Music, Royal Holloway, University of London. He is the founder and coordinator of the UK Latin American Music Seminar, Associate Fellow of the Institute for the Study of the Americas, and former Committee Member of the British Forum for Ethnomusicology. His doctoral research focused on the music of a Quechua-speaking herding and agricultural community of Northern Potosí, Bolivia. Following a research fellowship at Darwin College, Cambridge, he was appointed as the first lecturer in Ethnomusicology at Royal Holloway in 1999. His books include *Music and the Poetics of Production in the Bolivian Andes* (2006), the edited volume *The New (Ethno)musicologies* (2008), *Knowledge and Learning in the Andes: Ethnographic Perspectives* (co-edited with Rosaleen Howard, 2002), and the interdisciplinary volume *Sound* (co-edited with Patricia Kruth, 2000). His current research focuses on indigenous music VCD (DVD) production, music 'piracy' and cultural politics in the Bolivian Andes. He is also an active professional performer with the Early/World Music ensemble *SIRINU*, who have given hundreds of concerts and recorded on many European radio networks since their first Early Music Network tour in 1992.

Aaron Williamon is Professor at the Royal College of Music where he heads the Centre for Performance Science. His research focuses on music cognition, skilled performance, and applied psychological and health-related initiatives that inform music learning and teaching. His book, *Musical Excellence*, is published by Oxford University Press (2004) and draws together the findings of initiatives from across the arts and sciences, with the aim of offering musicians new perspectives and practical guidance for enhancing their performance. He is a Fellow of the Royal

Society of Arts and the UK's Higher Education Academy, and has been elected an Honorary Member of the RCM.

Clemens Wöllner is Interim Professor of Systematic Musicology at the University of Bremen, Germany. He holds a Masters degree in Psychology of Music (Sheffield) and a PhD in Systematic Musicology (Halle-Wittenberg). He was a full-time lecturer at Martin Luther University in Halle (Germany), and a Research Fellow in Psychology of Music at the Royal Northern College of Music in Manchester. In addition, he worked as a guest lecturer at the University of Oldenburg, the University of Applied Sciences in Magdeburg and the Stuttgart Academy of Music. His main research interests are in the fields of music performance, with a particular focus on musicians' skilled movements, orchestral conducting and ensemble coordination. In 2006, he received an ESCOM Young Researcher Award for his work on expressive conducting, and was recently appointed Associated Junior Fellow of the Hanse-Wissenschaftskolleg.

Series Editors' Preface

There has been an enormous growth over the past three decades of research into the psychology of music. SEMPRE (the Society for Education, Music and Psychology Research) is the only international society that embraces an interest in the psychology of music, research and education. SEMPRE was founded in 1972 and has published the journals *Psychology of Music* since 1973 and *Research Studies in Music* Education since 2008, both now in partnership with SAGE (see www.sempre.org.uk). Nevertheless, there is an ongoing need to promote the latest research to the widest possible audience if it is to have a distinctive impact on policy and practice. In collaboration with Ashgate since 2007, the 'SEMPRE Studies in The Psychology of Music' has been designed to address this need. The theme for the series is the psychology of music, broadly defined. Topics include (amongst others): musical development at different ages; musical cognition and context; culture, mind and music; micro to macro perspectives on the impact of music on the individual (such as from neurological studies through to social psychology); the development of advanced performance skills; musical behaviour and development in the context of special educational needs; and affective perspectives on musical learning. The series seeks to present the implications of research findings for a wide readership, including user-groups (music teachers, policy makers, parents), as well as the international academic and research communities. The distinguishing feature of the series is its broad focus that draws on basic and applied research from across the globe under the umbrella of SEMPRE's distinctive mission, which is to promote and ensure coherent and symbiotic links between education, music and psychology research.

Graham Welch
Institute of Education, University of London, UK
Adam Ockelford
Roehampton University, UK
Ian Cross
University of Cambridge, UK

Acknowledgements

We should like to thank the authors for their contributions to this volume. In addition, we wish to thank Heidi Bishop and Laura Macy at Ashgate for their constant support in enabling this volume to come together as well as the members of the production team for their assistance in the delivery of the book. We are grateful to the SEMPRE Series Editors, especially Ian Cross, for their enthusiasm and assistance in this project too.

Introduction

Elaine King and Helen M. Prior

The notion of familiarity is ubiquitous in our lives: it pervades everyday conversations, thoughts and activities. If one is familiar with someone or something, one might be described as being 'well acquainted', 'intimate' or 'close' to it. Idealistically, the Latin *familia* ('family') from which the word derives connotes a domestic, tightly knit unit. One might become more familiar with someone or something through repeated exposure, such as through meetings between people, frequenting a particular place, regular practice or dedicated study.

We often assume that familiarity is a dichotomous variable – we either know someone or something, or we do not – but we can think about familiarity on a bipolar continuous scale: we are more familiar with our nearest loved ones than our friends; more familiar with our close work colleagues than with acquaintances we have met on one or two occasions; more familiar with those acquaintances than with those whom we have read about, but not met; more familiar with the music recording we have listened to a hundred times than the one we have just encountered on a single occasion as background music. Yet familiarity is, in reality, still more complex, and is not adequately represented by this bipolar conception.

In some cases, we become familiar with someone or something without conscious effort or intention; in other cases, there is a deliberate desire, effort or need to increase (or decrease) our familiarity with someone or something.[1] There are of course countless reasons why an individual might wish to become more (or less) familiar with someone or something, but often it is simply because we want to know more (or less) about that person or thing than we already do.

In understanding the notion of familiarity, liking is a key variable. There is widespread belief that in the context of personal or professional relationships 'familiarity breeds contempt'[2] (that is, the more we get to know someone, the less we get to like them), yet this is by no means always the case: personal or

---

[1]   Whether or not it is possible to de-familiarise ourselves entirely with someone or something merits attention; arguably, a process of de-familiarisation could be achieved intentionally or unintentionally through distancing (in time or space). For example, when writing this Introduction, we drafted the text and then came back to it at a later stage: we tried to de-familiarise ourselves with the original draft so as to review it critically through a 'fresh' pair of eyes.

[2]   Apuleius, the Roman philosopher (124–70 AD) said 'familiarity breeds contempt; rarity wins admiration', while Aesop, the Greek writer, alludes to the proverb in his fables.

professional relationships can prosper over time. Indeed, the 'mere exposure' effect (Zajonc 1968), also known as the familiarity principle, is a psychological phenomenon which indicates that the more familiar we are with someone or something, the more we like it; in other words, we prefer something because it is familiar. However, our liking for someone or something will naturally fluctuate, and there could come a point at which familiarity with a person or thing leads to excessive comfort or boredom (or even contempt). There is potentially a fine line, therefore, between being 'under-' or 'over-' familiar with someone or something, and both states seem to be undesirable, as explained by an inverted U-shaped effect (this theory originated in the work of Wundt 1874). If 'optimum' levels of familiarity are desirable, then individuals might intentionally (or unintentionally) manipulate exposure to someone or something so as to 'control' this effect. The familiarity principle has been addressed in relation to liking and other factors in existing research over the past several decades and current thinking applied to music is presented in this volume (see overview below).

The way in which we become familiar with someone or something is influenced by our mode of interaction with it: familiarity gained through talking with someone on the telephone is not the same as familiarity gained through multi-modal interaction of a personal meeting. There are many ways in which musicians can become familiar with music, such as through performing, studying or listening to it, although the nature of engagement will affect the individual's experience and knowledge of it along with their sense of familiarity about it: a person listening to music while concentrating on some other activity will become familiar in a different way from a person listening to the same recording and giving it their full attention; a person listening to one recording of a piece will gain different knowledge about that piece from a person listening to another recording of the same piece; a musician performing a piece will experience it differently from someone listening to it.

Research in musicology, music psychology and music education often draws upon the notion of familiarity as it affects our understanding of and engagement with music. The largest body of research about familiarity and music focuses on listening. Philosophical, introspective accounts such as that provided by Cone (1977) have been supplemented by empirical research exploring the changes in the aspects of music that are understood by individuals as they become familiar with a specific piece of music (for two rare examples, see Deliège and Mélen 1997 and Pollard-Gott 1983). Attentive listening is not the only type of listening interaction known to be affected by familiarity, however: familiarity has been found to impact upon the effects of listening to background music while carrying out other tasks (Silverman 2010), something exemplified by numerous people on a daily basis. Other studies investigate familiarity as a variable influencing evaluative responses to music (Edmonston 1969; Gaver and Mandler 1987, Kinney 2009; Mull 1957; North and Hargreaves 1997, 2001, 2008; Ritossa and Rickard 2004; Schubert 2007; Tan, Spackman and Peaslee 2006), and such research often draws upon the inverted U-shaped curve described by Wilhelm Wundt (1874; see above) and elaborated by David Berlyne (1971), as do some of the authors in this volume.

Emotional responses to music are often considered to be a prime motivator for listening to music. Familiarity is sometimes considered as a dichotomous variable in studies of such responses and through in-depth examination by using immediate or delayed repetition of a piece (Ali and Peynircioğlu 2010; Iwanaga, Ikeda and Iwaki 1996). Some studies use music participants who have become familiar with pieces through performance (Fredrickson 1999; Sloboda and Lehmann 2001) and the bodily effects of familiar music are also explored (Lingham and Theorell 2009).

Familiarity is not restricted to exposure with a specific piece of music. Music may be more or less familiar in its language or genre, and this has been found to affect listeners' ability to memorise new pieces of music as well as the generation of musical expectations (Curtis and Bharucha 2009; Demorest et al. 2008). Technological developments have created opportunities for wider dissemination of music, perhaps increasing the idiosyncrasies of an individual's listening experiences. With such an influence on a wide range of variables relating to music listening, it is perhaps surprising that familiarity effects have only rarely been considered as a core subject of study, rather than a variable to be taken into account in the examination of other aspects of musical engagement.

Musicology has perhaps tacitly acknowledged the effects of familiarity on musical understanding. Donald Francis Tovey's guide to the Beethoven Piano Sonatas (Tovey [1931] 1998), for example, is aimed towards students encountering pieces of music for the first time, and therefore provides an aide to their engagement with, and understanding of, the music. Arguably, this type of music analysis promotes familiarity with an 'ideal' understanding of a piece of music. Indeed, in musicology, including music analysis, an 'ideal' listener is often assumed (Cook 1990; Dunsby 1995). That 'ideal' listener will of course be familiar not only with a wide range of music, but also with the networks of influence surrounding composers and the effects of these on their music, as identified by musicologists. Scholarly interpretations and readings of musical works and events are, without doubt, influenced by an individual's familiarity with particular texts or methods of enquiry (Bent with Drabkin 1987; Cone 1977) and such issues inevitably impact upon the ways in which music educators tackle the subject too.

For performers, the notion of familiarity underpins studies of performance preparation by musicians working at different levels, including novice (Frewen 2010) and professional (Chaffin, Imreh and Crawford 2002), while it also relates to research on specific aspects of performance, such as memorisation (Ginsborg 2004, Williamon and Valentine 2002) and ensemble playing (Ginsborg and King 2012; King and Ginsborg 2011; Williamon and Davidson 2002).

Despite a plethora of research about the notion of familiarity and music, there is no existing book or journal that focuses specifically on the subject. Following a successful conference of the Society for Education, Music and Psychology Research (SEMPRE) on 'Music and Familiarity' at the University of Hull in

October 2009, this edited volume draws together leading research showcased at the event along with invited contributions from colleagues to expose contemporary theoretical and empirical approaches to familiarity in relation to listening, studying and performing music.

Overview of *Music and Familiarity*

The 13 chapters in this book are conceived as a broad narrative trajectory, although they have been divided into three parts so as to highlight the notion of familiarity from three key perspectives: listening (Part I: chapters 1–4), musicology (Part II: chapters 5–9) and performance (Part III: chapters 10–13). In Part I, the chapters are driven by music psychologists who explore the influence of familiarity on our engagement with music through listening based on empirical enquiries, specifically how much we listen, and how much we like the music we listen to (Chapter 1), the process of getting to know music through regular listening (Chapter 2), how comfortable we feel when listening (Chapter 3), and music's efficacy as a pain-reliever (Chapter 4). The second group, Part II, exposes the notion of familiarity from varied musicological stances, including ethnomusicological (Chapter 5), analytical (Chapter 6), philosophical (Chapter 7), practical (Chapter 8) and educational (Chapter 9). In Part III, the effects of familiarity are explored in relation to different aspects of the Western art and popular performance process, specifically through memorisation (Chapter 10), rehearsal (Chapter 12) and performance itself (chapters 11 and 13).

There are numerous themes that emerge across the volume, providing important links across the three parts: the role of schemata in our cognitive understanding of music (chapters 2, 5, 9 and 13); responses to Berlyne's influential research (chapters 1, 4, 10 and 11); socio-cultural issues (chapters 3, 5, 6 and 8); group-music making (chapters 9, 11–13); memory and learning (chapters 9–10); and reflexivity in research (chapters 5–8 and 13).

In Chapter 1 ('Keeping it Fresh: How Listeners Regulate their own Exposure to Familiar Music'), Alinka Greasley and Alexandra Lamont address existing theoretical hypotheses concerning the ways in which we engage with music over long timespans, arguing that existing models such as the inverted U-shaped hypothesis that are derived from laboratory-based studies are too simplistic for a real-life, longer-term context. They reveal the complex ways in which listeners modify their music listening over a one-month period and over their lifespan, highlighting the differences between individuals' habits, as well as listeners' awareness of the effects of familiarity on their enjoyment of the music.

Although Greasley and Lamont take a broad approach to listeners' familiarity with music, they also advocate more detailed study of familiarity with specific pieces of music over relatively long timescales. Such tactics are adopted by Helen M. Prior in Chapter 2 ('Familiarity, Schemata and Patterns of Listening'), who undertakes a fine-grained examination of three listeners' perceptual responses to music by Clementi, Schoenberg and Berio over a fortnightly period of daily listening. Her presentation of qualitative data in representations of perceptual schemata allows some insight into the development of perceptual responses to these pieces over time, with such responses including descriptions of the listeners' understanding of the music as well as the fascinating connections made between music and their other experiences, knowledge and ideas.

Prior's focus on the effects of familiarity on specific pieces of music is continued to some extent in Chapter 3 ('The Effects of Repertoire Familiarity and Listening Preparation on New Audiences' Experiences of Classical Concert Attendance'), in which Melissa Dobson applies the concept of repeated listening to a real-life situation. Dobson reports the findings of a study in which she investigated the ways that repeated listening to specific pieces of music influenced novice concert attendees' enjoyment of concert performances of those pieces. Her findings demonstrate the complexity of our relationship with music as listeners.

In Chapter 4 ('Familiarity with Music in Post-operative Clinical Care: A Qualitative Study'), Katherine Finlay examines the use of music as an audio-analgesic through a study of music listening in a clinical setting. Although research indicates that music can support a standard clinical care regime, little research has investigated the impact of familiarity on the effectiveness of music in reducing pain. Finlay's chapter addresses this issue, exploring qualitative findings from a recent study of the use of music following knee surgery. Her findings underline the importance of familiarity as a variable with a strong influence on the benefits that may be gained from music listening.

In Chapter 5 ('Unfamiliar Sounds? Approaches to Intercultural Interaction in the World's Musics'), Henry Stobart contextualises and problematises the notion of familiarity as he reflects on our engagement with music, although his focus extends beyond the Western tradition to more or less familiar musical genres from other parts of the world. Stobart highlights the potential mismatch between familiarity and geographical proximity, the potentially perceptually narrowing effects of musical education, and the implications of perceptual mismatches across cultures, all of which provide important scope for consideration by all kinds of musicologists.

An analytical perspective on our engagement with music is demonstrated in Chapter 6 ('Well, What Do You Know? Or, What Do You Know Well? Familiarity as a Structural Force in Crumb's *Black Angels*') as Jonathan James Hargreaves considers the ways in which George Crumb manipulates familiar and unfamiliar musical materials to influence the listener's perception of meaning within his music. Hargreaves' analysis, however, takes into account not only those listeners familiar with the specific works quoted by Crumb in *Black Angels*, but also those who hear the work while they are unfamiliar with the other works quoted within the piece. As such, this chapter considers familiarity with music in multiple forms.

In Chapter 7 ('Familiarity, Information and Musicological Efficiency'), Vanessa Hawes shifts the focus from our engagement with pieces of music to our scholarly endeavour in musicology. She notes the complexities of the field, with its disparate sub-disciplines and concomitant methodologies, and suggests ways of assessing the efficiency of communication between researchers working with different backgrounds. These means, she suggests, will provide helpful ways for the relatively inexperienced (unfamiliar) researcher to assess the value of research from disparate fields. Hawes argues for a philosophy of musicology to provide scope for reflection on the discipline.

A similarly reflective approach is taken by Clemens Wöllner, Jane Ginsborg and Aaron Williamon in Chapter 8 ('Familiarity and Reflexivity in the Research Process'), although their focus is less on the field of musicology than on the researchers within that field. Specifically, they discuss findings from a recent questionnaire survey exploring music researchers' engagement with music, and the implications this familiarity with their subject of study may have on their research.

In Chapter 9 ('Familiarity in Music Education'), Susan Hallam explores the role played by familiarity in music education, examining existing research in relation to the enculturation of musical language, the development of musical skills, and teachers and teaching. Educational issues are further considered in Chapter 10 ('The Significance of Familiar Structures in Music Memorisation and Performance') as Artemis Apostolaki explores the development of memorised performances both theoretically and empirically in the light of cultural differences in learning music. Specifically, Apostolaki examines the effectiveness of the solfège system in aiding memorisation for performance.

Moving away from issues of pitch and memory, Rowan Oliver examines the performer's relationship with musical time in Chapter 11 ('Groove as Familiarity with Time'). Oliver interrogates notions of 'groove' in popular performance from a performer's perspective, arguing that groove stems not only from temporal understanding between two or more performers, but also from a solo performer's conception of musical time. As such, a performer's familiarity with a particular temporal frame of reference is seen to facilitate the experience of groove through examples of musical material used in varied contexts.

In Chapter 12 ('Social Familiarity: Styles of Interaction in Chamber Ensemble Rehearsal'), Elaine King looks beyond the musical interaction within and between performers to consider the socio-emotional behaviour arising in the chamber ensemble rehearsal context. She applies complementary analytical frameworks in the study of social interaction between musicians working in 'new' (unfamiliar) and 'established' (familiar) duo partnerships. Her research introduces transactional thinking into the analysis of musicians' socio-emotional behaviour in rehearsal and supports theories of group development. Ensemble work is also the focus of Chapter 13 ('Familiarity and Musical Performance'). Mine Doğantan-Dack documents the elusive effects of repeated performances of the same work, and hence the experience of familiarity through performance itself, by a philosophical reflective analysis of classical trio performances. She draws together issues of both musical and social familiarity in her account.

As a whole, this book is securely bound together by its overall theme of familiarity. The diversity of perspective and methodology enables valuable contributions to different disciplines of music research, notably psychology, musicology, education, analysis, theory and performance studies. We hope that readers from all of these fields, among others, will find the volume stimulating and enjoyable, while the material will be of interest to a range of readers, including students and experienced researchers.

References

Ali, S.O. and Peynircioğlu, Z.F. (2010). Intensity of Emotions Conveyed and Elicited by Familiar and Unfamiliar Music. *Music Perception* 27/3: 177–82.

Bent, I. with Drabkin, W. (1987). *Analysis. The New Grove Handbooks in Music.* London: Macmillan Press.

Berlyne, D.E. (1971). *Aesthetics and Psychobiology.* New York: Appleton-Century-Crofts.

Chaffin, R., Imreh, G. and Crawford, M. (2002). *Practicing Perfection: Memory and Piano Performance.* Hillsdale, NJ: Lawrence Erlbaum.

Cone, E.T. (1977). Three Ways of Reading a Detective Story – Or a Brahms Intermezzo. In R.P. Morgan (ed.), *Music: A View from Delft* (pp. 77–94). Chicago, IL: University of Chicago Press.

Cook, N. (1990). *Music, Imagination and Culture.* Oxford: Oxford University Press.

Curtis, M.E. and Bharucha, J.J. (2009). Memory and Musical Expectation for Tones in Cultural Context. *Music Perception* 26/4: 365–75.

Deliège, I. and Mélen, M. (1997). Cue Abstraction in the Representation of Musical Form. In I. Deliège and J. Sloboda (eds), *Perception and Cognition of Music* (pp. 387–412). Hove: Psychology Press.

Demorest, S.M., Morrison, S.J., Beken, M.N. and Jungbluth, D. (2008). Lost in Translation: An Enculturation Effect in Music Memory Performance. *Music Perception: An Interdisciplinary Journal* 25/3: 213–23.

Dunsby, J. (1995). *Performing Music: Shared Concerns.* Oxford: Clarendon Press.

Edmonston, W.E.J. (1969). Familiarity and Musical Training in the Esthetic Evaluation of Music. *The Journal of Social Psychology* 79: 109–11.

Fredrickson, W.E. (1999). Effect of Musical Performance on Perception of Tension in Gustav Holst's First Suite in E-flat. *Journal of Research in Music Education* 47/1: 44–52.

Frewen, K.G. (2010). Effects of Familiarity with a Melody Prior to Instruction on Children's Piano Performance Accuracy. *Journal of Research in Music Education* 57/4: 320–33.

Gaver, W.G. and Mandler, G. (1987). Play it Again, Sam: On Liking Music. *Cognition and Emotion* 1/3: 259–82.

Ginsborg, J. (2004). Singing By Heart: Memorization Strategies for the Words and Music of Songs. In J.W. Davidson (ed.), *The Music Practitioner: Exploring Practices and Research in the Development of the Expert Music Performance, Teacher and Listener* (pp. 149–60). Aldershot: Ashgate.

Ginsborg, J. and King, E. (2012). Rehearsal Talk: Familiarity and Expertise in Singer–Pianist Duos. *Musicae Scientiae*; doi: 10.1177/1029864911435733.

Iwanaga, M., Ikeda, M. and Iwaki, T. (1996). The Effects of Repetitive Exposure to Music on Subjective and Physiological Responses. *Journal of Music Therapy* 33/3: 219–21.

King, E. and Ginsborg, J. (2011). Gestures and Glances: Interactions in Ensemble Rehearsal. In A. Gritten and E. King (eds), *New Perspectives on Music and Gesture* (pp. 177–201). Aldershot: Ashgate.

Kinney, D.W. (2009). Internal Consistency of Performance Evaluations as a Function of Music Expertise and Excerpt Familiarity. *Journal of Research in Music Education* 56/4: 322–37.

Lingham, J. and Theorell, T. (2009). Self-reflected 'Favourite' Stimulative and Sedative Music Listening: How does Familiar and Preferred Music Listening Affect the Body? *Nordic Journal of Music Therapy* 18/2: 150–66.

Mull, H.K. (1957). The Effect of Repetition upon the Enjoyment of Modern Music. *The Journal of Psychology* 43: 155–62.

North, A.C. and Hargreaves, D.J. (1997). Experimental Aesthetics and Everyday Music Listening. In D.J. Hargreaves and A.C. North (eds), *The Social Psychology of Music* (pp. 84–103). Oxford: Oxford University Press.

North, A.C. and Hargreaves, D.J. (2001). Complexity, Prototypicality, Familiarity and the Perception of Musical Quality. *Psychomusicology* 17: 77–80.

North, A.C. and Hargreaves, D.J. (2008). *The Social and Applied Psychology of Music*. Oxford: Oxford University Press.

Pollard-Gott, L. (1983). Emergence of Thematic Concepts in Repeated Listening to Music. *Cognitive Psychology* 15: 66–94.

Ritossa, D.A. and Rickard, N.S. (2004). The Relative Utility of 'Pleasantness' and 'Liking' Dimensions in Predicting the Emotions Expressed by Music. *Psychology of Music* 32/1: 5–22.

Schubert, E. (2007). The Influence of Emotion, Locus of Emotion and Familiarity upon Preference in Music. *Psychology of Music* 35/3: 499–516.

Silverman, M.J. (2010). The Effect of Pitch, Rhythm, and Familiarity on Working Memory and Anxiety as Measured by Digit Recall Performance. *Journal of Music Therapy* 47/1: 70–83.

Sloboda, J.A. and Lehmann, A.C. (2001). Tracking Performance Correlates of Changes in Perceived Intensity of Emotion During Different Interpretations of a Chopin Piano Prelude. *Music Perception* 19/1: 87–120.

Tan, S.-L., Spackman, M.P. and Peaslee, C.L. (2006). The Effects of Repeated Exposure on Liking and Judgements of Musical Unity of Intact and Patchwork Compositions. *Music Perception* 23/5: 407–21.

Tovey, D.F. ([1931] 1998). *A Companion to Beethoven's Pianoforte Sonatas: A Bar-by-Bar Analysis of Beethoven's 32 Pianoforte Sonatas* (revised edition). London: The Associated Board of the Royal Schools of Music.

Williamon, A. and Davidson, J. (2002). Exploring Co-Performer Communication. *Musicae Scientiae* 6: 53–72.

Williamon, A. and Valentine, E. (2002). The Role of Retrieval Structures in Memorizing Music. *Cognitive Psychology* 44: 1–32.

Wundt, W. (1874). *Grundzüge der Physiologischen Psychologie*. Leipzig: Wilhelm Engelmann.

Zajonc, R.B. (1968). The Attitudinal Effects of Mere Exposure. *Journal of Personality and Social Psychology. Monograph Supplement* 9: 1–27.

PART I
Listening

Elaine King and Helen M. Prior

The first section of this volume explores our interaction with music through listening. The four chapters are bound together by their empirical approaches, while they also provide scope for a broader understanding of the effects of familiarity on music listening through their diverse settings and scales of approach. Firmly situated in everyday listening habits, Greasley and Lamont (Chapter 1) explore the ways in which listeners' listening habits change over time, describing the waxing and waning of listeners' liking for music during a month of normal listening and throughout their lives. In Chapter 2, Prior adopts a more music-specific approach, examining the effects of daily listening to pieces of tonal and atonal music on listeners' perceptions of those works. Case studies are used to exemplify the development of perceptual cues in listeners' conceptions of the music and the ways in which these change over a two-week period. The effects of repeated listening to specific pieces of music are also explored by Dobson in Chapter 3, who reports on her study of novice concert-goers and the effects of listening to recordings of concert works prior to a performance. Her fascinating findings point to the complex issues surrounding concert attendance, and situate the examination of the effects of familiarity on our engagement within music in a real-life context with particular contemporary relevance. The section is completed by the examination of the effects of familiarity with music in another real-life setting: that of a post-operative hospital ward (Chapter 4). Finlay reports on some of the qualitative findings of her mixed-methods study examining the use of music as an audio-analgesic. The section as a whole, therefore, provides a picture of the effects of familiarity on our engagement with music as listeners from the small to the large scale; from relatively controlled experimental settings to wholly ecologically valid real-life settings; from popular music to classical; and from listening purely for enjoyment to clinical applications of music. Although there will always be further questions to answer, this section of the book sheds some light on the role familiarity plays in an activity undertaken by the vast majority of those who engage with music.

Chapter 1

Keeping it Fresh: How Listeners Regulate their own Exposure to Familiar Music

Alinka Greasley and Alexandra Lamont

The purpose of this chapter is to explore current understanding of the ways in which familiarity with music shapes musical preferences and listening behaviour. It has been well documented that our liking for music increases as our familiarity increases, and decreases with repeated exposure (cf. North and Hargreaves 1997). However, most previous studies have based their conclusions on people's responses to experimenter-chosen music in laboratory conditions over short timeframes, and this explanation is too simplistic. More recent studies employing open-ended methods to explore engagement with music over time (Greasley 2008; Greasley and Lamont 2009; Lamont and Webb 2010) have shown that the process through which music becomes familiar and through which liking/disliking for certain styles of music develops, and the effect that familiarity has on listening behaviour, all vary from person to person and in response to the specific music a person is listening to. Research has thus only just begun to identify some of the key factors shaping the relationship between familiarity with music and subsequent uses of and responses to music. We draw here on literature from a range of music-psychological approaches to address the following questions: how does music become familiar? How do listeners understand the concept of familiar and 'favourite' music? How do listeners behave in order to maintain and protect familiarity in music? As well as outlining prevailing theories, we draw on research findings from two of our own recent studies.

Preference, Novelty and Familiarity

The process by which a piece of music becomes familiar is one of the best-theorised in music psychology. Within the framework of experimental aesthetics, different approaches have been formulated to explain listeners' responses to music. One of the earliest approaches was the mere exposure hypothesis (Zajonc 1968, 1980), which argues that, all other things being equal, exposure will make us like something more. This simple idea has received some empirical support in a range of domains, including music (Colman, Best and Austen 1986; Harrison 1977; North and Hargreaves 1997; Orr and Ohlsson 2001; Peretz, Gaudreau and Bonnel 1998; Sluckin, Hargreaves and Colman 1982, 1983). However, more complex theories have been developed to explain how this process might operate over time and with different kinds of stimuli, familiar as well as novel, and most of these draw inspiration from Berlyne's psychobiological theory of aesthetic preference.

Berlyne (1971) introduced the notion of arousal potential, arguing that preference results from the interaction between the listener's level of arousal (relatively stable) and the arousal potential of the music itself (which can vary). Properties of the stimulus that can evoke arousal include psychophysical properties such as loudness or pitch, ecological properties such as associations with positive or negative situations, and collative properties (novelty, complexity and familiarity). Berlyne used the inverted U-shape (or Wundt curve) to illustrate this relationship. Arousal potential (the sum of the above, with most emphasis on the collative properties) determines preference, so stimuli with a moderate level of arousal potential are preferred the most, while lower or higher than moderate levels of arousal potential lead to lower preference judgments.

In Berlyne's approach, familiarity is linked to complexity, both of which are necessarily subjective for a given listener at any time; as noted above, these are the most important determinants of preference. Much research has supported this theoretical explanation from a range of fields, finding that moderate levels of arousal lead to the highest levels of preference (Hargreaves 1984; Kellaris 1992; North and Hargreaves 1995, 1996b, c; Russell 1986; Schubert 2010). However, particularly in research into visual art, the importance of complexity and familiarity has been overshadowed by other features of the stimuli such as typicality (Martindale, Moore and Borkum 1990) or meaningfulness to the participants (Martindale and Moore 1989). In particular there has been some challenge to the notion of arousal as being too simplistic an explanation when considering complex stimuli like music that can be processed at a deep level (Martindale et al. 1990).

Berlyne's original approach did not include any explicitly temporal elements of preference, and research exploring its relevance has tended to take a piece of music and a listener at a given moment in time and assess their familiarity and levels of subjective complexity with the music alongside their liking (North and Hargreaves 1996b). However, familiarity is not a static concept. A development of Berlyne's ideas which attempts to capture this is Walker's (1973) psychological complexity and preference theory. This combines the principle of optimal complexity (a parameter of the individual) and situational psychological complexity to argue that experience with music alters its subjective complexity: with repeated experience, very complex pieces will be liked more, while very simple pieces will be liked less. This has received some empirical support (Heyduk 1975; Hunter and Schellenberg 2011; North and Hargreaves 1995; Schellenberg, Peretz and Viellard 2008; Szpunar, Schellenberg and Pliner 2004). For example, North and Hargreaves (1995) found that listeners' tolerance levels remained higher over successive repetitions of more complex music than simpler music. Schellenberg et al. (2008) found that liking ratings followed an inverted U-shape over successive exposures in focused listening situations in the laboratory. However, these studies found that liking when listening was not focused or when distractor tasks were being undertaken followed a more monotonic relationship with exposure. This could be because a lack of attention over such short time-spans does not allow subjective complexity to be reduced.

In more ecologically valid and realistic contexts, music preference has also been found to rise and fall in a similar way to the inverted U-shaped predictions (Russell 1986, 1987). Russell explored chart positions, radio airplay time and preference for current popular music, finding that preference for unfamiliar pieces increases with repeated exposure to a certain point, but then decreases as listeners become satiated with the music. This process of change in subjective complexity over time is illustrated in Figure 1.1.

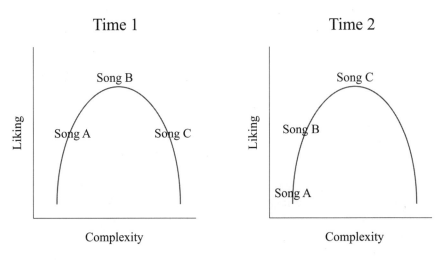

Figure 1.1   Patterns of changing complexity over time
Source: Hargreaves, North and Tarrant (2006), p. 141.

Russell's research raises the important question of choice over exposure to stimuli. In the laboratory studies that have dominated research in this field, most stimuli are artificial and exposure is highly controlled. In real-life settings many more factors play a role in shaping experience with naturalistic stimuli. When we can consciously control our degree of exposure to the stimulus, as is the case in most music listening situations, the preference-feedback hypothesis advanced by Sluckin et al. (1983) suggests that the peak of satiation is avoided through a process of self-regulation. By avoiding excessive levels of exposure, the relationship between familiarity/complexity and liking can proceed in a pattern of waxing and waning. This pattern has been demonstrated both in listeners' nominations of preferred music from different periods (North and Hargreaves 1996a) and in popularity rankings of composers and works over time (North and Hargreaves 2008; Simonton 1997). Music choices are also highly context-dependent (North,

Hargreaves and Hargreaves 2004), and greater levels of choice lead to more positive outcomes such as positive mood states (Sloboda, O'Neill and Ivaldi 2001) and lower levels of pain perception (Mitchell, MacDonald and Brodie 2006). This suggests that some concept of choice must be included in research attempting to explore familiarity. Much of the existing research draws on experimenter-chosen music in an attempt to control for factors such as subjective complexity, but by doing so omits important and interesting aspects of musical engagement.

The time course of musical preferences and familiarity as they might operate for an individual over weeks, months and years has not been studied very much at all. Most of the laboratory studies explore a limited number of repetitions within a single experimental session (for example, Szpunar et al. 2004 compare two, eight and 32 repetitions of tone sequences or 15-second extracts of music). Cross-sectional research has suggested that liking varies over the life-span as well as on an individual basis, and some generalisations are possible (cf. Hargreaves, North and Tarrant 2006). For example, younger children are more accepting of different musical styles or are 'open-eared'; adolescents are much less tolerant; and open-earedness partially rebounds in early adulthood and then declines in older age (LeBlanc 1991; LeBlanc et al. 1993). Complex music seems to be more preferred by older listeners (Hargreaves and Castell 1987), who also tend to report higher liking for classical music and jazz (Hargreaves and North 1999).

Subtle changes in people's patterns of response to music in the context of everyday music listening have not yet been satisfactorily addressed. How can listeners negotiate a path between their own self-chosen music listening and the music they hear on the radio and through other media when saturation is close? Does liking always decline in the second half of the inverted U, and how much time is required before a piece of music is sufficiently 'novel' or complex to permit re-engagement? Finally, how do everyday life situations and the role of listeners themselves, other people in their lives, and their situations, motivations and emotions affect this process? Studies of music in everyday life (Greasley and Lamont 2011; North et al. 2004; Juslin et al. 2008) have begun to map people's engagement with music across different settings (for example, contexts, activities, social situations) and increasingly account for the role of listener's personal musical choices (see Sloboda, Lamont and Greasley 2009). Recent studies investigating the influence of emotion on preference have also shown that, the stronger the emotions aroused in a listener, the more likely it is that the piece will be liked (Schubert 2007, 2010). However, these questions have received virtually no consideration in the literature on familiarity thus far, and so in the remainder of this chapter we present some detail of our own empirical investigations into some of these key issues.

Mapping Short-term Patterns of Preference

In our first study (Lamont and Webb 2010), we explored how musical preferences might ebb and flow over the relatively short time period of a month. This exploratory study centred on the relevance of the concept of musical 'favourites', asking participants to identify a daily favourite piece of music and exploring the situational factors around this as well as the representativeness of a daily favourite for more stable patterns of listening behaviour. The focus on early adulthood, although partly a matter of convenience, was also because early adulthood has been shown to be a period of consolidation of music preferences (LeBlanc 1991).

Nine participants (three men and six women, mean age 21 years 7 months, drawn from a university population) completed daily listening diaries for a week, followed by a two-week gap, and then again for another week. Only one participant was receiving musical training, but four others had learned instruments in childhood. The daily structured diaries were based on existing experience sampling research (Sloboda et al. 2001) and asked participants about key events in the day. This included nominating one favourite piece of music per day, with contextual information about whether that piece had been heard and if so whether it was through choice, for what purpose, and why. The purpose of the daily diary was to explore the face validity of identifying favourite pieces as a way into people's fluctuating patterns of involvement with music over time.

Each participant nominated at least two different favourite pieces of music over the course of the two weeks, with the average being six, although some nominated more than 14 in total (having chosen more than one on a particular day). The music chosen was very diverse, including contemporary pop (Christina Aguilera, Take That), older pop (The Cure, The Smiths), rock (Dire Straits, Eric Clapton, Led Zeppelin) and classical (Verdi, Pachelbel). Despite the homogeneity of the sample in terms of age, occupation and social environments (many were friends and four lived together), only 5 per cent of the nominations overlapped. This provided strong evidence for the unique nature of preferences and listening habits.

Almost all the favourite pieces were heard on the day in question, with just under two-thirds (61 per cent) being actively and deliberately chosen and a further 17 per cent being chosen through less explicit mechanisms (listening to the radio and shuffle mode on iPods). The data does not allow us to identify where the concept of the favourite begins: it is possible that participants have a favourite piece of music in mind and choose to listen to it, or conversely they may choose to listen to music they like and one of those pieces happens to 'stick' as a daily favourite. However, it does suggest that most memorable listening experiences are self-chosen and have a relationship to this concept of a daily favourite. It also gives some indication of the stability of this relationship: looking over the two weeks, purely random hearings of music that were nominated as a daily favourite were never repeated for the same participant, whereas deliberately chosen music appeared much more frequently in their weekly catalogues of favourites. Looking at the data when nominations were not heard on the day, although this only reflected a small number of the total data points, non-heard favourites were often in people's heads because they had been recently heard; out of the total of 125 nominations, only one daily favourite had not been heard or sung within the last month.

Open-ended responses showed that reasons for choosing the piece as a favourite were extremely varied. These included features of the music itself (for example, 'beautiful song'), emotional or physical responses in the listener ('makes me smile'), the appropriateness of the context ('good to drive to'), memories and links to other things ('reminds me of being little') and temporal factors ('first piece I heard, therefore was memorable'). People often gave more than one reason for listening to it on the day in question. This included liking (84 per cent of instances), mood (50 per cent), relaxation (38 per cent), enhancing an activity (32.5 per cent), creating an atmosphere (31 per cent) and distraction (26.5 per cent). Results thus chime with recent research exploring reasons for choosing music in daily life (Greasley and Lamont 2011; Juslin and Laukka 2004). As in many other studies, favourite music listening was either accompanied by or accompanied another activity, including studying (31 per cent), travelling (18 per cent) and socialising (16 per cent). More favourite music was heard alone (65 per cent) than other research suggests is typically the case when considering the totality of someone's daily listening experiences (for example, Juslin et al. 2008). This further suggests that daily listening to a favourite might reflect a more deliberate, conscious and reflective experience than the broader category of everyday exposure to music.

From the diary data, two types of listening and nomination behaviour were identified. 'Magpie' listeners (all female) nominated a moderate number of favourites (mean of eight across 14 days), most of which had been preferred for relatively short time periods of days to weeks and with several overlaps within the nominations. Conversely, 'squirrel' listeners (two female, three male), had a much larger number of favourites (mean of 14), with fewer overlapping tracks, preferred from a variety of time-spans from the day in question to several years previously. Their reasons for listening were broadly similar, although magpies selected more reasons and more frequently chose 'to pass the time' and 'out of habit', while squirrels chose fewer reasons and more frequently chose 'to create the right atmosphere'. This suggests two types of listening behaviour, which were further explored in interviews with Martha, a female magpie listener, and Simon, a male squirrel listener, to compare their engagement with music more carefully.

Both chose music that reflected the mood of the day and situations they were in on a number of occasions, as well as being influenced in their nominations by hearing music that was not self-chosen but which happened to fit. However, as suggested by the group differences, Martha tended to choose more of the same songs over the study weeks (selecting only seven songs out of a potential 14 nominations), often without considering the range of options open to her. She referred to the importance of repeated listening in maintaining short-term patterns of preference, and all her nominations were current chart hits over the study period, meaning that she could hear them from her own collection, at friends' houses, on the radio or in bars and clubs. Conversely, Simon was more varied and deliberate in his nominations (choosing 15 different songs), choosing music from a wider range of styles (including jazz, folk, sixties and current pop), and explained that he was aware of the need to revise and refresh his music listening habits regularly to avoid saturation with favourite pieces on a day-to-day basis.

Considering familiarity over much longer time-spans, both interviewees were asked about their long-term favourite pieces of music. Martha happened to hear hers by chance on the radio during the study, and she described this music ('Killing Me Softly' by The Fugees) as 'my old friend'. She went on to explain:

> Maybe it's just out of habit that of saying it's my favourite, I don't give anything else a chance any more, I like that one so much, um, I don't know … er, I guess like I've listened to it a lot more as well than like more recent things.

Simon was able to elaborate more about his personal associations with the music he claimed as his long-term favourites, discussing these as 'ingrained – they will not be shifted' and available for returning to at different moments in his life.

These results show that people have a catalogue of favourite music which they know, whose effects they are aware of (Batt-Rawden and DeNora 2005; DeNora 2000), which they are aware they need to revise and refresh on a regular basis, and from which they choose to listen to from day to day (Sloboda et al. 2001; North et al. 2004). These choices from the catalogue appear largely to determine nominations of daily favourites, and are highly context-specific (Sloboda et al. 2001). Magpie listeners have a relatively restricted and more ephemeral catalogue of favourites that they tend to draw on more heavily in the shorter term, while squirrel listeners have a larger catalogue and are more easily able to recall music from this without needing to directly experience it.

Considering the results in line with experimental aesthetics, a complex picture is found. The variety of music selected contradicts the mere exposure hypothesis (Zajonc 1980), since amongst this tightly knit group of friends who were all exposed to similar chart pop music there should have been much more commonality in their nominations. Magpie listeners showed a pattern of inverted U-shaped preference over time (cf. North and Hargreaves 2008). They explicitly noted the importance of familiarity in maintaining their preference, but indicated that, after a period of time – ranging from a few days to three weeks – music heard on a daily basis became boring and they regulated their exposure to it by not listening again. Squirrel listeners showed a pattern which reflected the waxing and waning described by Sluckin et al.'s (1983) preference-feedback hypothesis, choosing music drawn from much wider time-spans (with some pop music which was in the charts 30 or 40 years prior to the study), and returning to music that they had listened to in the past in addition to choosing music from their current catalogue. To add a final layer of complexity, our interview data suggest that longer-term favourite pieces of music might work in a different way to more ephemeral preferences created by familiarity and saturation, particularly for the magpie listener. Both types of listener had long-term favourites which held rich personal associations, although they might not choose to listen to these as frequently. These results are only exploratory, and our second study considers longer-term attachments to pieces of preferred music and what these might mean for listeners on a range of levels in more detail.

Mapping Long-term Patterns of Preference

Our second study (Greasley 2008; Greasley, Lamont and Sloboda 2013) was an in-depth qualitative investigation of everyday musical behaviour across the life-span. Twenty-three participants of varying ages (18–73) were interviewed at home with their music collections to explore past and present music preferences and patterns of listening behaviour. The study was designed to elicit as accurate accounts of musical behaviour as possible. Participants were asked to collect together all their vinyl/tapes/CDs/MP3s (in some cases retrieving music from the attic!) to facilitate a chronological discussion of the development of their collections. They were also encouraged to play preferred music in the context of the interview to act as an aide-memoire.

Interviewees' accounts highlighted the intensely personal nature of music listening behaviour. Whilst some of the participants conceptualised music as a 'soundtrack' to their everyday lives and described uses of preferred music in relation to daily contexts and activities, and mood and emotional states, others reported being less concerned about listening to music regularly. Similarly, whilst some listened to self-chosen music as much as possible, others predominantly listened to the radio. These differences relate to broader differences in levels of engagement with music in everyday life and are discussed in detail elsewhere (Greasley and Lamont 2006, 2011). Here we focus on the role of familiarity in shaping musical preferences and listening behaviour over time.

The Importance of Variety

Firstly, although interviewees described a great deal of music they consistently liked, and felt they would always like, our analysis shows that the one thing that is *constant* about musical preferences is that they are changing all the time, both on a day-to-day basis and over longer periods of time. All of the participants bar two or three (the less engaged listeners) reported that their breadth of preferences had increased steadily throughout their life-span and that they had gone through and were currently going through 'phases' of listening to different artists and styles. In their descriptions, they frequently described their likes and dislikes using the phrase 'at the moment', emphasising the ephemeral nature of preference.

Familiarity with music plays a central role in this process. A large proportion of the sample (20 out of 23) stated that they needed a constant renewal of what they were listening to and actively sought variety in music, emphasising the importance of listening to different styles and variety within music (particularly in terms of musical characteristics and cross-over in styles). For instance, Karl (male, aged 27) told us that there was not enough time to listen to the same music over and over again, and this was reflected in the breadth of his collection:

> the music collection I have now, it's, it, I go from reggae to hip-hop to country to pop, eighties pop, to techno to trance to hardcore, literally, erm, even some classical, Irish pub songs … I try to only buy stuff which I'm not gonna hear all the time.

When describing music that they disliked or no longer listened to, interviewees frequently referred to the music as 'sounding the same' or 'providing nothing new'. Chrissie (female, aged 48) said she had continued listening to Joni Mitchell, a favourite from early adulthood, throughout her adult life but she would often think: 'this is so familiar to me I want something different'. Rich (male, aged 25) explained that he purchased seventies reggae (harder to obtain) instead of mainstream reggae (for example, Bob Marley) because he had become too familiar with Bob Marley albums: 'it's not that I don't like it, it's just I've heard them so many times, I needed something more'. Often, as a result of listening to a wide variety of different music, many felt there was 'something in all styles' they could enjoy listening to, as Rich explained:

> after rave it was, erm, hip-hop, then it was heavy metal, so, and then it was rave again, then it was hip-hop, and then it was heavy metal and I went through phases and then eventually I got into liking hip-hop, heavy metal and rave, and then after that, when I moved up to Birmingham … it got to the point where I like all the kinds of music now, well most styles of music I'd say, I'd say I buy mostly drum 'n' bass, erm, but I don't always listen to drum 'n' bass most, I'd say it's a throw up between hip-hop and reggae most I'd say probably, at the moment.

Phases of Listening

Rich's account above illustrates the fluctuations in music listening behaviour that characterised most of the sample's listening behaviour. Interviewees reported going through phases of listening to different artists and styles and described how they self-regulated their exposure to preferred music over short- and long-term intervals. These phases were described at different levels of detail (for example, track, artist, style) and were strongly influenced by the importance participants placed on variety and the extent to which they listened to music repeatedly. Over half of the interviewees maintained that they listened to preferred music repeatedly, sometimes *ad nauseam*.

> I listened to it [Sarah McLachlan – "Adia"] and I liked it, so, and I played it over
> and over and over 'cause that's what I do with music, I play it over and over and
> over and over again until I'm kind of satiated with it, and then I put it away and
> don't listen to it for a while.

While reflecting on listening habits, most interviewees seemed acutely aware of approaching the point of saturation, when it was time to 'put away' the music and move onto something different. One or two talked about methods they employed to prolong this critical point such as playing preferred music on random/shuffle mode and avoiding radio stations that were 'plugging' tracks they liked (several spoke of the negative effects of radio plugging in bringing about premature disliking). Some appeared to be aware of how much time was necessary before they could listen again, be that days, weeks or even years, as Gerry (male, aged 29) suggested:

> after listening to them [the songs of John Lennon – Best Collection] a lot, I just
> got sick of him really, so I just need a period where I don't listen to it, probably
> for the next two or three years, and then I will start listening to it again.

Whether people can predict future listening behaviour so precisely is questionable, but the point remains that our interviewees reported self-regulating their exposure to familiar music over short and long time-spans. Sometimes phases resulted in a permanent disliking for music. Helen (female, aged 19) noted: 'I went through a stage where I really liked Eminem, like *really, really* liked Eminem, had all of his albums, and then, I don't know, I just couldn't stand to listen to it any more'. However in most of the examples participants drew on, phases of listening had resulted in a permanent liking for styles (as illustrated in Rich's account above). So what happens when participants return to preferred music after a period of self-regulated abstinence?

Coming Back to Music: Listening in a Different Way

Two points arose from interviewees' accounts of re-engaging with familiar music over time. The first relates to valence of response, in terms of both degree of liking and memories associated with music. Instances were recalled where previously liked music was subsequently disliked (as the case with Helen and her waning preference for Eminem), and instances were recalled where previously disliked music was subsequently liked. For example, Susan (female, aged 30) noted: 'I hated that album [*Scorpio Rising*, Death in Vegas] when I first got it … I didn't play it for ages and suddenly, I played it again and it was like, ahh this is quite good'. In other cases, liking/disliking for music was reported to wax and wane in a continuous cycle. For example, in her description of preference for Radiohead's *The Bends*, Nat (female, aged 26) noted that her relationships with the tracks on the album were in constant flux:

you can listen to it and like *those* songs and then you come back to it a few months later and you realize that it's *other* songs that, sort of, caught you, this definitely, I've just travelled through loving that one, then that one, then that one and it, again like, gone full circle and loved the original ones that I loved again.

Two of the participants also spoke of music which had previously been associated with negative or upsetting memories and which had become associated with more positive feelings over time. Kim (female, aged 40) described a song played at her husband's funeral; she said she had found it difficult to listen to for many years but that she could listen to it now without becoming upset. Similarly, Rich (male, aged 25) reported that the album *The Private Press* by DJ Shadow reminded him for a long time of a very unhappy point during his life, but that these negative memories had subsided over time: 'I listen in a different way now, I listen to it in a happy kind of way'.

The second main finding concerns changes in understanding and appreciation over time, which tended to stem from increased knowledge of the context surrounding music (for example, lyrics, artist's lives) and reflexive awareness of changing relationships with music. For example, Chrissie (female, aged 48), who emphasised her engagement with Joni Mitchell's music across her life-span, told us how as a young woman she would often sit in her room 'whipping herself with the music emotionally' and crying for cathartic reasons. Whilst asserting that the music was still a major form of catharsis in her life, she felt she responded differently now that she was older: 'I'm not as vulnerable emotionally so although it still moves me emotionally my response is far more, kind of, aware and knowledgeable'. Several interviewees reported returning to music and realising the lyrics were conveying a different message to that which was originally conceived.

Preference as a Process: Instant versus Gradual Liking

Preference for styles and artists was reported as being both an instantaneous and a gradual process. A few of the participants emphasised that it had taken a long time to 'get into' their most preferred styles of music (for example, months, years) and to appreciate them fully. Gerry (male, aged 27) said that his favourite artist for the last 12 years had been Bob Dylan, but that he had disliked Bob Dylan when he was younger. He remembered thinking: 'the man cannot play guitar and he can't sing, and his lyrics are daft, I'm not interested at all, I don't like it'. However, two experiences incited his interest in Dylan. The first was hearing the song 'Mr Tambourine Man' by The Byrds (written by Dylan) at which point he began to appreciate Dylan's lyrics, and the second was attending a Dylan concert in Athens many years later, which he explained 'helped me get used to the rest of the Dylan stuff'. Note the way in which Gerry emphasised that the experience had helped him to 'get used to' Dylan's other styles, highlighting that it took years for him to develop an appreciation of Bob Dylan's music in all its stylistic variety.

This kind of *gradual* liking was expressed by many of our interviewees. Nat (female, aged 26) explained that her appreciation for different artists and bands matured as she became more familiar with them. When describing her preference for the bands Echo and The Bunnymen and Cake, she acknowledged that one took her longer to get into than the other: 'I just love their sound ... it's probably not very challenging music, like Cake, for example, that takes a lot of getting into, but this just, on first hearing it's just nice music to me'. There were many examples of participants 'persevering' with disliked music. Susan (female, aged 30) reported that when she was younger, she had continued listening to Metallica even though she disliked it initially:

> Alinka: And why did you persevere?
>
> Susan: I have no idea, I don't, I just thought it was possibly, I don't know, good for the soul to kind of stick with something that's not necessarily immediately gratifying I suppose, I suppose that means, again it's another hypocrisy with classical, I, I don't stick with that, because it's not immediately gratifying, I think I've got lazy in my old age, I just want nice music, right now.

Note how Nat and Susan contrasted musical preferences that were immediately satisfying with those which had taken longer to 'get into'. Our interviewees also reported that they could learn to like music. Andy (male, aged 24) explained, regarding the *FabricLive* album series: 'even if you haven't heard of the artist, if it's not your type of music, you'll still end up liking it anyway'. Rich (male, aged 25), who was interested in a wide range of musical styles, argued that 'you can choose a different genre to listen to, and you get to like it'. The discourse of 'learning to like something' underlined the key role of familiarity in shaping preference over time.

Age-related Changes

As mentioned above, most of our interviewees reported a steady increase in musical preferences throughout the life-span, as they went through phases of listening to different styles over time. This contradicts LeBlanc's (1991) theory of the development of musical preferences which suggests that preferences narrow as individuals get older. To illustrate, Chrissie (female, aged 48) had what she described as an 'epiphany' experience one day whilst watching TV in her lounge when she saw the band Hanson for the first time (aged 44). She reported that she was captivated by their music and characters and was compelled to find out everything she could about them and get hold of their music. The experience sparked a fundamental change in the way she perceived and listened to music, and highlighted how she had previously become 'stuck in a rut' listening to over-familiar music:

it's [Hanson's music] got a lot of youthfulness about it but it's also got some quite serious underlying themes so you can kind of sit down and feel satisfied by it, it's not just a piece of bubble gum, but also it's light enough to kind of enjoy and just have fun with …what captured me about it because it was youthful at a time when I suppose I was seeing my youth, well, going, and I was thinking, where do I go from here, all I've got is bloody Cyndi Lauper and Joni Mitchell, you know, and that's not where I want to be forever, and in a sense, they [Hanson] kind of kick-started me into loving music in a more youthful exciting sort of way, not being so kind of heavy and serious about it, you know, just saying music's fun and you can just jump about to it and enjoy it, rather than it has to be this really serious encounter with life.

In summary, these results provide key insights into the nature of engagement with preferred music over time, and the role that familiarity plays in this. Musical preferences are reported to be in a constant state of flux and people are consciously aware of the ways in which they self-regulate their exposure to familiar music over short and long time-spans. The findings provide support for all four theories within the experimental aesthetics perspective. Our participants were confident that they could 'learn to like' styles if exposed to them, supporting the mere exposure hypothesis (Zajonc 1968); they emphasised variety within musical characteristics, started to dislike music very quickly if it was low in complexity, and referred to the way in which over-familiarity would result in steady decrease in liking, supporting the psychobiological (Berlyne 1971) and psychological complexity and preference (Walker 1973) theories; and were consciously aware of the peak of satiation proposed by the preference-feedback hypothesis (Sluckin et al. 1983). However, none of these theories are able to provide a complete explanation for the patterns of musical likes and dislikes seen here. People often self-regulate their exposure to familiar music before reaching satiation (that is, the 'peak' of the inverted U-shaped curve); they may return to previously disliked music and suddenly like it, or return to previously liked music and dislike it suddenly; and importantly, liking for certain tracks, artists and styles waxes and wanes continuously over time, which suggests that these explanations of preferences are too simplistic.

Revisiting Theory and Future Directions

Broadly speaking, our studies support the basic premise that familiarity affects liking. Over time, increased familiarity with music leads to greater liking up to a point of saturation at which the music begins to become disliked. As a fundamental principle, following the inverted U-shaped curve, this explains why people talk about a need to find new music to listen to and the importance of variety in their listening habits. People listen to preferred music frequently in daily life (North et al. 2004; Sloboda et al. 2009) and thus may become familiar with preferred music quickly. If music is low in complexity (for example, there is little variation between tracks on an album – a frequent reason given for disliking music), then liking decreases over a relatively short timeframe (our first study suggests that for pop music this happens over a period of weeks rather than months; cf. Russell 1986), and a renewal of what is being heard is necessary. In this way, our results provide support for the mere exposure (Zajonc 1968), psychobiological (Berlyne 1971) and psychological complexity (Walker 1973) theories.

However, our results also provide a number of key insights into engagement with familiar music that contradict these theories. Firstly, the process of preference takes different amounts of time, depending on both the listener and the music itself. While our first study suggests that some listeners become bored with music over a timeframe of a number of days or weeks, in our second study, some listeners reported taking a long time (that is, years) to 'get into' styles which do not seem to exhibit a great amount of objective complexity (for example, Bob Dylan), while others had music in their collections that could satisfy needs immediately (for example, Echo and The Bunnymen). This implies that the curvilinear function between familiarity and liking varies. Secondly, in both studies, many of our participants reported listening to music repeatedly on a daily basis, yet they also have a conscious awareness that excessive listening will lead to them becoming over-exposed to and bored with specific music and self-regulate their exposure to familiar music over both short and long timeframes to constantly refresh the process. This may be affected by the attention given to listening in different situations (cf. Szpunar et al. 2004) and underpinned by individual differences (Ladinig and Schellenberg, 2012). People halt the curve at different stages. Thirdly, we see evidence of listeners going through phases of preference for particular tracks, artists and styles – a waxing and waning of liking in response to familiarity over time. This last pattern would be more akin to a sine wave than an inverted U-shaped curve. In fact, the curve would potentially be different for every piece of music a listener engages with. Figure 1.2 depicts this diagrammatically in a simplified form. Each line on the graph represents a trajectory of liking (this can be a piece, artist or style), which illustrates a number of different patterns. These include instant liking followed by satiation, gradual liking developing over short and long time-spans, waxing and waning for different pieces over different timeframes and to different extents, and many other potential profiles.

Studying individuals' listening behaviour at a finer-grained level of detail is perhaps the only effective means of understanding how our relationships with music change over time. While this might look like an impossible enterprise (since a given piece could be listened to hundreds or thousands of times, and each piece is not experienced in a vacuum), there are some promising ways in which this can be tackled. One potential technique is mapping shorter-term engagement with music at a precise and fine-grained level over hours and days, drawing on diary methods such as the one used here (Lamont and Webb 2010) or methods that permit an even finer level of detail such as experience sampling methodology (Greasley and Lamont 2011; Juslin et al. 2008) to explore how listening behaviour changes and how familiarity affects the listener from moment

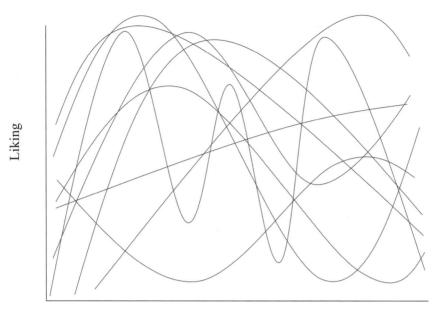

Time

Figure 1.2   Patterns of musical waxing and waning

to moment. Another fruitful approach, as illustrated here (Greasley 2008; see also Greasley et al. in press), is to explore the development of preference and its relationship to listening behaviour retrospectively. Our focus on the music collection as an 'aide-memoire' enabling participants to interact with their preferred music in a contextualised way is one means of achieving this at a level of detail. Other alternatives might include detailed biographical investigations that involve a great deal of non-musical information such as discussing specific friendship and social groups, fashions and leisure activities, with recourse to contextual and historical information in order to prompt people to reflect on their musical tastes and how familiarity might have shaped these over the longer term (as has been done for fashion by Woodward 2007).

These more detailed explorations also need to account for the wide range of listener characteristics that have been shown to affect the ways in which people listen. In addition to the well-known individual differences of age and gender, it appears there may be more subtle distinctions in the ways that people engage with music. For example, magpies and squirrels seem to explore and interact with music in different ways (Lamont and Webb 2010), as do listeners with lower and higher levels of musical engagement (and there may be some overlap with these categorisations). Highly musically trained listeners may have different ways of responding to repeated exposure (and different concepts of what counts as 'over-exposure') which have yet to be explored. Furthermore, personality types have been linked with musical preferences (for example, Rentfrow and Gosling 2003), self-reported reasons for music listening in general (Chamorro-Premuzic and Furnham 2007), and liking for unfamiliar music (see Hunter and Schellenberg 2011), and thus have some potential in explaining the moment-to-moment and longer-term patterns of engagement with music.

Concluding Remarks

Whilst experimental research exploring relationships between musical characteristics and liking for music has provided insights into the ways in which familiarity shapes preferences, our own recent social-psychological research has shown that explanations provided by this approach are too simplistic. Listeners are highly consciously aware of the levels of familiarity they have with the music they own, and self-regulate exposure in ways that open them up to new and unfamiliar styles and pieces, ensuring that their preferred music does not become disliked. Whilst exposure to familiar music cannot always be controlled for (as is the case with radio plugging), in many cases listeners are in control of their own music choices. Most listeners are constantly seeking a fresh renewal of what they are listening to, largely driven by the quantity and diversity of styles available today, and are proficient at managing relationships with preferred music over time. Our findings highlight the need to explore listening behaviour in fine-grained ways over shorter (for example, liking for music over days and weeks) and longer (for example, years) periods of time, as well as accounting for individual differences (for example, listener types). This is a substantial challenge for future research in this field, but essential if we are to fully understand the complex ways in which people engage with music over time.

References

Batt-Rawden, K. and DeNora, T. (2005). Music and Informal Learning in Everyday Life. *Music Education Research* 7: 289–304.

Berlyne, D.E. (1971). *Aesthetics and Psychobiology*. New York: Appleton-Century-Crofts.

Chamorro-Premuzic, T. and Furnham, A. (2007). Personality and Music: Can Traits Explain how People use Music in Everyday Life? *British Journal of Psychology* 98: 175–85.

Colman, A.M., Best, W.M. and Austen, A.J. (1986). Familiarity and Liking: Direct Tests of the Preference-feedback Hypothesis. *Psychological Reports* 58: 931–8.

DeNora, T. (2000). *Music in Everyday Life*. Cambridge: Cambridge University Press.

Greasley, A.E. (2008). *Engagement with Music in Everyday Life: An In-depth Study of Adults' Musical Preferences and Listening Behaviours.* Unpublished PhD dissertation, University of Keele.

Greasley, A.E. and Lamont, A. (2006). Musical Preference in Adulthood: Why do we Like the Music we do? In M. Baroni, A.R. Addessi, R. Caterina and M. Costa (eds), *Proceedings of the 9th International Conference on Music Perception and Cognition* (pp. 960–66). Bologna: University of Bologna.

Greasley, A.E. and Lamont, A. (2011). Exploring Engagement with Music in Everyday Life using Experience Sampling Methodology. *Musicae Scientiae* 15(1): 45–71.

Greasley, A.E., Lamont, A. and Sloboda J.A. (2013). Exploring Musical Preferences: An In-depth Qualitative Study of Adults' Liking for Music in their Personal Collections. *Qualitative Research in Psychology*; doi: 10.1080/147887.2011.647259.

Hargreaves, D.J. (1984). The Effects of Repetition on Liking for Music. *Journal of Research in Music Education* 32: 35–47.

Hargreaves, D.J. and Castell, K.C. (1987). Development of Liking for Familiar and Unfamiliar Melodies. *Bulletin of the Council for Research in Music Education* 91, 65–9.

Hargreaves, D.J. and North, A.C. (1999). Developing Concepts of Musical Style. *Musicae Scientiae* 3: 193–216.

Hargreaves, D.J., North, A.C. and Tarrant, M. (2006). Musical Preference and Taste in Childhood and Adolescence. In G.E. McPherson (ed.), *The Child as Musician* (pp. 135–54). Oxford: Oxford University Press.

Harrison, A.A. (1977). Mere Exposure. In L. Berkowitz (ed.), *Advances in Experimental Social Psychology*, Volume 10 (pp. 39–83). New York: Academic Press.

Heyduk, R.G. (1975). Rated Preference for Musical Composition as it Relates to Complexity and Exposure Frequency, *Perception and Psychophysics* 17: 84–91.

Hunter, P.G. and Schellenberg, E.G. (2011). Interactive Effects of Personality and Frequency of Exposure on Liking for Music. *Personality and Individual Differences* 50: 175–9.

Juslin, P.N. and Laukka, P. (2004). Expression, Perception, and Induction of Musical Emotions: A Review and a Questionnaire Study of Everyday Listening. *Journal of New Music Research* 33: 217–38.

Juslin, P.N., Liljeström, S., Västfjäll, D., Barradas, G. and Silva, A. (2008). An Experience Sampling Study of Emotional Reactions to Music: Listener, Music, and Situation. *Emotion* 8(5): 668–83.

Kellaris, J.J. (1992). Consumer Aesthetics Outside the Lab: Preliminary Report on a Musical Field Study. *Advances in Consumer Research* 19: 730–34.

Ladinig, O. and Schellenberg, E.G. (2012). Liking Unfamiliar Music: Effects of Felt Emotion and Individual Differences. *Psychology of Aesthetics, Creativity, and the Arts* 6/2: 146–54; doi: 10.1037/a0024671.

Lamont, A. and Greasley, A.E. (2009). Musical Preferences. In S. Hallam, I. Cross and M. Thaut (eds), *Oxford Handbook of Music Psychology* (pp. 160–68). Oxford: Oxford University Press.

Lamont, A. and Webb, R.J. (2010). Short- and Long-term Musical Preferences: What makes a Favourite Piece of Music? *Psychology of Music* 38: 222–41.

LeBlanc, A. (1991). Effect of Maturation/Aging on Music Listening Preference: A Review of the Literature. *Paper presented at the Ninth National Symposium on Research in Music Behavior*, Canon Beach, OR, 7–9 March.

LeBlanc, A. Sims, W.L., Siivola, C. and Obert, M. (1993). Music Style Preferences of Different-age Listeners. *Paper presented at the Tenth National Symposium in Research in Musical Behavior*, University of Alabama, Tuscaloosa, AL, April.

Martindale, C. and Moore, K. (1989). Relationship of Musical Preference to Collative, Ecological, and Psychophysical Variables. *Music Perception* 6: 431–46.

Martindale, C., Moore, K., and Borkum, J. (1990). Aesthetic Preference: Anomolous Findings for Berlyne's Psychobiological Theory. *American Journal of Psychology* 103: 53–80.

Mitchell, L.A., MacDonald, R.A.R. and Brodie, E.E. (2006). A Comparison of the Effects of Preferred Music, Arithmetic and Humour on Cold Pressor Pain, *European Journal of Pain* 10: 343–51.

North, A.C. and Hargreaves, D.J. (1995). Subjective Complexity, Familiarity, and Liking for Popular Music. *Psychomusicology* 14: 77–93.

North, A.C. and Hargreaves, D.J. (1996a). Responses to Music in Aerobic Exercise and Yogic Relaxation Classes. *British Journal of Psychology* 87: 535–47.

North, A.C. and Hargreaves, D.J. (1996b). The Effects of Music on Responses to a Dining Area. *Journal of Environmental Psychology* 16: 55–64.

North, A.C. and Hargreaves, D.J. (1996c). Situational Influences on Reported Musical Preferences. *Psychomusicology* 15: 30–45.

North, A.C. and Hargreaves, D.J. (1997). Experimental Aesthetics and Everyday Music Listening. In D.J. Hargreaves and A.C. North (eds), *The Social Psychology of Music* (pp. 84–103). Oxford: Oxford University Press.

North, A.C. and Hargreaves, D.J. (2008). *The Social and Applied Psychology of Music*. Oxford: Oxford University Press.

North, A.C., Hargreaves, D.J. and Hargreaves, J.J. (2004). Uses of Music in Everyday Life, *Music Perception* 22: 41–77.

Orr, M.G. and Ohlsson, S. (2001). The Relationship between Complexity and Liking across Musical Styles, *Psychology of Music* 29: 108–27.

Peretz, I., Gaudreau, D. and Bonnel, A.M. (1998). Exposure Effects on Music Preferences and Recognition. *Memory and Cognition* 26: 884–902.

Rentfrow, P.J. and Gosling, S.D. (2003). The Do Re Mi's of Everyday Life: The Structure and Personality Correlates of Music Preferences. *Journal of Personality and Social Psychology* 84: 1236–56.

Russell, P.A. (1986). Experimental Aesthetics of Popular Music Recordings: Pleasingness, Familiarity and Chart Performance. *Psychology of Music* 14: 33–43.

Russell, P.A. (1987). Effects of Repetition on the Familiarity and Likeability of Popular Music Recordings. *Psychology of Music* 15: 187–97.

Schellenberg, E.G., Peretz, I. and Viellard, S. (2008). Liking for Happy- and Sad-sounding Music: Effects of Exposure. *Cognition and Emotion* 22/2: 218–37.

Schubert, E. (2007). The Influence of Emotion, Locus of Emotion and Familiarity upon Preference in Music. *Psychology of Music* 35: 499–515.

Schubert, E. (2010). Affective, Evaluative, and Collative Responses to Hated and Loved Music. *Psychology of Aesthetics, Creativity and the Arts* 4: 36–46.

Simonton, D.K. (1997). Products, Persons, and Periods: Historiometric Analyses of Compositional Creativity. In D.J. Hargreaves and A.C. North (eds), *The Social Psychology of Music* (pp. 107–22). Oxford: Oxford University Press.

Sloboda, J.A., O'Neill, S.A. and Ivaldi, A. (2001). Functions of Music in Everyday Life: An Exploratory Study using the Experience Sampling Method. *Musicae Scientiae* 5: 9–32.

Sloboda, J.A., Lamont, A. and Greasley, A.E. (2009). Choosing to Hear Music: Motivation, Process and Effect. In S. Hallam, I. Cross and M. Thaut (eds), *Oxford Handbook of Music Psychology* (pp. 431–40). Oxford: Oxford University Press.

Sluckin, W., Hargreaves, D.J. and Colman, A.M. (1982). Some Experimental Studies of Familiarity and Liking. *Bulletin of the British Psychological Society* 35: 189–94.

Sluckin, W., Hargreaves, D.J. and Colman, A.M. (1983). Novelty and Human Aesthetic Preferences. In J. Archer and L. Birke (eds), *Exploration in Animals and Humans* (pp. 245–69). Wokingham: Van Nostrand Reinhold.

Szpunar, K.K., Schellenberg, E.G. and Pliner, P. (2004). Liking and Memory for Musical Stimuli as a Function of Exposure. *Journal of Experimental Psychology: Learning, Memory, and Cognition* 30: 370–81.

Walker, E.L. (1973). Psychological Complexity and Preference: A Hedgehog Theory of Behaviour. In D.E. Berlyne and K.B. Madsen (eds), *Pleasure, Reward, Preference: Their Nature, Determinants and Role in Behaviour* (pp. 65–97). London: Academic Press.

Woodward, S. (2007). *Why Women Wear What they Wear*. Oxford: Berg.

Zajonc, R.B. (1968). The Attitudinal Effects of Mere Exposure. *Journal of Personality and Social Psychology, Monograph Supplement* 9: 1–27.

Zajonc, R.B. (1980). Feeling and Thinking: Preferences need no Inferences. *American Psychologist* 35: 151–75.

Chapter 2
Familiarity, Schemata and Patterns of Listening

Helen M. Prior

The origins of this work, as in much research, lie in a personal experience. My first exposure to a selection of Schoenberg's piano works (which happened to be through a recording) had resulted in an ambivalent response, but upon hearing the same recording some months later for a second time, I found the experience more enjoyable and the music much easier to understand. Perhaps this was to be expected, but I was curious: what was it about the second listening that made it so much easier than the first? What was it that let me access the music in a manner akin to other piano music, and experience emotional responses to it, when I had not been able to do either of these things the first time? There may have been many situational factors, but becoming familiar with the music almost certainly played a role in this transformation.

This chapter explores ways of understanding the familiarity that listeners gain from multiple hearings of a single recording of a piece of music. I will begin by considering research in music and psychology concerning schemata as a means of understanding and representing knowledge. I will then outline some empirical work undertaken as part of a large-scale study of the effects of repeated music listening (Daynes 2007), and, finally, I will present representations of schemata as a means of understanding some of the details contained within a listener's perceptual responses to a piece, outlining what may be learned from this information.

Schemata

Schemata are highly versatile and flexible mental frameworks for representing knowledge that may be related hierarchically, associatively or in a time-dependent order (Eysenck and Keane 1995; Howard 1987; Sternberg 1996). They have been used by musicologists (Cone 1977; Pascall 1989) and music psychologists (Bharucha 1994; Deliège 1996, 2006, 2007; Deliège and El Ahmadi 1990; Deliège and Mélen 1997; Deliège et al. 1996; Gaver and Mandler 1987; Ockelford 1999, 2004, 2005) to aid the understanding of music perception. According to Deliège and Mélen (1997), the perception of a piece of music involves segmentation according to Gestalt rules. These segments are then categorised according to their 'sameness' or 'difference' to one another, or as Ockelford (2004) suggests, according to similarity relations. Once the segments are categorised, 'cues' (particularly salient features of the music that have been reinforced by repetition) develop, and act as memory triggers for the listener. These cues then fill the 'slots' in the schema, which forms the mental representation of the piece of music.[1]

Listeners' perceptions of these segments and cues, while important, are unlikely to be the only features perceived in the music that have the potential to form part of a musical schema. At a very basic level, music is heard as sound (Lavy 2001), and listeners are able to identify the source of that sound (Clarke 2005; Dibben 2001). Music also shares qualities with human utterance, and is always heard within a particular context (Lavy 2001). These three modes of listening are partly responsible for the ease with which listeners are able to identify emotion in music, and may trigger emotional responses to music either directly or indirectly through varied musical and non-musical associations (Sloboda and Juslin 2001). A further source of emotional response can be through the expectations listeners generate as they hear music (Huron 2007). According to Lavy (2001), such expectations can contribute to the creation of suspense and, through this, a narrative mode of listening. All of these features, and others, have the potential to contribute to a schema for a piece of music.

As a listener becomes familiar with a piece of music, their mental representation will change to reflect the new knowledge gained, which in turn serves to make the music 'more predictable' (Huron 2007, p. 241). Authors such as Deliège and Mélen (1997) and Pollard-Gott (1983) provide evidence to suggest that familiarity allows the listener to progress from the observation of readily accessible 'surface' features, such as dynamics, tempo and texture, to a deeper focus on thematic and structural features of the music. Their studies also showed some interesting effects of musical expertise: musicians showed recognition of the thematic or structural features more readily than non-musicians, suggesting that they were more efficient at creating perceptual schemata, but the non-musicians were able to 'catch up' as they became more familiar with the piece of music.

As Huron (2007) and Bharucha (1994) observe, listeners may be familiar not only with a specific piece, but with a particular musical form, style or language. Music that does not conform to familiar patterns of form, style or language is experienced as relatively unpredictable (Huron 2007). Free atonal music, for instance, is a musical language less familiar to most Western listeners than tonal music, and this lack of familiarity might be expected to have an effect on listeners' perceptual responses. Dibben (1994, 1996) suggests some important differences

---

[1]   Other music psychologists use the term 'schema' slightly differently. Bharucha (1994) for instance, considers schemata to be stylistically based mental representations that allow the formation of stylistic expectations for the music's continuation. Knowledge of the particular piece being heard is considered to be 'veridical', and forms a different representation. Huron (2007) also adopts this classification. I am using the term schema to encompass a mental representation of a specific piece, which may incorporate what Bharucha (1994) and Huron (2007) describe as veridical knowledge of a piece.

in the perception of tonal and atonal music. She suggests that the representational reductions used in the perception of tonal music are replaced by associative reductions in response to atonal music, and suggests important roles for both salience and semiotics in music perception. The implications of this are that our perceptual responses to atonal music may be somewhat different from those of tonal music.

Lerdahl (1988) discusses serial music, and suggests that the difficulty listeners generally have in identifying structural features of some serial music may be because of the significant differences between the 'grammatical rules' utilised by composer and listener. Although his work is not focused on free atonal music, but on serial music, his explication of the grammatical rules highlights some rules to which free atonal music does not conform. For instance, like serial music, free atonal music is not elaborational or hierarchical; it also compromises the relationship between spatial distance and cognitive distance. Music that does not conform to such constraints, Lerdahl argues, is hard to understand, or has an 'opaque' structure, so we might anticipate that listeners may have more difficulty identifying the musical structure of atonal pieces in comparison to tonal pieces.

The following study used empirical methods to examine the areas outlined above, investigating the perceptions of musicians and non-musicians listening to unfamiliar pieces of tonal and free atonal music repeatedly over a two-week period. The data were gathered as part of a larger enquiry that aimed to investigate the effects of familiarity on listeners' perceptual and emotional responses to tonal and atonal music. This chapter focuses on the listeners' perceptual responses, which addressed two research questions.[2]

Firstly, how do listeners' perceptual responses change with familiarity? Listeners' perceptual responses were expected to develop with familiarity, and show a greater awareness of deeper aspects of the musical structure. Musicians were expected to be more efficient in this development than non-musicians. Secondly, what are the similarities and differences between listeners' perceptual responses to tonal and atonal music? Existing research suggests both similarities and differences in listeners' perceptual responses to tonal and atonal music; there is some suggestion that listeners may find it harder to perceive the musical structure of an atonal piece than a tonal piece.

Method

Design

To investigate these research questions, a mixed methods design was used. Although both quantitative and qualitative data were gathered, the focus here will

---

[2] For information about other aspects of the study, see Daynes (2007, 2011).

be on the qualitative data. Three experiments were conducted, each of which took place over a two-week period. The experiments commenced approximately three months apart.

Participants

Nineteen participants were recruited from a British university. The group of 'musicians' were undergraduate music students ($n = 10$; three male, seven female; mean age = 20.2), and the other, 'non-musician' participants were recruited from other departments ($n = 9$; four male, five female; mean age = 25.9). The qualitative data from three carefully selected participants will be examined here.

Stimuli

Three short pieces of piano music were used in the study, one of which was tonal and the other two free atonal: the second movement of Muzio Clementi's Piano Sonata in F-sharp minor, Op. 25 No. 5 (tonal); the first of Arnold Schoenberg's Three Piano Pieces, Op. 11 (free atonal); and Luciano Berio's *Rounds* for Piano Solo (free atonal).

All three pieces were unfamiliar to the participants at the outset of the study. The Clementi was chosen for its tonal language, its clear structure and its conventional harmonic, melodic and rhythmic features.[3] Although not intended to be representative of all tonal music, this piece was used as a tonal baseline with which to compare participants' responses to atonal music. The Schoenberg is one of the composer's first free atonal works, but maintains relatively conventional features of form, melody, rhythm and texture.[4] The Berio is less conventional: although it has a ternary form structure, the piece is atonal (although one note, C-sharp, may be considered to be a pitch centre in the work), and has very complex rhythmic and textural features, as well as frequent silent pauses.[5]

A single recording of each work was used. Upon examination of Balázs Szokolay's rendition of the Clementi (Naxos 8.550452) with Audacity 1.2.4 (for more details see Cannam, Landone and Sandler 2010), it was clear that Szokolay used rubato to lengthen bars with phrase boundaries, and to shorten climactic moments that had thicker textures. His dynamic variation was reasonably predictable from the score, but had expressive freedom. For the Schoenberg, a

---

[3]   Few score-based analyses of this sonata exist in the English language; however Plantinga (2006) describes this particular sonata as having considerable 'expressive power'. A basic analysis is included in Daynes (2007).

[4]   This piece has been the subject of much discussion and analysis (Bailey 1998; Botstein 1999; Brower 1989; Frisch 1993; Kramer 1999; McCoy 2999; Neighbour 2007; Simms 1998; Webern 1999). For a brief review, please see Daynes (2007).

[5]   This work has been examined by Thow (1996); pertinent details are explored by Daynes (2007).

recording by Maurizio Pollini was used (Deutsche Grammophon 423 249-2). Like Szokolay, Pollini used rubato to indicate structural boundaries, and played with some dynamic freedom. David Arden's performance of the Berio (New Albion Records NA089CD) appeared to realise the score fairly accurately, even though the numerous pauses and subtle phrase boundaries made the timing relatively difficult to evaluate (see Daynes 2007, pp. 103–8, 179–84 and 259–64 for further details of the analysis of these recordings).

Procedure

Participants attended three experimental sessions (A, B and C) with the researcher over a two-week period; between these sessions, they were issued with a CD recording of the relevant piece and asked to listen to it on a daily basis and to complete a listening diary to record any thoughts, perceptions, feelings or emotional responses they experienced whilst listening to the music.[6] Within experimental sessions A, B and C, the participants recorded their current mood, their continuous emotional responses to each piece (twice)[7] and measures of familiarity and liking.[8] The participants were interviewed at the end of each session, during which they were asked to identify the triggers of their responses using a line graph based on the continuous response traces. The entire procedure was repeated with each piece of music at approximately three-month intervals, with as many of the existing participants as possible. Fourteen participants completed all three experiments. The qualitative data from the interviews and listening diaries of three participants will be reported in this chapter in the form of a case study for the reasons outlined below.

Case Study Participants and Data

Participants A, J and S completed all three experiments and provided comprehensive diary reports of their daily listening. They varied in musical experience, as well as in age, and for these reasons, they provide particularly interesting cases for

---

[6]   The exact text at the beginning of each week's listening diary was: 'Please use this booklet to write down any thoughts you have as you listen to the accompanying CD. I am interested in anything – so please don't restrict what you write down! These might include features of the music that come to your attention, any comparisons with other music, any pictures, memories or emotions you have in your mind, or apparently unrelated thoughts or feelings you have while listening' (Daynes 2007).

[7]   Participants recorded the 'intensity' of their emotional responses to the music using a specially designed computer program through which they used cursor keys to control a visual indicator of their intensity. Data were collected twice per second for the duration of the piece. Full details are reported in Daynes (2011).

[8]   These were measured on continuous scales from 'totally unfamiliar' to 'very familiar' and from 'I hate it' to 'I love it', respectively (see Daynes 2007 for full details).

comparison and were thus selected for in-depth analysis and cross-examination. Participant A was 18 years old, and an undergraduate music student in her first year. She had received more than five years of instrumental tuition, had achieved an A level in music, and Grade 7 or 8 in a musical instrument.[9] She also had absolute pitch. Participant J was 23 years old and also an undergraduate music student in his first year. He, too, had received more than five years of instrumental tuition, had achieved a qualification of the equivalent standard of an A level in music, and was of Grade 7 or 8 standard in a musical instrument. Participant S was 41 years old, and a postgraduate student in psychology. The only music tuition he had received was compulsory school music lessons to the age of 14. He did not possess any music qualifications.

These three participants generated a wealth of data alone: 27 interviews, all of which were transcribed verbatim, and 96 diary entries. These data were carefully examined and categorised using an inductive thematic analysis (Braun and Clarke 2006)[10] before the material was displayed in a largely data-led representation of participants' schemata as they developed over time. Schemata are rarely illustrated in literature – they are normally described in prose – but some useful examples as well as similar models were influential in the basic construction of these representations (Buzan 1995; Eysenck and Keane 2005; Howard 1987). Figure 2.1 shows the template for the schematic representations, which is partly associative and partly hierarchical.[11] Different types of information are indicated with different shapes and borderlines, specifically perceptual cues, large-scale structural features, musical and non-musical associations, judgments, mood responses, and narrative understandings. In order to represent perceptual changes over time, 'new' information (such as that provided in a current interview or diary entry during the study) is shown in black; as this information becomes 'older' with distance in time, it gets progressively smaller and lighter in colour in the schema (for example, by the schema for the third interview, the information provided in the first interview would be very small and grey, unless this cue had been reactivated in the current interview, in which case it would be black). Information from previous schemata is not deleted: this acknowledges the fact that, even when we come to a new way of understanding something (which may be simpler than a previous understanding), we can still be aware of earlier understandings. This

---

[9] Musical experience and qualifications were measured in fixed criteria that were reasonably broad and designed in part to avoid intimidating the non-musician participants. Participant A had, in fact, received considerably more than five years' of instrumental tuition, but this was the highest category.

[10] This is also similar in nature to Hseih and Shannon's (2005) conventional content analysis.

[11] It should be noted that, although this template is presented before any of the representations generated with the data, it did emerge during the analysis of the data, rather than being created as a theoretical schema through which data were interpreted. Nonetheless, some of the categories do reflect the existing literature discussed above.

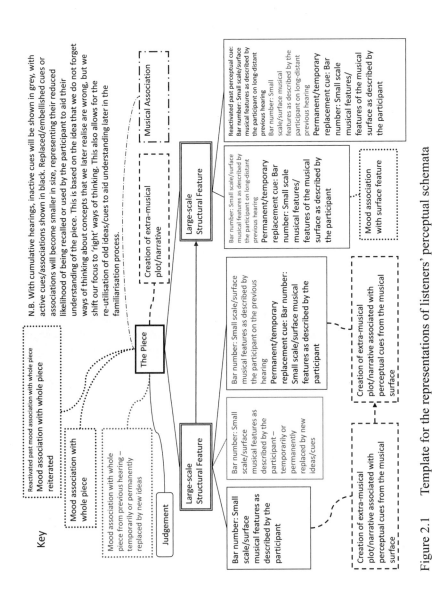

Figure 2.1   Template for the representations of listeners' perceptual schemata

also allows for participants to return to previous cues after a period of not using them. The series of schemata generated for each participant provides a profile of their perceptions; and the final schema in particular enables all of the data from an interview or diary entry to be displayed in one relatively compact space, and in the context of similar data from an earlier point in time, allowing the analysis of longitudinal trends.

Results

The schemata assembled for each of the three participants will be discussed in turn and reported according to each of the three pieces used in the experimental studies (Clementi, Schoenberg and Berio respectively). Selected representations of their schemata will be used to support these findings and, where appropriate, cross-comparisons will be made.[12]

Perceptions of Clementi's Piano Sonata in F-sharp minor, Op. 25 No. 5, II

Participant A   The schematic representations of participant A's responses to the Clementi are characterised by the presence of detailed descriptions of small-scale perceptual cues and the identification of large-scale structural features. She began in the very first interview to relate different parts of the music to one another, outlining a basic ternary form structure for the piece, and demonstrating a hierarchical understanding of the piece, which was reflected in the first schema. She also gave clear descriptions of various musical details, including chromatic movement and suspensions, and began to associate moods with specific features (for example, chromaticism with 'urgency'), and with the whole piece, which she described as 'mellow'. In the first diary entry, she gave a few more specific details, and invoked an overall narrative of 'an old man sitting in his chair remembering his past'. She also added some detailed descriptions of small-scale, correctly identified, musical features that appeared to serve as basic perceptual cues. This pattern continued in the second, third and fourth diary entries, but here, participant A corrected her conception of the musical structure by adding the recapitulation of the second section. Although she had invoked some narrative ideas in response to small details of the piece and to the piece as a whole, it was not until the sixth diary entry that she created a coherent overall narrative for the piece that was clearly tied to the formal structure, and to some small-scale details of the music.

The schemata for the second interview and seventh and eighth diary entries were once more focused on accurate additions or clarifications to the small-scale perceptual cues, with a few comments concerning mood. The schema for the ninth

---

[12]   A comprehensive set of all the schemata for these three participants is produced in Daynes (2007) and is freely available in electronic form, https://hydra.hull.ac.uk/resources/hull:753.

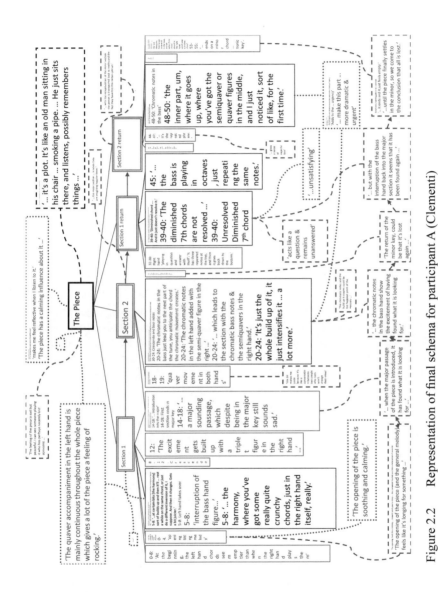

Figure 2.2   Representation of final schema for participant A (Clementi)

diary, however, returned to a narrative conception of the piece, this time linking the small-scale perceptual cues described in such detail with a quasi-musical narrative concerning longing, the joy of finding something (or someone) lost, and disappointment with the conclusion that it is lost after all. These narratives were clearly neither fixed nor temporary, as evinced by the schema for the third interview (see Figure 2.2), which shows participant A's return to the 'old man' narrative, whilst discussing details of the piece. Overall, participant A's schemata for this piece revealed a balance between accurate conceptions of small-scale perceptual cues and large-scale formal features and the use of narratives around which these cues and features were understood.

Participant J   Participant J's schemata were characterised by a high volume of comments, but also showed interesting development over the familiarisation period. In the first interview and first two diary entries, participant J focussed on detailed surface features; simultaneously, he began to develop narrative ideas whilst thinking about moods, judging the piece and comparing it with other music that he knew, such as Ravel, Liszt and Beethoven. His schemata for these interactions were almost entirely serial and associative, rather than showing a clear hierarchical conception of the piece. It was not until the third diary entry that participant J showed evidence of thinking about the large-scale structure; and at this point, he divided the piece into two sections.

The schema for the fourth diary entry showed a change of listening focus. Rather than considering the serial progress of the piece as before, participant J considered the harmony, rhythm, texture and narrative of the piece as a whole. He also identified a new section of the piece in terms of its mood ('sprightly, hopeful section', relating to the second subject of the movement). The fifth diary entry showed a further change of focus, with evidence of musical and non-musical associations and the use of a narrative understanding. These features were highlighted in the second interview, but here it also became clear that participant J understood the structure of the piece as having four sections, related to the themes of the music and recognised the repetition contained therein.

The remaining diary entries revealed a range of approaches to the music: in the sixth, and to some extent, the seventh, participant J focused on perceptual cues, whereas in the seventh, eighth and ninth diary entries, he began to critique the performance of the movement. This judgment of the performance continued in the third interview (see Figure 2.3), by which time he had an accurate conception of the musical structure, and linked it with composers closer to the right stylistic period (Mozart, Beethoven). He also continued to think about the moods or emotions portrayed in the music. Participant J's schemata revealed a varied listening stance, beginning with a focus on small-scale perceptual cues and associations, and with an increasing understanding of musical structure and tendency to critique the performance of the work. They also consistently showed consideration of the mood of the piece.

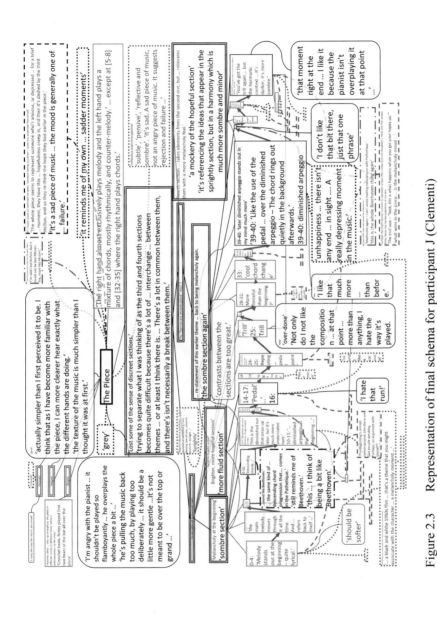

Figure 2.3   Representation of final schema for participant J (Clementi)

Participant S   Participant S's schemata contained a lower volume of comments than, in particular, participant J, and less of a tendency to focus on small-scale perceptual cues or large-scale structure. In the schema for his first interview, he talked a little about small-scale perceptual cues, and made an association with a film, but also invoked a narrative understanding of the music. A similar emphasis was shown in the schema of the first and second diary entries: only two additional perceptual cues from the surface of the music were added, and one of these was related to a judgment comment, but three potential narrative understandings of the music were outlined. The third, fourth and fifth diary entries revealed a greater number of small-scale perceptual cues, and the inclination to invoke a narrative understanding returned in the fifth, to be exploited further in the sixth. Participant S reported in his third diary that he had noticed some familiar patterns of notes returning during the piece, and also noted the point at which the music 'could have stopped' at bar 27. Although this might be taken to indicate an understanding of the two halves of the piece, the participant did not explicitly divide the piece into sections. This understanding was confirmed in the second interview, the schema of which was still serial and associative.

Brief judgments and narratives were highlighted in the schemata of the seventh and eighth diary entries, and this tendency towards narrative was used more in the ninth diary, and in relation to small-scale perceptual cues. Small parts of this were highlighted in the tenth and eleventh diary entries, and in the twelfth diary entry, participant S constructed a narrative based on the seasons, which he developed further in the third interview. He had also, by this time, divided the piece into two sections, although these did not correspond with the potential 'finishing point' he had discussed earlier. Although he recognised the cadence and the pause, he did not appear to understand this to indicate a structural boundary in the same manner as participants A and J. Overall, over the course of the diary entries, participant S used and developed his narrative conception of the piece, and added detail to the small-scale perceptual cues, spotting some larger-scale structural landmarks (see Figure 2.4).

It is evident that the three participants approached and understood the Clementi in different ways, with varying emphasis on small-scale perceptual cues, large-scale structure, associations, judgments of the piece, and the use of narrative. Given the relative familiarity of the tonal language in the Clementi, their approaches may diverge further, change independently, or remain relatively constant in response to the free atonal stimuli; either way, it was anticipated that they would find it more difficult to perceive the musical structures of the latter. The next section examines the participants' responses to the Schoenberg.

Perceptions of Schoenberg's Three Piano Pieces, Op. 11, I

Participant A   In response to the tonal stimulus, participant A had focused initially on her understanding of the small-scale perceptual cues and structure of the piece, before developing narratives in response to the movement. Her perceptions of

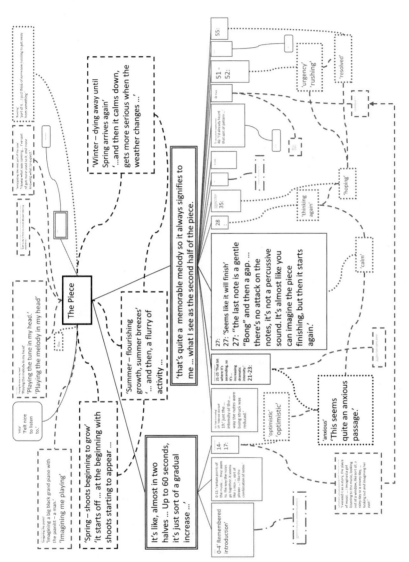

Figure 2.4   Representation of final schema for participant S (Clementi)

the Schoenberg showed a similar trend, although the schemata indicated greater use of narrative ideas to aid her understanding. In her first interview, she focused on small-scale perceptual cues and generated a mood association, describing the Schoenberg as 'quite a sad piece'. No large-scale structure was outlined here, nor in the first diary entry, which revealed a purely narrative understanding of the piece, with the music representing a three-stage development of a person's emotions over a period of time. This narrative perhaps triggered her tripartite conception of the structure of the piece, elucidated in the second diary entry, although the boundaries of the middle section of the piece differed slightly.

The schemata of the third and fourth diary entries elaborate on some of the existing small-scale perceptual cues, and include other associations and small-scale narratives (for example, in her third diary entry, participant A comments on a repeated note that 'makes it seem that someone's trying to get a point across'). The fifth diary entry revealed a four-part narrative understanding of the music, linked to small-scale perceptual cues, and later diary entries exploited this understanding with other narratives (for example, a storm). The schemata from the second interview and seventh diary entry revealed her increased understanding of the detail in the latter half of the piece, and in particular, her awareness of thematic unity and the return of the main musical idea. The ninth, tenth and eleventh diary entries reflected further judgment, small-scale cues, associations and small-scale narratives, before participant A returned to her larger-scale four-part narrative understanding, this time relating it to a person's mood as experienced over a day. This is highlighted in the schema of the final interview, when she labelled the structure of the piece as 'binary' among other details.

Overall, participant A's understanding of the Schoenberg appeared to be less immediate than her understanding of the Clementi, and she revealed her conceptualisation of the large-scale structure initially through narrative understanding, perhaps suggesting that this was a perceptual tool, rather than a creative by-product of listening. Furthermore, she continued to focus on small-scale perceptual cues throughout the familiarisation period, only identifying the structure of the movement after a few days of listening (see Figure 2.5).

Participant J As with his responses to the Clementi, participant J's schemata were characterised by a high volume of comments, which varied in nature over the familiarisation period. In the first interview, he made links with other pieces of music he knew, identified a large number of small-scale cues, and made several (mostly negative) judgments about the piece. He identified the piece as 'atonal', but did not provide a structure for it. The schemata from his first two diary entries revealed his search for structure in the piece, which he found, to some extent, in the schemata for the fifth and sixth diary entries as indicated by comments concerning motivic connections and musical repetitions. At this stage, he continued to make musical and non-musical associations, to use small-scale narratives, and to make judgments about the piece, and, in the second interview, to elaborate on some of the existing small-scale perceptual cues. He revealed further

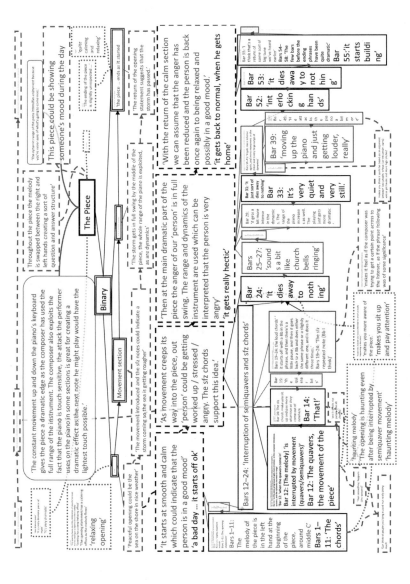

Figure 2.5   Representation of final schema for participant A (Schoenberg)

frustration in his seventh and eight diary entries at the apparent lack of structure of the piece, but his final schema, which related to the third interview, showed awareness of four structural sections and perception of the ways in which these related to repetition of the small-scale perceptual cues (see Figure 2.6). There is also a positive judgment comment at this stage. Overall, then, participant J expressed frustration at not being able to understand the structure of the piece, but this lessened when he finally identified repetition across the piece and divided it into sections. He thus searched for musical structure in his perception of the piece throughout the familiarisation period, and only articulated it in the final interview, after two weeks of daily listening.

Participant S   Compared with his responses to the Clementi, participant S provided slightly more information about his perceptions of the Schoenberg, particularly in the interviews. The schema from the first interview revealed numerous small-scale perceptual cues with narrative links and musical associations (for example, participant S revealed a perceptual cue by commenting that 'the contrast in the notes ... was almost ... they didn't fit together', and then created a narrative around this cue, stating, 'It's like he couldn't understand what to play, so he was doing lots of different notes'). These cues, associations and narratives were elaborated upon in each diary entry. The schema for the fifth diary entry, however, revealed preliminary formation of a larger-scale idea about the structure of the piece: he mentioned the return of an earlier tune at the end of the piece. This was not referred to again explicitly until the third interview. In the second interview, participant S referred twice to the use of narrative in his understanding, firstly suggesting that the music was not conducive to a narrative or imagery-led understanding, and secondly, that any associated narratives would be negatively valenced. His musical associations were related to a similarly negative 'murder' theme, reminding him of the music from a specific horror film, *Psycho*.

In subsequent diary entries, participant S developed other small-scale perceptual cues, and elaborated upon those he had discussed previously. He revealed further awareness of the thematic structure of the piece in the ninth and tenth diary entries, while four small-scale perceptual cues and related mood associations were established in the eleventh diary entry. In the final schema, developed from the third interview, participant S divided the piece into small chunks: the first two chunks corresponded with the first two sections of the piece outlined by participants A and J; the next two chunks were identified in a previous diary entry and reflected recognition of familiar melodic features; the fifth and sixth chunks were described as 'peaks', suggesting a move towards the climax of the piece; and the final chunk began with a return of the initial theme of the piece. Participant S suggested that this increased understanding of the structure of the piece increased his enjoyment of the piece (see Figure 2.7).

As before, participant S's conception of the piece was serial and associative, rather than hierarchical, until the final interview, when he revealed an understanding of the structure of the music based on themes. This was not identical to the

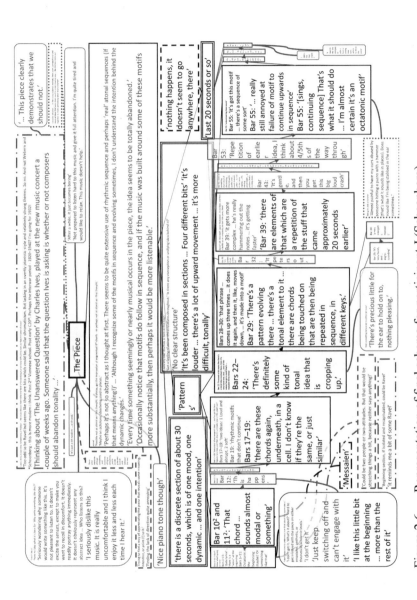

Figure 2.6   Representation of final schema for participant J (Schoenberg)

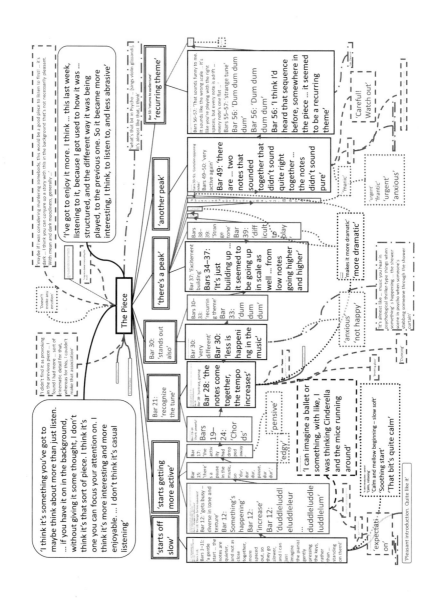

Figure 2.7   Representation of final schema for participant S (Schoenberg)

two musicians' conceptions, but the first two sections did overlap with their interpretations. It was interesting that participant S reported finding it easier to enjoy the piece once he had an understanding of its structure.

This piece prompted the three participants to consider, and in some cases, search for, the structure of the music. Participant A succeeded in identifying a structure relatively early on in the familiarisation period, and appeared to use her narrative conceptions of the music to aid her in this process. Participant J did not succeed in this until the final interview, and participant S only partially succeeded, although he did report an improved understanding. Participants J and S relied upon making musical and other associations in their continued exposure to the Schoenberg. The participants' responses to the Berio will shed light on whether or not their approaches will remain consistent when listening to another piece of free atonal music.

Perceptions of Berio's 'Rounds' for Piano Solo

Participant A   As with her previous responses, participant A's schemata for the Berio focussed on small-scale perceptual cues, this time with a few generalised features and narrative ideas concerning animal noises. By her third diary entry, she showed awareness of repetition within the piece: she noticed the return of the first section of the piece, clarifying the structure. Once this structure was observed, participant A combined it with a narrative idea: 'This could be a major catastrophe which all the busy "people" actually take notice of and perhaps try to improve it because we get similar material to what was at the beginning after that crash' (extract from the fourth diary entry). She continued to add small-scale perceptual cues as she noticed more features and, by the sixth diary entry, she began to make judgments about the piece (for example, 'The pauses for the most part are effective, but sometimes they are perhaps a little too long and make you wonder whether the piece has finished or not'). The schema for the second interview showed the addition of detailed descriptions of perceptual cues, and these increased in number in subsequent diary entries. A slightly different pattern emerged in the eighth diary entry, with increased general perceptual comments, mood associations and judgments, and the remaining diary entries showed a balance between these types of comments. The schema for the final interview revealed some new small-scale perceptual cues, participant A's clear understanding of the ternary form structure and a judgment about the piece that contained a slight hint of frustration over the many pauses (see Figure 2.8).

Overall, participant A's schemata for the Berio focused on small-scale perceptual cues (as with her responses to the Clementi and Schoenberg), but revealed a developing awareness of repetition as a feature of this particular stimulus. Her conception of the formal structure of the piece was clear and musically persuasive by the end of the experiment. Narrative was used to understand small-scale perceptual cues, while mood-related comments and judgments were also present.

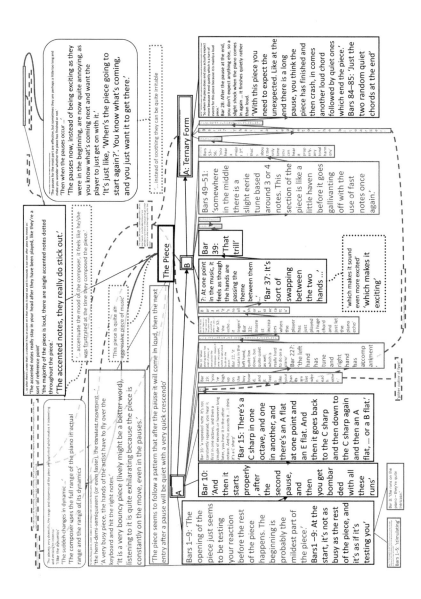

Figure 2.8   Representation of final schema for participant A (Berio)

Participant J   Participant J's responses to the Berio, like his responses to the Schoenberg, revealed his striving for an understanding of the musical structure. The schema for the first interview showed that he was already able to comment on repetition in the piece, although this was not transferred to an understanding of the form of the piece at this stage. He expressed intriguing ideas concerning the compositional structure of the music, suggesting that each section might incorporate the last. He commented, 'it seems like some kind of self referencing to different parts of what's come earlier. Maybe … each phrase incorporates every phrase that's gone before. Each section … incorporates all the elements that were in the previous sections.' Clearly he recognised the unified nature of the piece, even if his conception of the structure was inaccurate. As well as his extensive perceptual cues, he invoked a narrative understanding of the piece, highlighting the importance of suspense.[13] The urge to understand the structure continued across all of his diary entries, with a number of other comments reflecting musical associations and a gradual increase in small-scale perceptual cues.

Participant J's efforts to understand the structure resulted in his recognition of repetition in the work, and in the third interview, he suggested a formal structure with one repeated section. He identified the return of the first section of the piece, but did not notice that his first section was longer than his second. In this interview, he also discussed small-scale perceptual cues with a positive tone, and made an association with film music featuring elves and pixies. Participant J's schemata therefore reflected his developing clarity about the piece, from initial confusion to a conception with some accurate structural features and understanding of perceptual cues (see Figure 2.9).

Participant S   As in his responses to the Schoenberg, participant S struggled to create a narrative understanding of the Berio. The schema from his first interview showed evidence of small-scale perceptual cues about the first 42 bars and some general comments regarding the harmony and pauses across the piece. Additionally, he made a comparison between this piece and those used for the previous experiments. Regarding the pauses, he found it difficult to understand their purpose, and this manifested itself in subsequent diary entries as he tried to create a narrative understanding of the music. In the second interview, he reported that he was finding it impossible to identify a structure in the piece. He commented that there was no melody, and that 'there wasn't any structure, or normal structure as I would see it', adding, 'It didn't seem coherent, it didn't seem to be telling a story to me.' He elaborated on the 'unresolved notes' mentioned in his diary entries, and described some small-scale perceptual cues in some detail, such as those for bars 10–16: 'it got a bit more interesting with the use of the bass notes … they gave quite a deep sound … the percussive sound, but all seem to fit together'. His schema from the third interview incorporated perceptual cues, observation

---

[13]   Interestingly, one of his narratives involved a barking dog, as did one for participant A, perhaps suggesting a shared semiotic link.

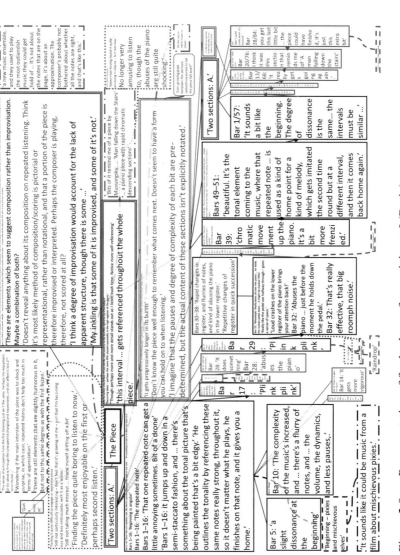

Figure 2.9   Representation of final schema for participant J (Berio)

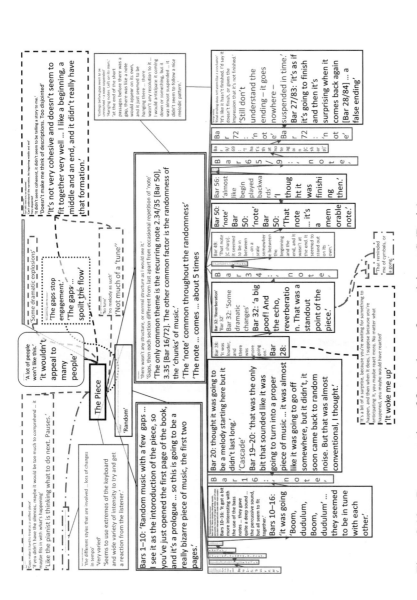

Figure 2.10   Representation of final schema for participant S (Berio)

of a repeated note across the piece, and a further disappointed attempt to use narrative to understand the piece. Participant S made an interesting suggestion about the potential for the piece to end at bar 56, the moment before the opening material returns: this indicated an understanding of the juncture as a large-scale structural boundary, although he did not perceive the repetition of the first section from that point onwards (he did, however, suggest that the material up to that point formed an introduction). Participant S provided some detail of his small-scale perceptual cues, and attempted to use narrative to aid his understanding of the piece, but found this difficult, if not impossible. He made a few musical and mood associations, as well as a judgment of the piece. Overall, participant S's responses to the Berio are marked by a striving for a narrative understanding of the music, with additional comments relating to small-scale perceptual cues, structural features and judgments of the piece (see Figure 2.10).

The three participants' schemata revealed considerable development over the familiarisation period in response to Berio's *Rounds*. They gradually generated conceptions of the musical structure, although only participant A was wholly successful in doing this. Both participants J and S struggled to invoke a narrative understanding of the music, and participant A only managed to do so via whole-piece or small-scale perceptual cues: unlike her responses to the other pieces, she did not tie her structural understanding to a narrative.

Comparison of the Participants' Perceptions

Overall, it is evident that the three pieces provoked both similar and varied responses in the three participants. Table 2.1 records the number of hearings preceding different types of responses to the three pieces by these participants. Various trends can be identified, both within and between participants, and within and between pieces. Participant A, for instance, identified large-scale musical structure most rapidly and accurately; participant S found this the hardest. All three participants identified structural features earliest in the Clementi, then in the Schoenberg, and then in the Berio. Participants' accuracy and detail in describing the music increased with familiarity, as revealed by the detailed perceptual cues and by their understanding of the musical structure. All participants tried to utilise a narrative understanding of the music from an early stage in the familiarisation process, and sometimes, as discussed, this was thwarted for them by the style of the music.

Readers might be tempted to use the data to generalise the features listeners heard to other people with similar levels of musical experience, or to responses to other pieces of music. This would be problematic, however, owing to the focus on a small number of participants and pieces. Perhaps a more appropriate way of working out the implications of these data is to generalise to the participant, following the example of Edward Stern (2004), who examines very small periods of time in people's routine behaviours (for example, having breakfast) to find patterns exhibited on a larger scale in their everyday lives. It may be possible,

Table 2.1   Number of hearings preceding different types of responses

	Clementi			Schoenberg			Berio		
	A	**J**	**S**	**A**	**J**	**S**	**A**	**J**	**S**
Small-scale perceptual cues	2	2	2	2	2	2	2	2	2
Overall conception of piece	2	3	2	2	2	2	2	2	2
Use of narrative	3	2	2	3	2	2	2	2	—
Mood association	2	2	3	2	2	2	12	2	2
Musical association	—	2	2	—	2	2	—	3	2
Judgment	—	2	4	14	2	2	3	4	2
Division of piece into sections	2	5	18	4	14	18	5	2	—
Recognition of thematic similarity	2	2	5	10	2	7	5	5	14
Accurate conception of large-scale musical structure	6	9	—	12	14	—	5/18	—	—

therefore, to use the data presented in the schemata to decipher a pattern of listening for each participant, that is, specific idiosyncratic listening behaviours in which each participant engages whilst listening to music. If there are sufficient commonalities between each listener's responses to the three pieces, these may enable us to predict the ways in which they might approach another piece of music.

Participant A, for instance, identified small- and large-scale musical structure quickly and accurately, and linked narrative ideas with specific aspects of this readily identified musical structure. She rarely made musical associations. In contrast, participant J made numerous and wide-ranging comparisons with other pieces of music. He also made value judgments about the music, and focused to some extent on performance features. He strove to understand musical structure, and became frustrated if he could not succeed in doing this. Participant S made non-musical associations easily, often with films or visual images. Rather than trying to identify musical structure, like participant J, he strove towards a narrative understanding of a piece. He had difficulty grasping the structure, which possibly reflected his lack of musical training or his listening strategy as narrative-based.[14] The participants did appear, therefore, to have idiosyncratic responses to perceiving music, which were consistent in some ways across the pieces, despite the contrasting musical languages of the stimuli.

---

[14]   It is impossible to identify cause and effect here.

Discussion and Conclusion

This study examined the effects of familiarity and musical language (tonal and atonal) on three listeners' perceptual responses to music. So, how do listeners' perceptual responses change with familiarity, and what are the similarities and differences between listeners' perceptual responses to tonal and atonal music? These questions will be considered together, before other findings are discussed.

Initial hearings of each piece of music seemed to allow participants access to small-scale perceptual cues, overall conceptions of the piece, and usually, mood associations and the use of narrative to understand the music, which supports existing research that suggests that these ways of listening are straightforward (Lavy 2001; Sloboda and Juslin 2001). Where musical associations were made, these were often made early in the familiarisation process. Judgments, too, appeared when participants were still relatively unfamiliar with the piece, as well as later on. Familiarity allowed the listeners access to thematic and structural understandings of the music, demonstrated by their division of the pieces into sections, their recognition of thematic similarity, and sometimes, their accurate conception of the large-scale structure of the pieces. This was expected, certainly within the tonal idiom, from previous studies (Pollard-Gott 1983); these findings suggested that responses to free atonal music were similar. Indeed, atonal stimuli may be especially useful in research with music listeners because the familiarisation process appeared to be slower with the atonal pieces, allowing more detailed access to its many facets. There were profound effects of musical language: the musical structure was more easily identified in the tonal piece, which supports and extends work by Lerdahl (1988) by providing empirical evidence. Although this evidence was strong for these three participants and pieces, the use of similar methods to explore a wider range of music would allow greater understanding of the influence of musical language on perceptual responses to music and their development with familiarity.

Additionally, this study revealed patterns or strategies of listening within participants, which were discernible through the visual representations of schemata generated as part of the data analysis process. These patterns were flexible: participants appeared to show a particular approach on a particular hearing, and at varied stages in the listening process. The patterns were, however, fairly consistent for each participant, suggesting that they are, to some extent, person-dependent. For some listeners, structure is important; for others it is musical associations, and for others still, narrative. It would be interesting to investigate the frequency of use of such approaches in listeners, and future studies could assess this with a sample sufficiently large to allow the generalisation of their findings. In particular, it would be interesting to explore the extent to which musical training or listening experience (and hence familiarity with the musical language) is related to particular listening approaches. Would the pattern observed here be replicated in other listeners, with those with less experience focussing on narrative listening, those with some experience using associations to guide their understanding, and those with most experience focussing on small-scale perceptual cues and musical structure?

It is inevitable that, even with qualitative data, participants' responses will be influenced to some extent by the method of data collection. Whilst this study made an attempt to allow participants freedom in their responses by avoiding leading questions in the interviews and by using an open-ended listening diary, it is acknowledged that the experimental situation and, in particular, the continuous emotional response mechanism, may have influenced participants' qualitative responses. Although it was made clear that there were no 'right' or 'wrong' answers, participants' responses may have been affected by demand characteristics. The schemata presented here were generated from the interview and diary entries, which were useful sources of data, but nonetheless were likely to be an incomplete representation of each participant's effable responses (not to mention those that were ineffable). Nonetheless, the data presented here are valuable, and show interesting trends.

The schemata presented above were not intended to be exact representations of participants' knowledge of the pieces of music under scrutiny. There will, quite naturally, be several layers of inaccuracy present between what listeners have heard, what they are aware of hearing, what they remember, what they say or write, and no matter how scrupulous the researcher, the researcher's interpretation of those data. Rather, the representations of schemata provided a means of depicting the data that resembled participants' knowledge gathered over time and after multiple interactions with a piece of music. It allowed the examination of changes in a listener's focus and in their understanding of a piece of music, and provides a potentially useful tool for other research studies investigating these issues.

It is perhaps worth noting that, whilst the dominant conception of perception in the field of psychology involves both bottom-up (sensation-based) and top-down (knowledge-based) processing, an ecological approach has been adopted in the past (Gibson 1979) and has recently been applied to music listening (Clarke 2005). This approach argues that perception is direct and does not involve a mental representation, which has been acknowledged as possible in some situations (Braisby and Gellatly 2005), but is otherwise rejected by mainstream psychologists (Eysenck and Keane 2005). As seen above, music psychologists have found evidence for mental representations; should, however, this evidence be superseded, the schemata presented here remain a useful means of understanding participants' knowledge of a piece of music as expressed verbally after listening.

The focused and detailed approach presented here allowed access to the familiarisation process for three listeners with varied backgrounds. As such, it cannot be representative of all listeners, or even all listeners with similar backgrounds. This in-depth approach, however, made it clear that listeners interact with music in complex and idiosyncratic ways as they become familiar with a piece via repeated exposure. It is these complexities, as well as the more straightforward and shared trends, which merit further study in research into music and familiarity.

References

Bailey, W.B. (1998). Biography. In W.B. Bailey (ed.), *The Arnold Schoenberg Companion* (pp. 11–39). Westport, CT: Greenwood Press.

Bharucha, J.J. (1994). Tonality and Expectation. In R. Aiello and J. Sloboda (eds), *Musical Perceptions* (pp. 213–39). Oxford: Oxford University Press.

Botstein, L. (1999). Schoenberg and the Audience. In W. Frisch (ed.), *Schoenberg and His World* (pp. 19–54). Chichester: Princeton University Press.

Braisby, N. and Gellatly, A. (2005). *Cognitive Psychology*. Oxford: Oxford University Press.

Braun, V. and Clarke, V. (2006). Using Thematic Analysis in Psychology. *Qualitative Research in Psychology* 3: 77–101.

Brower, C. (1989). Dramatic Structure in Schoenberg's Opus 11, Number 1. *Music Research Forum* 4: 25–52.

Buzan, T. (1995). *Use Your Head* (3rd edn). London: BBC Books.

Cannam, C., Landone, C. and Sandler, M. (2010). Sonic Visualiser: An Open Source Application for Viewing, Analysing, and Annotating Music Audio Files. *Paper presented at the 18th International ACM Multimedia Conference*, University of Trier, Florence, 25–29 October.

Clarke, E.F. (2005). *Ways of Listening: An Ecological Approach to the Perception of Musical Meaning*. Oxford: Oxford University Press.

Cone, E.T. (1977). Three Ways of Reading a Detective Story – Or a Brahms Intermezzo. In R.P. Morgan (ed.), *Music: A View from Delft* (pp. 77–94). Chicago, IL: University of Chicago Press.

Daynes, H. (2007). *Listeners' Perceptual and Emotional Responses to Tonal and Atonal Music*. Unpublished PhD dissertation, University of Hull, http://library.hull.ac.uk/record=b2088755~S3.

Daynes, H. (2011). Listeners' Perceptual and Emotional Responses to Tonal and Atonal Music. *Psychology of Music* 39/4: 468–502.

Delière, I. (1996). Cue Abstraction as a Component of Categorisation Processes in Music Listening. *Psychology of Music* 24: 131–56.

Delière, I. (2006). Analogy: Creative Support to Elaborate a Model of Music Listening. In I. Delière and G.A. Wiggins (eds), *Musical Creativity* (pp. 63–77). Hove: Psychology Press.

Delière, I. (2007). Similarity Relations in Listening to Music: How do they come into Play? *Musicae Scientiae. Discussion Forum* 4A: 9–37.

Delière, I. and El Ahmadi, A. (1990). Mechanisms of Cue Extraction in Musical Groupings: A Study of Perception on *Sequenza VI* for Viola Solo by Luciano Berio. *Psychology of Music* 18: 18–44.

Delière, I. and Mélen, M. (1997). Cue Abstraction in the Representation of Musical Form. In I. Delière and J. Sloboda (eds), *Perception and Cognition of Music* (pp. 387–412). Hove: Psychology Press.

Delière, I., Mélen, M., Stammers, D. and Cross, I. (1996). Musical Schemata in Real-time Listening to a Piece of Music. *Music Perception* 14/2: 117–60.

Dibben, N. (1994). The Cognitive Reality of Hierarchic Structure in Tonal and Atonal Music. *Music Perception* 12/1: 1–25.

Dibben, N. (1996). *The Role of Reductional Representations in the Perception of Atonal Music.* Unpublished PhD dissertation, University of Sheffield.

Dibben, N. (2001). What Do We Hear, When We Hear Music? Music Perception and Musical Material. *Musicae Scientiae* 5/2: 161–94.

Eysenck, M.W. and Keane, M.T. (1995). *Cognitive Psychology: A Student's Handbook* (3rd edn). Hove: Psychology Press.

Eysenck, M.W. and Keane, M.T. (2005). *Cognitive Psychology: A Student's Handbook* (5th edn). Hove: Psychology Press Limited.

Frisch, W. (1993). *The Early Works of Arnold Schoenberg 1893–1908.* Berkely, CA: University of California Press.

Gaver, W.G. and Mandler, G. (1987). Play it Again, Sam: On Liking Music. *Cognition and Emotion* 1/3: 259–82.

Gibson, J.J. (1979). *The Ecological Approach to Visual Perception.* Hillsdale, NJ: Lawrence Erlbaum.

Howard, R.W. (1987). *Concepts and Schemata: An Introduction.* London: Cassell Educational.

Hsieh, H.-F. and Shannon, S.E. (2005). Three Approaches to Qualitative Content Analysis. *Qualitative Health Research* 15/9: 1277–88.

Huron, D. (2007). *Sweet Anticipation: Music and the Psychology of Expectation.* Cambridge, MA: MIT Press.

Kramer, A.W. (1999). This Man Schönberg! In W. Frisch (ed.), *Schoenberg and His World* (pp. 312–15). Chichester: Princeton University Press.

Lavy, M.M. (2001). *Emotion and the Experience of Listening to Music: A Framework for Empirical Research.* Unpublished PhD dissertation, University of Cambridge.

Lerdahl, F. (1988). Cognitive Constraints on Compositional Systems. In J. Sloboda (ed.), *Generative Processes in Music: The Psychology of Performance, Improvisation and Composition* (pp. 231–59). Oxford: Clarendon Press.

McCoy, M. (1999). A Schoenberg Chronology. In W. Frisch (ed.), *Schoenberg and His World* (pp. 1–15). Chichester: Princeton University Press.

Neighbour, O.W. (2007). Schoenberg, Arnold, 1: Life up to World War I. *Grove Music Online*, http://www.grovemusic.com/shard/views.article.html?section=music.25024.1 (accessed 25 August 2007).

Ockelford, A. (1999). *The Cognition of Order in Music: A Metacognitive Study.* London: The Centre for Advanced Studies in Music Education, Roehampton Institute.

Ockelford, A. (2004). On Similarity, Derivation and the Cognition of Musical Structure. *Psychology of Music* 32/1: 23–74.

Ockelford, A. (2005). Relating Musical Structure and Content to Aesthetic Response: A Model and Analysis of Beethoven's Piano Sonata Op. 110. *Journal of the Royal Musical Association* 130/1: 74–118.

Pascall, R. (1989). Genre and the Finale of Brahms's Fourth Symphony. *Music Analysis* 8/3: 233–45.

Plantinga, L. (2006). Clementi, Muzio, 2: Works. *Grove Music Online*, http://www.grovemusic.com/shared/views/articles.html?section=music.40033.2 (accessed 19 June 2006).

Pollard-Gott, L. (1983). Emergence of Thematic Concepts in Repeated Listening to Music. *Cognitive Psychology* 15: 66–94.

Simms, B.R. (1998). Schoenberg: The Analyst and the Analyzed. In W.B. Bailey (ed.), *The Arnold Schoenberg Companion* (pp. 223–50). London: Greenwood Press.

Sloboda, J.A. and Juslin, P.N. (2001). Psychological Perspectives on Music and Emotion. In P.N. Juslin and J.A. Sloboda (eds), *Music and Emotion: Theory and Research* (pp. 71–105). Oxford: Oxford University Press.

Stern, D.N. (2004). *The Present Moment in Psychotherapy and Everyday Life.* New York: W.W. Norton.

Sternberg, R. J. (1996). *Cognitive Psychology.* Fort Worth: Harcourt Brace College Publishers.

Thow, J. (1996). *Sleeve Notes of Recording Arden, David, Luciano Berio: Complete Works for Solo Piano.* NA089 CD: New Albion Records.

Webern, A.V. (1999). Schoenberg's Music. In W. Frisch (ed.), *Schoenberg and His World* (pp. 210–30). Chichester: Princeton University Press.

Chapter 3

The Effects of Repertoire Familiarity and Listening Preparation on New Audiences' Experiences of Classical Concert Attendance

Melissa C. Dobson

When attending a classical concert for the first time, many aspects of the experience may seem unfamiliar or even unusual. This chapter reports on an exploratory study that sought to explore how new audience members respond to both familiar and novel musical experiences in the concert hall, and to question how familiarity with the repertoire performed might affect new audience members' concert experiences and enjoyment. While existing research has produced few answers to these questions, findings of research on this topic may have potentially useful implications, both for increasing our understanding of the roles of familiarity and novelty in the enjoyment of music listening, and for the marketing and audience development strategies of orchestras and concert organisations.

The lack of current knowledge about the role of repertoire familiarity on new audiences' experiences is important to note, as there is a prevailing idea that concert audiences are indeed often familiar with the music that they choose to see performed live. While classical audience members may of course have become familiar with a given work from recorded listening (Hennion 2001), the effects of orchestral programming may also play a role, where the repertoire that symphony orchestras present typically derives from a core number of symphonic works from the Western art music canon (Kramer 1995). Small (1998, p. 167) describes what he regards as the negative effects of repeated hearings of canonic symphonic works, arguing that these result in a 'loss of narrative meaning'. Similarly, Cone (1974, p. 116) writes that 'it is hard to make overfamiliar compositions yield vital experiences', yet conversely in another essay stresses the potential benefits of repeated hearings through gaining insight into the meanings and structure of a musical work (Cone 1989/1977).

Psychological research provides possible explanations for the phenomena Cone describes. There are several different theories that attempt to describe relationships between familiarity and liking of aesthetic stimuli. The simplest is the mere exposure effect (see Zajonc 2001), whereby an individual's liking for a novel stimulus increases with repeated exposure. This relationship between affective response and frequency of hearings has been demonstrated in numerous studies of a mere exposure effect for music (Peretz, Gaudreau and Bonnel 1998).

In the field of experimental aesthetics, Berlyne's (1971) arousal potential theory (which unlike the mere exposure effect relates specifically to aesthetic stimuli) proposes that liking of a stimulus is determined by the degree to which it induces physiological arousal. As Hargreaves and North (2010, p. 520) outline, 'Berlyne suggested that the listener "collates" the different properties of a given musical stimulus, such as its complexity, familiarity, or orderliness, and that these "collative variables" ... combine to produce predictable effects on the level of activity, or arousal, of the listener's autonomic nervous system.' Berlyne (1971) states that liking stands in an inverted U-shaped relationship with arousal, so that stimuli that create intermediate levels of arousal are preferred. This theory has been developed through the concept of subjective complexity (North and Hargreaves 1995), whereby liking is influenced by the listener's perception of the music's complexity. There is an optimal level of complexity for each individual listener, which depends on their degree of prior exposure. Repeated exposure increases familiarity and reduces the subjective complexity of the stimulus (North and Hargreaves 1995). Therefore, when inexperienced listeners are exposed to repetitions of complex music, their liking should increase, while experienced listeners who are exposed to multiple hearings of a simple piece should experience a decrease in liking (Hargreaves and North 2010, pp. 523–4).

Huron (2006) considers both the positive and negative effects of familiarity with music on our listening experiences. Familiarity can be inherently pleasurable, enabling us to make predictions about what we will hear next: an accurate prediction of musical events leads to a prediction response which 'serves the biologically essential function of rewarding and reinforcing those neural circuits that have successfully anticipated the ensuing events' (Huron 2006, p. 140). Moreover, while not all music is inherently predictable, with familiarity we can learn to 'expect the unexpected', and therefore reap a positively valenced prediction response even from music that violates expectations on a first hearing (p. 365). Identifying an 'extraordinary repetitiveness' inherent in music, both within and across individual works (p. 268), he theorises that this repetition contributes to our pleasure in listening to music (through the limbic reward effects of being able to approximately predict what will happen next), but also considers the negative effects of habituation in individuals with considerable exposure to music listening. Meyer (1967/1994, p. 48) notes similar effects, writing that 'the better we know a work, the more difficult it is to believe in, to be enchanted by, its action'.

A small number of studies have addressed the role of familiarity in the concert experiences of audiences in real-world settings, where it is likely that enjoying a piece in a concert performance is mediated by a range of variables (familiarity among them), from the listening environment to the listener's internal state (Hargreaves, Miell and MacDonald 2005; Thompson 2007). Thompson (2006) explored the effects of familiarity on the enjoyment of listening within the concert hall: aiming to identify factors that affect the enjoyment of a performance, Thompson's study found no relationship between enjoyment and prior familiarity with the repertoire performed. He proposes that familiarity with a piece may even exert a negative effect on the enjoyment of a performance, as comparisons with a recorded version of the work known to a listener may be inevitable (Thompson 2006, p. 233; cf. Cone 1974, p. 138).

This idea is supported by Roose's (2008, p. 247) large-scale survey of concert attenders, which indicated that, while experienced and frequent classical concert attenders prefer concert programmes that present works which are new to them, less frequent attenders exhibit a preference for hearing that music that they know, so that 'evaluating a concert according to the extent to which it contains familiar and easily recognizable tunes is negatively related with frequency of attendance'. In Pitts's (2005) research on audience experience at a chamber music festival, attendance decisions were strongly influenced by the audience members' familiarity with the repertoire on offer, with the desire to hear familiar pieces finely balanced with a 'cautiously openminded' willingness to explore unknown works (Pitts 2005, p. 263). However, as Pitts (2005, p. 264) points out, although the audience members in this setting had the benefit of previous positive experiences within the context of their festival attendance when they had chosen to 'take risks' with unfamiliar repertoire, they 'rarely transfer[ed] their more adventurous musical choices to other settings'.

Two studies addressing the concert experiences of audience members who do not usually attend classical performances suggest that, for these less experienced audience members, familiarity (or its lack) can also exert a strong influence on audience members' enjoyment of classical concert attendance. Radbourne et al. (2009, p. 23) interviewed first-time attenders after they had attended either a classical concert or theatre production, and found that both types of performance led participants to '[express] discomfort with not being sufficiently "in the know" to value what they were seeing'. In Kolb's (2000) study, three groups of students attended one concert each at London's Festival Hall: a classical 'pops' concert that contained many well-known pieces of classical music from film soundtracks; a 'traditional' classical concert of nineteenth-century orchestral music; and a concert of music composed for a science fiction film by composer Michael Nyman. While those attending the latter two concerts were concerned about their lack of knowledge, feeling that they did not know the music well enough to appreciate the performances properly, those at the classical pops concert were positively surprised by recognising much of the music performed. Kolb argues that this recognition increased their sense of belonging in the concert hall in comparison to those attending the other two concerts, who felt that 'to properly enjoy classical music you need to have extensive knowledge of the music' (Kolb 2000, p. 25).

This chapter reports on research that developed the design of Kolb's (2000) study. Given that infrequent classical concert attenders exhibit a preference for attending performances of works with which they are familiar (Roose 2008), an exploratory aim of the present study's design was to explore the effects of exposing new attenders in advance to the music they would hear live, as a potential means of providing them with existing experience with which to appraise the performances. This part of the study also aimed to gain further data on the experiences of those attending classical concerts for the first time *without* knowledge of the music provided in advance.

Method

Design

A group of participants was invited to attend three classical concerts in succession during a two-week period. The participants were all individuals for whom classical concert attendance was not the norm; the study aimed to explore their experiences and responses during and after this exposure to classical concert attendance (Dobson 2010, presents further information on the study's aims and design).

Participants

A total of nine participants aged 24–36 took part in the study (see Table 3.1). Potential participants were identified through acquaintances of the author and through advertising the study on a social networking site. A preliminary questionnaire was used to identify individuals who met the study's criteria. Participants were recruited on the basis that they frequently attended arts events (live music, theatre, art exhibitions and so on) but did not attend classical concerts: the recruitment criteria required that they had attended no more than one classical concert in the preceding 12 months. The participants reported listening to classical music with varying degrees of frequency; the majority of those who listened more frequently than 'rarely' exclusively chose to listen to classical music in the background when they were working, in order to aid concentration.

Procedure

The concerts took place in three different London concert venues (see Table 3.2). The participants' immediate responses were obtained in focus group interviews directly after Concerts 1 and 3. To elicit more detailed responses from the participants, one-to-one semi-structured interviews with each participant took place in the two weeks following the final concert.

To collect exploratory data on the effects of repertoire familiarity on the participants' enjoyment and experience of the concerts, half of the sample (labelled 'LP' – signifying 'Listening Preparation') was provided in advance with recordings of the music to be heard in performance two weeks before the first concert took place. They received three CDs (one relating to each concert) and were asked to listen to the relevant CD at least once before attending the corresponding concert. As much as possible, recordings were selected that would exhibit a similarity to

Table 3.1   Participant profiles

Participant	Age	Previous classical concert attendance	Frequency of classical music listening	Attended all three concerts?	Listening preparation status (see Procedure)
Carla	36	Once in past year, to see a friend performing in an amateur choral concert	Several times a week: background listening when working	Yes	LP
Dawn	35	None in past year; attended once on a school trip	Once a week: background listening when working	No: attended only Concert 1	LP
Dominic	27	Once in past year, when offered free tickets by a friend	Rarely	Yes	Non-LP
Emma	27	Once in past year, when offered free tickets by a friend	Rarely: background listening when working	No: attended Concerts 2 and 3	Non-LP
Kerry	30	Attended opera once in past year; never attended a classical concert	Rarely	Yes	Non-LP
Rachel	25	None	Once a week: when teaching, playing music to her school class	Yes	Non-LP
Stuart	25	None	Every so often	Yes	LP
Tara	24	None	Rarely; previously listened when a work colleague played classical CDs in the office	Yes	LP
Toby	27	Attended opera twice in last 12 months when offered free tickets by a friend; never attended a classical concert	Every so often: background listening when working	No: attended Concerts 1 and 3	Non-LP

Table 3.2   The programmes of the three concerts attended

Concert	Date, performers and venue	Programme
1	13 February 2008	[Joseph Phibbs: Shruti – not included in the listening preparation task]
	London Symphony Orchestra	Rachmaninov: *Rhapsody on a Theme of Paganini*
	Barbican Hall, Barbican Centre, London	Shostakovich: Symphony No. 15
2	19 February 2008	[Mozart: Overture to *Der Schauspieldirektor* – not included in the listening preparation task]
	The Night Shift, Orchestra of the Age of Enlightenment Queen Elizabeth Hall, South Bank Centre, London	Mozart: Piano Concerto No. 21, movements II and III Beethoven: *Coriolan* Overture
3	20 February 2008	Strauss: *Die Fledermaus* Overture
		Schumann: Piano Concerto
	London Chamber Orchestra St John's, Smith Square, London	Brahms: Symphony No. 1

the live performances the participants would hear. For example, the CD relating to Concert 2 (performed by the Orchestra of the Age of Enlightenment, a period instrument orchestra) comprised recordings made by other period instrument ensembles. A work which had not been listed in the available promotional literature about the performances was performed in Concerts 1 and 2, respectively; in these cases the CDs did not completely represent the combination of works that the participants were to hear live. The quantitative analysis that follows is therefore based on the seven pieces which the LP participants had listened to in advance.

The first four participants to agree to take part in the study were provided with the recordings. The participants were instructed that, as long as they had listened to each CD once (and before the concert to which it pertained), they were free to listen to the CDs as many times as they wished. They were informed that they could listen to the CDs in any situation (including listening while focusing on other tasks), and were free to transfer the tracks from the CD to their computer or mp3 player should they wish. They were provided with a rating card for each CD and were asked to record the date of each occasion when they listened to some or all of each recording.

One LP participant (Dawn) ceased her involvement in the study on the day of the second concert. There was not time for her replacement to be provided with recordings and asked to listen to them before attending the performances. Dawn's data relating to Concert 1 has nonetheless been used in the analysis. With the inclusion of Dawn's replacement (Emma), the group of participants who did not participate in the listening preparation task (labelled 'non-LP') thus increased from four to five. It was decided that the possibility of the non-LP participants having heard some of the repertoire prior to attending the concerts could not be controlled: they were not, for example, instructed *not* to listen to the music in advance. Rather, it was assumed that they were unlikely to prepare for attending the concerts by seeking out recordings of the works. Their perceived levels of familiarity with the works were obtained using a ratings sheet at each concert and were discussed in the focus group and individual interviews.

Data Collection and Analysis

The participants were asked to complete a ratings sheet immediately after each concert. The ratings sheet recorded their perceived familiarity with each piece performed; their enjoyment of each piece; and an enjoyment rating for the concert overall. All ratings employed a seven-point scale, where 1 represented 'not at all', and 7 represented 'very much so'. The interview schedules for the focus groups and for the individual interviews each contained a series of questions about the participants' degrees of familiarity with the music performed. The LP participants were asked whether they thought that hearing the music in advance had affected their live listening experiences. The non-LP participants were asked whether there were points in the music that they did in fact recognise; they were also asked if they would have liked the opportunity to hear the repertoire from recordings before attending the concerts.

The focus group and interview data were analysed thematically using a grounded theory approach. The transcripts were each read repeatedly, with summaries, interpretations and initial theme titles noted in the margin. A list of theme titles was devised; the themes were then grouped together using axial coding (Strauss and Corbin 1998) to produce a number of higher-order concepts. The focus group interviews were analysed first; the individual interview data were then analysed, using the existing themes created from the focus group analysis where these were appropriate, but also creating new themes when necessary. Once the set of individual interviews had been analysed, the focus group data were returned to, checking for instances of new themes that had been subsequently identified.

The analysis which follows firstly presents quantitative findings on the effects of the listening preparation task on the participants' enjoyment ratings. These findings are then contextualised with qualitative data from the focus groups and interviews, exploring the participants' responses to participating in the task and investigating its effects on their experiences within the concert hall. The participants are identified by pseudonyms; quoted interview data is labelled by the participant's status in the listening preparation task (LP or non-LP), and by the interviewing occasion (FG1 – focus group following Concert 1; FG2 – focus group following Concert 3; I – individual interview).

Results and Discussion

The participants' mean enjoyment ratings for the seven pieces for which ratings were attained were reasonably high, with mean ratings all equal to or exceeding the rating scale's mid-point of 4. The Mozart (Concert 2) and Rachmaninov (Concert 1) received the highest enjoyment ratings, with mean scores greater than 6, while the Shostakovich symphony (Concert 1) received the lowest mean enjoyment rating, of 4 (see Figure 3.1).

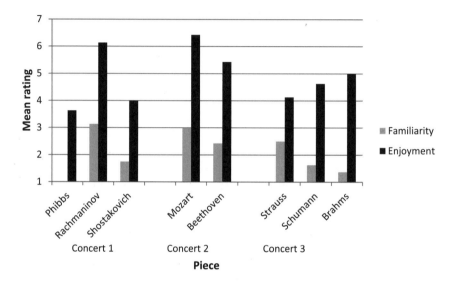

Figure 3.1   Mean familiarity and enjoyment ratings for each piece in the three
                concerts attended by the participants

Note: Error bars represent 1 standard deviation.

As expected, the group of LP participants (who had received and listened to recordings of the music in advance) gave higher mean familiarity ratings than the non-LP group for all seven pieces for which ratings were obtained (see Figure 3.2). When the familiarity ratings from all seven pieces were aggregated, the mean familiarity rating for the LP group (mean, $M = 3.30$; standard deviation, SD = 1.40; $n = 23$) was higher than the mean familiarity rating for the non-LP group ($M = 1.45$; SD = 0.85; $n = 31$). A Mann–Whitney U-test showed this difference to be significant ($U = 81$; $p < 0.01$ one-tailed).

As shown in Figure 3.3, the LP group produced higher mean enjoyment ratings than the non-LP participants for four of the seven pieces for which ratings were obtained. When the enjoyment ratings from the seven pieces were aggregated, the mean enjoyment rating for the LP group ($M = 5.26$; SD = 1.36; $n = 23$) was

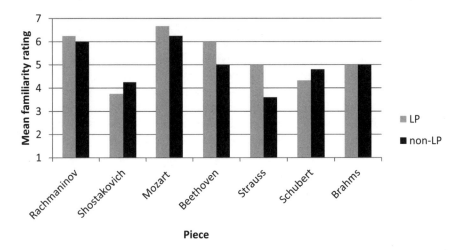

Figure 3.2   Mean familiarity ratings for each piece from the LP
and non-LP groups

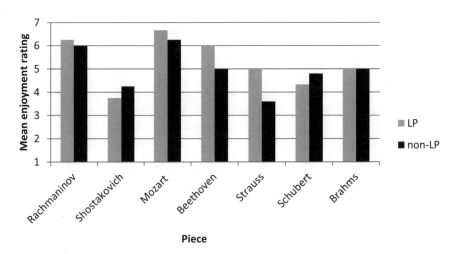

Figure 3.3   Mean enjoyment ratings for each piece from the LP
and non-LP groups

marginally higher than the mean enjoyment rating for the non-LP group (M = 4.94;
SD = 1.18; n = 31). A Mann–Whitney U-test showed that the difference in mean
ratings between the two groups was not significant (U = 299; p = 0.15, one-tailed).

Additionally, when the *overall* enjoyment ratings for the three concerts were
aggregated, the mean overall enjoyment rating for the LP group (M = 5.40; SD =
0.84; n = 10) was marginally lower than the mean overall enjoyment rating for the

non-LP group (M = 5.46; SD = 0.78; n = 13). A Mann–Whitney U-test showed that the difference in mean overall enjoyment ratings between the two groups was not significant (U = 61.5; p = 0.45, one-tailed). These results therefore indicated that, while the listening preparation task did exert an effect on the LP participants' perceived familiarity with the works, it did not exert an effect on the enjoyment ratings provided by the LP and non-LP groups. This lack of a difference in enjoyment ratings between the two groups is surprising given the strong empirical evidence for a mere exposure effect in music, whereby repeated exposure to a novel stimulus leads to an increase in liking (Peretz et al. 1998; cf. Huron 2006).

Difficulties Remembering or Recognising

Notably, the two symphonies that the participants listened to over the course of the study (Shostakovich, Concert 1; and Brahms, Concert 3) were the pieces that received the two lowest mean familiarity ratings from the LP group, and were the two pieces in which the difference between the mean familiarity ratings from the two groups was the smallest (see Figure 3.2). With the recordings of these pieces lasting 46 minutes and 44 minutes, respectively, these were also the longest works the participants were asked to listen to. In contrast, the two shortest works (Beethoven, Concert 2, and Strauss, Concert 3, both lasting around 7.5 minutes) were the two pieces in which the difference between the mean familiarity ratings from the two groups was greatest.

These observations raise the question of whether the listening preparation task was effective in increasing the LP participants' perceived familiarity with symphonic works of considerable length. Two LP participants (Stuart and Dawn) felt that the lengths of even the individual movements of classical works hindered their ability to become acquainted with the music from listening in advance. Stuart, when discussing the Shostakovich symphony, described how:

> even though I've listened to it sort of several times, apart from the few really recognisable bits that probably people who haven't even heard the CDs would recognize … some of the songs [movements] are whatever, fifteen minutes long and so listening to it a few times it was impossible to remember them, so it was just as surprising as if I hadn't heard the CDs before. … But I think I would have had to listen to them a lot of times before I was able to remember, you know, most of the piece. [LP Stuart, FG1]

Because of this difficulty, some of the participants concurred with Stuart that listening to the pieces only a few times in advance did not (or would not) provide an adequate level of familiarity. This problem was accentuated for the other two LP participants, Carla and Tara, both of whom noted difficulties finding time to listen to the recordings, especially as they did not own a portable music device (for example, an mp3 player), and so were only able to listen to the CDs when at home.

Similarly, Stuart described a sense of begrudging obligation to complete the listening preparation task, again relating this difficulty in part to the length of the works:

I definitely sort of had to go out of my way to listen to them, I had to sort of set aside time … There was an element to them which was sort of a little bit arduous having to listen to them all the way through. [LP Stuart, I]

Supporting this idea, when asked if they would have liked to have heard the music in advance, some non-LP participants expressed the belief that prior exposure would have only aided their enjoyment of the performances if they had enjoyed listening to the music first from recordings. Emma (non-LP) made a distinction between listening to styles of music she already holds an affinity with and listening to classical music, describing how the lengths of the works would necessitate many hearings in order to reap any benefits when listening live:

I don't think it's the same as listening to, you know, before you go and see a pop concert you listen to the album just so you remember all the words so you can sing along. … [Classical music's] not that easy to remember, it's not that simple, it's more complicated. [non-LP Emma, I]

One interpretation of Stuart and Emma's responses is that they are willing to prepare for concert attendance for other styles of music but when considering doing so for a classical concert, preparation is viewed in more negative terms as 'work'. This is understandable, given that they were asked to devote time to listen to music of which they had no prior knowledge, and little information with which to contextualise what they were hearing. This is in contrast to the way they might listen to music they already like and feel an emotional connection towards, where listening in advance of seeing a live performance may form a valued part of the ritual involved in attending a concert (Cavicchi 1998). In addition, with popular styles of music the participants have the benefit of a cumulative exposure to – and thus schemata of – the music and its conventions, making new music they are exposed to within these styles easier to assimilate, remember and recognise (Meyer 1967/1994, p. 287).

Overall, all four LP participants were in agreement that listening in advance was not always rewarded by an ability to recognise the music in live performance. They particularly noticed this discrepancy in relation to the Shostakovich symphony in Concert 1, where Dawn and Stuart talked of *only* recognising the quotations from Rossini's *William Tell Overture*, despite having listened to the work more than once. While some were merely surprised about this, two participants in particular (Carla and Dawn) interpreted their lack of recognition as an almost personal failing. Carla described her lack of recognition, as well as being surprising, as 'a bit embarrassing', finding that despite listening to each CD at least once she did not recognise any of the music she heard in the concerts:

I don't know whether it's my music memory or whatever. Because it's like it was completely different music. I couldn't say that what I have listened [to] today was exactly what I have listened to on the CD. I don't link it at all. [LP Carla, FG1]

A sense of personal failure in relation to listening and appreciation at the concerts was a broader theme that extended beyond the LP participants' perceived familiarity with the works (see Dobson 2010). In Dawn's case, however, the provision of the music to listen to in advance seemed to reinforce a perceived ignorance about classical music. She spoke of deliberately listening to the CD of Concert 1 more than she had anticipated because she found the Shostakovich 'hard to remember' and wanted to 'become more familiar before I went' [LP Dawn, I]. Rather than providing a sense of confidence through prior exposure and knowledge, the listening preparation instead instilled a sense of pressure:

It was useful, but I think it was also … it also kind of made you think "Oh gosh, I don't know this piece. I know that I don't know this piece; I'm going to a concert, I don't know the piece, and I can't possibly get to know it that well in the short time I've got before I go." Because obviously, you know, when you like pieces, you find pieces you like – you play them continually over a long period of time and you get to know them. It becomes subconscious, almost … what you know about them. So I think in some ways, listening to pieces I didn't know didn't make me look forward to it so much. Yeah, that's probably a very ignorant way of seeing it. [LP Dawn, I]

By increasing Dawn's awareness of her lack of knowledge of the music performed, it is plausible that the listening preparation task heightened negative perceptions about her competence as a listener. The inclusion of the task in the study may have reinforced (or even instigated) participants' expectations that classical audiences know the music well (data on perceived lack of knowledge is discussed further in Dobson 2010; Dobson and Pitts 2011). Participating in the listening preparation task therefore did not appear to alleviate problems in knowing how to appreciate classical performances: all of the LP participants demonstrated difficulties in appraising the performances, either by attributing a lack of enjoyment to an internal fault, or by noting difficulties in understanding which aspects of classical performance they should be showing appreciation for (Dobson 2010).

The Effects of Recognition on a First Hearing

Because of the time constraints of the task, the average number of occasions on which the LP participants listened to each CD decreased as the study progressed: the majority listened to the CD of Concert 1 three times, Concert 2 twice, and Concert 3 once. The LP participants' perceived familiarity with the music (and its perceived accessibility) was mediated by whether they recognised the music when first listening to the recordings. Pieces they did not instantly recognise on the first hearing were still perceived as music they 'didn't know' even *after* they had undertaken repeated hearings. In Stuart's case, not being able to recognise a piece when first listening to the recording seemed to be an overriding factor in the degree to which he perceived the music to be accessible and enjoyable:

> The third CD I think I enjoyed least; I didn't like the music as much and there was very little of it which I recognized or had heard before. It wasn't very easy to listen to, I didn't find much that was easy to enjoy. … Yeah, it was less accessible … I suppose the classical concerts were the first time I'd gone to see something which, whilst I'd listened to the CDs, it wasn't something I'd listened to before, before that, you know. I tend to go to gigs of things I know of, and like. [LP Stuart, I]

Three LP participants recognised Rachmaninov's *Rhapsody on a Theme of Paganini* when first listening to the recordings, and with this piece they cited a sense of anticipation during the live performance of the work as a positive effect of the familiarity developed through the listening preparation task. They talked of being 'genuinely excited' [LP Stuart, FG1] about hearing it live, describing how knowing 'how it was going to start' and 'what was possibly going to happen next' [LP Tara, FG1] increased their enjoyment of the performance. Combined with the primacy effect of the Rachmaninov being the first work that the LP participants would have heard on the first CD, this aspect may also have been accentuated by prior exposure. Participants in both groups described recognising two distinct 'hooks' in the Rachmaninov: the theme (which most recognised as the television theme tune to *The South Bank Show*) and Variation 18 (which has appeared in various film soundtracks). Notably, the theme occurs within 30 seconds of the opening of the piece, meaning that the participants would have heard something they knew almost immediately on their first hearing of the work. The appearance of Variation 18 later on provided the LP participants with a degree of schematic geography with which to approach the work, meaning that from the beginning of the live performance they were expecting – and looking forward to – this later recognisable section (Meyer 1956). Similarly, three of the non-LP participants felt that familiarity would have heightened their enjoyment, precisely because prior exposure to a piece 'acts as a bit of a mental guide' [non-LP Toby, I]. For those in the LP group, therefore, this unexpected recognition on their first hearing may have helped create an enjoyable first experience of encountering the entire piece, engendering positive expectation about subsequent hearings and the live performance that ensued (Huron 2006, p. 327).

Non-LP Participants' Responses to Unexpected Recognition

As experiences of the Rachmaninov demonstrate, despite not being asked to purposively listen to the works in advance, most of the non-LP group did coincidentally recognise some sections of music during at least one of the concerts. In Concert 2, some had a general sense of familiarity with the Mozart piano concerto. They were not, however, able to identify specifically where they had heard the work previously; one participant described how 'because Mozart sort of is so ubiquitous … you feel like even if you haven't heard that particular piece, sometimes you feel like you've heard that [before]' [non-LP Dominic, I]. In Concert 1 the non-LP participants all directly recognised small sections of the Rachmaninov and/or Shostakovich. In general, recognition was viewed positively; they noted that moments of recognition put them at ease within the concert situation and provided confidence. For some of the non-LP group, recognition appeared to instil a sense of emotional security: '[it's] like a sort of comfort toy … It was quite a nice feeling, like thinking "oh I recognize that from so and so" or "oh I've heard that before"' [non-LP Rachel, I]. Perhaps because the non-LP group had no reason to assume that they would know the music, recognition exerted stronger positive effects on these participants compared with their LP counterparts, whose expectations about how much of the music they would recognise were frequently not met.

Unexpectedly, however, the effects of recognition in the non-LP group were not exclusively positive: two participants described recognition as a distracting feature, rather than finding that it increased their engagement in the performance. Kerry, for example, didn't 'know whether it enhanced my enjoyment. … I think I spent more time trying to work out where I'd heard it before, rather than going "ok, yeah, this is quite nice to have heard it"' [non-LP Kerry, I]. In Toby's experience, meanwhile, recognition 'actually cheapens it slightly, because you just end up thinking "Ah, it's a Direct Line advert" or whatever it might be' [non-LP Toby, I]. Both of these accounts highlight a difficulty for non-attenders, whose only prior exposure to classical music may have been through popular media and music used in advertisements (where it is re-appropriated, creating new meanings and associations; cf. Tota 2001).

Recognition in this sense perhaps served to reinforce their perceived limited experience, highlighting a disparity between their levels of knowledge and exposure to classical works, and the greater depth of knowledge that they assumed other audience members to possess (see Dobson 2010; Dobson and Pitts 2011). This seemed to be the case for Rachel, who describes her positive surprise at recognising some of the Rachmaninov, but ends by stressing the 'low-art' setting from which she assumes she has encountered it previously, demonstrating a tension between the positive effects of recognition and her perceived level of knowledge:

> When that bit came on [Rachmaninov , Variation 18], I was like "Oh! Is that what it is" [laughter] "Oooh!" … I don't know whether it was just a surprise

thinking "Oh, I recognize this". Like because apart from that thing I don't know very many ... er ... many tunes or anything. I mean listening to this one, I was like "Oh, I have heard that before". But as you say it's probably off some ... popular culture show. [non-LP Rachel, FG1]

Perhaps because of this tension between recognition and knowledge, one participant [non-LP Dominic] in particular spoke of valuing the opportunity to assimilate new information from the context provided during the concert (through programme notes or spoken introductions) and to then 'give yourself a pat on the back' at identifying particular elements of the music during the performance, or 'spot[ting] ... this is where this movement cuts off and this movement starts' [non-LP Dominic, I]. This account highlights the importance of recognising that increased confidence can be instilled in new audience members through the provision of verbal descriptions as well as through auditory exposure.

Conclusions

It was surprising that participating in the listening preparation task did not increase the LP group's enjoyment ratings in comparison to those of the non-LP group, as the theories of mere exposure and subjective complexity would dictate. It is plausible that enjoyment ratings from the LP group may have been higher had they been able to recognise a greater proportion of the music performed. The design of the study may have played a part in this respect, as the participants were required devote full attention to the music when listening: they therefore most frequently engaged in 'incidental' (that is, background) rather than 'active' or focused listening. Szpunar, Schellenberg and Pliner (2004, p. 370) found that listening strategies were critical in determining subsequent recognition, with those who engaged in focused listening improving their ability to subsequently recognise musical stimuli in comparison to those who listened 'incidentally'. However, given the small number of hearings undertaken even on an incidental basis, asking the LP participants to engage only in focused listening would have most probably diminished the likelihood that they would complete the task at all.

Silva and Silva (2009) found that mere exposure did not improve an unfamiliar song's appeal, but that providing information about the artist or an endorsement from an authority figure did. In alignment with Silva and Silva's findings, the piece rated most highly for enjoyment by both the LP and non-LP groups was the Mozart piano concerto in Concert 2, about which the greatest degree of accessible context was delivered, in the form of an extensive spoken introduction and demonstration from the pianist (see Dobson 2010). Importantly, the participants' responses indicated that this provision of context was more instrumental in increasing their enjoyment than prior exposure. However, it is also possible that prototypicality played a role in their preference for this work: Martindale and Moore's (1988) findings indicate that aesthetic preference is influenced by the degree to which a stimulus conforms to mental schemata, so that 'typical instances of any category should be preferred because they give rise to a stronger activation of the relevant cognitive representations than atypical instances' (Hargreaves and North 2010, p. 525). Of all pieces, the Mozart received the highest mean familiarity rating from the non-LP group, but unlike other works they felt some familiarity with, the participants did not identify particular passages or sections that they recognised, nor could they identify a specific context in which they had heard the Mozart previously. Rather, they noted feeling familiar with Mozart 'in general', suggesting that the piano concerto conformed to their preconceptions of what Mozart, and possibly even *classical music*, should sound like.

For some LP participants, the task produced the opposite of its intended effects. Instead of instilling confidence, it appeared to make them even more aware of their status as novices in the concert hall. Some of LP participants noted the differences between the somewhat artificial demands of the listening preparation task and the more gradual, organic process of becoming acquainted with music of their own choice that typically characterised their usual listening practices. Szpunar et al. (2004) ran three experiments on the effects of exposure on liking and memory, each using stimuli with a different level of ecological validity. They found that 'repetition led to greater liking as well as increased recognition' in the more ecologically valid contexts (p. 378). By the standard of Szpunar et al.'s experimentally controlled stimulus contexts (which ranged from short tone sequences to 15 second excerpts from orchestral recordings), the listening preparation task was highly ecologically valid. The LP participants did recognise its artificiality, however, and so it is possible that, in line with Szpunar et al.'s results, liking and recognition may have increased if there had been a more ecologically valid means of providing the LP participants with prior exposure (cf. Hargreaves and North 2010).

These findings indicate that, when considering listening in real-world situations, the mere exposure theory does not adequately provide explanations for listeners' responses, as the effects of exposure are inevitably mediated by cultural and/or social factors. This was especially evident in some of the LP participants' negative responses to undertaking the listening preparation task. One of the most important findings to emerge from the non-attender data was the significant role of recognition on a first hearing, and the way in which, for some participants, this was a greater determinant of enjoyment than repeated exposure. This finding is concordant with Peretz et al.'s (1998, p. 898) identification of a preference bias for melodies that participants already knew prior to testing, in comparison with novel melodies which participants were exposed to repeatedly in the experimental situation.

While unexpected, these results do hold potentially useful applications, suggesting that it may be better for new audience members to have heard short, recognisable extracts in advance, rather than being advised to listen to a whole work. Orchestras including the Philharmonia and the London Symphony Orchestra already present short audio clips amongst the concert listings on their websites, while the City of Birmingham Symphony Orchestra, in its section on advice for first-time attenders, highlights works in their upcoming season that listeners may recognise, noting in which film or television soundtracks the music has featured. A more integrated approach, using both of these methods (so allowing prospective audience members to hear short extracts in advance, but also pointing out where they might have heard them before) would provide prior exposure, but would also serve to legitimise the state of being familiar with classical music from 'popular' sources alone. Further work might build on this exploratory research by replicating the study with larger comparison groups and systematically considering the effects of both repertoire familiarity and information about the music provided verbally, allowing further exploration of how these two related sources of contextualising new listeners' experiences may interact.

References

Berlyne, D.E. (1971). *Aesthetics and Psychobiology*. New York: Appelton-Century-Crofts.

Cavicchi, D. (1998). *Tramps Like Us: Music and Meaning Among Springsteen Fans*. Oxford: Oxford University Press.

Cone, E.T. (1974). *The Composer's Voice*. London: University of California Press.

Cone, E.T. (1989/1977). Three Ways of Reading a Detective Story – Or a Brahms Intermezzo. In E.T. Cone and R.P. Morgan (eds), *Music: A View from the Delft: Selected Essays* (pp. 77–94). Chicago, IL: The University of Chicago Press.

Dobson, M.C. (2010). New Audiences for Classical Music: The Experiences of Non-attenders at Live Orchestral Concerts. *Journal of New Music Research* 39/2: 111–24.

Dobson, M.C. and Pitts, S.E. (2011). Classical Cult or Learning Community? Exploring New Audience Members' Social and Musical Responses to First-Time Concert Attendance. *Ethnomusicology Forum* 20/3: 353–83.

Hargreaves, D.J. and North, A.C. (2010). Experimental Aesthetics and Liking for Music. In P.N. Juslin and J.A. Sloboda (eds), *Handbook of Music and Emotion: Theory, Research, Applications* (pp. 515–46). Oxford: Oxford University Press.

Hargreaves, D.J., Miell, D.E. and MacDonald, R.A.R. (2005). How do People Communicate using Music? In D.E. Miell, R.A.R. MacDonald and D.J. Hargreaves (eds), *Musical Communication* (pp. 1–25). Oxford: Oxford University Press.

Hennion, A. (2001). Music Lovers: Taste as Performance. *Theory, Culture and Society* 18/5: 1–22.

Huron, D. (2006). *Sweet Anticipation: Music and the Psychology of Expectation*. Cambridge, MA: MIT Press.

Kolb, B.M. (2000). You Call this Fun? Reactions of Young First-time Attendees to a Classical Concert. *Journal of the Music and Entertainment Industry Educators Association* 1/1: 13–28.

Kramer, L. (1995). *Classical Music and Postmodern Knowledge.* Berkeley, CA: University of California Press.

Martindale, C. and Moore, K. (1988). Priming, Prototypicality, and Preference. *Journal of Experimental Psychology: Human Perception and Performance* 14/4: 661–70.

Meyer, L.B. (1956). *Emotion and Meaning in Music.* Chicago, IL: The University of Chicago Press.

Meyer, L.B. (1967/1994). *Music, the Arts, and Ideas: Patterns and Predictions in Twentieth-Century Culture* (2nd edn). Chicago, IL: The University of Chicago Press.

North, A.C. and Hargreaves, D.J. (1995). Subjective Complexity, Familiarity, and Liking for Popular Music. *Psychomusicology* 14: 77–93.

Peretz, I., Gaudreau, D. and Bonnel, A. (1998). Exposure Effects on Music Preference and Recognition. *Memory and Cognition* 26/5: 884–902.

Pitts, S.E. (2005). What Makes an Audience? Investigating the Roles and Experiences of Listeners at a Chamber Music Festival. *Music and Letters* 86/2: 257–69.

Radbourne, J., Johanson, K., Glow, H. and White, T. (2009). The Audience Experience: Measuring Quality in the Performing Arts. *International Journal of Arts Management* 11/3: 16–29.

Roose, H. (2008). Many-voiced or Unisono?: An Inquiry into Motives for Attendance and Aesthetic Dispositions of the Audience Attending Classical Concerts. *Acta Sociologica* 51/3: 237–53.

Silva, K.M. and Silva, F.J. (2009). What Radio can do to Increase a Song's Appeal: A Study of Canadian Music presented to American College Students. *Psychology of Music* 37/2: 181–94.

Small, C. (1998). *Musicking: The Meanings of Performing and Listening.* Middletown, CT: Wesleyan University Press.

Strauss, A. and Corbin, J. (1998). *Basics of Qualitative Research: Techniques and Procedures for Developing Grounded Theory* (2nd edn). London: Sage.

Thompson, S. (2006). Audience Responses to a Live Orchestral Concert. *Musicae Scientiae* 10/2: 215–44.

Szpunar, K.K., Schellenberg, E.G. and Pliner, P. (2004). Liking and Memory for Musical Stimuli as a Function of Exposure. *Journal of Experimental Psychology: Learning, Memory, and Cognition,* 30/2: 370–81.

Thompson, S. (2007). Determinants of Listeners' Enjoyment of a Performance. *Psychology of Music* 35/1: 20–36.

Tota, A.L. (2001). 'When Orff meets Guinness': Music in Advertising as a Form of Cultural Hybrid. *Poetics* 29/2: 109–23.

Zajonc, R.B. (2001). Mere Exposure: A Gateway to the Subliminal. *Current Directions in Psychological Science* 10/6: 224–8.

Chapter 4

Familiarity with Music in Post-Operative Clinical Care: A Qualitative Study

Katherine A. Finlay

Music is a highly engaging stimulus that occurs in every culture (Sloboda 2002). From an early age, music listening is an everyday experience and as such, it is a uniquely familiar and privileged art-form. Universal, cross-cultural appreciation for music and the perceived 'power of music' has rarely, if ever, been in question (Staricoff 2004). Over the past 20 years, research in the realm of music psychology has increasingly been widened to include clinical uses of music listening. Music is advantaged in a clinical setting owing to its non-invasive nature (Standley 2002), ease of accessibility and administration (Lim and Locsin 2006), relatively low financial outlay (Good, Al-Dgheim and Cong 2004; Good et al. 2005) and flexibility, and its close association with the desires and preferences of the patient. Music listening is an effective intervention that can be both patient-centred and patient-administered. Patients may select their own music, either from an offered library or from their own collections. They may listen on their own personal music players without interrupting others or jeopardising their own standard of medical care. Music can be used as directed, or when wanted, day or night. The flexibility of music as an intervention is unusual and, as such, is highly relevant for potential inclusion in a multi-modal model of patient care.

Significant progress has been made in relation to the contribution of music to a pain management programme. Audio-analgesia, the ability of music to affect and attenuate pain perception (Mitchell, MacDonald and Brodie 2006) is at the forefront of clinical music-related research. The use of music *as* medicine has been found to be associated with resultant physiological and psychological benefits (see Cepeda, Carr and Lau 2005 or Maratos and Gold 2005). That music may reduce pain levels, alongside other benefits such as anxiety reduction, mood alteration, distraction and relaxation, renders music listening a potentially important contribution to pain management (see Mitchell and MacDonald 2006; McCaffrey and Locsin 2002; Macdonald et al. 2003; Mitchell et al. 2006; Mitchell, MacDonald and Knussen 2008).

The use of music in a clinical setting may be appropriate in three possible contexts: (1) in conjunction with the normal treatment of a patient (standard care); (2) in situations in which medication may be 'less effective' and therefore an adjunctive treatment may be beneficial; or (3) on occasions when medication does not have time to take effect or is not desired, such as in childbirth. Physiologically, the benefits of music have been demonstrated intra-operatively, post-operatively and with laboratory-induced pain, acute pain and chronic pain. Music has been found to limit the amount of analgesia required post- and intra-operatively, and can lower the quota of rescue analgesic requested by patients. Music may therefore be considered a significant contributor to pain management and patient care.[1]

Conceptualising and Theorising Pain

Pain is a multi-factorial phenomenon that is highly individuated, and which incorporates sensory, affective, motivational and cognitive factors (Melzack and Casey 1968). Essentially, 'pain is whatever the experiencing person says it is, existing whenever he says it does' (Melzack 1983, p. 71). Traditionally, pain has been viewed as a simple sensory stimulus–response model (the Cartesian view of pain). Pain, according to this viewpoint, is the directly proportional response to a nociceptive stimulus.[2] Yet pain is undoubtedly more complex: Engel's (1977) Biopsychosocial Model of Pain revealed the reductionistic nature of this perspective, and it is now impossible to view pain as a uni-dimensional phenomenon. Pain management programmes must reflect all dimensions of the pain experience: any treatment for pain must be multi-modal in order to maximise potential benefits. 'Standard care' for all medical conditions must be widened to ensure potential to ameliorate pain across biological, psychological and sociological dimensions.

The efficacy of music in attenuating pain has been attributed to a number of theoretical approaches. Melzack and Wall (1965) proposed Gate Control Theory, in which the conscious apprehension of pain is modulated through a gate control mechanism located in the dorsal horns of the spinal cord. As a pain signal passes through neural networks to/from the central nervous system (CNS), neural activity in the dorsal horns of the spinal cord acts as a 'gate', facilitating or inhibiting signal transmission (Trout 2004). Importantly, Gate Control Theory stated that pain could result from or be modulated by psychological and sociological factors. It argued that there are cyclical interactions between pain and its causes, both internal (personal) and external (environmental). Through the multi-dimensionality of this model, music can be perceived as a potential modulator of afferent impulses, reducing the transmission of nociceptive activity to the CNS.

Gate Control Theory has since been supplemented by the addition of a 'neuromatrix' (Melzack 1999, 2001). The neuromatrix is theorised as a system of parallel and

---

[1]   For supporting research, see Nilsson, Rawal, and Unosson (2003a); Good (2002); Good et al. (1999, 2002, 2004, 2005, 2010); Hekmat and Hertel (1993); Mitchell and MacDonald (2006); Mitchell et al. (2006); Cepeda et al. (1998); Ikonomidou, Rehnström and Naesh (2004); MacDonald et al. (2003); Nilsson (2008); Nilsson et al. (2003b); Tse, Chan and Benzie (2005); Voss et al. (2004); McCaffrey and Freeman (2003); Mitchell et al. (2007); Schorr (1993); Zimmerman et al. (1989); Koch et al. (1998).

[2]   Nociception is a process by which pain signals are detected, encoded and processed.

cyclical processing loops in the brain, which provide a convergent output, dictating each individual's pain perception. The processing loops are thought to be sensory-discriminative (somatosensory), affective-motivational (limbic) and cognitive-evaluative (thalamocortical) neural networks. The neuromatrix is a genetically constructed matrix of neurons – a synaptic architecture – which produces characteristic nerve impulse patterns for the body. This patterning is thought to be uniquely personal: a neurosignature (Melzack 2001). The neurosignature specifies individual pain perception and (in)voluntary 'action systems' in which a person responds to their pain via personal pain-related coping strategies (Trout 2004). Therefore a person's historical and present pain experiences (personal and environmental) contribute to their dynamic perception of pain. Where the neurosignature is altered, potentially through a music listening intervention or an enhancement/development of coping strategies, then pain perception is also impacted.

Understanding the Psychological Impact of Pain

Pain and Attention

Just as audio-analgesia may impact upon neural patterning and pain impulse transmission, pain intensity can be further modulated by the control of attention by the CNS. Pain can disrupt behaviour and thought (Melzack and Casey 1968; Melzack and Torgerson 1971). Pain demands attention and interrupts and interferes with daily activity (Eccleston 2001; Eccleston and Crombez 1999); the processing of pain signals is prioritised by the CNS above other attentional demands (Eccleston and Crombez 1999). This is theorised through the traditional model of attentional capacity. Baddeley (1986) posited that attentional capacity is finite and is divided and allocated based upon task demands. Attention is a limited resource which is distributed according to the importance of tasks (Shiffrin 1988). The appropriate distribution of attentional resources can become flawed when the combined activity required by the tasks exceeds the limits of the available attentional resources. The processing of information is bi-directional: bottom-up and top-down. Automatic, bottom-up processing is unconscious and occurs without invoking attention and is driven by stimulus attributes. Effortful, top-down processing requires deliberate goal-oriented control of attention and cognitive capacity as the subject drives information processing through deliberate strategies and intentions (Hammar 2003; Hammar, Lund and Hugdahl 2003). Music has the potential to engage attention in both bottom-up and top-down processing.

Pain signals impose a new action priority on the organism to escape. Pain signals grab attention in order to motivate avoidance behaviour, which may reduce or remove the noxious stimulus (Eccleston and Crombez 1999). Owing to the limited availability of attentional resources, the attentional system becomes selective in processing and filtering complex information when in a situation of heightened attentional demand, such as that of a complex task or in response to pain. Conversely, if attention is absorbed by other demands (such as complex tasks), then there are fewer attentional resources available to attend to pain. If attention is persistently directed elsewhere, the ability to attend fully to pain is inhibited: attention engaged in non-pain demands cannot be allocated to pain processing (McCaul and Malott 1984). Selective attention alerts the prefrontal cortex to the distractor rather than to the noxious stimulus and therefore inhibits pain (Good et al. 2005). This is an example of top-down processing.

Leventhal (1992) extended the attention-diversion concept by suggesting that, owing to the affective nature of pain, cognitive-coping strategies with an emotional content might be most salient. An emotionally resonant stimulus has the potential to operate on pain affect and pain sensation simultaneously through invoking personalised action systems operant in pain circuitry and the pain neurosignature. Leventhal and Everhart (1979) argued that distracting attention away from the pain stimulus can prevent the activation of pain distress schemata, which minimises the negative affect naturally associated with pain signals (Christenfeld 1997). It is in this context that music may be most valuable. Music has been found to have particular efficacy in engendering emotional engagement with the sound (Mitchell et al. 2006). It is a statement of the self through musical preference and personal identification with the music; it is played for mood enhancement or mood alteration (Ali and Peynircioğlu 2010); and is an attention-diversion strategy. Music is immediately emotionally engaging and listeners have strong evaluative responses following emotional identification with music (Hargreaves 1986; Sloboda 2002).

Music and Attention

Beyond emotionality, music is thought to be a uniquely attentionally demanding and distracting stimulus (Mitchell et al. 2007), in part due to the complexity of harmonic and melodic interaction (Williams 2005). The processing of harmonic expectations when listening to music is highly absorbing for both musicians and non-musicians (Bigand, McAdams and Foret 2000). Music listening necessitates attentional focus through psychological and sociological engagement, as, for example, the listener dynamically attends to any or all of melody, harmony, genre, structure, lyrics, their personal history with the music, their listening context, associations with the music and their developing relationship with the music. Music has been found to be more effective than humour or arithmetic (Mitchell et al. 2006) or guided relaxation training (Good 2002) in invoking audio-analgesia and distraction away from laboratory-induced pain. Additionally, music has proven to be more effective in these measures and in relaxation than teaching patients about pain management (Good 2008). Music is a highly familiar stimulus that has an innate ability to hold attention (Mitchell et al. 2006).

Music as a Distractor

One of the mechanisms postulated to explain the effects associated with audio-analgesia is to implicate music as a distractor. Distraction of attention is a commonly used coping strategy to control pain in everyday situations (Van Damme et al. 2008). Distraction is the deliberate replacement of a (noxious) stimulus with a more pleasant focus of attention. Research into distraction has shown that distraction can facilitate pain reduction (Boyle et al. 2008; Eccleston 1995; Eccleston et al. 2002) and increase pain tolerance (Van Damme et al. 2008). By distracting and diverting attention from the pain experience, patients are actively engaging in a focus external to their pain.

Music-related distraction methods used in pain research to date have included music and relaxation, white noise, guided imagery, music therapy, music and therapeutic suggestions, experimenter-chosen music listening and patient-selected music.[3] The results of these studies have been variable, with some showing positive benefits, but others not.

MacDonald et al. (2003) conducted two studies researching the impact of music on pain with two patient populations. Participants were requested to bring their own choice of CD to the hospital for use as much as possible in the post-operative period. The first study looked at pain after foot-surgery and results supported the use of music as a method of anxiety reduction, but not pain reduction. Results from the second study with hysterectomy patients found that, although pain levels and anxiety scores diminished across the course of the study, the music group was not advantaged above the control group. This was hypothesised to be a result of the high level of emotional significance surrounding hysterectomy surgery, fostering group cohesion that was preferred to the isolation of individual music listening. Similarly, Ikonomidou et al. (2004) found that music did not induce audio-analgesia after laparoscopic gynaecological surgery. This was true despite the fact that 74 per cent of the patients stated that they enjoyed the music and found it beneficial. Consequently, it is not possible to conclude definitively whether music-induced distraction is a viable, effective and consistently beneficial methodology, and further research is needed.

Familiarity and Audio-analgesia Research

Familiarity research commonly cites Berlyne (1971), who advocated the concept of an inverted U-shaped relationship between liking for musical stimuli and their arousal potential (as cited elsewhere in this volume). The more moderate the arousal potential, the more the music is preferred. When arousal is minimal

---

[3]   For supporting research, see Hirokawa (2004); Boyle et al. (2008); Janata (2004); Lee et al. (2004); Maratos and Gold (2005); Wigram, Saperston and West (1995); Nilsson et al. (2001, 2003a); Good (1996); McCaffrey and Freeman (2003); McCaffrey and Locsin (2002); MacDonald et al. (2003); Mitchell et al. (2006).

or maximal, liking is thought to decline. The subjective complexity of music is important and, with more exposure to music, subjective complexity decreases, with moderate, optimal complexity levels leading to greater liking for the stimulus (North and Hargreaves 1997). Liking, complexity and arousal are interlinked in their relationship with musical perception and emotion. The relationship is a dynamic trade-off: a highly familiar piece can be disliked owing to 'over-playing' and a reduction in subjective complexity, in which case it is liked less and its arousal potential is diminished. The emotions that participants experience when listening to music are predictable by the extent to which they like and are aroused by the musical stimulus (Ritossa and Rickard 2004). Whilst noting that the inverted U-shaped model is not without its critics (see Martindale, Moore and Borkum 1990), it provides a helpful model for the conceptualisation of dynamic changes to preference and taste (Rawlings and Leow 2008).

To extend this model to pain research, just as the arousal potential of music alters the listener's preference for music, it may also impact upon the efficacy of music as a pain-reducing stimulus. Limited capacity models of attention (Shiffrin 1988) argue that distraction from pain must be attentionally demanding and emotionally engaging, and must displace the noxious pain stimulus from the focal point of attention. The efficacy of an audio-analgesia intervention may parallel the inverted U-shape, with peak analgesia occurring simultaneously with peak familiarity/arousal/subjective complexity in response to the music listening stimulus. The pain-relieving effects of music may be augmented until the listener becomes satiated with the stimulus, at which point pain levels rise again and the efficacy of the intervention declines. Research to date has not thoroughly investigated this hypothesis, although some initial research suggests that this extension may be warranted.

For instance, McCaffrey and Freeman (2003) asked 66 older people with chronic osteo-arthritic pain to listen to 20 minutes of relaxing classical music daily in the morning for 2 weeks and compared this group against participants who spent that time in quiet relaxation. Results showed that the music listening group had a significantly better reduction in pre- to post-test pain on all days of testing (days 1, 7 and 14). Scores also showed a cumulative decrease in pain across the 2 weeks of testing for the music group, but not for the control group. This suggests that music listening may have an immediate pain-relieving effect, but also a cumulative pain-relieving effect that enhanced the level of audio-analgesia for at least the 2 weeks tested – potentially as optimal complexity, arousal and familiarity were attained. Similarly, Mitchell et al. (2007) found that chronic pain participants who reported listening to music regularly demonstrated reduced pain and improved quality of life than those survey respondents who did not value music or listen to music regularly. The role of familiarity in pain relief enabled by music listening has been neglected by pain research. Consequently, this study was intended to investigate whether familiarity effects do indeed occur in audio-analgesia.

It is possible also that familiarity effects may explain some of the differential findings in audio-analgesia research (see Macdonald et al. 2003). Audio-analgesia may be mediated by the efficacy of the music, the familiarity of the listener to the stimulus and also the ability of the person to apply the cognitive-coping strategy. Fundamental to this is the level of absorption in the intervention. If the participant is not adequately absorbed in the coping strategy, then the benefits of the strategy may be minimal. The power of music is not a 'pharmaceutical property of the sound stimulus' (Sloboda 2002, p. 384), but is a tool that can either be used effectively or remain untouched. Therefore, audio-analgesia may be differentially beneficial as a consequence of the challenges inherent in maintaining absorption in music as a cognitive-coping strategy. To date, research has rarely assessed exactly what has occurred *during* the music listening intervention. Further research is needed to investigate whether it is viable to use music regularly for pain control and whether familiarity has an effect on audio-analgesia.

This research therefore aimed to investigate the effect of familiarity on pain management facilitated by music listening. Patients undergoing knee replacement surgery listened to 15 minutes of music on the ward for each day of their in-patient stay (Finlay 2009). An exploratory qualitative design was used to investigate the multi-dimensional experience of music listening as a pain management intervention. Participants provided daily qualitative feedback, from which themes were drawn to illustrate the impact of music listening on post-operative pain, alongside broader physiological and psychological factors.

Method

Design

An exploratory qualitative research design was used and participant responses were analysed using conventional content analysis (Hsieh and Shannon 2005). To minimise inter-observer variation, the principal investigator carried out all assessments and interviews and was present daily for passive capture of information (see Dworkin et al. 2005).

Participants

Seventy-eight patients (32 males, 46 females; mean age = 68; standard deviation, SD = 7.8 years) scheduled to undergo primary total knee arthroplasty in the Orthopaedics Department of the Royal Infirmary of Edinburgh, took part in the study. Thirty-seven participants were undergoing surgery on their right knee and 41 participants had surgery on their left knee. Recruitment took place over an 18-month period. All participants suffered from arthritic pain and radiographic arthritis and had experienced knee problems for around 9 years (mean = 8.97 years; SD = 7.78 years; range 1–40 years).

In order to avoid complications owing to the elevated levels of mortality and surgical failure, patients with any history of previous total knee arthroplasty surgery were not accepted. Patients were excluded if they had contraindications to central neural blockades, were unable to give informed consent or co-operate with pain assessment, had a history of allergy to local anaesthetics, suffered from non-osteo-arthritic chronic pain, or had been recruited for other research. Patients who felt that their hearing negatively affected their ability to listen to music were also excluded.

Apparatus and Materials

The use of headphones for research into music and pain has been recommended by Carroll and Seers (1998) and Nilsson et al. (2003b) in order to reduce environmental noise. In this study, Bose Noise Cancelling Headphones and a portable CD-Walkman were used.

A battery of musical examples was created (see Table 4.1). All examples were of approximately 12–15 minutes in length. This duration was deemed appropriate as short-term bursts of music listening have been proven effective in an acute pain setting (following Lee, Henderson and Shum 2004; McCaffrey and Good 2000).[4] Extracts were recorded at a sample rate of 16 bits and 44.1 kHz. Where multiple tracks were used to create a combined length, Pro Tools LE 6.9.2 was used to merge between individual extracts. All musical extracts were commercially available, did not include lyrics and represented a wide range of possible genres. All record companies and copyright holders were contacted and permission to use the music was granted.

A short pre-admissions clinic demographic questionnaire containing information about age, gender, medical history, pain history, musical background, musical preferences and music listening habits was created. A standardised anaesthetic regimen was used for all participants. All participants received combined spinal, femoral and sciatic anaesthesia and intravenous sedation.

Qualitative Data

Qualitative feedback was collected using a semi-structured interview and was, where possible, subject-directed. The experimenter used prompts to facilitate the patient, asking open-ended questions such as 'How is your pain today?' Prompts

---

[4] This time period of listening was selected through interaction with nurses, doctors and consultants, who considered it appropriate for demonstrating the efficacy of the treatment, yet minimising intrusion into clinical care and the daily activity that is required to successfully and efficiently operate an orthopaedic ward.

Table 4.1   Musical examples

No.	Composer/artist	Title	Length
1	Ali Khan and Purna	Emptiness is Form	16.21
2	Pärt	*Festina Lente*, Cantus in Memory of Benjamin Britten	15.26
3	Shakuhachi	Akita No Sugagaki, Gekko Roteki	12.42
4	Tommy Smith	Into Silence, tracks 8, 9, 12, 15, 25	14.52
5	Mahler	Symphony 5, *Adagietto Sehr Langsam*	11.54
6	Vaughan-Williams	Fantasia on a Theme by Thomas Tallis	14.41
7	Glass	Low Symphony, Movement 1, Subterraneans	15.07
8	Mike Oldfield	Tubular Bells: Part 1	15.00
9	Keith Jarrett	(If the) Misfits (Wear it)	13.15
10	Pat Metheny	Sirabhorn and Unity Village	14.39
11	Miles Davis	Miles Runs the Voodoo Down	14.01
12	Vivaldi	*Le Quattro Staggioni, La Primavera*	11.16
13	Dvořák	Slavonic Dances, Op. 46	16.18
14	Smooth Jazz Collection	Tourist in Paradise, Drop Top, Kari	11.47
15	Kartsonakis and Bonar	Vacation in the Sun, Return of the Dove	12.38
16	Trad. Arr. Williamson	Selection of traditional harp music	13.44
17	Stan Getz	I can't get started	11.27

covered five broad areas for discussion: pain, mood, sleep, physiotherapy/activity and music/relaxation, alongside any additional themes that the patients wished to talk about. In order to ensure the validity of the qualitative data, reflective notes and patient-generated themes were taken each day. Daily themes were sorted into categories by constant comparison through conventional content analysis (Hsieh and Shannon 2005).

Procedure

Pre-operative assessment   The study was approved by NHS Lothian Regional Ethics Committee (approval no. 06/S1101/5). On-site permissions were granted by the Royal Infirmary of Edinburgh Research and Development Office (reference 2006/R/AN/06). Patients who were scheduled for primary total knee arthroplasty were identified from the hospital waiting lists. Following this, participants eligible for the study were approached by letter approximately 2 weeks before their attendance at their Pre-admissions Clinic.

Upon arrival at the Pre-admissions Clinic, participants were approached by the principal investigator; the details of the study were further outlined and patients were asked whether they wished to participate. If so, informed consent was taken.

Participants then undertook a 20 minute preliminary assessment. Participants were asked to complete the pre-admissions demographic questionnaire.[5] They then listened to 1 minute musical excerpts from the sample CD and selected their favourite. The chosen extract was later used on the ward post-surgery in its entirety (approximately 15 minutes). Participants were informed that they would receive standard care in addition to the music-listening intervention. All patients were requested to refrain from listening to their own music and were advised that they could withdraw from the study at any time. Patients' general practitioners were informed of their participation by letter.

Post-operative assessment   On the first day of assessment, participants were met on the ward by the researcher. After completing the non-intervention-specific qualitative questions, the music/control equipment was set up appropriately and given to the patient for use. During the 15 minute period of music listening, participants were asked to refrain from doing any other activity external to the music listening. Participants were asked to relax, listen and focus on the music. The researcher left the curtained area for 15 minutes and ensured that the participant was not interrupted. Post-intervention, participants commented qualitatively on their music, mood, pain or other thoughts. Qualitative data was recorded daily (5 days maximum). All completed, full days of data were included in the analysis.

Results

Musical Background

The majority of participants had not played a musical instrument (67.9 per cent). Of those who did, most had played the piano (17.9 per cent), voice (5.1 per cent) or folk instruments (3.8 per cent). Those who reported playing an instrument had predominantly played for between 5 and 10 years (11.5 per cent) or less than 1 year (6.4 per cent). Only three participants reported still playing a musical instrument (3.8 per cent). Twenty-three participants had undertaken formal instrumental musical tuition (29.5 per cent), predominantly for 5–10 years (12.8 per cent) or 3–4 years (6.4 per cent).

Musical preferences were varied (see Table 4.2), with participants showing a broad preference for classical music (23.1 per cent) or country and western music (17.9 per cent). Pop music was most often disliked (34.5 per cent), followed by jazz (17.9 per cent). In all, 74.4 per cent of participants reported regularly listening to music. Music listening most often occurred whilst driving (26.9 per cent) and the radio was the most popular form of music listening (51.3 per cent). Participants reported a wide variation in amount of time spent listening to music

---

[5]   Quantitative pain, mood state, cortisol and physiological assessments were also taken (Finlay 2009) but are not reported in this chapter.

per week (mean = 19.62 hours; SD = 24.6 hours; range 0–112 hours). Music was most often listened to through 'passive' music listening – music playing in the background whilst the participant was engaging in another activity (74.4 per cent). By contrast, deliberately focussed, 'active' music listening was reported by 12 participants (15.4 per cent). No participants reported having previously heard the musical excerpt that they selected for use during the research study.

Table 4.2   Participant music listening preferences according to genre

Genre	Liked		Disliked	
	No. of participants	Percentage	No. of participants	Percentage
Classical	18	23.1	1	1.3
Country and Western	14	17.9	2	2.6
Easy listening	9	11.5	1	1.3
Choral/opera	8	10.3	6	7.7
Pop	7	9	27	34.6
Folk	5	6.4	0	0
60s/70s	5	6.4	0	0
Jazz	4	5.1	14	17.9
Rock	3	3.8	12	15.4
New age	1	1.3	0	0
Rap	0	0	8	10.3
Metal	0	0	3	3.8
No particular preferences	4	5.1	4	5.1

Qualitative Results

Ten key themes were identified in the data.

1. Pain relief: Actual versus perceived   Some participants were confident that their intervention had made a significant difference: 'It's taken away the pain when I was listening to the music' (female, aged 70) and 'It reduced my pain quite considerably – on the first day it was quite dramatic' (male, aged 54), while other participants disagreed, stating 'My pain levels weren't changed' (male, aged 69) or 'My pain levels did not change at all' (female, aged 62).

Participants seemed to differentiate between 'actual' pain reduction and 'perceived' pain reduction. 'Actual' pain relief was defined by patients as durable pain relief achieved through pharmacological analgesia. One patient commented that 'I don't know whether it's helped or not. The pain is bad, but it comes and goes. One minute it's good, and you think it's getting better, and then the next you have a sharp pain. When you're in pain, you're in pain. I only know that painkillers work' (male, aged 71). Some patients felt that the intervention did not provide any 'actual' pain relief: 'I wouldn't say that it affected my pain at all, though I think it probably could help' (male, aged 77).

The contrasting viewpoint was that of 'perceived' pain relief. 'Perceived' pain relief refers to the themes in which patients felt that their pain had been reduced or eliminated, but that this was not an absolute, biological reduction of their pain, but a diversion of attention away from their pain. Responses regarding 'perceived' pain relief suggested that, 'It actually seemed to numb it. I just lay there and thought "this is wonderful". You shut off from the rest of the ward, you listen to your music and you don't feel the rest of your pain' (female, aged 60), and 'Your pain was gone, I didn't think about the pain, I just enjoyed it (the music)' (male, aged 62). Some patients consequently struggled to differentiate between 'actual' and 'perceived' pain relief and were not sure how the intervention had helped: 'Once, it felt very pleasant, but otherwise my pain was not really affected ... because you come at the end of the 1 p.m. cocktail [of pain medication]' (female, aged 81).

2. Mood modulation: Positivity and agency Participants reported feelings of positivity and an increased sense of control over their treatment. Positive change in mood was key, with patients commenting that 'It makes you happier' (male, aged 71) and 'It lifts your mood right up and makes you more relaxed, less tense' (male, aged 57). The intervention was found to be uplifting and promoted equilibrium by balancing out moods, for example, 'I would think that if you're upset, you'd feel much calmer' (female, aged 59) and 'It depends on what mood you are to start with. If you're low to begin with, it keeps you steady. If you're feeling great, then it lifts you even more' (female, aged 57). Mood improvement was also associated with having the ability to increase the sense of agency in treatment, a serious concern for patients, for example, 'There are times when you feel anxious, especially at the loss of control' (male, aged 66). In a clinical setting where the patient has minimal involvement in decision-making and choice of care, using a music/noise-reducing intervention allowed the patient to impact upon their own treatment. Patients found that 'It was good because you're doing something to help yourself' (female, aged 76), 'it made me feel good and useful' (male, aged 80) and 'It made me think and say "get a grip of yourself"' (female, aged 69).

3. Relaxation: Learned associations and active facilitation Psychologically, patients found that the intervention induced a relaxation effect: 'It made me more relaxed' (male, aged 73) and 'The music was relaxing in itself' (female, aged 82). This took the form of specific feelings of relaxation such as calmness and tranquillity, shown in responses such as 'The music was soothing. It was useful for me' (male, aged 67) and 'The relaxation was really good. It was really tranquil' (female, aged 69). In some cases, the relaxation effect was deliberately enabled by the patient, as patients stated that, 'It's listening for a different reason. As you're doing it to relax, it's totally different to at home' (female, aged 56). A theme of learned-association emerged; as participants perceived benefits in the early sessions, they facilitated relaxation in later sessions, thus learning to associate the intervention with the relaxation response. One patient commented regarding that: 'There's a link between the body and music – it's sleepy time! It's like a wave, the same today as yesterday. It's just something you look forward to actually' (male, aged 71). This indicates some active involvement in their intervention on the part of the patient.

4. Anxiety-reduction: Relaxation and muscular tension   Anxiety reduction was key as patients associated the intervention with an anxiolytic effect: 'Anxiety was the main change. You just get total relaxation and don't feel so sad' (female, aged 57). Anxiety was cross-associated with physiological themes of relaxation and muscular tension reduction: 'It made me less anxious and a bit more relaxed and less tense' (male, aged 63). Anxiolytic effects were commonplace, however some found that their anxiety had not changed or were unsure about any change: 'There's been no change' (male, aged 69) and 'It maybe changed' (female, aged 74). For those patients who did experience an anxiolytic effect, such anxiety reduction enabled a change in mood in response to the intervention, and one patient commented that 'It lessened your anxiety levels, because if you and your mind are more relaxed, you're less anxious … It definitely relaxed me, therefore my mood lifted' (female, aged 71).

5. Durability of effects: Residual and restricted   Patients had a variety of thoughts about the durability of the 'perceived' pain relief, with some feeling that; 'In the short-term I didn't have any pain. I feel a lot better after listening to the music for some strange reason, until the nurses come round and then you're back to square one again' (male, aged 69). It was thought that the durability of the improvement could be residual: 'There is an aftermath, it [pain relief] carries on, carries forward. If I was quietly sitting at home, I never tire of listening and the effect would have carried on for some time' (female, aged 72). Others considered that the pain control was effective during the intervention and had a short residual effect, for example, 'The pain relief lasted during the music and then for 5–10 mins afterwards until you moved again' (female, aged 76). By contrast, durability was restricted to exactly the duration of the intervention, for example, 'I didn't feel much pain during my relaxation, but now the relaxation has stopped, the pain has started to come back, very sudden and severe' (female, aged 73).

6. Absorption: Distraction and practice effects   The degree of absorption in the intervention session was considered important:

> When you link up with the music, it pervades the body and it really is lovely, like how you would think about a nice dream. I had two interruptions though today, and my train of thought was aggravated about three times and I lost it. I never relaxed, right from the word go. You and the music need to be together. (male, aged 71)

Some patients found that they were easily able to focus on their intervention and were not distracted: 'I was quite focused on the music' (male, aged 65); 'I didn't get distracted' (female, aged 62); and 'I wasn't aware of anything that was going on around me. You can go into your own space as I call it' (female, aged 57). Other participants stated that they were distracted by external events: 'I have been quite distracted by football or by the other people in the ward' (male, aged 71) and 'I was distracted by the general noise' (male, aged 75). This was thought to have a direct impact upon their degree of absorption: 'I got distracted on Day 3. I just opened my eyes and that was it, I was not into it' (female, aged 62). Some patients specifically related that they struggled with concentration and that this was a concern for them: 'The music would have been okay if I'd been feeling better. I just wanted it to stop as I couldn't concentrate on it' (female, aged 62) and 'I would find it hard to concentrate were I in more pain and trying to listen to the music' (female, aged 82). It was deemed important to concentrate when listening; 'It helps to concentrate on music and to take your mind off [your pain]' (male, aged 62).

Without some involvement in the intervention and without being given ample opportunity to specifically relax, perhaps owing to interruption, patients felt that the psychological benefits of the intervention could be lost. This sentiment that the relaxation could be improved or, conversely, lost, was embodied in further themes of relaxation, those stating that the potential for relaxation was not uniform in its effect. Some patients reported that their ability to relax changed from day to day, 'I enjoyed the music today. It was more relaxing than yesterday; I really got into it' (female, aged 60). This was both positive and negative, with some participants finding the intervention progressively more relaxing, for example, 'I was getting to the stage where the more I was doing it, the more I was relaxing when I was listening' (male, aged 65), whereas others found the relaxation effect diminished; 'The first day was very relaxing, Day 2 was not so relaxing, and today [Day 3] was not relaxing at all' (female, aged 78).

7. Imagery: Spontaneous and self-initiated   The intervention enabled some participants to use imagery: 'I was thinking about the music and trying to describe what would go best with the music. For the first part it would be dolphins and for the second part it would be lovers running through fields with not a care in the world. I felt like it was me doing that' (male, aged 69). Patients were actively guiding their thoughts to intervention-generated images, such as 'I imagined being in Latin America and drinking cocktails' (male, aged 77) and also to personally generated and personally resonant images, for example, 'I was thinking of the walk to Silvernowes ... I walked further [along the beach] today' (female, aged 77).

8. Music preferences: Liking and anticipation   Of first priority for the participants was their liking for or disliking of the music. Participants who enjoyed their music commented: 'Lovely, lovely music, it was very soothing' (female, aged 78) and

'I enjoyed the music. It was beautiful' (male, aged 67). Patients who liked their music chose to verbally validate their choices, asserting their feelings that they chose correctly: 'The selection was chosen by myself and was very good, not too heavy' (male, aged 57), 'I've heard that type of thing before. I picked it because I think it is the type of music you would pick to listen to if you were in pain' (female, aged 59) and 'It's definitely the type of music I'd pick. I like to listen to music, even at home – soothing music' (female, aged 57). Those patients who enjoyed their music referred to the theme of anticipation – they looked forward to their music listening sessions: 'I was really looking forward to my music today' (male, aged 71). By contrast, a few patients stated that the music was not 'their type' of music: 'It's not something I'd buy, but I wouldn't turn it off if it came on the radio! I wouldn't close the door to it' (male, aged 64).

9. Familiarity/complexity   Participants reported changes in their enjoyment/ dislike of the music as the study progressed. As patients became more familiar with the music, it became less difficult to understand and more accessible, enjoyable and beneficial: 'You understand the music a bit better' (female, aged 67). As they heard too much of the music, they became over-familiar with the music and it became simplistic and minimally stimulating for them, thus boredom ensued. Patients expressed this theme in two directions, as expected. Patients liked the music better as the study progressed: 'I was apprehensive at the start about whether I would enjoy it or not. I don't normally like modern jazz. I picked it though, and after the first time it got better and better', and 'I enjoyed it today [Day 3]. I had a chance to get into it today and was picking out the sound of the harp. The first two times it kinda washes over you. Then as you know more you see new things, like if you watch a film a few times' (male, aged 73). Similarly, patients commented that: 'I think it's "growing on me"' (male, aged 72) and 'The more you get to know it, the more you drift away' (female, aged 74). When overly familiar with the music, patients noted that: 'I only liked it on Day 1 … My liking for the music got gradually less each day' (female, aged 78), 'The music sounded soothing, but I was really quite indifferent today; I suppose I've heard it quite a few times now' (male, aged 56) and 'I got a bit bored of the music actually' (male, aged 72). Finally, patients suggested that they may have enjoyed a change in the music, for example, 'I might have liked a change of music everyday' (male, aged 64) or to have been given the opportunity to choose their own music: 'If you bring your own music in, or the type of music you prefer, it would help' (female, aged 62).

10. Deeper listening: Structural recognition   With repeated hearings, participants increasingly commented on the musical form of the piece which they were listening to and stated that their knowledge of the music increased as the number of hearings rose. Participants referred to instrumentation, such as 'I like the instruments in it actually, the strings and the piccolo' (female, aged 81), to the composers, for example 'I like virtually everything Mahler has composed' (female, aged 78), and to the structural design of the music, such as, 'Some of it was nice, but other sections were a wee bit thumpy' (male, aged 65). As they got to know it better, they commented more perceptively about the composition: 'I liked the variety and change of keys and moods. I became more familiar with it as the week went on, and I was waiting for entries (of musical instruments)' (female, aged 72).

Discussion and Conclusion

The results of this study offer an insight into patient perceptions of the use of music for pain management. As modelled by previous studies, qualitative responses showed equivocal audio-analgesia. Patients differentiated between actual and perceived pain relief and held strong convictions about whether music listening would affect their pain. The attribution of 'real' pain relief to pharmacology demonstrates a need for patient education about pain mechanisms. Post-surgical pain is severe and highly salient for the patient as it represents their first benchmark for their recovery after their emergence from anaesthesia. Pain at the outset of this study therefore had a very high threat value, which may have interfered with the patient's abilities to effectively utilise the music as a distractor (following Van Damme et al. 2008).

Results demonstrated that music facilitated complete attentional diversion from pain for some patients and this had concomitant benefits in inducing mood alterations and relaxation. Following Chafin et al. (2004) and Khalfa et al. (2003), the distraction participants experienced both (a) diverted attention from pain and (b) enabled patients to avoid ruminating on their pain, thereby reducing psychological distress, pain intensity and pain-related disability, and promoting greater success in recovery (Pavlin et al. 2005; Turner et al. 2002). This minimised pain catastrophising, a thought process characterised by an excessive focus on pain sensations, with an exaggeration of threat and the self-perception of not being able to cope with the pain situation (Sullivan et al. 2006). Audio-analgesia may not only be the result of diverted attention away from the stimulus; distraction may also prevent negative rumination about the noxious stimulus or stressor (Chafin et al. 2004). This can be termed a 'response shift' as participants used the intervention to separate themselves from their previous pain sensations and promote a renewal in terms of positive psychological state (Razmjou et al. 2006).

Listening to music in the post-operative recovery period enabled participants to improve their mood state and promoted positivity and emotional stability. That music impacted upon participant emotions demonstrated the role of emotional engagement in increasing the music's efficacy and the listener's absorption. This supports Leventhal's (1992) finding that greater emotional engagement in a distraction task enhances the efficacy of the cognitive-coping method used. In particular, emotional reactions to music may activate those areas of the brain that are thought to be related to pain modulation (Blood and Zatorre 2001; Roy, Peretz and Rainville 2008).

In addition to heightened positivity, participants reported that the music listening intervention improved their sense of agency, giving them the opportunity to participate in their own treatment. Using music to accompany tasks or to support a patient in a clinical setting is a way of returning autonomy and personalisation back to the patient and to their treatment, avoiding learned helplessness. These findings replicated McCaffrey and Freeman (2003), who found that music listening improved motivation, elevated mood and emphasised feelings of responsibility and control. When patients engage with music, or listen to music in an effort to reduce their pain, they are actively participating in changing their pain state. Enabling patients to develop a sense of agency in their treatment, internalising their locus of control, can assist recovery and heighten patient satisfaction with treatment (McCaffrey and Freeman 2003).

Key themes of relaxation and anxiety reduction were demonstrated in this study. Relaxation demonstrated a dynamic patterning comparable to a conditioning mechanism. Participants experienced feelings of relaxation in early listening sessions and this consolidated the close association between the intervention and the relaxation response. This association led to active anticipation of relaxation and it suggests that audio-analgesia has the potential to reinforce classical conditioning. Familiarity with the musical stimulus and repetition of the intervention enhanced the likelihood of audio-analgesia and relaxation at later hearings. Relaxation served to attenuate the affective distress component of pain and this may have operated in conjunction with distraction away from pain, which reduces pain sensation (Stevensen 1995).

Differential results were demonstrated for anxiety-reduction: some participants experienced anxiolytic effects and others no benefits. The variability in this finding replicates the work of Macdonald et al. (2003), who found that music was not universally anxiolytic. According to the cognitive theory of anxiety (Eysenck 1992), anxiety acts to enhance and quicken the detection of noxious stimuli. The organism becomes hypervigilant to threat and somatosensory signals such as pain, elevating pain intensity. Music as a distraction intervention may have been effective as it offers an alternative source of sensory input upon which hypervigilant patients may focus. It therefore prevents pain distress schemata from activating (Leventhal and Everhart 1979) and moderates the pain signals experienced and the anxiety felt.

Research into the role of catastrophisation in modulating the efficacy of cognitive-coping strategies may explain the opposite finding of no anxiety reduction. Campbell et al. (2010) showed that pain catastrophising elevated pain levels, regardless of the pain management intervention. Patients who ruminated on their pain and thought negatively about its intensity and their self-efficacy disrupted the distraction task and analgesia was reduced, particularly at the start of the session. Catastrophising led to minimal distraction analgesia at the outset of the distraction process. Comparably, in this study, some patients showed some degree of pain catastrophising, which may have affected the initiation of audio-analgesia. Patients predominantly reported the worst pain on the first day following surgery.

The efficacy and normal patterning of audio-analgesia over time may thus have been reduced in early sessions by a combination of catastrophisation, increased pain and unfamiliarity with the music devaluing the music listening intervention.

Participants felt that music listening offered a short-term pain-relieving intervention. This was thought to be restricted to the duration of the musical stimulus, or to have a short-term residual effect. This finding validates the limited-capacity model of attention (Shiffrin 1988) as during the musical intervention pain is minimised as attentional focus is redirected towards the music listening. At the conclusion of the intervention, however, pain signals provoke greater attentional focus than residual physiological or psychological benefits from the music listening. The duration of audio-analgesia may be delineated by absorption in the music listening intervention. A deeper focus and concentration on the music and loss of awareness of external events were found to be fundamental to the efficacy of the intervention. With disruption of attention through break-through noise or distraction, absorption was reduced and benefits depleted. Engagement with the stimulus showed individual differences in its patterning over time. Qualitative results demonstrated varying degrees of attentional focus on the music across different days of testing, leading to differential reporting of the efficacy of the intervention. This finding can be explained in reference to effortful engagement with the stimulus and familiarity effects.

Research has suggested that the degree of absorption and engagement in music mediates the efficacy of that music as a cognitive-coping strategy (Leventhal 1992). Music is not an 'auditory vitamin pill' (Sloboda 2002, p. 241), but is more effective if the patient is actively focused on the intervention (Good et al. 1999). Active focus necessitates a desire to engage with treatment. Verhoeven et al. (2010) compared three groups who underwent the Cold Pressor Task:[6] a distraction group; a distraction and financial motivation group; and a control group. Participants in both distraction groups reported less pain than the control group. Importantly, engagement in a tone-detection task was higher in the distraction-motivation group than in the pure distraction group, demonstrating the importance of motivation. Where some patients in the present study reported engaging with the music, others were cynical about psychological interventions and therefore had low distraction motivation. Future research could aim to assess the level of motivation to engage in pain management interventions in order to provide patients with an optimum and appropriate level of clinical care. It is likely that motivation would interact with the familiarity effect, potentially elevating or reducing audio-analgesia. Low motivation and absorption may result in poor compliance with study methodology owing to boredom/frustration, leading to reduced responsivity (Phumdoung and Good 2003).

---

[6]   The Cold Pressor Task is a laboratory-based pain induction intervention that is widely used for distraction research. Participants submerge their dominant arm in ice-cold water and are assessed for pain intensity, threshold and endurance (see Mitchell, MacDonald and Brodie 2004 for protocol).

That cognitive-coping strategies can be effortful was shown through the frequency with which participants supplemented the music listening with supplementary coping techniques. Participants reported using active guidance of thoughts to imagery and found that this boosted absorption. Music provided a temporal space in which patients could reinforce and re-use the cognitive-coping skills they had already learned (Hekmat and Hertel 2003). Research by Forys and Dahlquist (2007) criticised the 'one-size-fits-all' approach to using cognitive coping strategies for pain management. They suggested that this can serve to artificially minimise the benefits of a psychological pain management intervention and that natural coping styles can be suppressed. Music allows for individuation on a multiplicity of levels – through the music selected and through the space that it allows for the patient to invoke adjunctive personally appropriate coping strategies such as the imagery reported in this study.

Preference for music is highly influential: an initial liking led to positive anticipation, verbal validation and enhanced receptivity for the intervention. As expected, preferences did show familiarity effects. Participants felt that they knew and enjoyed the music better with repeated playings. For some participants hearing the same music again helped increase the benefits of the intervention, with each day of listening better received than the day before – a positive conditioning mechanism. Others described reaching a peak on a certain day of testing, after which point their liking of and benefits from the music declined. The variability in pain attenuation and enjoyment may be related to the inverted U-shaped function described by Berlyne (1971). On the first day of music listening the music was essentially unfamiliar, suggesting that its complexity on first hearing would be high and arousal therefore low. With repeated playings, progressing towards optimal subjective preference, familiarity with the music increased, leading to greater musicological/structural knowledge of the extract, decreased subjective complexity, greater liking for the music and enhanced audio-analgesia. Where audio-analgesia fluctuated, the extension of the familiarity model appears justified as subjective complexity/familiarity was reported to vary on different days of testing between different participants, with some reaching satiation before others.

It is not known how long a hypothesised inverted U-shape in audio-analgesia would take to reach maximal pain reduction, nor how quickly music would lose its potency as a distractor. In addition, it would be beneficial to know whether regular changes in the chosen musical stimulus would enhance or diminish pain reduction. This research was restricted by the focus on acute, post-operative pain. Future research could aim to work with chronic pain participants for an opportunity to map familiarity effects over a longer time period. Additionally, this research utilised music that was unfamiliar to the participants. Recent research has suggested that preferred music may enhance audio-analgesia further (Macdonald et al. 2003; Mitchell and MacDonald 2006; Mitchell et al. 2006); therefore, studies should aim to compare the changes in audio-analgesia over time with music of differing initial levels of familiarity. The issue of patient choice is also important for future research, including when and how often music should be used, as well as what music is optimum.

This research has provided an insight into the qualitative experience of music listening for pain management. Results have clearly demonstrated that familiarity significantly impacts upon treatment responsiveness, particularly in terms of facilitating learned relaxation responses, learning to apply the intervention and maximising pain-relieving effects. The durability of audio-analgesia is undoubtedly connected to participant absorption in the sound stimulus, with the inverted U-shape between familiarity and complexity dynamically altering these processes. Further research is needed, quantitatively and qualitatively, to determine the time-course and patterning of familiarity effects in audio-analgesia. It seems, however, that the ability of music to modulate pain, improve patient satisfaction with treatment and offer a potentially powerful adjunct to standard care is significant and worthy of integration into clinical pain management programmes.

Acknowledgements

The author would like to acknowledge Professor Ian Power, Dr John Wilson, Mr Paul Gaston (University of Edinburgh) and Dr Emad Al-Dujaili (Queen Margaret University).

References

Ali, S.O. and Peynircioğlu, Z.F. (2010). Intensity of Emotions Conveyed and Elicited by Familiar and Unfamiliar Music. *Music Perception* 27/3: 177–82; doi: 10.1525/mp.2010.27.3.177.

Baddeley, A.D. (1986). *Working Memory*. Cambridge: Cambridge University Press.

Berlyne, D.E. (1971). *Aesthetics and Psychobiology*. New York: Appleton-Century Crofts.

Bigand, E., McAdams, S. and Foret, S. (2000). Divided Attention in Music. *International Journal of Psychology* 35/6: 270–78; doi: 10.1080/002075900 750047987.

Blood, A.J. and Zatorre, R.J. (2001). Intensely Pleasurable Responses to Music Correlate with Activity in Brain Regions Implicated in Reward and Emotion. *Proceedings of the National Academy of Sciences of the United States of America* 98/20: 11818–23.

Boyle, Y., El-Deredy, W., Martínez Montes, E., et al. (2008). Selective Modulation of Nociceptive Processing due to Noise Distraction. *Pain* 138/3: 630–40; doi: 10.1016/j.pain.2008.02.020.

Campbell, C.M., Witmer, K., Simango, M., et al. (2010). Catastrophizing Delays the Analgesic Effect of Distraction. *Pain* 149/2: 202–7; doi: 10.1016/j.pain. 2009.11.012.

Carroll, D. and Seers, K. (1998). Relaxation for the Relief of Chronic Pain: A Systematic Review. *Journal of Advanced Nursing* 27: 476–87.

Cepeda, M.S., Diaz, J.E., Hernandez, V., et al. (1998). Music does not Reduce Alfentanil Requirement during Patient-controlled Analgesia (PCA) use in Extracorporeal Shock Wave Lithotripsy for Renal Stones. *Journal of Pain and Symptom Management* 16/6: 382–7.

Cepeda, S., Carr, D.B. and Lau, J. (2005). Music for Pain Relief (Protocol). *The Cochrane Library* 2: 1–5.

Chafin, S., Roy, M., Gerin, W. and Christenfeld, N. (2004). Music can Facilitate Blood Pressure Recovery from Stress. *British Journal of Health Psychology* 9: 393–403.

Christenfeld, N. (1997). Memory for Pain and the Delayed Effects of Distraction. *Health Psychology* 16/4: 327–30; doi: 10.1037/0278-6133.16.4.327.

Dworkin, R.H., Turk, D.C., Farrar, J.T., et al. (2005). Core Outcome Measures for Chronic Pain Clinical Trials: IMMPACT recommendations. *Pain* 113/1–2: 9–19; doi: 10.1016/j.pain.2004.09.012.

Eccleston, C. (1995). Chronic Pain and Distraction: An Experimental Investigation into the Role of Sustained and Shifting Attention in the Processing of Chronic Persistent Pain. *Behavioural Research Therapy* 33/4: 391–405.

Eccleston, C. (2001). Role of Psychology in Pain Management. *British Journal of Anaesthesia* 87/1: 144–52.

Eccleston, C. and Crombez, G. (1999). Pain Demands Attention: A Cognitive-Affective Model of the Interruptive Function of Pain. *Psychological Bulletin* 125/3: 356–66.

Eccleston, C., Morley, S., Williams, A., et al. (2002). Systematic Review of Randomized Controlled Trials of Psychological Therapy for Chronic Pain in Children and Adolescents, with a Subset Meta-analysis of Pain Relief. *Pain* 99: 157–65.

Engel, G. (1977). The Need for a New Medical Model: A Challenge for Biomedical Science. *Science* 196/4286: 129–36; doi: 10.1126/science.847460.

Eysenck, M.W. (1992). *Anxiety: The Cognitive Perspective*. Hillsdale: NJ: Erlbaum.

Finlay, K.A. (2009). *Audio-analgesia and Multi-disciplinary Pain Management: A Psychological Investigation into Acute, Post-operative Pain*. Unpublished PhD dissertation, University of Edinburgh.

Forys, K.L. and Dahlquist, L.M. (2007). The Influence of Preferred Coping Style and Cognitive Strategy on Laboratory-induced Pain. *Health Psychology* 26/1: 22–9; doi: 10.1037/0278-6133.26.1.22.

Good, M. (1996). Effects of Relaxation and Music on Postoperative Pain: A review. *Journal of Advanced Nursing* 24: 905–14.

Good, M. (2002). Relaxation and Music Reduce Pain after Gynecologic Surgery. *Pain Management Nursing* 3/2: 61–70; doi: 10.1053/jpmn.2002.123846.

Good, M. (2008). Effects of Relaxation/Music and Patient Teaching for Pain Management on Salivary Cortisol. *The Journal of Pain* 9/4: 54.

Good, M., Al-Dgheim, R. and Cong, X. (2004). Cognitive/Behavioral Approaches: Relaxation and Music Reduce Pain after Abdominal Surgery in Older Adults. *The Journal of Pain* 5/3: S92.

Good, M., Albert, J.M., Anderson, G.C., et al. (2010). Supplementing Relaxation and Music for Pain after Surgery. *Nursing Research* 59/4: 259–69; doi: 10.1097/NNR.0b013e3181dbb2b3.

Good, M., Anderson, G.C., Ahn, S., et al. (2005). Relaxation and Music Reduce Pain Following Intestinal Surgery. *Research in Nursing and Health* 28/3: 240–51.

Good, M., Anderson, G.C., Stanton-Hicks, M. and Makii, M. (2002). Relaxation and Music Reduce Pain After Gynecologic Surgery. *Pain Management Nursing* 3/2: 61–70.

Good, M., Stanton-Hicks, M., Grass, J.A., et al. (1999). Relief of Postoperative Pain with Jaw Relaxation, Music and their Combination. *Pain* 81/1–2: 163–72; doi: 10.1016/S0304-3959(99)00002-0.

Hammar, A. (2003). Automatic and Effortful Information Processing in Unipolar Major Depression. *Scandinavian Journal of Psychology* 44: 409–13.

Hammar, A., Lund, A. and Hugdahl, K. (2003). Long-lasting Cognitive Impairment in Unipolar Major Depression: A 6-month Follow-up Study. *Psychiatry Research* 118: 189–96.

Hargreaves, D.J. (1986). *The Developmental Psychology of Music*. Cambridge: Cambridge University Press.

Hekmat, H.M. and Hertel, J.B. (1993). Pain Attenuating Effects of Preferred Versus Non-preferred Music Interventions. *Psychology of Music* 21/2: 163–73; doi: 10.1177/030573569302100205.

Hirokawa, E. (2004). Effects of Music Listening and Relaxation Instructions on Arousal Changes and the Working Memory Task in Older Adults. *Journal of Music Therapy*, 41/2: 107–27.

Hsieh, H.F. and Shannon, S.E. (2005). Three Approaches to Qualitative Content Analysis. *Qualitative Health Research* 15/9:1277–88.

Ikonomidou, E., Rehnström, A. and Naesh, O. (2004). Effect of Music on Vital Signs and Postoperative Pain. *Association of Operating Room Nurses Journal* 80/2: 269–74, 277–8, http://www.ncbi.nlm.nih.gov/pubmed/15382598.

Janata, P. (2004). When Music Tells a Story. *Nature Neuroscience* 7/3: 203–4; doi: 10.1038/nn0304-203.

Khalfa, S., Dalla Bella, S., Roy, M., et al. (2003). Effects of Relaxing Music on Salivary Cortisol Level after Psychological Stress. *Annals of the New York Academy of Science* 999: 274–376.

Koch, M.E., Kain, Z.N., Ayoub, C. and Rosenbaum, S.H. (1998). The Sedative and Analgesic Sparing Effect of Music. *Anesthesiology* 89/2: 300–306.

Lee, D., Henderson, A. and Shum, D. (2004). The Effect of Music on Preprocedure Anxiety in Hong Kong Chinese Day Patients. *Journal of Clinical Nursing* 13: 297–303.

Leventhal, H. (1992). I Know Distraction Works Even Though It Doesn't. *Health Psychology* 11: 208–9.

Leventhal, H. and Everhart, D. (1979). Emotion, Pain, and Physical Illness. In C.E. Izard (ed.), *Emotions in Personality and Psychopathology* (pp. 263–99). New York: Plenum.

Lim, P.H. and Locsin, R.C. (2006). Music as Nursing Intervention for Pain in Five Asian Countries. *International Nursing Review* 53/3: 189–96; doi: 10.1111/j.1466-7657.2006.00480.x.

MacDonald, R.A.R., Mitchell, L.A., Dillon, T., et al. (2003). An Empirical Investigation of the Anxiolytic and Pain Reducing Effects of Music. *Psychology of Music* 31/2: 187–203; doi: 10.1177/0305735603031002294.

Maratos, A. and Gold, C. (2005). Music Therapy for Depression. *Cochrane Library* 2: 1–7.

Martindale, C., Moore, K. and Borkum, J. (1990). Aesthetic Preference: Anomalous Findings for Berlyne's Psychobiological Theory. *The American Journal of Psychology*, 103/1: 53–80.

McCaffrey, R. and Freeman, E. (2003). Effect of Music on Chronic Osteoarthritis Pain in Older People. *Journal of Advanced Nursing* 44/5: 517–24.

McCaffrey, R. and Good, M. (2000). The Lived Experience of Listening to Music While Recovering from Surgery. *Journal of Holistic Nursing* 18/4: 378–90.

McCaffrey, R. and Locsin, R.C. (2002). Music Listening as a Nursing Intervention: A Symphony of Practice. *Holistic Nursing Practice* 16/3: 70–77.

McCaul, K.D. and Malott, J.M. (1984). Distraction and Coping with Pain. *Psychological Bulletin* 95: 516–33.

Melzack, R. (1983). *Pain Measurement and Assessment*. New York: Raven Publishing.

Melzack, R. (1999). From the Gate to the Neuromatrix. *Pain* 82: S121–S126.

Melzack, R. (2001). Pain and the Neuromatrix in the Brain. *Journal of Dental Education* 65/12: 1378–82.

Melzack, R. and Casey, K.L. (1968). Sensory, Motivational and Central Control Determinants of Pain. A New Conceptual Model. In D.R. Kenshalo (ed.), *The Skin Senses* (pp. 432–43). Springfield, IL: Charles C. Thomas.

Melzack, R. and Torgerson, W.S. (1971). On the Language of Pain. *Anesthesiology* 34: 50–59.

Melzack, R. and Wall, P.D. (1965). Pain Mechanisms: A New Theory. *Science* 150: 971–9.

Mitchell, L.A. and MacDonald, R.A.R. (2006). An Experimental Investigation of the Effects of Preferred and Relaxing Music Listening on Pain Perception. *Journal of Music Therapy* 43/4: 295–316.

Mitchell, L.A., MacDonald, R.A.R. and Brodie, E.E. (2004). Temperature and the Cold Pressor Test. *Journal of Pain* 5/4: 233–7.

Mitchell, L.A., MacDonald, R.A.R. and Brodie, E.E. (2006). A Comparison of the Effects of Preferred Music, Arithmetic and Humour on Cold Pressor Pain. *European Journal of Pain* 10/4: 343–51.

Mitchell, L.A., MacDonald, R.A.R. and Knussen, C. (2008). An Investigation of the Effects of Music and Art on Pain Perception. *Psychology of Aesthetics, Creativity, and the Arts* 2/3: 162–70.

Mitchell, L.A., MacDonald, R.A.R., Knussen, C. and Serpell, M.G. (2007). A Survey Investigation of the Effects of Music Listening on Chronic Pain. *Psychology of Music* 35/1: 37–57; doi: 10.1177/0305735607068887.

Nilsson, U. (2008). The Anxiety- and Pain-Reducing Effects of Music Interventions: A Systematic Review. *Association of Operating Room Nurses Journal* 87: 780–807.

Nilsson, U., Rawal, N., Enqvist, B. and Unosson, M. (2003b). Analgesia following Music and Therapeutic Suggestings in the PACU in Ambulatory Surgery: A Randomized Controlled Trial. *Acta Anaesthesiologica Scandinivica* 47: 278–83.

Nilsson, U., Rawal, N., Unestahl, L.E., et al. (2001). Improved Recovery after Music and Therapeutic Suggestions during General Anaesthesia: A Double-blind Randomized Controlled Trial. *Acta Anaesthesiologica Scandinivica* 45: 812–17.

Nilsson, U., Rawal, N. and Unosson, M. (2003a). A Comparison of Intra-operative or Postoperative Exposure to Music: A Controlled Trial of the Effects on Postoperative Pain. *Anaesthesia* 58/7: 699–703; doi: 10.1046/j.1365-2044.20 03.03189_4.x.

North, A.C. and Hargreaves, D.J. (1997). Liking for Musical Styles. *Musicae Scientiae* 1/1: 109–28.

Pavlin, D.J., Sullivan, M.J.L., Freund, P.R. and Roesen, K. (2005). Catastrophizing: A Risk Factor For Postsurgical Pain. *Clinical Journal of Pain* 21/1: 83–90.

Phumdoung, S. and Good, M. (2003). Music Reduces Sensation and Distress of Labor Pain. *Pain Management Nursing* 4/2: 54–61.

Rawlings, D. and Leow, S.H. (2008). Investigating the Role of Psychoticism and Sensation Seeking in Predicting Emotional Reactions to Music. *Psychology of Music* 36/3: 269–87.

Razmjou, H., Yee, A., Ford, M. and Finkelstein, J.A. (2006). Response Shift in Outcome Assessment in Patients Undergoing Total Knee Arthroplasty. *The Journal of Bone and Joint Surgery. American Volume* 88/12: 2590–95; doi: 10.2106/JBJS.F.0028.

Ritossa, D.A. and Rickard, N.S. (2004). The Relative Utility of 'Pleasantness' and 'Liking' Dimensions in Predicting the Emotions Expressed by Music. *Psychology of Music* 32/1: 5–22.

Roy, M., Peretz, I. and Rainville, P. (2008). Emotional Valence Contributes to Music-induced Analgesia. *Pain* 134/1–2: 140–47.

Schorr, J.A. (1993). Music and Pattern Change in Chronic Pain. *Advances in Nursing Science* 15/4: 27–36.

Shiffrin, R.M. (1988). Attention. In R.C. Atkinson, R.J. Herrnstein, G. Lindzey and R.D. Luce (eds), *Stevens' Handbook of Experimental Psychology* (2nd edn) (pp. 739–811). New York: Wiley.

Sloboda, J.A. (2002). The 'Sound of Music' versus the 'Essence of Music': Dilemmas for Music-emotion Researchers (Commentary). *Musicae Scientiae* (special issue 2001–02), 237–55.

Standley, J. (2002). A Meta-analysis of the Efficacy of Music Therapy for Premature Infants. *Journal of Pediatric Nursing* 17/2: 107–13; doi: 10.1053/jpdn.2002.124128.

Staricoff, R. (2004). *Arts in Health: A Review of the Medical Literature* (36th edn). London: Arts Council England.

Stevensen, C. (1995). Non-pharmacological Aspects of Acute Pain Management. *Complementary Therapies in Nursing and Midwifery* 1/3: 77–84, http://www.ncbi.nlm.nih.gov/pubmed/9456714.

Sullivan, M.J.L., Martel, M.O., Tripp, D., et al. (2006). The Relation between Catastrophizing and the Communication of Pain Experience. *Pain* 122/3: 282–8; doi: 10.1016/j.pain.2006.02.001.

Trout, K.K. (2004). The Neuromatrix Theory of Pain: Implications for Selected Nonpharmacologic Methods of Pain Relief for Labor. *Journal of Midwifery and Women's Health* 49/6: 482–8; doi: 10.1016/j.jmwh.2004.07.009.

Tse, M.M., Chan, M.F. and Benzie, I.F. (2005). The Effect of Music Therapy on Postoperative Pain, Heart Rate, Systolic Blood Pressures and Analgesic Use Following Nasal Surgery. *Journal of Pain, Palliative Care and Pharmacotherapy* 19/3: 21–9.

Turner, J.A., Jensen, M.P, Warms, C.A. and Cardenas, D.D. (2002). Catastrophizing is Associated with Pain Intensity, Psychological Distress, and Pain-related Disability among Individuals with Chronic Pain after Spinal Cord Injury. *Pain* 98/1–2: 127–34.

Van Damme, S., Crombez, G., Van Nieuwenborgh-De Wever, K. and Goubert, L. (2008). Is Distraction Less Effective when Pain is Threatening? An Experimental Investigation with the Cold Pressor Task. *European Journal of Pain* 12/1: 60–67; doi: 10.1016/j.ejpain.2007.03.001.

Verhoeven, K., Crombez, G., Eccleston, C., et al. (2010). The Role of Motivation in Distracting Attention Away from Pain: An Experimental Study. *Pain* 149/2: 229–34; doi: 10.1016/j.pain.2010.01.019.

Voss, J.A., Good, M., Yates, B., et al. (2004). Sedative Music Reduces Anxiety and Pain during Chair Rest after Open-heart Surgery. *Pain* 112/1–2: 197–203.

Wigram, T., Saperston, B. and West, R. (eds) (1995). *The Art and Science of Music Therapy: A Handbook. The Art and Science of Music Therapy: A Handbook.* Chichester: Harwood Academic Publishers.

Williams, L.R. (2005). The Effect of Music Training and Musical Complexity on Focus of Attention to Melody or Harmony. *Journal of Research in Music Education* 53/3: 210–21.

Zimmerman, L., Pozehl, B., Duncan, K. and Schmitz, R. (1989). Effects of Music in Patients who had Chronic Cancer Pain. *Western Journal of Nursing Research* 11/3: 298–309.

PART II
Musicology

Elaine King and Helen M. Prior

Part II of this volume approaches the notion of familiarity through reflections on the discipline of musicology and its researchers from philosophical, analytical and empirical perspectives. The section opens with Stobart's consideration of the familiar and unfamiliar within music from our own and other cultures, and his exploration of the influences upon our perception of musical sounds and pieces as more or less familiar (Chapter 5). His dissection of traditional notions of familiarity – such as those based on geographical proximity, or the presence or absence of particular musical sounds in a person's environment – reveals the ever-changing and complex perceptual relationship we have with music of our own and other cultures. A further dissection of similarly complex perceptual relationships is undertaken in Chapter 6, although here, Hargreaves analyses the familiar and unfamiliar within a single musical work: George Crumb's *Black Angels*.

A reflective stance is adopted by the authors of chapters 7 and 8. Hawes, in Chapter 7, explores musicology and its many sub-disciplines, noting the inevitable lack of familiarity an individual graduate student will have with some of those sub-disciplines. She argues for efficient communication between researchers using different approaches, offering various criteria through which this may be judged. Such a model, she argues, will be highly beneficial to the field as a whole and its individual researchers. The focus remains on researchers in Chapter 8, as Wöllner, Ginsborg and Williamon examine the impact of musical engagement on research practice. Music researchers are privileged in that they can gain a range of musical experiences, such as through learning an instrument and/or composing music, prior to and alongside their research endeavour, and this has an unavoidable and sometimes valuable effect on the ways in which they approach their work. Finally, the section is completed by Hallam's discussion of the role of familiarity in music education (Chapter 9). Familiarity, she notes, is ubiquitous in the process of learning, and the complex nature of our relationship with music is highlighted here.

Overall, this section raises many valuable questions from a number of different musicological perspectives, and calls attention to some of the myriad ways in which familiarity influences musical research.

Chapter 5

Unfamiliar Sounds? Approaches to Intercultural Interaction in the World's Musics

Henry Stobart

The notion of 'familiarity' in the title of this book suggests associations with the known, secure, embodied or predictable.[1] 'Familiar', in early English usage (for example, by Shakespeare), was also sometimes synonymous with 'domesticated' or 'tame'. Its etymology clearly relates to the idea of 'family', and similarly evokes ideas of shared experience, cultural competency, social intimacy or nostalgia. These ideas might be seen to lie at the heart of identity and culture. Meanwhile, the familiar also carries connections with the mundane, dull or everyday, suggesting a potential lack of challenge, excitement or imagination. In contrast, the 'unfamiliar', which always exists in dialogue or tension with the 'familiar', typically evokes connections with the unknown, unpredictable or 'other'. In turn, this often entails associations with danger, insecurity, anxiety or alienation. Yet, unfamiliarity also suggests excitement, challenge, novelty, innovation and liberty, and is often associated with creativity and the imagination. In this context, it is hardly surprising that throughout music history musical creativity and inspiration have widely been connected with danger and unpredictability. (By music history I refer to all musics, in any time or place, and whether documented or not.) Although socially powerful, beautiful or affective musical expressions must necessarily include familiar elements, their very potency is often associated with dark, mysterious and unpredictable realms – unfamiliar territories of the body and psyche encapsulated, for example, in the imaginary figures of the siren (Austern and Naroditskaya 2006), jinn (Neuman 1990, p. 64) or spiritguide (Roseman 1991).

The effective balancing of the familiar and unfamiliar seems to be at the heart of most successful music making and communication. Indeed, might this even be a universal? Accordingly, Steven Feld observes (in conversation with Charles Keil) that 'as music grooves, there is always something new *and* something familiar' (Keil and Feld 1994, p. 23). It is often precisely the distinctive, unanticipated or even idiosyncratic elements an individual or group brings to a performance,

---

[1]   I use familiarity and unfamiliarity in this chapter as 'thinking tools'. From a more critical etymological perspective, their Latinate root and its histories of usage can, of course, be seen to be shape the semantic space they occupy in culturally specific ways.

or composer to a piece, that mark it out as musically engaging.[2] However, let us not forget that the degree to which the introduction of unfamiliar elements is welcomed varies immensely according to context and tradition. It is also important to stress that such elements – which may be seen to reflect the identity of the performer(s), whilst possibly representing a form of alterity to certain listeners – must be adequately framed within the familiar. We often welcome, and even celebrate, unfamiliar elements when introduced into a familiar environment in which we feel secure, in control, and are able to orientate ourselves. However, predominantly alien environments (musical and other), in which we are unable to recognise, reproduce or respond appropriately to structures, patterns or modes of expression, or in which our ignorance or powerlessness is made manifest, may provoke feelings of insecurity, disorientation and anxiety, as well as negative evaluations.

Perceptions of the unfamiliar depend on subject position and it is important to consider how other people might perceive or conceive of the music that you or I consider familiar, as well as how you or I might perceive or conceive of music that we consider culturally unfamiliar. Clearly, each reader will have their own graduated, shifting and context sensitive conceptions and perceptions of familiar and unfamiliar musics, certain aspects of which I probably share and others which I do not. I wish to avoid reducing familiarity and unfamiliarity to a binary, equivalent to the problematic 'self/other' dichotomy, because musics are rarely, if ever, entirely familiar or unfamiliar. Rather, our musical engagements tend to consist of points on a continuum (or even on a more complex 3D kind of matrix), with particular dimensions of any given musical encounter being more or less familiar. In other words, there are aspects of any musical performance, however culturally close or distant, with which we are likely to feel degrees of both unfamiliarity and familiarity.

In this chapter I want to suggest that the notions of familiarity and unfamiliarity are not only fruitful ways for thinking about music more generally, but that – when applied to relations between musical expressions from around the world – they throw up a range of challenges to commonplace assumptions. In particular, I wish to use this as an opportunity to question how we might approach the idea of 'unfamiliar sounds'; this is the focus of the first half of the chapter, which is divided into two main parts (each consisting of several shorter sections). What might make certain musical sounds unfamiliar? A simple, initial, answer is likely to be that they have not been part of a given listener's (musical) environment. Yet, this immediately throws up more complex questions about the agency, opportunities and motivations of the listener, which inevitably involve, for example, identity, power, politics, ideology and gender. Are all musics approached in the same way or is there a tendency for some, especially those viewed as

---

[2]   This point is sometimes more easily appreciated in the case of performance from notation, as in the case of Chinese Guqin music (Yung 1987, p. 85) or a Chopin piano Prelude (Cook 1992, p. 124).

culturally distant, to be rendered unfamiliar or unknowable by a presupposition of difference? As we keep asking these questions, we are in turn drawn to consider the cultural construction of musical aesthetics, and the challenges presented by aesthetically unfamiliar sounds, both from distant locations and from close to home. I end the first part of the chapter by considering the historical development of communications and especially audio technology. Has the ubiquity of World Music (following its rise as a marketing phenomenon in the late 1980s) rendered the idea of musically unfamiliar cultures obsolete? The second part of the chapter looks more specifically at music perception and paradoxically suggests that, from such a perspective, musical unfamiliarity is a consequence of cognition. It goes on to consider some of the perceptual challenges involved in learning to perform culturally unfamiliar musics. Finally, whilst acknowledging the powerful sense of cultural and perceptual consensus and stability that often surrounds musical performance, I explore the relevance of what I call *creative misperception* to histories of music making. A key point to emerge from many discussions in this chapter is that a neat bracketing of features common to some notion of 'non-Western' musics, which somehow differentiate them from the equally problematic idea of 'Western' musics, is entirely inappropriate. Indeed, I begin Part 1 of the chapter by contesting the use of the all too familiar term 'non-Western', followed by an overview of ethnomusicology's complex relationship with notions of the familiar and unfamiliar.

Part I: Conceptualising Unfamiliar Sounds

Resisting the Familiar and Ethnomusicology's Paradox

My invited remit for this chapter was to focus on 'non-Western' music and to provide an ethnomusicological perspective. However, like many other terms whose very familiarity and apparent utility deter critical reflection, I wish to start out by resisting the term 'non-Western'. Albeit a handy, catch all, expression which is widely used in 'Euro-American'-derived discourse (including quite regularly by ethnomusicologists), it is ambiguous, problematic, anachronistic and unhelpful. Even if, arguably, well employed as a weapon of postcolonial critique of former decades, its relevance for the 2010s is doubtful. Evidently the significance of 'non-Western' resides in its duality with the 'West', a malleable but increasingly destabilised imaginary and target, often historically linked with 'whiteness', Christianity and claims of superiority and rationality – it is 'always a fiction, an exercise in global legitimation' (Trouillot 2003, p. 1). The 'West' has been deployed and employed in a multiplicity of geo-political projects and, despite its anachronistic connection with scientific, epistemological or economic superiority, will probably retain its currency for a good many decades to come (Bonnett 2004, pp. 163–4). Even if too much is invested in the idea of the 'West' for it to be superseded any time soon, a reality check on the relevance of the term 'non-Western' to the study of music is a more realistic goal, and one with considerable benefits for a more holistic music scholarship.

Maintaining the duality Western/non-Western encourages us to overlook, or simply exoticise or fetishise, the interpenetration of the world's various music histories and to ignore the musical realities that surround us (Taylor 2007). What relevance do such terms have to conceptualising a Taiwanese concert pianist, a British Asian Bhangra artist, or the sounds of the Istrian *doojkinje* (Croatian double flute) which – although European – would strike most British listeners as culturally remote? Another uncomfortable aspect of the totalising tendency of the word 'non-Western' is that it implies an identity defined by exclusion, absence or deficiency rather than contribution. In his book *Beyond Exoticism*, Timothy Taylor (2007, p. 7) contests the idea of a unitary and essentialist 'musical Other', arguing instead for closer historical, cultural and social examination of how people construct their 'others', sometimes through music. Similarly, Michel-Rolph Trouillot (2003) observes that:

> The "us and all of them" binary, implicit in the symbolic order that creates the West, is an ideological construct … There is no Other, but multitudes of others who are all others for different reasons. (p. 27)

In short, rather than encouraging us to engage with people around the world as equals, starting from the position of common humanity, the term 'non-Western' presupposes difference, re-inscribes the boundary line in us/them, and often presumes an implicit hierarchy.[3] Instead of lumping familiar and unfamiliar cultural expressions into well-worn binaries such as 'Western' and 'non-Western', we need to learn to be much more specific in the ways we articulate experience and knowledge about the world's various musics and our own individual and subjective relationships to them. An important point to emerge here, as regards this volume, is that familiarity can easily become complacency. Accordingly, the unfamiliar can pose not only perceptual, aesthetic and intellectual challenges, but also potent political ones.

Following the adoption of the title Ethnomusicology in the 1950s, this field of scholarship was almost exclusively dedicated to the study of 'culturally unfamiliar' musics. Indeed, my own extensive research in the Bolivian Andes (Stobart 2006), over the past 25 years, might be seen to fit this traditional paradigm. However, since the late 1980s, and in part responding to the reflexive turn in anthropology (Clifford and Marcus 1986), ethnomusicological research has increasingly included case studies that are both more familiar and geographically closer to home. For example, Bruno Nettl (1995) researched his home music institution and Stephen Cottrell (2004) studied the London freelance music scene in which

---

[3]   See Agawu (2003, pp. 151–71) on the theme of contesting difference. However, strikingly throughout this highly critical chapter, he maintains the use of the term 'non-Western'.

he was professionally involved. Accordingly, many involved in ethnomusicology today are likely to insist that their field of study is defined by methodology rather than object of study (Stobart 2008). Indeed, the research of a high proportion of current ethnomusicology graduate students focuses on musical phenomena that are relatively familiar, close to home or easily encountered on the internet; what they share is the use of ethnographic approaches. This methodology typically involves acquiring familiarity with particular musical practices and with the people involved in them through close participation and observation. In short, it might be argued that, despite its stereotypical and ongoing association with culturally distant musics, ethnomusicology in the 2010s is more defined by familiarity (also in the sense of rapport) – as its central methodological contribution – than unfamiliarity.

Nonetheless, ethnomusicology – alongside anthropology – has not found it easy to shrug off its historical associations with the unfamiliar or exotic; nor up to now has it been entirely in its interests to do so. Indeed, ethnomusicology has occupied a paradoxical position, on the one hand contributing to the deconstruction of difference by rendering unfamiliar musics familiar, but, on the other, constructing difference – often unintentionally – by presenting the musics studied as unfamiliar, exotic or 'other'.[4] This is hardly surprising given the common human tendency to notice or dwell on the unfamiliar, remarkable and memorable; to catch people's imagination and interest, or even make an 'original contribution to knowledge' (as doctoral students are required) through identifying things that are different, rather than examining what is the same, already known or familiar. As Roger Keesing (1989, p. 460) has observed for the case of anthropology, 'the reward structures, criteria for publishability, and theoretical premises of our discipline [mean that] papers that might show how un-exotic and un-alien other people's worlds are never get written or published'.

Even though, as noted above, the primary research of many ethnomusicology doctoral students focuses on familiar musics from close to home, the employment of ethnomusicologists in university music departments usually remains predicated on the teaching of 'culturally unfamiliar' or 'world' musics. Undoubtedly, the exposure of students to a variety of the world's musical cultures is immensely valuable in terms of widening musical and other horizons and raising awareness of new possibilities. However, another key benefit is its potential to encourage students to reflect on the familiar – the particularities, constraints and conventions of their own musical practices and experiences. Yet, perhaps this is also the greatest challenge. For example, whilst usually happy to write about the ritual nature of musical practices in Papua New Guinea, Indonesia, Africa or South America, many students find it hard to accept that, for example, the conventions of a European classical concert are also rituals. In other words, students often struggle to see the continuities between their own familiar musical practices or experiences and those of geographically distant cultures. Yet of course, they are

---

[4]   In part this reflects the methodology of balancing so-called 'emic' and 'etic', or insider and outsider, perspectives (Nettl 2005, p. 228).

far from alone in experiencing such – at least, initial – blindness or even resistance to the idea of cultural continuity. They simply reflect a much wider tendency to presuppose difference (Agawu 2003).

The Politics of Particularity: Cultural Meaning and Identity

In Renaissance and later accounts of global exploration, European travellers and the indigenous people they encountered are periodically reported to have used music as a means to establish contact. Where no mutually intelligible language was available, performing music to one another sometimes contributed to the initiation of peaceful relations (Woodfield 1995, p. 101). While these musics undoubtedly took radically different forms, their very existence and expression presumably communicated to both parties a sense of common humanity. However, such musical exchanges did not always function as a universal language; they sometimes led to fatal misunderstandings. For example, during Abel Tasman's voyage of exploration to New Zealand in 1642, his Dutch crew failed to recognise that the Maori's chants in a 'rough hollow voice' and trumpet sounds (probably a *putatara*) performed to them from their canoes were intended as a challenge to the strangers and an invitation to fight (see Figure 5.1). Neither did the Maoris appreciate that the welcoming tunes played by the Dutch trumpeters in reply were of peaceful intent. Only were the consequences of this mutual musical misunderstanding made apparent when a rowing boat, containing seven unarmed Dutch sailors, was launched a few days later and immediately attacked by two Maori canoes. Four sailors were killed and the three others only escaped by swimming back to the ships (Lodge 2009, pp. 626–7). Ethnographic detail does not permit us to be sure how much this dramatic incident can be attributed to a lack of musical familiarity, but the particularity of the respective musics and cultural traditions is clear. Growing up within and acquiring familiarity with particular musical environments evidently lead us to hear, interpret and contextualise sounds in particular ways.

This encounter presents the two musics as if they were mutually unintelligible, much in the same way as languages. However, the sounds of music, unlike those of language, are primarily concerned with expression rather than communication, and are neither arbitrary signs nor usually effective means for conveying specific or propositional meanings. Rather, a key aspect of music's power lies precisely in its semantic ambiguity and flexibility, which enable it to 'mean different things to different people, different things at different times, or even to mean many things at once' (Slevc and Patel 2011, p. 111; Cross 2008). Musics that are culturally unfamiliar can sometimes provide us with immense aesthetic, sensory and emotional enjoyment and afford a range of meanings, even if these are quite different from those experienced by a person for whom such music is deeply familiar. Yet, unfamiliar music can also potentially communicate hostility, and provoke a sense of confusion, fear or alienation.

Figure 5.1   Isaac Gilsemans, *A View of the Murderers' Bay* (1642)
Source: Reproduced with the permission of the Alexander Turnbull Library, Wellington, New Zealand, PUBL-0086-021.

It is unclear from the account of Tasman's encounter whether the Maoris heard aggression and challenge in the sonorities of the Dutch trumpets. Unlike visual images, musical sounds literally surround or immerse the listener, imposing their temporalities, and invading the subject's physical and temporal space. We can close our eyes, but not our ears, and even so-called 'civilised' societies continue to use music for torture (Cusick 2006). Unfamiliar musics can signal alternative or conflicting temporalities, values and ways of ordering or understanding the world, which may be heard as noise, 'disorder', a threat to order or alterity. Sometimes such musics can instil terror in the hearer, whereas for people who identify with or who are familiar with these sounds they may be perceived as beautiful and convey a sense of solidarity. For example, in his 1609 *Royal Commentaries of the Inca*, Garcilaso de la Vega includes a description of the historical defeat of the Andean Huanca people by the Inca king Pachacuti. According to Garcilaso, the Huancas worshipped dogs, took immense delight in consuming their flesh, and even made a form of horn (*bocina*) out of their heads. These dog head horns were played during the Huanca's feasts and dances, and produced a music that was sweet (*suave*) to their ears, but when used in battle stunned and terrified their enemies. As Garcilaso observes, 'they said that the power of their [dog] god brought about these two contrary effects: to them it sounded good, because they honoured him, but it bewildered their enemies and caused them to flee' (Garcilaso de la Vega 1609, Book 6, Chapter 10, my translation). It is notable that Garcilaso specifically relates this contrast in perception to belief or ideology – belief or non-belief in the Huanca's 'dog god'.

More generally, the association of musics with particular ideologies, belief systems, values or identities can sometimes provoke a powerful sense of aversion, alienation or negative evaluation among those who do not identify or empathise with them, which may involve a suspension of listening or disengagement (Stokes 1994). Such processes of avoidance can lead certain musics, even when produced close to home, to be rendered unfamiliar or simply to be heard as noise. Heavy metal music is an obvious example of a sub genre that is deeply appreciated and closely identified with by some, but because of its associations and sonic exterior, is often avoided or dismissed as 'noise' by others, for whom it thereby remains unfamiliar. However, musics have also often been rendered unfamiliar through processes of exclusion. For example, the development and practice of art music genres in, for example, India, China and Europe, were usually – historically at least – restricted to the privileged classes and their musician employees. This privileged position, and relative freedom from concerns with subsistence, provided opportunities for cultivating familiarity with highly elaborated and extended musical forms and sometimes dilettante study and performance. Yet, for a broad spectrum of the population such musics remained unfamiliar (Booth and Kuhn 1990, p. 423), just as class or caste affiliation ensured that much vernacular music remained unfamiliar to elite groups. Thus, rather than being a given (as it is often perceived), musical unfamiliarity can be, and often is, a product of wider social and cultural processes, involving, for example, identity, politics, ideology, economics, gender, ethnicity, class and religion.

Attitudes to intercultural engagement are also inevitably influenced by some of these factors, alongside the power dynamics accompanying global and historical processes, such as colonialism. The perception that music is 'unfamiliar' may have much less to do with cultural distance than a lack of motivation to engage with it, sometimes in turn reflecting the ascription of low or undesirable cultural values. For example, the number of Europeans and North Americans motivated to acquire performance skills in the musical traditions of other cultures is very small compared with that of people from elsewhere who dedicate themselves to European–American classical and popular music traditions – sometimes to become world-leading exponents. It is also notable that the local or indigenous musics of many of these latter musicians remain deeply unfamiliar to them. Often, only later, after spending time away from their country of origin, do such musicians begin to engage with the music of their homeland and come to view it as interesting, valuable or part of their cultural identity. On the one hand, the motivation to engage with unfamiliar or culturally alien musics often concerns the cultural capital and aspirations with which these musics are associated. On the other hand, such motivation may reflect political agendas or a wish to explore alternatives, new challenges or creative opportunities. In turn, this may reflect dissatisfaction with more familiar music, or a desire to objectify, develop or reinvigorate it. Indeed, the culturally unfamiliar sometimes comes to represent a key creative resource, possibly stimulating a kind of collector mentality or the reification of musical phenomena which then stand out, or 'figure', against the 'ground' of the familiar. Thus, whilst cultural unfamiliarity may sometimes motivate a tendency to suspend listening or to disengage, at other times it may focus and intensify the listener's attention. As noted previously, it is often unfamiliar elements in an otherwise familiar piece of music that emerge as the most salient and remarkable features, commanding attention and interest, and perhaps provoking a sense of excitement or emotional potency.

The world's diverse music histories abound with, and might be said to be characterised or motivated by, the practice of acquiring more or less familiar musical elements, ideas or resources from other people or cultures and modifying them to conform or appeal to the receiver's tastes. When the musical resources in question are considered culturally distant or unfamiliar and adopted from a position of power, this process is often dubbed 'exoticism' or 'orientalism' (for example, Locke 2009; Said 1978). In these cases, the faithfulness and extent to which unfamiliar cultural resources are imitated or incorporated vary immensely. For example, the forms taken by 'exoticist' European classical music to invoke unfamiliar or exotic cultures may have little or no basis in the actual music of the culture represented. The sounds index the unfamiliar or exotic, but in reality they are often familiar and stereotyped semiotic codes used to evoke exotic imaginaries that may themselves have little basis in cultural reality. In other words, familiar codes may come to represent the idea of cultural unfamiliarity.

In turn, these same codes for the exotic, alongside stereotypical or homogenised versions of the culture that stress the unfamiliar – sometimes labelled by outsiders as 'true' or 'authentic' (Diamond, Cronk and von Rosen 1994, p. 44) – may come to be adopted by the people they are purported to represent, as a form of 'strategic essentialism' (Spivak 1987). Thus, for example, indigenous Americans whose own cultural traditions do not include notably exotic or unfamiliar features are likely to appear to dominant settler society as disappointing, inauthentic or corrupted (Conklin 1997; Dueck 2005, pp. 170–71). Once again, this reflects a widespread tendency, noted above, to presuppose difference (Agawu 2003), especially by metropolitan populations when they encounter or conceptualise culturally distant peoples. Nonetheless, with the rise of identity politics (especially since the early 1990s), many self-identified ethnic or cultural groups have been deeply occupied in defining their cultural uniqueness and exploring the benefits, especially in terms of economics and rights, that such distinct resources might hold for them (Comaroff and Comaroff 2009; Brown 2003). Indeed, communities who lack adequately distinctive and marketable musical or other forms of cultural resources or heritage – in other words, who appear too familiar – may be seen to be at a disadvantage.

Challenging Aesthetics

Our social and cultural environments inevitably lead us to become attentive and receptive to particular forms and ways of structuring musical sounds, which in turn underlie the ongoing development of our aesthetic appreciation and values. Although, as noted above, social processes often contribute to the unfamiliarity of particular musics, it is also evident that certain musical phenomena provide greater challenges for intercultural listening than others. For example, the polyphonic singing and yodelling of Central African forest peoples, such as the Mbuti and BaAka, is immediately appealing to many Europeans and North Americans. The minor divergences of such music's pitch intervals and vocal sonorities from Euro-American models are usually heard as charmingly unfamiliar, rather than as unpleasant or aesthetically challenging. However, instrument sonorities, vocal timbres and tuning systems developed in other parts of the world are sometimes much more demanding for Euro-American ears. Indeed, such musics are notable for their absence, at least in an unmodified form, from most World Music marketing. For example, the strident high-pitched women's singing of some indigenous groups of the Bolivian and Peruvian Andes is not always perceived as immediately attractive by North Americans and Europeans – although many of my own students greatly enjoy emulating these vocal sonorities. Similarly, the various forms of panpipes and flutes played in this region tend to be blown strongly, exploiting the upper register, to produce what outsiders typically characterise as 'harsh' or 'dissonant' timbres. During field research in the Andes, I found that these rural musicians had little aesthetic appreciation for the lyrical and virtuosic recorder music I played to them, pieces that had received acclaim in the UK. My Andean friends were unimpressed with the gentle and expressive 'fluty' sound, suggesting instead that I 'blow more strongly'.

What these same musicians particularly appreciated when they played their own *pinkillu* flutes (see Figure 5.2) – which were probably modelled on European Renaissance recorders – was the strong vibrant timbre that they referred to as *tara* (Stobart 1996b). From a European scientific perspective, the beating sound of *tara* is literally 'dissonant' or inharmonic. It results from a combination of instrument construction and performance practice where, as my hosts put it, the flute speaks 'with two mouths', producing a double (multiphonic) timbre consisting of two sounds pitched approximately an octave apart, but not precisely, so as to create a beating effect (Stobart 2006, p. 215). Significantly, the concept of *tara* was closely associated with notions of social harmony and abundance, and was contrasted with a much less appreciated thin 'fluty' sound, referred to as *q'iwa*, also produced by these flutes. Whereas the acoustically 'dissonant' *tara* sound was widely connected with social harmony, the 'fluty' *q'iwa* sound was linked with social dissonance (Stobart 2006, p. 216). For example, the word *q'iwa* was applied to string instruments that would not stay in tune, awkward-shaped objects that did not fit or people who were mean or selfish, whereas *tara* was related to two people walking together, double objects or things that were in balance (Stobart 1996a).

Figure 5.2   A *pinkillu* flute consort performing in a llama corral during the
            feast Carnival, Wak'an Phukru community, Macha, Bolivia
Source: Photo by author.

The simple point I wish to stress here is that musical expressions of 'harmony' and 'dissonance' are socially constructed. Accordingly, such constructions, and the types of aesthetic preferences they entail, do not necessarily communicate across culture. What might be heard by an Andean rural musician as harmonious, vibrant and abundant, may well be perceived by an outsider acculturated into a different musical environment as harsh, 'dissonant' and even 'unmusical'.

This sense of cultural particularity, as mutually unintelligible musical aesthetics or taste, was baldly expressed by Charles Darwin in *The Descent of Man* (1871):

> so different is the taste of the several races, that our music gives not the least pleasure to savages, and their music is to us in most cases hideous and unmeaning. (p. 333)

Darwin's conception of the music of 'savages' was probably largely based on his encounters with the indigenous people of Tierra de Fuego during his voyage to South America on the *Beagle* in 1832, an experience that was critical to the development of his theory of evolution (Desmond and Moore 1992, p. 133). We cannot be entirely sure what Darwin heard, but a series of recordings of songs performed by Lola Kiepja, who is presented as the last of the traditional Selk'nam people of Tierra del Fuego, may at least give us a flavour. The songs were recorded by Anne Chapman and appear on two LP discs entitled *Selk'nam Chants of Tierra del Fuego, Argentina* (1972, 1978). In his review of the discs, Dale Olsen (1980, p. 289) observes that the 'casual listener will undoubtedly find the songs in these two albums hopelessly repetitive'. Although close listening is rewarded by the revelation of musical intimacy, subtle and varied tone colours, rhythmic complexity and the use of microtonal intervals, there is little in the vocal quality, melodic gestures or rhythmic cadence of these unaccompanied songs that the average Euro-American listener will find immediately attractive or meaningful. Indeed, I can well imagine less culturally open-minded (or politically correct) individuals today dismissing these sounds as 'hideous' – as did Darwin. Nonetheless, many Euro-American listeners might well find the sounds of certain experimental or atonal musics produced close to home even more sonically unfamiliar – or more 'hideous and unmeaning' – than, for example, many of the world's indigenous musics. In other words, we need to be careful not to assume that cultural distance can be neatly mapped onto musical aesthetics or sonic unfamiliarity.

The challenging aesthetics and sense of unfamiliarity provoked by much experimental music is often deeply political in motivation. They intentionally defy familiar musics, as sounds that are perceived to represent the status quo, convention or complacency. Yet, it is striking that people do not usually refer to such unfamiliar sounds as 'exotic'; their unfamiliarity comes from within rather than from outside. According to Jacques Attali, music is used by power to create an illusion of harmony in the world, making people forget the general violence. It is also used to *silence* or censor all other human noises 'by mass-producing a deafening, syncretic kind of music' – in other words, music that is mass mediated and familiar to the majority (Attali 1985, p. 19). Thus, rebellious musics may often be seen to exploit and celebrate precisely those unfamiliar sounds or 'noises' that familiar hegemonic musics attempt to censor. Examples abound, including the British punk movement or the diverse inharmonic sounds incorporated into the compositions of Brazilian composer and instrumentalist Hermeto Paschoal (Lima and Costa 2000). The unfamiliar sounds of experimental or rebellious music may be seen – among those who identify with them – to provoke what might be characterised as a kind of political or ideological listening. Thus, to create, perform or listen to music that is constructed as aesthetically 'difficult' or unfamiliar, in relation to accepted or mainstream musics, may sometimes be seen as an expression of solidarity with (or rejection of) particular ideologies, politics or values. Indeed, for those committed to such projects or ideologies, these 'unfamiliar sounds' may come to symbolise integrity and to be experienced as beautiful. Some of these kinds of processes are beautifully charted in the documentary film *Hanoi Eclipse: The Music of Dai Lam Linh* (2010) by Barley Norton, which features the controversial Vietnamese band Dai Lam Linh.

The kinds of ideological listening described above also partly explain the large audience that developed in Europe during the late 1970s and 1980s for the previously unfamiliar (but aesthetically approachable) sounds of Andean neo-folklore music, especially in the form of Chilean 'New Song'. For many people in Europe, engaging with this music was a way to express solidarity with Chilean victims and exiles following the 1973 coup by Augusto Pinochet, the dictatorship's censorship of such sounds affording them special potency (Morris 1986). However, I also want to suggest that the sounds of culturally unfamiliar musics, especially when encountered in audio recordings, are sometimes attractive precisely because they do *not* entangle the listener in the music's ideologies, politics or broader contexts, dynamics which sometimes constrain our listening nearer to home. Paradoxically, this means that, despite the social inequalities that so often surround the production and circulation of culturally unfamiliar musics, they are often heard as novel, fresh and free of ideological baggage. They liberate listeners from the constraints of the familiar and offer a sense of utopia or innocence. Thus, in recordings of the polyphonic singing of the BaAka Forest People (Central African Republic), Euro-American listeners tend to hear a sense of peace, naturalness, humour and community, rather than the jealousies, disease, hunger, anxiety and conflicts with agriculturalist villagers that underlie such people's lives and performance (Locke 1996, p. 130). Paucity of contextual or linguistic knowledge may also mean that problematic aspects of unfamiliar musics, such as unacceptably homophobic or racist lyrics, may pass unnoticed. In short, unfamiliar musical sounds can, in some contexts, be intensely political – explicitly employed to challenge the familiar or conventional – but in others they may be perceived as apolitical, as offering a sense of ideological liberation or nostalgic return to an imagined innocence.

From Unfamiliar to Ubiquitous?

The sensation of unfamiliarity associated with musics of distant cultures has undoubtedly been, and sometimes still remains, heightened by the challenges and dangers of travel. Furthermore, interpretations of encounters with such cultures have often been informed by existing imaginaries of the unfamiliar, such as the utopian and fantastical accounts of Amazons from *Mandeville's Travels*, which in turn drew on Greek myth (Klarer 1993). Yet, these early musical encounters, even if sometimes with fatal consequences, were characterised by direct human contact. By contrast, the majority of encounters with culturally unfamiliar musics over the past century have occurred via recorded media. Although such mediated listening vastly reduces the risks associated with travel (Taylor 2007, p. 206), the separation of sounds from their sources also entails an absence of direct human engagements. While such so-called 'acousmatic' (Chion [1983] 2009) or 'schizophonic' (Schafer [1977] 1994, p. 90) listening might be seen to open up new spaces for the imagination, beyond the constraints of lived human interactions, it might also be seen to restrict opportunities to develop mutual understanding. Indeed, Ross Daly advocates replacing recordings of World Music with a greatly increased quantity of live performance. For him, 'any attempt to approach the various musical traditions of the world has to involve an appreciation of the musicians themselves' (Daly 1992 cited in Aubert 2007, p. 55). Arguably, however, this problematic space for the imagination or re-interpretation offered by audio recordings is in part offset by other factors surrounding developments in communications. Migration, travel, tourism and the circulation of knowledge about other cultures enable vastly expanded opportunities for intercultural interaction and understanding that could not have been imagined several decades, let alone a century, ago. In this context it is interesting to consider two statements about the reception of unfamiliar or exotic musics made respectively in 1934 and 1990:

> Nowadays, with ample gramophone records of exotic music and recently published books of musical research, our sources of information are many and inexpensive. There is no longer the necessity for the appalling ignorance that darkens our musical life and for the prejudices that arise out of that ignorance. (Grainger 1934 in Blacking 1987, pp. 151–2)

> It may be a natural human tendency to label unfamiliar musical traditions as primitive, barbarian, the chirping of birds, or meaningless, but the continually increasing speed and intensity of intercultural communication in the twentieth century has shown such ethnocentric value judgments to be indefensible. (Booth and Kuhn 1990, p. 411)

Grainger's words date from the time that 78 rpm discs of music from unfamiliar cultures were just becoming available in the UK, while Booth and Kuhn were writing soon after World Music marketing had become big business and the internet was in its infancy. The variety of culturally diverse music available to us in the 2010s, at the click of an internet search engine or from an online store, has grown exponentially, but does this mean that people have become more tolerant of culturally unfamiliar musical sounds or that ethnocentric value judgements and prejudice have diminished?

Dependent on context, it would be possible to argue both for and against the proposition that culturally unfamiliar musics are more accepted today. Developments in communications mean that an extraordinary range of culturally diverse musics, and information about them, is certainly widely available. However, although an almost unimaginable multiplicity of localised musical expressions from around the world is available on Youtube, finding particular examples usually requires local or specialist knowledge. For example, to locate the music videos of Gregorio Mamani that feature indigenous music from the Northern Potosí region of the Bolivian Andes (Stobart 2011), specific key words, such as the name of the artist, genre or instruments would be required. My simple point is that internet searches tend to be confined to familiar linguistic and cultural territory – and it is in the economic interests of internet search engines to keep it this way. While in many respects musical sounds from around the world have become ubiquitous, the versions that dominate mass media tend to be carefully selected so as not to offend or disturb generalised sensibilities.[5] While some are

---

[5]   Key actors here include artists, record companies, distributors and broadcasters.

immediately approachable (as mentioned above), others that might challenge mainstream sensibilities are sometimes blended with familiar musical resources; in this form they may then potentially add spice to the music or evoke other people, times or places. For example, a World Music hit – which might seem almost as unlikely as the Selk'nam songs of Tierra del Fuego – was created when the New Age pop group Enigma combined sampled excerpts from a 1989 recording of a Taiwanese aboriginal song with synthesiser sounds and a disco beat to create *Return to Innocence* (1993). The song, which explicitly used unfamiliar vocal sounds to invoke nostalgia for modernity's loss of innocence, was selected as the theme song for the 1996 Atlanta Olympics and received very wide circulation (Tan 2008, p. 222). Anahid Kassabian has argued that such blends of exotic and familiar sounds, as also offered by the Putumayo recordings heard in Starbucks coffee shops, do not – as some scholars of postmodernity would have it – collapse the distinction between 'here' and 'there'. Rather, she suggests, they enact a kind of *distributed tourism* (or *distributed subjectivity*) where the difference between 'here' and 'there' is actively maintained, to create a kind of 'entangled space' of being 't/here'. However, this experience is much the same whichever part of the world the coffee shop is located in (Kassabian 2004, pp. 219–21). We might also question the degree to which such ubiquitous forms of background music, which often create ambience without receiving close listening, can be said to be truly 'familiar'.

Part II: Perceiving Unfamiliar Sounds

Music Perception and the Unfamiliar

Although perhaps counter-intuitive, it would appear that perceptions of musical unfamiliarity intensify with cognitive development. Infants exhibit an extraordinary ability to discriminate speech sounds, including almost every phonetic contrast. Yet, with cognitive development over the first year this ability declines as infants increasingly focus their attention on familiar acoustic dimensions relevant to the linguistic environment (Maye with Werker and Gerken 2002). According to Lynch and Eilers, similar culturally specific perceptual reorganisation for musical tuning also begins to affect infants' perception between 6 and 12 months (Stevens 2004, p. 434). Just as adults' perception of speech sounds becomes constrained by the phonetic organisation of the speaker's native language (Maye et al. 2002), so it seems their ability to perceive tonal relations that do not match the 'tonal schemata' developed during musical acculturation is reduced (Lynch and Eilers 1991, p. 122). In his theory of auditory scene analysis, Bregman (1990, p. 641) suggests that we actively and constructively process auditory information using 'schemas that incorporate our knowledge of familiar sounds' (in Clayton 2008, p. 139; also see Chapter 2). For the case of both music and speech, such schemata may be seen, on the one hand, to greatly increase our ability to perceive and process patterns of sound encountered in familiar cultural environments, but, on the other, to constrain our ability to distinguish unfamiliar patterns of sound encountered in other cultural environments.

Such learned perceptual dispositions would seem to impact on a variety of the ways in which people perceive culturally unfamiliar music. For example, besides the effects of pitch recognition mentioned above, various experiments have demonstrated that subjects' recognition and memory of musical structures, and their ability to predict or reproduce them, are more effective in culturally familiar than unfamiliar musics (Ayari and McAdams 2003, p. 191; Eerola et al. 2006; Morrison, with Demorest and Stambaugh 2008). Also, while certain aspects of temporal processing appear to be universal, such as working memory and biases towards particular periodicities, attentional foci or behavioural timing (Cross 2008, p. 150; Nan, with Knösche and Friederici 2006, p. 179), the structuring and perception of rhythmic relations vary considerably across culture. For example, the inability of outsiders to identify an underlying beat or pulse, which is often unmarked acoustically, in certain African or African-derived musics is well documented (Temperley 2000, p. 71). This has led artists such as Yousou N'Dour from Senegal to include an acoustic cue in his recordings of Mbalax music for the international market to enable foreign audiences to identify and move in time with the dance beat. This helps avoid the sense of disorientation, sometimes encountered by African musicians on international tour, when audiences perceive and dance to a different pulse from that structuring their performance.

Issues concerning cross-cultural differences in the perception of pulse have also emerged from my research in the rural Andes, analysed in a paper with Ian Cross (Stobart and Cross 2000). When European subjects were asked to tap along with Quechua language 'Easter Songs' from Northern Potosí in Bolivia, with very few exceptions they treated the songs as anacrustic: as if in 6/8 time with a quaver upbeat. By contrast, the tapping of all Quechua-speaking Bolivian subjects, even when unfamiliar with the genre, treated the songs as non-anacrustic: as if in 2/4 starting on the first beat (see Example 5.1). Interestingly, several of the Bolivian subjects also tapped along with recordings of anacrustic (6/8) British folk songs, perceiving the anacrustic upbeat quaver as a downbeat. (To my ears this tapping initially sounded random, but measurement using computer software demonstrated that it was regular.)

Our analysis also considered the role of production, such as the strumming of the mandolin-like *charango* that accompanies Easter songs (see Figure 5.3). This sometimes created rhythmic ambiguity by stressing the up-stroke of the strum, while the down-stroke, which coincides with the footfalls of the dance (or tactus), is sometimes silent, not striking the strings. Both the charango up-stroke and the sung rhythms of the songs also tended towards asymmetric proportions (averaging a 2:3 ratio), adding further rhythmic instability that we interpreted in terms of keeping the performance exciting and 'on the edge'. The Andean participants' rhythmic perception, and their ability for rhythmic 'play' that

Example 5.1   Easter song perception: (a) by European subjects (6/8 upbeat);
(b) by Bolivian (Quechua-speaking) subjects and performers
(2/4 on beat)

Figure 5.3   *Charango* players of Easter songs in Sacaca, Bolivia
Source: Photo by author.

often seemed to contradict acoustic cues, may also have been facilitated by the
stress rules of the Quechua language. The primary stress in Quechua falls on the
penultimate syllable (and is thus moveable, depending on word length), but an
unmarked secondary stress falls on the first syllable, acting, we have proposed,
as a kind of perceptual anchor.

An important point to emerge here is that the same musical sounds can not
only afford different associative meanings for different people, but also that –
according to the perceptual dispositions of the hearer – the ways in which the
patterns and structures of the sounds are heard may be strikingly divergent. In turn,

this can lead to differences in bodily responses as well as to broader conceptions of how the music is organised. Indeed, might listeners be said to be hearing quite different musics? According to John Blacking:

> Music can communicate nothing to unprepared and unreceptive minds, in spite of what some writers have suggested to the contrary. The power of music *as music* must depend in the last resort on people's perceptions of specific patterns of melody, rhythm and texture, and on the bodily sensations and responses that these elicit. (Blacking 1987, p. 30)

Even if questions about the ontology of *music* and its presentation as almost exclusively aural might arise from this statement, Blacking is clearly right to stress the body as the primary locus of our engagements with music. The only problem with the first part of Blacking's statement is that culturally unfamiliar musics almost always communicate *something*. This may not be the same as would be perceived by people who are deeply familiar with the music in question, but it is *something*, not *nothing*. I will return to this point in more detail below. However, at the time Blacking was writing, there were good historical, ideological and political reasons to make such a statement. It encouraged serious consideration of the perceptions, knowledge and musical meanings shared by marginalised and discriminated groups, such as the Venda with whom Blacking undertook extensive research under South African apartheid, against which he was an outspoken critic (Byron 1995, p. 16). It also challenged ethnomusicologists not to rely on their own subjective perceptions or presumed meanings, but instead to become deeply familiar with the musical cultures they studied through extended ethnographic research. As Titon (1996, p. xxiii) puts it, 'as much as possible, an unfamiliar music should be understood at the outset in its own terms: that is, as the people who make the music understand it'; an objective that may involve considerable perceptual challenges.

Challenging Perceptions and Performing the Unfamiliar

In ethnomusicology, the acquisition of familiarity with the musical culture under study through learning to perform has long been viewed as a crucial research methodology (see also Chapter 7). The issues surrounding the development of such performance skills have received considerable discussion, especially in the wake of Mantle Hood's classic 1960 essay 'The Challenge of Bi-Musicality'. While for the casual listener perceptual divergences in the reception of culturally unfamiliar music may pass almost unnoticed, for those attempting to engage more closely with, participate in or reproduce such musics, mismatches in perception may throw up considerable challenges. Many examples of these perceptual problems and the sense of disorientation they sometimes provoke are found in the ethnomusicological literature. For example, David Locke describes his own disorientation when first trying to achieve polyrhythmic synchrony with other parts in (African) Ewe drum music, and expresses sympathy for his US students who, when trying to do the same, 'would break down in tears of frustration and self-doubt' (Locke 2004, p. 169). Similarly, Ali Jihad Racy reflects on his US students' problems when playing microtonal intervals in Arabic music, and the tendency for beginning students to play 'neutral' intervals too flat:

> Are they in fact hearing the neutral interval as a flat note, as being "minor-ish"? Or do they hear it properly but cannot play it as such because some acquired control mechanism is holding them back? Are their minds "correcting" the intervals by fitting them to a familiar intonational paradigm? This might be like learning French or any foreign language and realizing you are speaking with an accent. You know what the native speakers sound like and probably can imitate them if you really try. But somehow your self-consciousness is standing in the way of your loosening up. You need to get rid of the inhibitions. I wonder if taking a course in acting would help my students play microtones better. (Racy 2004, pp. 160–61)

Long ago, Mantle Hood (1960, p. 56) also observed that the 'tendency of Westerners to "correct" unfamiliar intervals, usually without being aware of doing so, can itself be corrected only by repeated exposure to listening and by singing'. To 'correct' unfamiliar acoustic patterns in this way highlights a general cognitive propensity, noted above, to assimilate sounds (and other information from our environment) into familiar categories or schemata, which are sometimes very durable. Gerhard Kubik even goes so far as to suggest that:

> Hearing habits in the field of the recognition of note systems, once learned, are apparently irreversible. Someone who has "grown up" into a given note system from childhood onward perceives the note material of a foreign musical culture always in relation to his own patterns. Musicians brought up in Western musical culture, for example, hear the equiheptatonic scales of Africa instinctively in relation to the known diatonic scale, and equipentatonic systems as C, D, E, G, A. Even a major effort of will cannot change this perception process. (Kubik 1979, p. 242)

Nonetheless, it would also seem that with adequate repetition and immersion – and possibly even acting classes! – certain of these perceptual challenges can be, at least partially, overcome. This has obvious parallels with learning foreign languages after childhood, where varying degrees of competency can be achieved according to context, motivation and aptitude. However, even when immersed exclusively in a second language environment over many decades, it is quite rare for mother-tongue characteristics to be entirely undetectable in a person's second language.[6]

---

[6]   It should also be noted that long-term immersion in a second language environment may also modify aspects of speech in the person's mother tongue.

According to Hood (1960, p. 56), when learning foreign musics, students without previous musical training probably have an advantage over those who have received advanced proficiency in their home musical culture. Aubert (2007, pp. 75–6) even presents such prior expertise as a 'handicap'. As perception and cognition are apparently influenced even by passive exposure to music from childhood (Drake and El Heni 2003), presumably music specialisation into adulthood leads to increased culture-specific adaptation and refinement. Might this mean a corresponding reduction in the cognitive potential to process certain unfamiliar parameters of foreign music cultures? Or, alternatively, might prior investment in specialised skills, which inevitably incorporate (competing) musical values, serve to reduce the student's flexibility in approach or attitude? Students without prior specialisation may have less existing commitment to particular musical values and thus less to lose. In short, the challenges to music learning for those students who have already acquired specialist music skills may be both perceptual *and* motivational, albeit sometimes unconsciously.

In this context it is notable that, for Hood, learning to perform musics from other cultures was ultimately a means to develop musicality (Hood 1960, p. 59). John Baily (2008, p. 118) has suggested replacing Hood's concept of 'bi-musicality' with the notion of 'intermusability', which helpfully escapes the implication of two musics learned from childhood (as suggested by 'bilingual') and shifts the focus to abilities. What is also significant about this term is the way that, rather than implying two 'mutually unintelligible' musical languages, it suggests the possibility of overlaps and continuities in skills, competencies and other forms of musical experience. Might this mean a more flexible attitude to intercultural engagement and participation, and a move away from discourses of cultural exclusivity? As Laurent Aubert observes:

> one meets performers all over the world fully qualified in Western classical music, jazz or rock. But the reciprocal, if it is true, is only rarely accepted. As soon as it is about *flamenco* or Gypsy violin, African percussion or Indian *sitar*, someone will retort that "they have it in their blood", that "it is necessary to have been born in that place" to play like that. What is going on here? Is there a universal music accessible to all, and yet other music that is "intransmissible" because it emerges from innate predispositions? (Aubert 2007, p. 7)

Histories of Misperception, Transmission and Creativity

Modes of musical behaviour and perception are often reproduced and maintained by a consensus that divergence from the culturally familiar is 'wrong'. Music learners everywhere experience social pressure to conform to such accepted norms, where transmission is sometimes underscored by formal pedagogy. From this perspective, the perception of Bolivian Easter songs by Europeans as anacrustic (mentioned above) is 'incorrect' or a form of 'misperception'. However, listeners may be entirely unaware that their perceptions do not match those of the people producing the music.[7] Nor does this necessarily diminish their appreciation of the music or restrict its reproduction or transmission, albeit in a perceptually reinterpreted form (for example, Manuel 1988, p. 21). Thus, from a broader perspective we might question the degree to which a given mode of cultural hearing can always be said to be 'correct' or 'incorrect'. I would like to take two aspects into consideration here: power dynamics and cultural mixing or adaptation.

Firstly, as Agawu (2003, p. 164) reminds us, 'categories of perception are made, not given. Every act of perception carries implicit baggage from a history of habits of constructing the world'. Thus, we need to consider how notions of power and authority impact on hierarchies of perception and misperception during transmission processes. Whose perceptions dominate in given contexts? As noted above, musicians reproducing elements from unfamiliar musical traditions will often automatically 'correct' sound patterns to fit their existing acculturated schemata or categories. With continued transmission such reinterpreted perceptions can become accepted, potentially even leading to the marginalisation or demise of particular ways of hearing. The wide dissemination of the equally tempered diatonic scale is an obvious example, its transmission in part underwritten by music technology. However, if attempts to reproduce unfamiliar intervals are guided by a figure of authority (such as Ali Jihad Racy in the case of Arabic music), alternative types of perceptual sensibility and values may develop. As regards rhythm, I have encountered several published transcriptions of Andean music where the transcriber has misperceived the metric organisation of the music. Performances based on such notations, doubly imbued with authority through both notation and publication, would doubtless be based on the transcriber's misperceptions. Similarly, I taught an Andean song to British students for many years before realising that I had misperceived it as anacrustic.

Secondly, as regards cultural mixing, it is inevitable that the diverse borrowings, appropriations, parodies, assimilations and other forms of interaction, participation and transmission that have characterised music history, have involved direct or indirect exchanges between music makers with differing perceptual dispositions. Shifting our focus away from unequal power relations between participants, I wish to speculate on the 'creative' dimension of many intercultural encounters and reinterpretations, what I will term *creative misperception*. When musical elements are acquired from another cultural source they tend to be reconfigured (to a greater or lesser extent) according to the performers' perceptions, discrepancies sometimes passing unnoticed or emerging as foci for creative elaboration. Thus, in the context of hip-hop samples, Michael Krimper (2010) observes how, when

---

[7]   Arguably, a lack of awareness of such perceptual discrepancies is increased by recording technology's schizophonic separation of sounds from their sources. However, from another perspective, the potentially limitless repetition enabled by such technology (especially when digital) may be seen to fix and give authority to particular performances in turn limiting divergences (or perceptual mismatches) in the transmission process.

engaging with unfamiliar sound materials, his 'most creative ideas emerge in ... uncanny moments of misperception'. Also, perceptual discrepancies between musicians during intercultural performances, whilst sometimes potentially disturbing interactions, may in other contexts fuel creativity and provoke imaginative responses. Charles Keil's notion of 'participatory discrepancies' seems relevant here (although his discussion of the concept is not directly related to perceptual dispositions): 'Music, to be personally involving and socially valuable, must be "out of time" and "out of tune"' (Keil 1987, p. 275). Both the 'culture of correction' and the cult of the composer, which often accompany performance from music notation and formal music pedagogy, may sometimes be seen to work against the 'participatory discrepancies' that result from interacting subjectivities and perceptions in performance. For example, in collaborative compositional processes (found, for example, among many rock bands) individuals may find the musical ideas they offer misperceived by others. Such musicians often realise that relinquishing authority over an idea – and their perception of it – is often necessary for it to be taken up by the rest of the group and incorporated into a collaborative composition. Similarly, when I demonstrated a melody to a group of rural Andean panpipe players with whom I was playing during a feast, they misperceived various aspects. Rather than correct them, which would have seemed pedantic, I relinquished authority and a new and very satisfactory piece emerged.[8]

The diverse mismatches in perception that accompanied the ways African, European and Indigenous peoples borrowed, appropriated or assimilated one another's mutually unfamiliar musical resources during the colonisation of the Americas provide a broader historical perspective on such processes. For example, the famous Argentine musicologist, Carlos Vega, observed that, when indigenous South American singers applied their own 'rhythmic system' to the performance of a Spanish song, a 'hybrid' would have resulted (Vega 1941, pp. 495–6). Just such a hybrid song genre is to be found in the *mestizo huayño* of the Bolivian Andes, which Ellen Leichtman (1987, p. 170) has described as 'a blending of Indian [Indigenous] and European rhythmic understanding'. Although musical fusions and adaptations have been widely discussed in the literature – especially when accompanying racial and cultural *mestizaje* or 'mixing' (Moehn 2008; Wade 2000) – much less attention has been dedicated to the mismatches of perceptions that surely shaped such processes. Albeit largely speculative and difficult to demonstrate empirically, *creative misperception* has surely been fundamental to music history and has much potential for future research.

---

[8]   In this context it is notable that such musicians refer to the creation or acquisition of new music with the Quechua verb *q'iwiy* – 'to twist, stir or remix' (Stobart 2006, p. 244).

Conclusion

This chapter has taken the form of a reflection on the notion of musical unfamiliarity, and its repercussions, from an intercultural perspective. A chapter of this length can only hope to scratch the surface of this hugely complex theme, but I hope nonetheless to have opened up lines for future enquiry and debate. Although specifically invited to write from the perspective of 'non-Western' music and ethnomusicology, I have argued that a more holistic music scholarship – which better reflects global reality in the 2010s – would benefit from dispensing with the polarising term 'non-Western'. I have also observed that, although ethnomusicology is widely identified with the study of the unfamiliar, its methods are in many respects more characterised by the acquisition of familiarity, achieved through close ethnographic observation and participation. My approach has been both to explore the unfamiliar in music – its challenges and opportunities – and to question what social, cultural and cognitive processes lead music to be perceived as unfamiliar.

An important point to emerge is unfamiliarity's internal tension; its potential to provoke, on the one hand, anxiety, insecurity or fear and, on the other, a sense of excitement, novelty or revelation. It is also clear, and perhaps a musical universal, that for music making to be engaging and socially meaningful, the unfamiliar must be adequately framed within, or balanced by, the familiar. Cultural isolation or separation undoubtedly lies behind much experience of musical unfamiliarity, as especially evident before the rise of global communications systems. Indeed, it was questioned how much the ubiquity of global sounds resulting from modern technologies impacts upon musical familiarity and intercultural understanding. However, I also suggested that a range of social processes, including class, race, gender, ideology, politics and economics, or the presupposition of difference, often contribute to rendering certain musics unfamiliar. In addition it was shown that unfamiliar musical sounds may used to political ends, but conversely may be perceived as liberating precisely because they do not entangle listeners in politics, ideologies or the concerns of daily life. In line with my resistance to the idea of the 'non-Western', I have questioned the assumption that cultural distance can somehow be neatly mapped onto musical or aesthetic unfamiliarity. Thus, many British people may find aspects of certain musical phenomena from close to home in the UK less familiar than specific genres performed, for example, by indigenous people in the Amazon.

The second half of the chapter focused more specifically on music perception and began by noting that an infant's ability to identify a diversity of sounds diminishes with cognitive development; in other words, with such development (or cultural specialisation), the perception of unfamiliarity increases. It is evident, from a range of studies of intercultural music perception, that subjects' musical operations and processes (such as pitch, rhythmic or structural recognition, or memory) are more restricted in culturally unfamiliar musical environments than familiar ones. I also considered, in some detail, a study of intercultural rhythmic perception focusing on a specific Andean song genre. The tendency of British and Andean (Quechua speaking) subjects to perceive, and tap along to, the pulse of these songs in fundamentally different ways was related to a range of factors, including the stress patterns of the Quechua language. This led to discussion of the perceptual challenges that ethnomusicologists face when learning to perform culturally unfamiliar musics. Students sometimes experience a sense of confusion, owing to mismatches of perception, or may 'correct' unfamiliar intervals or other patterns to fit their existing perceptual dispositions. Finally, I briefly speculated on how mismatches of perception might have impacted on intercultural musical borrowings, especially in the context of broader histories of colonisation and migration. Power relations undoubtedly shaped whose perceptions were accepted in given contexts, but I also make a case for what I call *creative misperception*, the idea that perceptual mismatches in intercultural contexts may have provided, and continue to provide, an important creative focus for the development of musical styles and genres.

Overall, I hope this chapter contributes to a more nuanced approach to understandings of the familiar and unfamiliar in music, and, in particular, that it encourages an attitude of openness and enquiry to the challenges and creative opportunities of musical unfamiliarity. Rather than approaching the musically unfamiliar as 'Other', and presupposing difference, it should perhaps provoke us to listen more carefully, to engage and seek understanding. Whilst this may involve perceptual disorientations and challenging aesthetic terrains, if such engagements are treated as encounters between equals, they are likely to be deeply enriching – both for ourselves and for our relations with many other people around the world ...

Acknowledgements

I am grateful to Hettie Malcomson for her insightful comments on an earlier draft of this essay, to the editors for their helpful suggestions, and to Tina K. Ramnarine, who first brought the problems surrounding the term 'non-Western' to my attention.

References

Agawu, K. (2003). *Representing African Music: Postcolonial Notes, Queries, Positions*. New York: Routledge.

Attali, J. (1985). *Noise: The Political Economy of Music*. Trans. B. Massumi. Manchester: Manchester University Press.

Aubert, L. (2007). *The Music of the Other: New Challenges for Ethnomusicology in a Global Age*. Trans. C. Ribeiro. Aldershot: Ashgate.

Austern, L. and Naroditskaya, I (eds) (2006). *Music of the Sirens*. Bloomington, IN: Indiana University Press.

Ayari, M. and McAdams, S. (2003). Aural Analysis of Arabic Improvised Instrumental Music (Taqsīm). *Music Perception* 21/2: 159–216.

Baily, J. (2008). Ethnomusicology, Intermusability, and Performance Practice. In H. Stobart (ed.), *The New (Ethno)musicologies* (pp. 117–34). Lanham, MD: Scarecrow.

Blacking, J. (1987). *A Commonsense View of all Music: Reflections on Percy Grainger's Contribution to Ethnomusicology and Music Education.* Cambridge: Cambridge University Press.

Bonnett, A. (2004). *The Idea of the West: Culture, Politics and History.* Basingstoke: Palgrave Macmillan.

Booth, G. and Kuhn, T.L. (1990). Economic and Transmission Factors as Essential Elements in the Definition of Folk, Art, and Pop Music. *Musical Quarterly* 74/3: 411–38.

Bregman, A. (1990). *Auditory Scene Analysis: The Perceptual Organization of Sound.* Cambridge, MA: MIT Press.

Brown, M. (2003). *Who Owns Native Culture?* Cambridge, MA: Harvard University Press.

Byron, R. (1995). The Ethnomusicology of John Blacking. In J. Blacking (ed.), *Music, Culture and Experience: Selected Papers of John Blacking* (pp. 1–28). Chicago, IL: University of Chicago Press.

Chion, M. ([1983] 2009). *Guide to Sound Objects: Pierre Schaeffer and Musical Research.* Trans. J. Dack, http://juancantizzani.wordpress.com (accessed 1 July 2011).

Clayton, M. (2008). Toward an Ethnomusicology of Sound Experience. In H. Stobart (ed.), *The New (Ethno)musicologies* (pp. 135–69). Lanham, MD: Scarecrow.

Clifford, J. and Marcus, G. (1986). *Writing Culture: The Poetics and Politics of Ethnography.* Berkeley, CA: University of California Press.

Comaroff, J. and Comaroff, J. (2009). *Ethnicity, Inc.* Chicago, IL: University of Chicago Press.

Conklin, B. (1997). Body Paint, Feathers and VCRs: Aestetics and Authenticity in Amazonian Activism. *American Ethnologist* 24/4: 711–37.

Cook, N. (1992). *Music, Imagination and Culture.* Oxford: Oxford University Press.

Cottrell, S. (2004). *Professional Music Making in London: Ethnography and Experience.* Aldershot: Ashgate.

Cross, I. (2008). Musicality and the Human Capacity for Culture. *Musicae Scientiae* 12/1: 147–67.

Cusick, S. (2006). Music as Torture/Music as Weapon. *TRANS: Transcultural Music Review* 10 (article 11), http://www.sibetrans.com/trans/a152/music-as-torture-music-as-weapon (accessed 12 August 2011).

Darwin, C. (1871). *The Descent of Man, and Selection in Relation to Sex*, Volume 2. London: John Murray.

Desmond, A. and Moore, J. (1992). *Darwin.* Harmondsworth: Penguin.

Diamond, B., Cronk, M.S. and von Rosen, F. (1994). *Visions of Sound: Musical Instruments of First Nations Communities in Northeastern America*. Chicago, IL: Chicago University Press.

Drake, C. and El Heni, J.B. (2003). Synchronising with Music: Intercultural Differences. *Annals of the New York Academy of Sciences* 999: 429–37.

Dueck, B. (2005). *Festival of Nations: First Nations and Métis Music in Public Performance*. Unpublished PhD dissertation, University of Chicago.

Eerola, T., Himberg, T., Toiviainen, P. and Louhivuori, J. (2006). Perceived Complexity of Western and African Folk Melodies by Western and African Listeners. *Psychology of Music* 34/3: 337–71.

Garcilaso de la Vega (1609). *Commentarios Reales que Traten del Origen de los Incas (Primera Parte)*. Lisbon: Pedro Crasbeeck.

Hood, M. (1960). The Challenge of 'Bi-Musicality'. *Ethnomusicology* 4/2: 55–9.

Kassabian, A. (2004). Would You Like Some World Music with your Latte? Starbucks, Putumayo, and Distributed Tourism. *Twentieth-Century Music* 2/1: 209–23.

Keesing, R. (1989). Exotic Readings of Cultural Texts. *Current Anthropology* 30/4: 459–79.

Keil, C. (1987). Participatory Discrepancies and the Power of Music. *Cultural Anthropology* 2/3: 275–83.

Keil, C. and Feld, S. (1994). *Music Grooves*. Chicago, IL: University of Chicago Press.

Klarer, M. (1993). Woman and Arcadia: The Impact of Ancient Utopian Thought on the Early Image of America. *Journal of American Studies* 27/1: 1–17.

Krimper, M. (2010). Imaginative Misperception: A Study of Hip-Hop Vocal Samplings. *Hydra Magazine*, 5 February, http://www.hydramag.com/ (accessed 20 September 2011).

Kubik, G. (1979). Pattern Perception and Recognition in African Music. In J. Blacking and J.W. Kaeliinohomoku (eds), *The Performing Arts: Music and Dance* (pp. 221–50). The Hague: Mouton.

Leichtman, E. (1987). *The Bolivian Huayño: A Study in Musical Understanding*. PhD dissertation, Brown University.

Lima, N. and Costa, L. (2000). The Experimental Music of Hermeto Paschoal (1981–93): A Musical System in the Making. *British Journal of Ethnomusicology* 9/1: 119–42.

Locke, D. (1996). Africa/Ewe, Mande, Dagbamba, Shona, BaAka. In J.T. Titon (ed.), *Worlds of Music* (pp. 71–143). New York: Schirmer.

Locke, D. (2004). The African Ensemble in America: Contradictions and Possibilities. In T. Solís (ed.), *Performing Ethnomusicology: Teaching and Representation in World Music Ensembles* (pp. 168–88). Berkeley, CA: University of California Press.

Locke, R. (2009). *Musical Exoticism: Images and Reflections*. Cambridge: Cambridge University Press.

Lodge, M. (2009). Music Historiography in New Zealand. In Z. Blazekovic and B. Dobbs Mackenzie (eds), *Music's Intellectual History* (pp. 625–52). New York: Répertoire International de Littérature Musical.

Lynch, M. and Eilers, R. (1991). Children's Perception of Native and Nonnative Musical Scales. *Music Perception* 9/1: 121–31.

Manuel, P. (1988). *Popular Musics of the Non-Western World.* Oxford: Oxford University Press.

Maye, J. with Werker, J. and Gerken, L. (2002). Infant Sensitivity to Distributional Information can Affect Phonetic Discrimination. *Cognition* 82/3: 101–11.

Moehn, F. (2008). Music, Mixing and Modernity in Rio de Janiero. *Ethnomusicology Forum* 17/2: 165–202.

Morris, N. (1986). Canto Porque es Necesario Cantar: The New Song Movement in Chile, 1973–1983. *Latin American Research Review* 21/2: 117–36.

Morrison, S. with Demorest, S. and Stambaugh, L. (2008). Enculturation Effects in Music Cognition: The Role of Age and Music Complexity. *Journal of Research in Music Education* 56/2: 118–29.

Nan, Y. with Knösche, T. and Friederici, A. (2006). The Perception of Musical Phrase Structure: A Cross-cultural ERP Study. *Brain Research* 1094: 179–91.

Nettl, B. (1995). *Heartland Excursions: Ethnomusicological Reflections on Schools of Music.* Urbana, IL: University of Illinois Press.

Nettl, B. (2005). *The Study of Ethnomusicology: Thirty-one Issues and Concepts.* Urbana, IL: University of Illinois Press.

Neuman, D. (1990). *The Life of Music in North India.* Chicago, IL: University of Chicago Press.

Olsen, D. (1980). Selk'nam Chants of Tierra del Fuego, Argentina (Record Review). *Latin American Music Review* 1/2: 289–91.

Racy, A.J. (2004). 'Can't Help but Speak, Can't Help but Play': Dual Discourse in Arab Music Pedagogy (Interview with Ali Jihad Racy by S. Marcus and T. Solís). In T. Solís (ed.), *Performing Ethnomusicology: Teaching and Representation in World Music Ensembles* (pp. 155–67). Berkeley, CA: University of California Press.

Roseman, M. (1991). *Healing Sounds from the Malaysian Rainforest: Temiar Music and Medicine.* Berkeley, CA: University of California Press.

Said, E. (1978). *Orientalism.* New York: Routledge and Kegan Paul.

Schafer, R.M. ([1977] 1994). *The Soundscape: Our Sonic Environment and the Tuning of the World.* Rochester, VT: Destiny Books.

Slevc, R. and Patel, A. (2011). Meaning in Music and Language: Three Key Differences. (Comment on 'Towards a Neural Basis of Processing Musical Semantics' by S. Koelsch.) *Physics of Life Reviews* 8: 110–11.

Spivak, G. (1987). *In Other Worlds: Essays in Cultural Politics.* New York: Taylor and Francis.

Stevens, C. (2004). Cross-cultural Studies of Musical Pitch and Time. *Acoustic Science and Technology* 25/6: 433–8.

Stobart, H. (1996a). *Tara* and *Q'iwa*: Worlds of Sound and Meaning. In M. Baumann (ed.), *Cosmología y Música en los Andes* (pp. 67–82). Frankfurt am Main and Madrid: Vervuert and Iberoamericana.

Stobart, H. (1996b). The Llama's Flute: Musical Misunderstandings in the Andes. *Early Music* 24/3: 470–82.

Stobart, H. (2006). *Music and the Poetics of Production in the Bolivian Andes.* Aldershot: Ashgate.

Stobart, H. (ed.) (2008). *The New (Ethno)musicologies.* Lanham, MD: Scarecrow.

Stobart, H. (2011). Constructing Community on the Digital Home Studio: Carnival, Creativity and Indigenous Music Video Production in the Bolivian Andes. *Popular Music* 30/2: 209–26.

Stobart, H. and Cross, I. (2000). The Andean Anacrusis? Rhythmic Structure and Perception in the Easter Songs of Northern Potosí, Bolivia. *British Journal of Ethnomusicology* 9/2: 63–92.

Stokes, M. (ed.) (1994). *Ethnicity, Identity and Music.* Oxford: Berg.

Tan, S.E. (2008). Returning To and From 'Innocence': Taiwan Aboriginal Recordings. *Journal of American Folklore* 121/480: 222–35.

Taylor, T. (2007). *Beyond Exoticism: Western Music and the World.* Durham, NC: Duke University Press.

Temperley, D. (2000). Meter and Grouping in African Music: A View from Music Theory. *Ethnomusicology* 44/1: 65–96.

Titon, J.T. (1996). Preface. In J.T. Titon (ed.), *Worlds of Music* (pp. xxi–xxiv). New York: Schirmer.

Trouillot, M.-R. (2003). *Global Transformations: Anthropology and the Modern World.* New York: Palgrave Macmillan.

Vega, C. (1941). *Música Popular Argentina – Canciones y Danzas Criollos: Fraseología.* Buenos Aires: Institute de literatura Argentina/Universidad de Buenos Aires.

Wade, P. (2000). *Music, Race, and Nation: Música Tropical in Colombia.* Chicago, IL: University of Chicago Press.

Woodfield, I. (1995). *English Musicians in the Age of Exploration.* Hillsdale, NJ: Pendragon Press.

Yung, B. (1987). Interdependency of Music: A Case Study of the Chinese Seven-string Zither. *Journal of the American Musicological Society* 40/1: 82–91.

Chapter 6

Well, What Do You Know? Or, What Do You Know Well? Familiarity as a Structural Force in Crumb's *Black Angels*

Jonathan James Hargreaves

George Crumb's quotation of Schubert in the middle of *Black Angels* (1970) means different things to different people. To some, it might merely be sad music, to others, ('Big-C') classical music, or perhaps Renaissance music, to others still, it might be the second movement of Schubert's String Quartet in D Minor (D810), the 'Death and the Maiden' quartet (1824). It is, of course, all of these things, although the level of specificity at which listeners recognise it can have important implications for its meaning: what listeners know is a function of how well they know it.[1] *Black Angels* makes a particularly interesting case-study for investigating the role played by familiarity in musical communication. The fundamental question, therefore, is how the information posited by its composer can be understood to be meaningful by listeners, although in order to situate this discussion more generally, the relationship between music and familiarity should be discussed.

Familiarity plays a fundamentally important role in what and how music means. In an absolutist sense, it is evoked merely when something is called 'music'; in more everyday terms, it is manifest in the perception of almost all musical structures. Motivic development, thematic transformation, harmonic, serial and tonal structures, indeed tonalities themselves (both as individual key centres and as holistic bases for pitch organisation) are all essentially forms of repetition, and

---

[1]   The issue of how *a priori* knowledge relates to and influences musical perception is – it goes without saying – extremely complex, and so it cannot be discussed in full here. For canonic in-depth discussions, see Clarke's *Ways of Listening* (2005) and Cook's *Music, Imagination and Culture* (1990). Isolating the essential character of different types of listening leads authors to draw different lines of distinction between listeners according to their analytical slant, and thus to use different terminologies. Presently, the issue is how well listeners know the repertory under discussion, and in simplified terms, this might be expressed as an opposition between 'expert' and 'novice'; for a discussion of the differences between such hearings, see Hargreaves, Hargreaves and North (2012, pp. 164–6). Accordingly, I use the term 'qualification' later in this chapter in its literal sense: readers should not infer that this refers to any sort of educational certification!

thus, the implications for relationships between sounds – the musical meanings – that they bring about are dependent on listeners' recognition and familiarity. Even where repetition is avoided or presented as insignificant, such as in Stockhausen's moment form or Cage's chance music, the influence of familiarity is important: in these cases, conscious steps are taken by the composers to prevent it from becoming a haven for meaning. At the other end of the spectrum, composers can play upon the (un)familiar using quotations, potentially ascribing meaning to their music in doing so.[2] Importantly, as a product of memory, familiarity itself resides in listeners; just as composers cannot dictate what their music means, so they cannot write familiarity into it. It is worth considering those parts of the communication process that are the provinces of the composer, and of the listener, with particular reference to quotation, before focussing on *Black Angels*, with its many allusions to other music.

Quotation for Composers and Listeners

Quotation is an extreme type of musical reference; other examples include sampling, exotic instrumentation, arrangement, pastiche and stylistic allusion. In each case, composers (similarly, DJs, sound designers and other exploiters of music) import specific or general aspects of other music to the present context. Paradoxically therefore, a musical reference is both an expression of otherness and also a form of repetition: *alien* styles and/or pieces are *reiterated*. Importantly, this does not empower composers to bring about familiarity, although by using references they can assert *familial* relations with foreign contexts. By framing the relationships between their chosen materials in different ways – as growing out of each other, for example, or alternatively perhaps, as interrupting one another – they can play on both the extent and the nature of belonging: 'just *how* similar/ different *is* the quotation from its original context?' or, for less direct references, 'just *how is* the quotation similar/different to its original context?'; further, 'what does that difference imply?' As a platform for sameness (repetition) and difference (importation), reference provides composers with a powerful expressive tool, and should their assertions be understood – should listeners recognise a reference as

---

[2]   Straus discusses the reworking of materials from the past – which he calls 'misreadings' – by twentieth-century composers in his *Remaking the Past* (1990) in which, fittingly, he applies pre-existing ideas from Bloom's *The Anxiety of Influence: A Theory of Poetry* to music. In turn, in his *Writing through Music: Essays on Music, Culture and Politics* (Pasler 2008, pp. 52–3), Pasler refers to Straus's and Bloom's notion of 'creative misreadings'. Straus's analyses are performed at the level of pitch content, moving 'upwards' later in the book, to formal concerns. In the present chapter, discussion is concerned less with notes-on-the-page details, and is more akin to that found in Metzer's *Quotation and Cultural Meaning in Twentieth-Century Music* (2003).

belonging in a context other than the present – then familiarity has played a part in the communication that has occurred.[3]

Shifting the focus from composers to audiences, responses to musical references can vary enormously from one listener to the next; inevitably, the context into which information is received – memory – differs for each individual, because of previous experiences, habits and so on. Metaphorically, everyone has their own musical geography – a conceptual map of their musical memory – and in accordance with listening tastes and habits, particular areas might be more familiar than others: a person might be aware of the exact placement of two similar pieces that they know, whereas highly contrasted, unfamiliar styles would be situated approximately, although far apart. Thus, while on the one hand, listeners *construct* mental representations of music, on the other, they also *situate* them, and familiarity plays a particularly important role in the latter process. Situation occurs in all musical listening, and quotation can invest it with added layers of meaning: interplay between the imported material and its new context might imply that either one of the two is predominant, growing, disintegrating and so on; or, for example, the difference between sudden and gradual recognition of quoted materials can be loaded with implications. Such relationships might be appreciated without any knowledge of the extra-musical associations of the referent. However, those connotations can be played upon to great effect.

The underlying means by which music connotes extra-musical phenomena is simple enough: music presents listeners with a 'bundle of generic attributes in search of an object' (Cook 1998, p. 23), and should one of its strands be attached to (mnemonically associated with) a particular object (meaning), the connotation is manifest. Reference, therefore, is a doubly potent means of communication, as it furnishes composers (and listeners) with two sets of attributes – those belonging to the quoted material and those of the new context – which can be played off against each other. Thus, in referring to other music, composers do not merely present an opposition between old and new. Rather, they present a *context* for meaning, in the form of a three-part network, as shown in Figure 6.1.

Moving around the diagram, firstly, the connection between the sounds heard in the present context and their intra-musical referents is automatic, provided that the listener hears given parts of the musical texture as having originated elsewhere. This need not involve exact, 'name-that-tune'-type knowledge of the referent in question; merely recognising a different style can have implications. Secondly, the link between the sounds heard and their extra-musical connotations works,

---

[3]   This leads to the question of what constitutes the requisite conditions for such recognition to occur. Taking a semiotic approach to this matter, Tagg (2004) evokes Nattiez in explaining that the difficulty of defining the minimal unit of musical meaning – the 'museme' – across different contexts lies in its simultaneously syntagmatic and paradigmatic nature. Indeed, that musical perception involves an ongoing negotiation between that which is fundamental and that which is ornamental suggests that to attempt to identify an absolute level at which musemes function would be to work against the material under discussion.

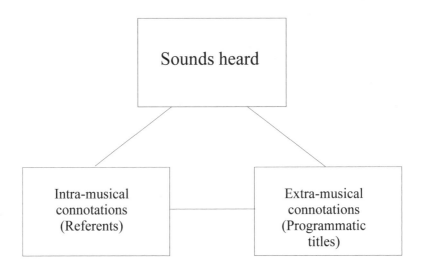

Figure 6.1   Three-part network of associations in a musical reference

essentially, by analogy and association: the qualities of the sounds and the ways in which they change over time might relate somehow to those of the phenomenon that is connoted, or as explained above, that bundle of musical attributes may become associated with a particular (or non-particular) object. The third and final relationship in the network, between the intra- and extra-musical, is particularly interesting, since it constitutes a 'meta-' level of situation – the extra-musical connotations of the original referent can be imported and, being placed in a new context, commented upon.

A classic example can be found in film music: the image of an empty playground swing has potentially tragic, worrying connotations when accompanied by a music box. Amongst the attributes commonly attached to that sound, innocence is particularly salient. By extension, therefore, to perceive the *lack* of a child on the swing implies an unexplained loss of innocence. Thus, the musical referent imparts an added layer of meaning to its filmic context, and of course, vice versa. This chapter is about concert music rather than films, however, and that begs the question of how this meta-situation process is used compositionally: where, unlike in films, extra-musical implications are not specified, how can they be commented upon?

In music, listeners unconsciously and sometimes consciously use mnemonic labels to rationalise their experience. There is, of course, not usually a kind of expert internal 'naming ceremony' in which different parts of the texture are labelled 'motif x', 'second subject', 'false recapitulation' and so on. Rather, the process is very simple: just by recognising something as different, one labels it –

this might apply to hearing a chorus rather than a verse, 'the weird bits' rather than 'the good bits', a C-major triad rather than a guitar, or even something as vague as 'this' rather than 'that'. Thus, merely by hearing something as a quotation, or as belonging to another style, listeners apply labels. Of course, this can be played upon in musical reference: the name given to the foreign material – the basis on which it is 'other' than the context into which it is imported – has a great influence on how it is heard.

By playing on situation, musical reference opens up listeners' musical geographies as a context for meaning. As materials are presented in simple 'same/different' oppositions, their connotations and those of their contexts can be loaded with implication. For example, such relationships might be inflected using carefully chosen names. Familiarity is the basic resource for this, one that is exploited deliberately, carefully and skilfully in *Black Angels*.

Crumb's *Black Angels*

The American composer George Crumb (b. 1929) has enjoyed a highly decorated career, winning the Pullitzer Prize for Composition (1968) and a Grammy Award (2001) alongside numerous honorary degrees. His pieces often use reference: music by and inspired by earlier composers sit alongside styles from elsewhere in the world, and sometimes pieces by Crumb himself, providing a rich referential resource. Clearly, familiarity is fundamental here, enabling listeners to make perceptual connections between references and referents, and to appreciate extra-musical connotations, in the hope that meaningful networks might develop. Indeed, referential networks other than the musical are often evoked; Crumb frequently uses rhetoric and numerology in his compositions. Therefore, in considering how this music communicates, it is important to try to separate out these symbolic layers of implication from those that are grounded in musical experience; it is easy to read meanings into it that may not be there.

The score of *Black Angels* (1970) is marked 'George Crumb (in tempore belli)' [in time of war], and perhaps as a direct result, the work is often associated with the Vietnam War. Indeed, Arnold cites it as the only well-known piece of art-music 'to have emerged from or about the war', although he later explains that in fact, it is 'an outgrowth of the Vietnam age and not particularly about it' (Arnold 1991, p. 326). As far as Vietnam is concerned, the ideas that are evoked offer listeners a means to reflect upon the tragic nature of military conflict: as explained below, this is an *anti*-war piece, and suitably, Crumb seeks to communicate by presenting oppositions. The paradoxical title *Black Angels* encompasses both elements of a binary opposition – black and white – and accordingly, the materials in the work belong to contrasting styles. Listeners are invited to find meaning in stark contrasts, as titles are used to ally them with extra-musical phenomena.

Calling Names ...

> The image of the "Black Angel" was a conventional device used by early
> painters to symbolize the fallen angel. (The Official George Crumb Home Page,
> Compositions)

Since Crumb's inspiration in naming this piece lies in the symbolism of a bygone
age, it comes as little surprise that the same is true of the music he has written.
Tritones abound, representing the *diabolus in musica*, although the present purpose
is to consider how the choice of title serves to inform the listening experience,
rather than be sidetracked into discussing the history of such rhetoric.

In full, the work is called *Black Angels (Thirteen Images from the Dark
Land) – [Images 1]*. In addition to the specific visual-arts motif of the main
title, the notion 'image' is in both subtitles. Clearly, rather than providing
information about musical coherence, the composer's intention in naming this
piece is to conjure up pictures in the minds of listeners, by manipulating the
perception of the musically familiar. These subtitles do inform the listening
experience, however, offering information as to how the work might be situated.
The dramatic (even theatrical) tension implied by the subtitle *Thirteen Images
from the Dark Land*, in combination with an elaborate programme note showing
the name and numerological concept (as opposed to percept) underpinning each
one, suggests affiliation to the nineteenth-century tradition of programme music.
Indeed, at points there are references to Saint-Saëns and other music of that era.
The present focus, however, is on the programmatic element in *Black Angels* as
a whole.

... And Programming Numbers

> *Black Angels* was conceived as a parable on our troubled contemporary world.
> The work portrays a voyage of the soul. The three stages of this voyage are
> Departure (fall from grace), Absence (spiritual annihilation) and Return
> (redemption). (Crumb cited in Yaple 1990, n.p.)

The purpose here is definitely not to speculate as to how the soul might travel from
one stage to the next, nor even to reflect on how that voyage is portrayed musically;
rather, it is to consider the impact of this narrative on the listening experience.
Crumb's comment implies that there is a single thread that runs through the work,
and in turn this raises the issue of unity in so discontinuous and diverse a piece. In
the programme note at the front of the score, this conceptual 'voyage' is broken
down further into 13 smaller stages (see Figure 6.2).

PROGRAM

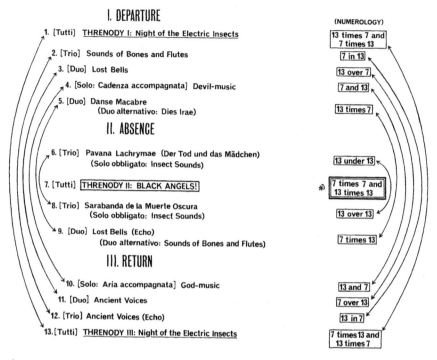

I. DEPARTURE

1. [Tutti] THRENODY I: Night of the Electric Insects
2. [Trio] Sounds of Bones and Flutes
3. [Duo] Lost Bells
4. [Solo: Cadenza accompagnata] Devil-music
5. [Duo] Danse Macabre
    (Duo alternativo: Dies Irae)

II. ABSENCE

6. [Trio] Pavana Lachrymae (Der Tod und das Mädchen)
    (Solo obbligato: Insect Sounds)
7. [Tutti] THRENODY II: BLACK ANGELS!
8. [Trio] Sarabanda de la Muerte Oscura
    (Solo obbligato: Insect Sounds)
9. [Duo] Lost Bells (Echo)
    (Duo alternativo: Sounds of Bones and Flutes)

III. RETURN

10. [Solo: Aria accompagnata] God-music
11. [Duo] Ancient Voices
12. [Trio] Ancient Voices (Echo)
13. [Tutti] THRENODY III: Night of the Electric Insects

(NUMEROLOGY)

13 times 7 and 7 times 13
7 in 13
13 over 7
7 and 13
13 times 7
13 under 13
7 times 7 and 13 times 13
13 over 13
7 times 13
13 and 7
7 over 13
13 in 7
7 times 13 and 13 times 7

* This central motto is also the numerological basis of the entire work

Figure 6.2   Programme for *Black Angels*

As Crumb's arrows indicate, the work is organised according to two patterns of numbers, implying a symmetrical arch made up of paired movements. The arrows on the left, for example, connect sections involving the same number of players, indicating the following undulating pattern, although in performance, most sections involve the entire quartet:

4–3–2–1–2–3–4–3–2–1–2–3–4

The arrows on the right-hand side illustrate an elaborate numerological scheme based on the 'magic' numbers 7 and 13, with their connotations of good and bad luck. Numerology underlies compositional decisions at various structural levels, although its effect on the listening experience is limited. As Crumb says, 'The numerological symbolism of *Black Angels*, while perhaps not immediately perceptible to the ear, is nonetheless quite faithfully reflected in the musical structure' (Crumb cited in Yaple 1990, n.p.).

Consider all of this from the point of view of communication: the composer actually displays *two* numerical patterns in support of his formal scheme, although neither of them serves to unify *Black Angels* as a perceptual context for musical meaning. There is an arc-figure that recurs, although the stylistic contrasts between the 13 sub-sections prevent any perceptible motivic structure; this motif 'functions' as a picture, merely being displayed on a wall, rather than as a brick within it. Moreover, the establishment and manipulation of perceptual patterns – the basis for most musical forms – seem to have played, at most, an unimportant role here. Inevitably, this begs the question of how – if at all – meaning can be perceived. Continuing the analogy of the visual, *Black Angels* functions like an art gallery: its meaning(s) lie in the things that it points to, which are suggested by the names of the tableaux in the programme. By exhibiting a number of styles, this music brings about meaningful intra- and extra-musical connotations; familiar musical models are evoked and enhanced.

Visual Enhancement

Enhancement is vital to communication in *Black Angels*; many of the things in this work are presented as exaggerated versions of their conventional selves. Amplification transforms the ensemble into an 'Electric String Quartet', for example, and in addition to playing their standard instruments, the performers are required to shout and whisper in a number of languages, to whistle, and to play various percussion instruments: tam tams and pitched crystal glasses are bowed and struck in the course of a performance. A number of extended string techniques are also used: players use glass rods to stop their strings, as well as tapping them with thimbles; noisy 'pedal tones' are generated by exerting extreme pressure with the bow. In addition to their timbral implications, these enhancements have a significant impact upon visual presentation.

It is clear from a single glance at the stage that this is not a conventional ensemble: the familiar, traditional string quartet is 'exploded', and this plays a role in communication with the audience. Even before the quartet enters the room, the presence of added resources – the tam tams in particular, with their dramatic, theatrical connotations (of being struck) – provides visual cues, intensifying the expectations of the audience. The players' movements in and around the performance space lend the music a visual layer of significance. For example, when the two violinists and the violist bow crystal glasses in the tenth subsection of the form ('God Music'), their graceful, unified motion reinforces the musically serene atmosphere. Thus, in a very real sense, part of the significance of this piece is derived from the 'thirteen images' it presents.

Black Angels plays upon the theatrical element of the concert format in a way that more conventional pieces do not, to the extent that, at times, it might even be argued that watching is as integral as hearing. Undoubtedly, the clearest example of this occurs at the two points where the players are required to bow on the 'wrong side' of the left hand, whilst holding their instruments 'in the manner of viols' (these are: 6. 'Pavana Lachrymae', and 8. 'Sarabanda de la Muerte Oscura'). The resultant viol–consort sonority, and pose, emphasise the importance of the relationship between sound and image in this work: this sounds and looks like Renaissance music. This first occurs in the Schubert quotation, and the implications of this moment are discussed in more detail below.

Absence and Echoes: 'Pavana Lachrymae'

'Pavana Lachrymae' opens the second large-scale section of the form, 'Absence', and it is marked to be played – or perhaps heard – as if it were 'A fragile echo of ancient music'. Rhetorically, therefore, it is doubly absent: 'ancient music' denotes one level of remove from the present, and the intimation that this is an 'echo' implies a second. As discussed above, the 'Pavana' even 'looks' antiquated on stage, and its viol-like timbre is far thinner and less pronounced – less acoustically 'present' – than that produced ordinarily by (amplified) modern string instruments. This demonstrates the link between the twentieth-century electric string quartet and its ancestor, the Renaissance viol consort, in as immediate and effective a manner as possible. Crumb's intention goes beyond merely pointing out the history of string ensembles, however. Given the stylistic implications and extra-musical connotations of the material – Schubert's 'Death and the Maiden' quartet – this is an allusion to a tragic loss of innocence. By implication, present-day instruments themselves are fallen (*Black*) *Angels*.

'Pavana Lachrymae' is a peculiarly powerful title with strong connotations, not only of a particular Renaissance dance form but also, more specifically, of a tradition associated with the sixteenth- and seventeenth-century English lutenist and composer John Dowland. 'His most popular piece was the pavan *Lachrimae*. He turned it into the song *Flow my tears*, and it occurs in about 100 manuscripts and prints in many different solo and ensemble arrangements' (Holman and O'Dette 2001). Indeed, there is a long historical line of pieces that take the *Lachrimae* theme as source material. Even today, it continues to be rearranged and quoted, appearing in ever more diverse contexts,[4] ensuring and sustaining the ubiquity of its opening 'falling tear' motif. Having appeared in print so many times under *Lachrimae*, that melodic figure is virtually synonymous with the word.

Simply by using this title, Crumb evokes in his (appropriately qualified) listeners a specific expectation of hearing this familiar melodic line. However, having aroused their anticipation, he denies its satisfaction, quoting Schubert instead. Arguably, the peculiar power of this moment arises from the interaction

---

[4] Indeed, the influence of that song is still felt in the twenty-first century. Recently, the pop star Sting was inspired to record the song (Sting 2006) after having heard a version by the tenor John Potter accompanied by jazz musician John Surman on bass clarinet, lutenist Stephen Stubbs, Baroque violinist Maya Homberger, and jazz musician and avant garde composer Barry Guy (Potter 1999).

between listeners' knowledge of the referents and this abstract 'familiarity-dynamic': the subtitle, 'Pavana Lachrymae' evokes so specific an expectation in qualified listeners that a reaction stemming directly from familiarity with the Dowland is inevitable on recognising its absence.[5]

This begs questions about potential experiences of an unqualified listener, who is familiar neither with *Lachrimae* nor with the 'Death and the Maiden' quartet. Arguably, the implication of special significance can be perceived independent of musical knowledge at this point, because of the pose adopted by the players. Further, the quietness and minor-tonality suggest a historically generated mood of sadness. Where knowledge comes into play is in the recognition that the visual and timbral allusion to 'Early music' connotes lost innocence. (Indeed, owing in no small part to the tradition that grew around the original *Lachrimae* pavan, viol-consorts themselves have strong intimations of melancholy.) Arguably, both Dowland's melody and Schubert's chorale are iconic of melancholy and tragedy, having provided reference points for many musicians in the centuries since they were written; they are pillars of what we in the West know to be 'Sad Music'. Thus, even those who are not familiar with this material are bound to, and might expect to, hear it as melancholy and anachronistic, or, put more poetically, to hear the 'echo of a fragile, ancient music'.

As well as imposing a different name on the Schubert, Crumb made a few alterations: it is scored for trio, for example, and major tonality, with its positive rhetorical implications purposefully removed. However, these changes are relatively slight in comparison with the points of similarity between the quotation and the original. As well as the musical material itself, Crumb imports aspects of the context in which it is presented: in its nineteenth- and twentieth-century incarnation it opens the second movement, and in each instance it has similar dynamic function, marking a subdued point of contrast. This may (literally) come as no surprise in *Black Angels*, owing to the holistic changes from one formal segment to the next, although even in this stylistically volatile context, the introduction of a regular pulse and conventional tonal harmony is a significant change. Similarly, in Schubert's quartet, this material forms a gesture at the level of style: its regular four-bar phrases and repeated long–short–short rhythms are purposefully made to sound markedly distinct within their early-Romantic context. Like Crumb, the nineteenth-century Austrian composer makes reference to a stylistic *other*; the Renaissance pavan.

---

[5]   It is impossible to speak for other audience members of course, and their experiences may well have been different, although I remember my own first experience of this moment distinctly: I was qualified to expect the Dowland, but had never knowingly heard the Schubert quartet. So, my confidence in predicting the falling tear motif was juxtaposed with deflation at its absence, and puzzled bewilderment (hence this chapter) as I heard quiet, tragic Renaissance-type music played.

Example 6.1   Echo effect in Crumb's *Black Angels*, 'Pavana Lachrymae'

Source: *Black Angels (Thirteen Images from the Dark Land)* by George Crumb. Published by C.F. Peters Corporation, New York (EP 66304). © Copyright by C.F. Peters Corporation, New York. All Rights Reserved. Reproduced by permission of Peters Edition Limited, London.

Crumb's 'Pavana Lachrymae' is a quotation of a quotation. As well constituting a reference to the pavan genre, in its own turn, this excerpt of the 'Death and the Maiden' quartet consists of material taken from Schubert's own 1817 setting of Matthias Claudius's poem, 'Der Tod und Das Mädchen'. Thus, by hinting at Dowland specifically, and imitating Renaissance music more generally, Crumb quotes Schubert-quoting-himself-imitating-Renaissance music – clearly, the situation (process) is complex, offering a multi-levelled perspective on events. Example 6.1 sets out the network of connections implied by 'Pavana Lachrymae'.

At every stage in this multi-levelled reference, the material is dependent for its identity on other, older music. Even at the immediate, theatrical level, there is a connection between Crumb's electric quartet and the Renaissance viol consort; even visibly, this quotation can be understood as a doubly absent 'fragile echo of ancient music'. This is a wonderfully poetic concept, implying notions of pieces contained within pieces; the intangibility of the present; that the past itself harks back to other, more distant pasts. Cutting through those indulgent mists of time, this quotation offers significance from a number of perspectives, as it can be situated in relation to a number of pieces. The careful choice of referent opens up intermediate levels of meaning for more highly qualified – more 'familiar' – listeners.

In this instance, to infuse the listening experience with such historical depth is clearly the composer's intention, although, arguably, most pieces have the potential to be heard in this way, given the ubiquity of familiarity in listening, and the perpetual cross-fertilisation of musical styles. This, indeed, helped make the Dowland and the Schubert iconic in the first place. Effectively, this quotation is a comment on the traditions associated with its referents, which give rise to a distinctive means of significance. For a fuller understanding, it should be considered how that distinction is brought about – how communication in 'Pavana Lachrymae' functions in the context of the work as a whole.

Blocking the Network: Formal Structure

Black Angels is a work concerned with contrasts, and Crumb uses the programme to assert musical and extra-musical oppositions. At one level, section-names and contents offer listeners networks of expectation in which to situate the sounds they hear: by naming extra-musical connotations ('God Music', 'Devil Music'), Crumb invites listeners to perceive music in particular ways; conversely, by naming music ('Danse Macabre', 'Pavana Lachrymae'), he invites listeners to expect and appreciate particular extra-musical associations. (Note that every formal subsection is named after a sound event of one kind or another; see Figure 6.2.) However, at a higher level, the composer contends that the work is a symmetrical arch. Notionally therefore, that shape is the underlying platform for communication, housing the section-to-section relationships and providing a context in which they can signify. Inevitably however, the many drastic changes of style challenge this formal unity: in the context of such stylistic diversity it is highly questionable that the global arch shown in the programme can actually be perceived. This might seem to lessen the value of analysing form – why consider music that may not even be there?

Communication involves the composer's conception, as well as the listeners' perception, and it is worth considering Crumb's large-scale thinking here. Numerology provides a global framework for the piece, although as the composer himself admits, it is imperceptible for listeners. One of the more novel ways in which the programme itself – the 'voyage of the soul' – is articulated is using physical materials. The emphasis on 'wooden' sounds (*col legno*, 'knuckles on wood') in the first movement, and 'glass' sounds (bowed crystal glasses, the use of glass rods) in the third, corresponds with programmatic associations: opacity and transparency equate to black and white, which, in turn, are analogous to good and evil, or 'Departure' (fall from grace) and 'Return' (redemption). Ultimately, however, this is merely another layer of rhetoric; it does not give rise to an inherently perceptible musical shape. Paradoxically, there are diverse ways of considering – as opposed, perhaps, to hearing – the work as a whole.

An arch is a gradual and continual increase in a single intensity, followed by an equivalent decrease, and as a piece built out of references, the musical significance of *Black Angels* lies at the level of style (see Figure 6.3). Unsurprisingly, therefore, that large-scale shape is realised in the distribution of historical references. The first section of the form, 'Threnody I', refers to Crumb's contemporary, Penderecki, as discussed below. Thus, the musico-historical arch starts near the present (of 1970), and becomes more absent as Saint-Saëns and Dowland are passed, before a reference to Messiaen's Second World War-associated *Quartet for the End of Time* initiates the return to Crumb's own 'tempore belli'. This overarching pattern gives rise to trends in the behaviour of material from one block to the next: harmony is increasingly less chromatic towards the middle of the form, and rhythms are increasingly measured and regular. These cannot be perceived as aural threads of continuity,

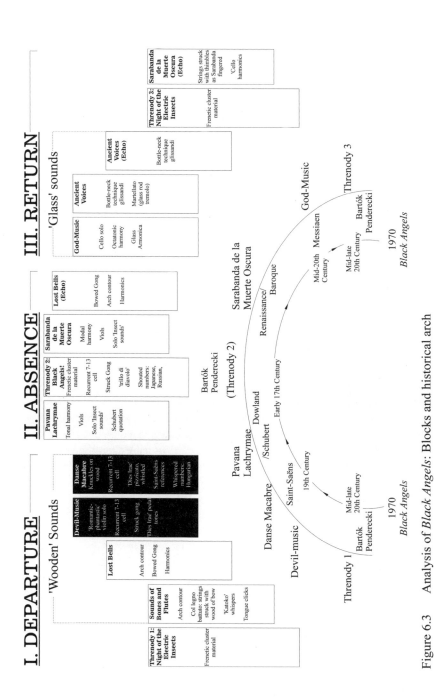

Figure 6.3   Analysis of *Black Angels*: Blocks and historical arch

however; because the style changes with the passing sub-sections, there is no musical constant against which to perceive change. Instead, the ongoing theme of presence and absence is used as a notional extra-musical constant, manifest in the historic-stylistic arch.

The highest-level assertion in George Crumb's *Black Angels* is implicit: that familiarity itself is an aspect of musical material, and therefore that it might act as a compositional resource. Accordingly, the piece is conceived and organised within that 'parameter', and presented as an arch: the referents are older and arguably better-known in the middle of the piece than at either end ('Threnody II: Black Angels!' merely serves as a structural pillar). As explained above, the echo effect in 'Pavana Lachrymae' marks the apex of this contour: famous material is heard, which might be familiar to many people in a range of ways. It is logical, therefore, to balance this out by considering communication at one of the ends of the arch, 'Threnody I: Night of the Electric Insects'.

Threnody I: Night of the Electric Insects

'Threnody I' refers most directly to Penderecki's famous *Threnody – To the Victims of Hiroshima* (1960), and this sub-section has associations with other sounds and other music(s). In addition to this titular reference, familiarity is invoked by other means, such as imitation and stylistic allusion. Indeed, nowadays, Crumb's 'Threnody' has historical connotations of its own, having familial relations with other music(s) of the 1960s and 1970s. All of this can be explained by taking the words of the title in reverse order. Example 6.2 shows the opening bars.

The opening section of *Black Angels* mimics the sound of a swarm of insects. The silence preceding a concert performance is interrupted by this explosive sonority, and that initial contrast is indicative of the extreme nature of the material: it is frantically fast, alternately very loud and very quiet indeed, and scored in the highest register for each member of the ensemble. The swarm-like quality lies in its internal characteristics and its behaviour through time. As an example, take the way in which the bipolar changes of dynamic are emphasised; any reader who has suffered the misfortune of a persistent in-ear mosquito will appreciate the effect – and referential significance – of the bracketed *crescendi* in the violins!

The simple answer to the question of how the proposed insects are 'electric' is that the instruments are amplified. The resulting timbres 'are intended to produce a highly surrealistic effect' (The Official George Crumb Home Page, Compositions), and despite denying a direct link between the genesis of this work and the Vietnam war, Arnold explains that such ugly sounds are typical of 'War Music and the American Composer during the Vietnam Era' (Arnold 1991): 'To deal with the Vietnam War in music, composers intensified the horror in its portrayal of war. Composers used electronic machine gun fire, sounds of actual bombs exploding, indeterminant sections with singers shouting and screaming, and other realistic sounds' (p. 322). Where Crumb speaks of the timbrally surreal, Arnold writes of the sonically real; essentially, both are referring to the extension of the timbral

Example 6.2   Crumb's *Black Angels*, opening

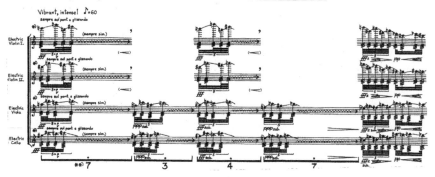

Source: *Black Angels (Thirteen Images from the Dark Land)* by George Crumb. Published by C.F. Peters Corporation, New York (EP 66304). © Copyright by C.F. Peters Corporation, New York. All Rights Reserved. Reproduced by permission of Peters Edition Limited, London.

palette beyond concert-hall norms. Notably, however, the electrification (and other extensions) of the string quartet in *Black Angels* creates familial relations with the psychedelic rock music at the time, which undoubtedly contributed to the broad appeal of this work.

The use of the word 'night' in the subtitle is a reference to the 'night music' idiom of the Hungarian composer, Bela Bartók. Certain generic features of 'Threnody I' conform to that style, albeit in a subversive fashion. The name of this particular brand of Bartók derives from the title of the fourth movement of his *Out of Doors* Suite for Piano (1926), 'Night Music', although the idiom also appears in many of his other works (notably the string quartets). Besides the presence of birdsong transcriptions (especially of the nightingale), and other symbolic motifs, it is characterised by 'quiet, tiny, delicate textures … set against a background of "stylized noise" suggested by a semitone cluster' (Harley 1995, p. 331; also see pp. 331–3). 'Crumb acknowledges his debt to … Bartók' (Borroff 1986, p. 552; also see Moevs 1976, pp. 294–6; Steinitz 1978, p. 845), and it is clear that, although the texture in 'Threnody I: Night of the Electric Insects' is explosive rather than delicate, the use of that title is an invitation to listeners to situate this section in relation to this Bartókian 'nocturnal' style.

Regarding the associations evoked by the first word of the title, a threnody is 'a poem, or its musical setting, expressing a strong feeling of grief for the dead; the term has much the same meaning as "lament"' (Boyd 2001, p. 433). Arguably, the word has stronger connotations of Penderecki than of this definition for most listeners of twentieth-century music. The Polish composer's synonymity with this type of piece stands as testament to the associative power of titles. Indeed, the *Threnody – to the Victims of Hiroshima* was only so-named 'following the comments of a conductor who felt the original title – a simple description of the work's duration – would not attract sufficient interest' (Lee 1975, p. 585). Certainly, that reference to nuclear war lends his work an unforgettably intense emotional atmosphere. In recognition of the sheer scale of the devastation caused by the bomb, the sonorities in the Penderecki are more suggestive of pained screams of anguish than a wailing lament.

The connection between *Black Angels* and Penderecki's *Threnody* is easy to appreciate for qualified listeners: both works start with loud, high-register chromatic clusters, with an emphasis on timbre over pitch: 'stylized noise'. Although this is not an exact quotation, the title invites listeners to situate this material in direct relation to the earlier, more famous *Threnody*. Crumb's output as a whole has a great deal in common with that of Penderecki. For the American, whose concern for sonority is demonstrated by his 'explosion' of the string quartet, the textural music of the Pole holds obvious attractions; similarly, unconventional notation characterises both their outputs; finally, of particular relevance here, there is a strong Bartókian influence on the music of both composers.

Overall, 'Threnody I: Night of the Electric Insects' communicates in a similar way to 'Pavana Lachrymae', with referential implications on a number of levels. By making particular titular allusions, Crumb hopes to evoke extra-musical connotations shared by his piece and others to which it refers. Specifically, 'threnody' is intimately associated with war and death; 'night' and 'insects' with blackness and gothic, macabre imagery; and 'electric' might be construed as alluding to a potentially dangerous force. Most simply, the sounds in this section mimic the behaviour of insects, although they also have more subtle, stylistic implications relating to Bartók and Penderecki. Discussion of each of the words of the title has led away from musical aspects and towards extra-musical ones, illustrating the extra levels of 'distance' from the musical surface that come into play for more highly qualified listeners to *Black Angels*.

Conclusion: Pause and Reflect

The basic communicative mechanism asserted in *Black Angels* is commonly accepted and experienced in everyday life: the bundles of attributes presented by music can become – indeed, cannot escape being – associated with extra-musical meanings as its familiarity grows and spreads. Thus, by carefully naming his work, the composer suggests 13 different connotations to his listeners. However, there is an added dimension to the compositional contention: if this ordered placement of historical styles can be perceived as a global arch, then familiarity can be used as a structural force, in this case fuelling a *hierarchy* of musical meaning.

There are those who would call Mr. Crumb's music theatrical, and not in the flattering sense. They would add that the attempt is not to move and satisfy but to amaze … as if the composer were … bent on the thrill of the moment but devoid of any deeper sense of order. Believers might argue that there are other kinds of order than the arrangements of notes into linear and vertical groups. They would propose that the knockout sonority can be every bit as profound as Mozartian sonata form. (Holland 2002)

As shown above, Crumb uses the theatrical aspect of his music as a means of communication, in this case presenting the string quartet as a visual convention onto which alternative meanings and implications can be projected. Holland refers to theatricality of a rather different nature, however; the implication is that, underneath all its layers of rhetoric and references, *Black Angels* is nothing more than an act – this collage of quotations merely results in a cosmetic imitation of music, rather than the real thing, whatever that may be.

The comparison between Mozart and Crumb could refer to many aspects of their musics, although by setting the profundity of sonata form in opposition with 'the thrill of the moment', Holland exposes a particular contrast. The issue here lies in the idea that music is somehow 'deeper' when order is perceived over extended time periods (thus, inevitably, the order in question has influence over more music; by extension, it is more meaningful). Where, in Mozart, indubitably perceptible musical aspects form patterns of coincidence (tonality, rhythmic patterns, phrase boundaries and others), such that notions like 'Sonata Form' can unify large-scale shapes, *Black Angels* has no such single musical thread. Continual changes of style mean that the 'Thirteen Images' of the subtitle each exhibit different means of organising pitch and rhythm, for example. Instead, the bases for unity are extra-musical: a programmatic narrative provides listeners with a means of negotiating significance, and thus 'navigating' through the piece, notionally reinforced by numerology. To listeners, the number scheme is purely rhetorical and, by and large, imperceptible. The programme, however, is supported by references of different kinds, and, as it were, 'strengths': it is dependent on familiarity; on what individual listeners know, and how well they know it.

Other than its resonance with psychedelic rock, an idea that has not been discussed so far is that of familiarity with *Black Angels* itself, and its implications for meaning. Arguably, the very idea of networks based on musical references is already familiar to listeners; some would say it has become over-familiar. In the spirit of meta-quotations, the following captures this point.

Richard Steinitz has remarked that: "The direct quotations from Bach, Schubert or Chopin, heard through Crumb's strange and unworldly soundscape, acquire an amazing aura of distance both cultural and temporal. Surrealist museum exhibits, their mummified beauty seems utterly remote, like a childhood memory of warm, homely security." It worked. But it worked only so long as tonal and atonal were strictly separate categories, implying a similarly strict separation

> between ancient and modern. Once composers began re-establishing tonality, and working again in traditional genres ... such quotations as Crumb's lost the shock, the inadmissibility, on which their sentimental effect depended. (Griffiths 1995, p. 162 citing Steinitz 1975, p. 11)

Both Griffiths's and Holland's Crumb-sceptics' criticisms concern the lack of musical order and function. To echo Holland, believers might argue that, in a piece about disorder and dysfunction, understanding (how or whether) Crumb's music 'works' is not the issue. This is the place for opining neither as to how and whether *Black Angels* is effective, nor as to the reasons behind that efficacy; the purpose has been to recognise and discuss the mechanisms facilitating communication. This is a piece of extremes, and fittingly, its means of communication could scarcely be more referential: rather than use tonality or other means to generate temporal momentum and musical unity, Crumb uses a sophisticated network of references to divert attention elsewhere. *Black Angels* does not offer an experience of musical cause and effect; instead it invites listeners to pause and reflect. However sceptical one may be, this composition cannot fail to play on listeners' situation process; its stylistic and other oppositions present a carefully considered context for meaning, rather than a standalone, arbitrary assertion of extra-musical association. Musical references are set in a fixed order – a network of familial relationships – that might function as a basis for signification, depending on the individual listener.

Inevitably, there are limits as to what can be surmised from discussion of two out of 13 formal segments. By implication, however, as extremes of various oppositional continua within which the work is conceived (quiet/loud; slow/fast; consistent/changeable), 'Pavana Lachrymae' and 'Threnody I' constitute those limits. Thus, comparison of the processes by which they evoke extra-musical connotations – rather than of their musical contents – might provide higher perspectives on the role played by familiarity in the work as a whole. 'Threnody I: Night of the Electric Insects' guides listeners more directly than 'Pavana Lachrymae': whereas the former title provides numerous referential frames, the latter denotes a particular one, only for the sounds heard to be drawn specifically from another. Of course, one cannot speak for the composer, and in turn, he cannot speak for listeners, but the implication seems to be that 'Pavana Lachrymae' is better known than 'Threnody I: Night of the Electric Insects'; because Schubert (along with Dowland) is notionally more familiar than Penderecki, Bartók, and so on, listeners do not need to be told how to understand it. Thus, their appreciation can be taken for granted, and manipulated.

Familiarity plays a role in all musical communication, whether directly or indirectly, as explained above. To use it as the basis for a global arch of styles prompts questions, however; the idea that a piece might be 'more familiar' in the middle raises the issue of whether familiarity has a quantitative or qualitative influence within musical listening. Is the familiarity arising from the 'Pavana' of a different *kind* to that of the 'Threnody', or simply more of the same? The answer to this lies beyond the scope of this chapter, and of this kind of analysis: on the qualitative side, there is no accounting for individuals' musical tastes, listening habits and musical geographies; on the quantitative side of course, there is no standard unit for measuring familiarity. Ultimately, because of the way Crumb presents his materials, the two interact: those qualities of music that listeners know (the 'what'), and the extent to which ('how well') they know it, are placed in an ongoing relationship that is fundamental to communication in *Black Angels*, and the role played by familiarity within it.

One final issue arises from the notion of familiarity as a structural force. An important difference between 'Threnody' and 'Pavana' is that they play on expectation in different ways. The long title of the opening casts the referential net wide; listeners are in a sense given licence to expect far more referents than in the sixth segment. Attention is focussed on familiarity in this way to different extents throughout the form, and this manipulates their perception of time. Thus, *Black Angels* highlights one final opposition that might be inverted. Focussed as it is on a single piece, this chapter has examined the role played by familiarity within musical communication. Conversely, consideration of the referential meanings in this programmatic work also illuminates the role played by this music in familiarity.

References

Arnold, B. (1991). War Music and the American Composer during the Vietnam Era. *The Musical Quarterly* 75/3: 316–35.

Bloom, H. (1973). *The Anxiety of Influence: A Theory of Poetry*. Oxford: Oxford University Press.

Borroff, E. (1986). George Crumb. In H. Wiley Hitchock and S. Sadie (eds), *The New Grove Dictionary of American Music* (pp. 551–3). London: Macmillan.

Boyd, M. (2001). Threnody. In H. Wiley Hitchock and S. Sadie (eds), *The New Grove Dictionary of Music and Musicians* (2nd edn). London: Macmillan, www.grovemusic.com (accessed 13 July 2008).

Clarke, E.F. (2005). *Ways of Listening: An Ecological Approach to the Perception of Musical Meaning*. Oxford: Oxford University Press.

Cook, N. (1990). *Music, Imagination and Culture*. Oxford: Oxford University Press.

Cook, N. (1998). *Analysing Musical Multimedia*. Oxford: Oxford University Press.

Griffiths, P. (1995). *Modern Music and After: Directions Since 1945*. Oxford: Oxford University Press.

Hargreaves, D.J., Hargreaves, J.J. and North, A.C. (2012). Imagination and Creativity in Music Listening. In D.J. Hargreaves, D. Miell and R. MacDonald (eds), *Musical Imaginations: Multidisciplinary Perspectives on Creativity, Performance, and Perception*. Oxford: Oxford University Press.

Harley, M.A. (1995). 'Natura Naturans, Natura Naturata' and Bartók's Nature Music Idiom. *Studia Musicologica Academiae Scientiarum Hungaricae* T. 36,

Fasc. 3/4. *Proceedings of the International Bartók Colloquium* (pp. 329–49). Szombathely, 3–5 July (JSTOR; accessed 3 March 2008).

Holland, B. (2002). Bowed Gongs in Lurid Red Light Set a Festival's Tone. *New York Times*, 5 October, http://query.nytimes.com/gst/fullpage.html?res= 9B09E5D8173BF936A3575C1A9649C8B63 (accessed 13 September 2007).

Holman, P. and O'Dette, P. (2001). Dowland, John, §2: Works, (ii) Lute music. In S. Sadie and J. Tyrrell (eds), *The New Grove Dictionary of Music and Musicians* (2nd edn). London: Macmillan, www.grovemusic.com (accessed 13 July 2008).

Lee, D.A. (1975). Penderecki and Crumb at Wichita State. *The Musical Quarterly* 61/4: 584–8.

Metzer, D. (2003). *Quotation and Cultural Meaning in Twentieth-Century Music*. Cambridge: Cambridge University Press.

Moevs, R. (1976). Review: *George Crumb: Music for a Summer Evening (Makrokosmos III). For Two Amplified Pianos and Percussion* by Gilbert Kalish; James Freeman; Raymond DesRoches; Richard Fitz; George Crumb. *The Musical Quarterly* 62/2: 293–302.

Pasler, J. (2008). *Writing Through Music: Essays on Music, Culture and Politics*. Oxford: Oxford University Press.

Potter, J. (1999). *Flow my Tears*. On *Dowland: In Darkness Let Me Dwell*. John Potter (tenor), John Surman (soprano saxophone) and Stephen Stubbs (bass clarinet), Stephen Stubbs (lute), Maya Homberger (Baroque violin), Barry Guy (double bass). CD ECM 1697.

Steinitz, R. (1975). The Music of George Crumb. *Contact* 11: 14–22.

Steinitz, R. (1978). George Crumb. *The Musical Times* 119/1628: 844–7.

Sting (2006). *Flow My Tears*. On *Songs from the Labyrinth*. CD Deutsche Gramophone 170 3139.

Straus, J.N. (1990). *Remaking the Past: Musical Modernism and the Influence of the Tonal Tradition*. Cambridge, MA: Harvard University Press.

Tagg, P. (2004). Musical Meanings, Classical and Popular, www.tagg.org (accessed 21 March 2012).

The Official George Crumb Home Page, Compositions, www.georgecrumb.net/ comp/black-p.html (accessed 13 July 2008).

Yaple, C. (1990). Black Angels. Liner Notes to *Black Angels*. Kronos Quartet. CD Elektra Nonesuch 79242-2.

Chapter 7

Familiarity, Information and Musicological Efficiency

Vanessa Hawes

Musicology, as a discipline, is based neither wholly on scientific methods, nor on methods from the humanities. The conceptual space of the discipline is complex, incorporating many sub-disciplines, each with their own structures and methods. This complexity makes it difficult for a graduate student in musicology to become familiar with and to navigate sub-disciplines that can vary widely from those which the student is familiar with from their previous experience or supervisor's influence. However, becoming familiar with approaches from other sub-disciplines affords new perspectives on the student's specialist topic and opportunities for interdisciplinary work.

All musicological outputs communicate some new knowledge or idea to a reader or listener. In the sciences, tools derived from an information-theoretical approach have been proposed for the exploration and evaluation of theories, concentrating on issues of communication and efficiency (Brillouin 1962, 1964; Goertzel 2004). Similarly, approaches from performance analysis, critical theory and systematic musicology suggest an approach to the philosophy of musicology that addresses the fragmentation of musicology's discourse using concepts of effective communication between scholars and musicians (Cook 1999; Korsyn 2003; Parncutt 2007b). Building on these points of view from very different sub-disciplines within musicology, approaches to assessing the efficiency of musicology, independent from its specific topic or eventual purpose, are discussed.

Three questions are proposed here as the starting point for an examination of musicology with regards to efficient communication: questions about scope, predictive accuracy and simplicity. The interaction between perceived 'familiar' and 'unfamiliar' elements are then used in examples from computational analysis (Temperley 2009), ethnomusicology (Browner 2000) and popular music studies (Björnberg 1994) to explore the three questions and to examine issues of communication within those three example outputs. This chapter proposes the use of this kind of thinking for a systematic approach to the study of musicology as a discipline, with the purpose of providing people with tools to help them negotiate their way through musicology and developing a systematic approach to a philosophy of musicology.

Themes in Musicology

In recent years, many musicology conferences in the UK have tended to be organised around broad themes.[1] These types of conferences emphasise music's relationship to other spheres of thought and the importance of relationships and connections within musicology, by their use of the word 'and'. Themed conferences provide the opportunity for graduate students, early career researchers and more established researchers to present a paper on a topic they are familiar with (their specialty – most likely the PhD topic for a graduate student), and place it into the context of the theme of the conference using approaches and methodologies suggested by that theme.[2] The mix of familiar and unfamiliar in relation to topic and approach provides the foundation for shared understanding, leading to efficient communication. The scholar can communicate new ideas about a topic (unfamiliar to the audience) – most often the 'music' of the left-hand side of the conference title – using the frames provided by the concepts suggested by the word(s) on the right-hand side of the 'and' (familiar to many in the audience).

Themed musicology conferences create new identifiable research threads, which, depending on their relationship to musicological sub-disciplines and extra-musicological discourse, either form a category related in a specific way but which also belongs to other disciplinary-framed categories, or go on to form a new sub-discipline (or sub-sub-discipline).[3] The themed conference allows musicologists to make far-reaching connections through efficient communication in a conference context, and it provides the conditions necessary for the creation of a new musicological category.

Placing the familiar in the context of the unfamiliar (but not completely unknown), or vice versa, to provide the conditions necessary for efficient communication, echoes Richard Parncutt's model for interdisciplinary work in which communication between two researchers in different areas is built upon each knowing a little of the other's specialty but not having to become an absolute

---

[1]   For example: Music and Consciousness (University of Sheffield, July 2006); Music and Gesture (University of East Anglia, August 2003; Royal Northern College of Music, Manchester, July 2006); Music and Language as Cognitive Systems (University of Cambridge, May 2007); Music and Evolutionary Thought (University of Durham, June 2007); Music and the Melodramatic Aesthetic (University of Nottingham, September 2008); Music and Morality (University of London, June 2009); Music and Numbers (Canterbury Christ Church University, May 2010); Music and Familiarity (University of Hull, October 2009).

[2]   Examples can be found in the programmes for any of these conferences, and, indeed, this chapter is an example. 'Music and Familiarity' provides a broad and inclusive topic area, with the opportunity for each contributor to define the relationship between their research and the theme in a variety of ways.

[3]   For example, the success of 'Music and Gesture' as a new and fruitful category of musicological work.

expert in the other's area (Parncutt 2007b, pp. 21–2). It also echoes an information-theoretical approach to communication, in which effective communication is achieved only because of the foundation of shared contextual knowledge between the sender of a message and its receiver (Kraehenbuehl 1983; Potter et al. 2007, p. 300). In the discipline of musicology, another informational concept – feedback (Weiner 1961)[4] – is important for the continual evaluation and re-evaluation of the discipline: what can be assumed to be familiar to musicologists, and what will be unfamiliar in the current climate.

Music is often characterised as a form of communication drawing on the interaction of familiar and unfamiliar elements. Novel musical material can be presented within a familiar (established) form or structure, or familiar musical material (obeying a familiar prescriptive musical theory, such as common practice harmony) can be presented in a novel form or structure: 'large forms develop through the generating power of contrasts' (Schoenberg 1967, p. 178); 'always the same but not in the same way' (Schenker 1979, p. 6). Conferences without a specific theme might be organised around the theories and methods of a specific sub-discipline. In this context, the frames within which new material is presented are already formalised and institutionalised, creating a situation where more assumptions can be made about audience familiarity and, therefore, information might be communicated more efficiently because less time is needed to explain the sub-discipline-specific norms within which the work is contextualised because those connections have already been rendered familiar within the sub-disciplinary framework.

However, there are several more specific scholarly frameworks within which musicological work at a sub-disciplinary conference might exist. Some of these will be reflected in the titles of conference sessions (for example, phenomenological analysis; structural analysis; computational analysis), and the delegate's paper must conform or refer to the models and methods of these areas. Other scholarly frameworks are less explicit, but still affect delegates' communication with their audience, for example, the constantly shifting framework of fashions within the sub-discipline. If students are not familiar with this discourse, their work will be judged unfavourably by an audience who is. This kind of framework has many influences: which approaches have famous musicologists endorsed; which have been criticised or judged to have come to a dead end; which approaches are attracting the most grant money?

To study the relationship between specific musicological work and scholarly frameworks, the work itself must be defined in some way. Categorising musicological projects, like categorising music, is problematic. Projects within a sub-discipline can work at any level of musical or musicological detail, addressing

---

[4]   Weiner's *Cybernetics*, first published in 1948, emphasises the importance of feedback in information-based models of communication, where Shannon and Weaver's (1964) *Mathematical Theory of Communication* concentrated on modelling communication as a one-way process from sender to receiver.

points to do with a single note, up to a century's worth of music. Topics can be any 'thing' (and what, come to that, *is* music?), a person, a written note, a heard sequence of notes, and so on. Contexts can then be defined in a number of ways and use or omit any available data. When starting with the content of musicological work, it is very difficult to place a musicological project into any category other than one containing only itself.

All musicology is interdisciplinary, incorporating methods, theories and strategies from several established fields of thought, and crossing disciplinary boundaries to get nearer to its topic. Within musicology there is inter-sub-disciplinary work, in that topics and methods simultaneously exist in different established disciplinary frameworks. For example, Schenkerian theory might be seen as a self-contained method for the analysis of masterpieces of certain periods of Western art music. However, within musicology as a whole it has been applied not only to Western art music, but also to other groups of musical works in other disciplinary areas, such as popular music (see Gallardo 2000), or in ethnomusicological projects (Larson 2010). Mirroring these possibilities, one piece of music, oeuvre or genre can be explored using methods that evolved within different sub-disciplines.

Graduate students in musicology in the UK have a huge amount of choice concerning their thesis topic. They do not have to work within one particular musicological sub-discipline and can often engage with a number of different approaches and methodologies in a single project.[5] They could, for example, decide that their historical/theoretical research might benefit from the inclusion of some computational, systematic or empirical research, and they would find themselves faced with significant boundaries within musicology: those between humanities-based and science-based musicology. Crossing the boundary from one field to another, graduate students have to navigate the unfamiliar structures of scientific academia, involving the submission of papers to conferences whose processes and origins are not humanistic or musicological, but scientific. The rules, conventions and structures of scientific scholarly activity, without being entirely alien, are likely to be unfamiliar to these students, and so they not only have to familiarise themselves with the type of rigour required for framing work within the conventions of the scientific method, but they also must familiarise themselves with new, and often unspoken, extra-musicological frameworks for scholarly communication. How can someone studying *musicology* build a picture of the discipline as a whole while acknowledging these different philosophies, frameworks and approaches? How can musicology be presented as a unified

---

[5]   This situation is mirrored by or, perhaps, results from, the structure of many undergraduate music courses in the UK where a student engages simultaneously with many different aspects of musical education (performance is taught alongside analysis, for example), whereas often a student of the visual arts will choose between either practice or history as a focus of their whole programme of study.

discipline, while maintaining an awareness of its internal variation? What options are there for developing this kind of philosophy of musicology?

Performing Musicology and Performing Science

In 'Analysing Performance and Performing Analysis', Nicholas Cook talks about the closed form of paradigms in music analysis, that 'what they cannot say, they deny' (Cook 1999, p. 257–8). The same problem is present in musicology. Different musicological views exist: what a science-based musicological paradigm cannot say, it denies; what a humanities-based musicological paradigm cannot say, it denies. Just as Cook observes the closed form of music-analytical paradigms, Léon Brillouin observed the same closed forms in science, in his work on an information theory-based philosophy of science:

> Science is not a mere accumulation of results. It is essentially an attempt at understanding and ordering these results ... This rather artificial procedure [the selection of facts that fit a certain theory, while ignoring those that do not] is our own invention and we are so proud of it that we insist its results should be considered as "laws of nature". (Brillouin 1964, pp. vii–viii)

In the development of a scientific theory or paradigm, 'facts' that do not fit the logic of the theory are simply discarded.

Cook suggests a performative perspective for musical analysis as an alternative to the closed form of paradigms: 'If we think of analysis, or for that matter any musicology, in terms of what it does and not just what it represents, then we have a semantic plane that can accommodate any number of metaphorical representations of music' (Cook 1999, p. 258). For musicology a performative perspective could create a semantic plane that can accommodate any number of musicological ideals, taking into account what a work of musicology does, rather than what it is.

The musical object is difficult to define and categorise, and the purposes and definitions of musical activity vary widely. The musicological object, however, has one primary purpose – communication – because it is academic and disciplinary, and its purpose is to disseminate new knowledge. This is true for the musicological object itself, rather than the content, which may have many different purposes and aims. The object itself (research paper, journal article, seminar presentation) has to communicate efficiently and effectively. Efficient communication, then, is the goal of the musicological object and how its performance can be initially assessed. Developing a general performance perspective with which to examine musicology as a human activity can be based on the idea of efficient and effective communication.

Brillouin suggested that using concepts from information theory as meta-analytic tools would allow an 'unbiased and free discussion of the greatness and shortcomings of theories' (Brillouin 1964, p. ix). In the hands of the philosopher of science, Léon Brillouin (1962, p. xii) claimed, these tools can be useful in investigating the efficiency of experiments and observation as well as accuracy and reliability. The reason he finds information theory, and the related concept of entropy, useful for this purpose is because:

> Every physical system is incompletely defined. We only know the values of some macroscopic variables, and we are unable to specify the exact positions and velocities of all the molecules contained in a system. We have only scanty partial information on the system, and most of the information on the detailed structure is missing. Entropy measures the lack of information; it gives us the total amount of missing information on the ultramicroscopic structure of the system. (Brillouin 1962, p. xii)

Any scientific observation or experiment will result in an increase in the entropy of its environment (the more you know, the more you know you do not know), while at the same time increasing information in the environment (you have results available to be interpreted into meaningful knowledge; you know more about the specific object of investigation than you did before the experiment). For scientific theories where the ratio of information to entropy is positively stacked in favour of information, efficiency is high. Of course, there is a further process of translation of the information necessary to actually produce new and useful knowledge in the real world, to give it meaning and present it in a form in which it can be communicated and participate effectively within musicological discourse, but this process is one that can be examined separately. The human element is absent from the information theoretical approach proposed by Brillouin, because he wished to examine specific observations and experiments in a context in which they could be compared, without the complications of the meaning of the content or the bias of the scientist. Robert Morgan, in the first issue of the *Journal of Musicology*, talked about the difficulty of approaching the study of music in this way:

> The problem is that music, while containing quantifiable elements that lend themselves to generalization (and which, indeed, can be treated efficiently *only* through generalization), is created – like other arts – by human beings who differ in ways that are not quantifiable: in personality, social and historical environment, inherited stylistic conventions, artistic aims, and in other ways too numerous to list. (Morgan 1982, pp. 16–17)

Taking into account the difficulty of isolating generalisable results in musicology like music, from their human contexts, how can Brillouin's idea of information and entropy be put into practice for the examination of the performance of a musicological work without disregarding their human contexts? One approach would be to take a top-down viewpoint of the generalisations of musicological communication, using generalised concepts from an informational point of view in order to suggest what to isolate and analyse within a musicological output, but adjusting those decisions as a result of informal reasoning about musicological communication and specific research contexts.

The idea of assessing the information of a system and its communicational efficiency in these broad terms was applied to science by Ben Goertzel (2004). He attempts to 'do justice to both the relativism and sociological embeddedness of science, and the objectivity and rationality of science' (Goertzel 2004), drawing on aspects of a general information theory in probabilistic terms, and on a generalised Carnot principle. The Carnot principle, used in thermodynamics, states that, in making any heat engine, the motive power of heat is independent of the material of the working substance, and depends solely on the temperatures between which the heat particles are transported (Gribbin 2002, p. 382). In other words, the power of a system can be examined independently of the content and can be investigated through the relationships between identifiable elements.

Critical and Systematic Musicology

The idea that the structure of a physical system ultimately depends on the relationships between its working parts is a useful metaphor for an approach to a philosophy of musicology. Because no musicologist can be a true musicological polymath with an in-depth knowledge of every approach and every extra-musical and extra-musicological influence, examining and comparing the conditions of effective communication (or elements participating in the successful communication of new knowledge) is a way of situating disparate musicological projects within a unified whole. This idea is echoed, again, in Morgan's consideration of the study of music in his preface to the *Journal of Musicology*:

> We need to learn (relearn?) to be more responsive to the varied ramifications of music, to see the individual work as part of a vast and only partly penetrable network of connections and associations that encompass other compositions, other musics, and other spheres of thought. (Morgan 1982, p. 17)

In his 2007 article for the first *Journal of Interdisciplinary Music Studies*, Richard Parncutt makes a statement similar to that of Morgan 15 years earlier about musicology:

> The future development, and perhaps survival, of musicology will depend on the degree to which musicological institutions can achieve a balance between musicological sub-disciplines, celebrate their diversity and promote constructive interactions between them. (Parncutt 2007a, p. 2)

For Parncutt, using a 'top-down' systematic view of musicology, communication between researchers in different sub-disciplines is the key to the health of the discipline. Systematic musicology with its emphasis on music in general, rather than the music of the particular (as the object of study would be in, for example, ethnomusicology), provides some clues about how to approach musicology in general, rather than the musicology of the particular (sub-discipline). A systematic study of musicology might take a top-down view of it, with relationships, interdisciplinary information exchange (based on the balance of the familiar and the unfamiliar) and effective communication as the basis for observations, generalities and, eventually, theories.

Systematic musicology is not the only sub-discipline in which researchers are exploring ways of characterising and improving musicological communication. Kevin Korsyn's aim in his 2003 book is to understand musical research as an 'institutional discourse', and he draws on Mark Poster's 'mode of information' in postindustrial society (Korsyn 2003, p. 8):

> The electronically mediated exchanges that pervade our lives today, for example, are not merely tools at our disposal; they also disperse the self, placing in question "not simply the sensory apparatus but the very shape of subjectivity". (Korsyn 2003, p. 8)

In other words, methods of information exchange shape the understanding and structure of that information and the knowledge that is derived from it. The structure of a system depends on relationships between working parts. This gives new importance to the research output as an object of study in itself: 'how music is discursively framed will affect our reactions and may even determine whether we regard something as music at all' (Korsyn 2003, p. 9).

Research into the nature of musicology filters down into the disciplinary framed notion of what *music* is. As with any kind of analysis, the true value of the work can only be judged when the results affect action. Any methods for examining musicology separate from its context would only be one stage in a broader process. One thing shared by both Parncutt's and Korsyn's approaches is the emphasis on musicology as something that is done by people, and that communication between people is of great importance. These two approaches, one derived from empirical and scientific investigation, and the other from literary theory, come to this point of agreement. By stressing that musicology is a creative act with a higher level, social function – the communication of knowledge – science-based and humanities-based musicology might not seem to have such impenetrable boundaries between them.

Three Questions about Efficient Communication

Goertzel (2004) states three questions, derived from general Carnot principles and information theory, which can be used for assessing the efficiency of theories in informational terms:

1. How wide is its applicability?
2. How simple is it?
3. How accurate are its predictions?

These questions, even though they are derived from science, are general enough to be independent of any specific topic or data set and can be applied to data that are not framed scientifically. They are certainly not the *only* conditions of efficient communication, and do not rule out the consideration of any others.

Scope can be examined through observed applicability, which is often explicitly stated by researchers through what they choose as their objects of study. Predictive accuracy can be examined by making comparisons between what a work predicts will happen, and what actually does happen, or comparing how much a reader/listener knows at the end compared with the beginning: if predictions are accurate the reader will know more at the end, since an implicit prediction on the part of an author in any academic work is that they will communicate understandable information to their audience efficiently and effectively. Simplicity can be measured by making comparisons between size in both time and space.

These are relatively straightforward comparisons to make in scientific musicology, where sub-disciplinary output requirements, inherited from the hard sciences, are that every research output has an explicit prediction (hypothesis) at its outset, and a conclusion, referring to the hypothesis, at its end. Assessing predictive accuracy in non-scientific musicology would require more creative investigation, and a more flexible approach. In traditional musical analysis, predictability can be assessed through considering whether the analytical system being demonstrated provides new knowledge about a specific piece of music. The process of exploring the body of work to identify the information required to make these kinds of evaluations would be a valuable process, and one that would form the basis for interesting outputs in itself.

Identifying statements within the research outputs of non-scientific musicology for assessing scope, simplicity and predictive accuracy would not be impossible since every piece of academic writing requires a certain amount of familiar structure – Gaddis's thin, perilous condition imposed on the basic reality of chaos to enable communication.[6] Informal reasoning is often used to track relationships

---

[6]   With a quote from William Gaddis Jr, Joe Moran illustrates the link between the organisation of information and education, while simultaneously hinting at the resistance researchers might have to applying the necessary communication principles to their own work: 'Before we go any further here, has it ever occurred to you that all of this is simply one grand misunderstanding? Since you're not here to learn anything, but to be taught so you can pass these tests, knowledge has to be organized so that it can be taught, and it has to be reduced to information so it can be organized, do you follow that? In other words, this leads you to assume that organization is an inherent property of the knowledge itself, and that disorder and chaos are simply irrelevant forces that threaten it from outside. In fact it's

between aspects of real-world phenomena (the object of research) and the structure of a theory or approach.

I have chosen three examples to demonstrate the scope of this kind of approach for different kinds of musicology. These examples are by no means exhaustive, but are presented for the purpose of demonstrating some of the processes and issues representative of the approach suggested here. The examples are chosen not only for their self-reflective nature, but also for their common emphasis on communication; the three sub-disciplines from which the examples come have been chosen for efficiency: I approach them with a certain amount of familiarity with those areas of study. For the purposes of the communication of the points I wish to make with these examples, they remain general and short.

Computational Analysis

The first example is an examination of David Temperley's project developing probabilistic models for music analysis. This project (which consists of several outputs from 2001 to the present, see Temperley 2001, 2007, 2009) can be categorised as computational analysis, or as American music theory. His stated aim does not vary between outputs: 'My aim in this study is to gain insight into the processes whereby listeners infer basic kinds of structure from musical input' (Temperley 2001, p. 6).

His aim to 'gain insight' directly outlines his prediction that at the end more information will have been generated by the system (and, presumably, the entropy of the system – the system in this case being information exchange between Temperley and his audience, rather than the analytic model – will have decreased). He creates complex computational models designed to predict a 'correct' structural analysis of musical components (defined by indications in a score or by an 'experienced' human listener), such as beat, metre and melodic phrasing. These models are then tested on real music encoded into MIDI sequences.

Even though the system in question here is communication between Temperley and his audience, his analytical model, being the primary subject of the research output, provides malleable results about scope, predictive accuracy and simplicity. The scope is stated in Temperley's description of the development of his model over time. The 2001 model deals with predictions of note-patterns in terms of pitch only, whereas the 2009 model can predict metre as well as pitch. Additionally, the 2001 model only deals with monophonic music, whereas the 2009 model deals with polyphonic music, incorporating different lines and the relationships between them. This is an increase in scope in two conceptual directions. There is, therefore, an increase in the efficiency of the model between 2001 and 2009, increasing

---

exactly the opposite. Order is simply a thin, perilous condition we try to impose on the basic reality of chaos' (Gaddis 1976, p. 20 cited in Moran 2010, p. 1).

the amount of information that can be generated by the model and therefore communicated through the related output.

The predictive accuracy of the model also improves between 2001 and 2009. The average success for predicting metre, for example, in 2001 was 88.22 per cent, whereas the average success in 2009 was 90.29 per cent. Temperley ascribes the improved performance of the second model (albeit a small improvement) to the fact that it took into account more of the connections between different features of the music analysed. One example he gives is the improved prediction of beat owing to the combination of harmonic and metric analyses. By including harmony in the data the model was trained on, Temperley was able to improve its performance for predicting different, but connected, structural components. More aspects of the tested music participate in correct predictions – a definite increase in efficiency.

Simplicity is the most difficult aspect to measure in this case, and Temperley gives us no convenient numerical summary. The 2009 model is considerably more complex than the 2001 model, so a judgment would need to be made by the investigator about whether the increase in complexity is made up for by increases in scope and predictive accuracy. A decrease in simplicity for the 2009 model has less of an effect, resulting in an overall increase in efficient communication over the three questions asked for readers who are already familiar with the 2001 model (and, indeed, the 2007 model – which I have left out here for the sake of simplicity). These readers will already have a familiar context within which to place the newest features of the model, and so will participate in efficient communication when interacting with later research outputs. In this case, simplicity can only be judged with recourse to what the audience is or is not familiar with.

Additional analysis of the connections between the structure of the two models would yield information upon which to base predictions about the future development of the model – what its simplicity, predictability and scope might be in 2011 or 2013. In effect, analysing the structures of the models, just as Temperley examined the structure of various music, could result in the development of a computational model whose purpose would be to predict what Temperley's next modifications to his model would be, illustrating the recursive nature of musicology as it evolves.

Ethnomusicology

The second example considers Tara Browner's research on North American pow-wow dance music. Browner's article (2000) uses pow-wow as an example to showcase a specific analytical methodology and method of communicating the results for working with non-Western music: the 'bi-analytical' approach, which 'presents the understandings and interpretations of both realms [Western musical scholarship and native understanding of music making] as equally valid, and allows each to critique the other' (p. 216). Unlike Temperley, Browner has no facts and figures to report in order to allow an elementary survey of scope, predictive accuracy and simplicity, and so the process of identifying indicators of these in the text is subjective, depending on what investigators choose to examine in the article, and how they examine it. However subjective the identification and evaluation of these three, the top-down process of searching for these indicators within the output provides a new perspective on the work as a communicative entity.

The scope of the article is, in terms of the analytical object, pow-wow dance music of the North American Indians. When taking into account the article's purpose of showcasing the bi-analytical approach, however, Browner explicitly demonstrates the broad scope of the article and the methodology in terms of the number of people who can gain useful knowledge from it. Browner deliberately writes for two sets of readers – academic and native – so the purpose of the article is to increase the scope of readership (Browner 2000, p. 216). An increase in scope with regard to the kinds of music to which the method can be applied is suggested, but the article itself concentrates primarily on increasing the potential for a large audience.

Like much ethnomusicology, the article assumes two things of its reader: firstly, that they are not intimately familiar with the music being discussed (of course, there will be many interested experts who will seek the article out, but most of the readers of *Ethnomusicology* will not be experts in this particular type of music); and secondly, that they are familiar with the norms and structures of Western music analysis. The author situates what she assumes to be unfamiliar music within what she assumes to be the familiar context of Western musical analysis. A reader for whom these assumptions are true will gain much from reading the article (much information will have been efficiently communicated to them). Of course, these assumptions are not always true and, increasingly, people who are likely to read *Ethnomusicology* will not necessarily have a background in Western music analysis. Browner also draws on the comparison of unfamiliar and familiar at the data-gathering stage of her research: while interviewing native Indians, she asks them to choose terms from different sets of vocabularies, and so builds relationships between sets of terms used in one tradition and those used in another, creating the foundation for more efficient communication.

It is difficult to directly identify and assess the predictive accuracy of the article, since it is an early output of a broader project putting the analytical methodology into practice. The predictions made by the author are that the bi-analytical approach will be acceptable both to Western scholars and to native Indians, therefore enabling more efficient information exchange between those two groups and increasing the efficiency of communication by only requiring the use of one system of analysis for both audiences, rather than the two required to 'reach' the same number of people without using Browner's approach. Browner provides a concluding passage in which the reader is assured that this is the case:

> My hope is that by introducing all relevant sources relating to music history, analysis, composition/making, and performance, I can mediate the sometimes immense distance between Western scholarship and native understanding. (Browner 2000, p. 231)

A Western academic who has understood Browner's ideas about how to analyse pow-wow songs will certainly be of the opinion that, from the evidence she gives, communication between Browner and native Indians using the bi-analytical approach will be successful and efficient. Even if it has not been proved with empirical evidence, the reader still experiences a high level of perceived predictive accuracy.

Simplicity can also be assessed in relation to Browner's contextualisation of unfamiliar elements within familiar ones. A reader familiar with the difficulties of writing about non-Western music will also be familiar with the issues Browner tackles, perhaps, while being unfamiliar with the specific example used. Therefore communication will be efficient between Browner and these ethnomusicologists. For these scholars, the article is simple because it does not introduce completely new data *and* a completely new theoretical framework, but draws on existing analytical methodologies.

Popular Music Studies

The third example derives from popular music study and focuses on Alf Björnberg's research about structural relationships of music and images in popular music video (Björnberg 1994). This is similar to Browner's article in that it presents information about a particular act of analysis the author has carried out, but it also contextualises that analysis within an evaluation of other approaches – demonstrating that outputs in disparate areas often have the same self-critical character, critiquing musicology at the same time as music and other subjects. Like the Temperley example, this article is based on a previous published work, so anyone who is familiar with those outputs will benefit from relatively high perceived simplicity. Readers familiar with Björnberg's work on the characteristics of popular music syntax and their consequences for the analysis of music video will have the basis for very efficient communication of the new material of this article.

The scope of the article is stated as part of the author's aims, in that he feels that the method is appropriate for the analysis of any music video. This could be seen as a relatively limited scope, since the music video only populates a very small area in the overall concerns of musicology (but a somewhat larger area in popular music studies and the semiotics of music), but, like Browner, Björnberg's article has an increase in scope with regard to who will find it interesting, since he incorporates in his methodology an openness to varying kinds of reception and interpretation (Björnberg 1994, p. 51). Anyone familiar with the methods explored by Björnberg will have a conceptual comparative basis for efficiently understanding his proposed approach.

Accuracy of predictions is, again, something that the investigator must glean from the author's summary. Björnberg's aim is to align musical features with video features and show their relationship as being more significant than other authors have stated, thereby assuming the reader's prior knowledge of related work. The journey on which Björnberg takes the reader through his examples to his conclusion is certainly convincing of this point, carefully aligning features of musical narrative with visual narrative, connecting them using the frameworks and disciplinary norms of semiotics (drawn from past musical and visual work); in effect, he uses music and video as a way of aligning the areas of narrative study and semiotics. The organisation and structure of the work give a strong impression of predictive accuracy. If the information had been organised differently, the reader may not be *as* convinced by the 'correct' prediction. The author tells the reader that they know more at the end of the article than they did at the beginning – and who are we to argue?

Conclusions

Information as a concept defined and explored by Brillouin does not involve any notion of value or meaning (Brillouin 1962, pp. 9–10). Why, then, would a concept void of the very lifeblood of musicological investigation be useful in such an investigation? Surely the point of studying any body of musicology is to appraise its value and meaning? Surely to discard the notion of criticism altogether is very dangerous in a post-Kerman world?

The value of such an exercise as suggested here is to do with the aims and goals of the person doing the analysis. In suggesting these ways of thinking about musicology, my aim is to begin to address how those who engage with musicology, particularly those who are trying to understand its procedures from a position of little experience, can work within musicology as a whole and navigate the various sub-disciplines within it. I suggest that there is a process for organising our understanding of musicology, which can be sought through an examination of efficient communication.

A systematic approach to the communicative performance of musicology might eventually lead to philosophies or theories about musicology. Any general theory or philosophy would have to have a large scope, have accurate predictions and be simple enough to enable readers to familiarise themselves with it efficiently and effectively. If not, it becomes just another translation of one set of facts into another with no generation of new knowledge. The purpose and use of any new knowledge, however, is something that can only be examined when put back into its original context. A theory of musicology as communication would have the same purpose as any theory. These purposes are stated by Lawrence Zbikowski in *Conceptualizing Music* (2002, pp. 206–7):

1. Theories guide understanding and reasoning.
2. Theories provide answers to conceptual puzzles.
3. Theories simplify reality.
4. Theories involve a number of conceptual models.
5. Theories are dynamic.

If they provide testable predictions at their outset, they can have predictive power. Developing approaches to a philosophy of musicology would mean addressing the following three points. Firstly, our understanding of and reasoning about musicology could be more easily guided, perhaps to the point where a clear picture of its processes and connections as a whole would help scholars (particularly graduate students) identify their place in musicology and what directions and collaborations would be most fruitful. Secondly, patterns in musicology could be explored by processes of summarisation and the simplification of ideas into describable models or maps. Thirdly, it might become easier to assess the effect of research programmes on the larger structures of musicology as a discipline, as well as the effect of external disciplines. These are grand claims for this speculative approach to investigating musicology, but in predicting a fruitful outcome of such an endeavour, I hope to provide the conditions for more efficient subsequent communication about this topic in the future.

Any historian or critic of musicology will say that it is a relatively new discipline (compared with, for example, some of the literary disciplines), and perhaps the rise of systematic musicology and postmodern and poststructuralist ways of thinking about musicology are signs that it has changed to the point where a top-down, self-aware approach is something that is needed for the success of the discipline in the future, to counter its outward growth into increasingly disparate sub-disciplines, and to prevent different sub-disciplines from developing completely different understandings of what music itself is. Contrasting that which has been ordered or processed within musicology (rendered familiar to those working within the discipline) with that which is yet to be ordered (the disciplinarily unfamiliar, whether it be a concept from another discipline, a new piece of music or a new collection or connection) enables the integration of the unfamiliar into the disciplinary discourse, and in this way a discipline increases its scope, the accuracy of its predictions, and at some level, its simplicity. This suggestion raises questions about how musicology as a research activity is tracked and judged. A common point for institutional debate is the degree to which very new types of work within musicology as a whole count as 'research', and how an area or subject that is unfamiliar in disciplinary terms might be integrated into an evaluative framework or system (such as the UK's Research Excellence Framework, or even the peer review system for publication). The assumption that research should fall into a 'familiar' category or within 'familiar' disciplinary boundaries, in order to 'fit' the remit of a particular research framework or outlet so that it might be judged at all, might be something that prevents the efficient integration of unfamiliar areas into musicological discourse and possibly artificially slows the development of the discipline and research activity.

References

Björnberg, A. (1994). Structural Relationships of Music and Images in Music Video. *Popular Music* 13/1: 51–74.

Brillouin, L. (1962). *Science and Information Theory* (2nd edn). New York: Academic Press.

Brillouin, L. (1964). *Scientific Uncertainty and Information*. New York: Academic Press.

Browner, T. (2000). Making and Singing Pow-wow Songs: Text, Form, and the Significance of Culture-based Analysis. *Ethnomusicology* 44/2: 214–33.

Cook, N. (1987). *A Guide to Musical Analysis*. Oxford: Oxford University Press.

Cook, N. (1999). Analysing Performance and Performing Analysis. In N. Cook and M. Everist (eds), *Rethinking Music* (pp. 239–61). Oxford: Oxford University Press.

Gallardo, C.L.G. (2000). Schenkerian Analysis and Popular Music. *Transcultural Music Review* 5, http://www.sibetrans.com/trans/trans5/garcia.htm (accessed 1 July 2009).

Goertzel, B. (2004). Science, Probability and Human Nature: A Sociological/ Computational/Probabilist Philosophy of Science. *Dynamical Psychology*, http://wp.dynapsyc.org/ (accessed 16 June 2009).

Gribbin, J. (2002). *Science: A History*. Harmondsworth: Penguin.

Korsyn, K. (2003).*Decentering Music: A Critique of Contemporary Musical Research*. Oxford: Oxford University Press.

Kraehenbuehl, D. (1983). *Communication: Process and Purpose*. MSS 79, The David Kraehenbuehl Papers in the Irving S. Gilmore Music Library of Yale University.

Larson, S. (2010). Path and Purpose in Raga. *Paper presented at the 1st International Conference on Analytical Approaches to World Music*, University of Massachusetts, Amherst, MA, 19–21 February.

Moran, J. (2010). *Interdisciplinarity* (2nd edn). London: Routledge.

Morgan, R.P. (1982). Theory, Analysis and Criticism. *Journal of Musicology* 1/1: 15–18.

Parncutt, R. (2007a). Systematic Musicology and the History and Future of Western Musical Scholarship. *Journal of Interdisciplinary Music Studies* 1/1: 1–32.

Parncutt, R. (2007b). Can Researchers Help Artists? Music Performance Research for Music Students. *Music Performance Research* 1/1: 1–25.

Potter, K., Wiggins, G.A. and Pearce, M.T. (2007). Towards Greater Objectivity in Music Theory: Information-dynamic Analysis of Minimalist Music. *Musicae Scientiae* 11/2: 295–322.

Schenker, H. (1979). *Free Composition. Vol. III (New Musical Theories and Fantasies)*. Trans. E. Oster. London: Longman.

Schoenberg, A. (1967). *Fundamentals of Musical Composition* (ed. G. Strang). London: Faber and Faber.

Shannon, C. and Weaver, W. (1964). *The Mathematical Theory of Communication*. Urbana, IL: University of Illinois Press.

Temperley, D. (2001). *The Cognition of Basic Musical Structures*. Cambridge, MA: MIT Press.

Temperley, D. (2007). *Music and Probability*. Cambridge, MA: MIT Press.

Temperley, D. (2009). A Unified Probabilistic Model for Polyphonic Music Analysis. *Journal of New Music Research* 38/1: 3–18.

Weiner, N. ([1948] 1961). *Cybernetics: Or Control and Communication in the Animal and the Machine* (2nd edn). New York: MIT Press.

Zbikowski, L. (2002). *Conceptualizing Music: Cognitive Structure, Theory and Analysis*. Oxford: Oxford University Press.

Chapter 8
Familiarity and Reflexivity in the Research Process

Clemens Wöllner, Jane Ginsborg and Aaron Williamon

The assumption that researchers are highly familiar with the subject of their investigations has not been much debated. Many years of secondary and tertiary education, as well as continuous engagement with new developments in their respective fields, form a basis for accumulated expert knowledge. Yet there are disciplines in which certain skills can only be acquired by extensive training outside the academic sphere. In particular, areas involving precise control of the body such as sports, dance or music performance require sustained physical training before individuals acquire expert-level knowledge (Ericsson 1996). In this regard, these disciplines differ from other academic fields, and it can be asked whether a certain level of skill and familiarity is necessary or at least beneficial for carrying out research in these areas.

Familiarity in research can be defined as having particular insights into specific areas of investigation. The concept of familiarity between members of a research team, with the potential for increased coordination (Moreland, Argote and Krishnan 1998) or trust (cf. Edmondson, Bohmer and Pisano 2001), will not be discussed in detail here. In fields such as music psychology or performance science, the types of insights based on familiarity with a given subject may vary according to the scope and specific approaches of the research carried out. Researchers investigating the fundamental principles of pitch perception, for example, primarily need a psychological understanding of auditory processing. If more complex expectations in listening to musical examples are investigated, then expert knowledge of music-theoretical concepts of counterpoint or harmonic progressions is advantageous. Researchers addressing aspects of applied music psychology, on the other hand, such as performance anxiety or fingering in piano playing (for overviews, see Parncutt and McPherson 2002; Williamon 2004) may benefit from the experience of performing in public, or from having struggled themselves to find the best fingering in a Bach fugue. Thus intentions of whether or not research should be applicable to practitioners can influence notions of familiarity. It may be the case that researchers need to be more familiar, personally, with the subject of their investigations when their findings have practical applications than when they do not. If so, familiarity in applied music research can further be defined as the researcher's close relationship to practice as well as personal experience of music performance (examples of practitioners carrying out research are provided in Davidson, 2004).

A question arises, nevertheless, as to the necessity for researchers to maintain 'critical distance' from their topics (see, for example, Clases and Wehner 2009; Linstead 1994). In what ways do individual experiences and personal involvement shape or even in some cases undermine research? What is the difference, for example, between the ways two critics view a work of art when one is a creative artist and the other is not? If researchers carry out (qualitative) studies in their own environments, Hanson (1994) warns that there is a potential for role confusion, such that the researcher's objectivity might be compromised. Yet a detailed knowledge of 'cultural norms and values' can also be beneficial for the research process (p. 940). In the field of music studies, researchers are most likely to be experienced listeners with particular attitudes to music and listening habits. Thus it may be particularly difficult for those who carry out research into music perception to discount their own preferences when attempting to keep a 'professional' distance from their subject of study.

The question as to how researchers should tackle their personal experience in the fields they are investigating has been addressed, for example, by sociologists and ethnographers. Some sociologists, sceptical of 'scientific' knowledge and taking the view that, like other kinds of knowledge, it is socially constructed, began more than two decades ago to urge researchers to be more reflexive (Ashmore 1989). In other words, they are required to acknowledge consciously (Davies et al. 2004) the roles they play themselves in choosing their topics and methods of investigation, and the perspectives they take when interpreting their findings. Linstead (1994), rejecting the possibility that results can ever be seen as having been obtained 'objectively' and claiming that scientific knowledge should be treated as 'fiction' (p. 1321), argues that self-knowing as well as self-declaration are paramount in reflexive research. In ethnography, researchers maintain that the researcher's perspective, which is likely to arise from his or her cultural background at both micro- and macroscopic level, needs to be made explicit (Barz and Cooley 2008). The reasons for this claim are that the interpretation of other individuals' behaviours is potentially biased, and there are limitations to the generalizability of the researcher's viewpoint. Furthermore, relationships between field researchers may influence the research process (Pasquini and Olaniyan 2004) and should be reflected, thus revealing the 'situatedness' of some of the gained knowledge (p. 24), and more generally, the 'positionality' of researchers particularly in cross-cultural studies (cf. Twyman, Morrison and Sporton 1999). In addition, it is important for researchers to reflect on their relationships with research participants and the roles they play besides that of investigator (Jootun, McGhee and Marland 2009).

In the remainder of this chapter we reflect on the backgrounds of music researchers (including musicologists, music psychologists and performance scientists) so as to shed some light on the perspectives they take in their research. Familiarity in research is considered under four main headings: researchers' career transitions; their decisions to work in the field; their experiences as music performers; and their listening habits. Findings from a recent online survey of more than 100 music researchers are presented, providing an empirical basis for discussion (Wöllner, Ginsborg and Williamon 2011). We see our approach as being in line with the increasing awareness of the importance of reflexivity in research described above. In our own field, a recent special issue of the journal *Psychomusicology* included short autobiographies of 13 influential music researchers on the basis that 'to understand the creation, one must also understand the creator [and that] autobiography ... contributes context to this understanding by suggesting why certain questions were asked at a particular time, and why they were addressed in a particular way. Autobiography may reveal the force driving the research' (Cohen 2009, p. 11). Similarly, exploring the context of research may further our understanding of well-established as well as new approaches, and could be beneficial in developing future directions.

Findings from Four Domains

Are Music Researchers' Careers Different from those in Other Fields?

Music researchers are rather different from other kinds of researchers, in that their interest in their subject is likely to have predated their development as researchers *per se*; they are likely to be music-lovers and indeed to have had considerable experience of music-making at an amateur or professional level.

Influences on career choice have been widely studied. Recent articles report the reasons individuals choose to become accountants (Myburgh 2005; Sellers and Fogarty 2010), entrepreneurs (Carter et al. 2003; Cools and Vanderheyden 2009), engineers (Loui 2005), managers (Malach-Pines and Kaspi-Baruch 2008), medical students and doctors in a range of specialties (Goldacre, Laxton and Lambert 2010) including psychiatry (Lamb, Evans and Baillie 2006) and dentistry (Mariño et al. 2006), restaurateurs (Inal and Karatas-Ozkan 2007) and hospitality and tourism managers (Wong and Liu 2010), school teachers (Bayer et al. 2009) including physical education teachers (Metzler, Lund and Gurvitch 2008) from teaching assistants (Dunne, Goddard and Woolhouse 2008) to school principals (Stevenson 2006), veterinary surgeons (Ilgen et al. 2003) and writers of fiction (Kaufman and Kaufman 2007). Influences on women's career choices, in particular, first began to be studied in the 1980s (Betz and Fitzgerald 1987; Fitzgerald and Crites 1980) and continue to be of interest to researchers (Gottfredson and Lapan 1997; Coogan and Chen 2007). They have been investigated in several professions that have traditionally been the preserve of men: the construction industry (Johns and Johnson 2006), information technology (Quesenberry and Trauth 2007), policing (Irving 2009) and science (Besecke and Reilly 2006).

For the purposes of this chapter, the professions in which influences on career choice are perhaps most relevant are music (Simonton 2009; Vuust et al. 2010; Zwaan, ter Bogt and Raaijmakers 2010) and teaching and research in higher education (Bennion and Locke 2010; Kornhauser et al. 2008), although Simonton (1997, 2000) investigated career trajectories in the creative professions more generally, and Fournier and Bujold (2005) report on the career trajectories of 62 individuals – several with bachelors' degrees and PhDs – who had worked in a series of jobs for up to three years. Fournier and Bujold refer to these as 'non-standard' careers, and in many ways they resemble those of the freelance and therefore insecure classical musicians described by Harper (2002), as well as the 'boundaryless' careers (Arthur and Rousseau 2006) of the pop musicians studied by Zwaan et al. (2010).

The earliest accounts of the psychology of career choice (as opposed to explanations made from the perspective of organisational behaviour), published during the first half of the last century, derive from the seminal work of Parsons (1909). This has come to be known as the 'matching-men-and-jobs' approach (Schreuder and Coetzee 2006), whereby people are encouraged to choose their career not only on the basis of their knowledge of a range of occupations, in terms of what they involve and what they have to offer, but also taking into account their own personal strengths and weaknesses, and having considered carefully the relationship between personality and occupation. Personality traits play a part in more recent theories (Holland 1959, 1985, 1997). According to a review by Hackett, Lent and Greenhaus (1991), Ginzberg et al. (1951) were the first to put forward a developmental theory of occupational choice, followed by Super (for example, 1953) and his colleagues (Super et al. 1957) and Tiedeman and O'Hara (1963); the 1960s also saw the introduction of psychodynamic (for example, Bordin, Nachmann and Segal 1963) and vocation decision-making models (for example, Hershenson and Roth 1966). Lifespan developmental theories continued to be influential into the last decade of the twentieth century (Gottfredson 1981; Levinson et al. 1978; Super 1980, 1990), particularly Gottfredson's theory of circumscription, compromise and self-creation, which takes into account the social and economic circumstances of individuals (Gottfredson 2002).

Social learning theory – which evolved into social cognitive theory (Bandura 1986, 2009) – has had, perhaps, the most impact on the understanding of career choice, particularly with its emphasis on self-efficacy (Bandura 1977; Hackett and Betz 1981). Social Cognitive Career Theory (SCCT) (Lent, Brown and Hackett 1994, 2000) 'focuses on self-efficacy, expected outcomes, goal mechanisms and how they may interrelate with other person (for example, gender), contextual (for example, support system), and experiential/learning factors' (Lent et al. 1994, p. 79). On the basis of a meta-analysis of previous findings Lent et al. suggest, firstly, that self-efficacy and outcome expectations influence choice of career, and secondly, that as outcome expectations change, so do career choices. This model has been used to explore the four categories of career self-efficacy expectations set out by Lent and Brown (2006): content- or task-specific self-efficacy; coping self-efficacy; process self-efficacy, which includes the skills needed to explore careers, make decisions and implement them; and self-regulatory self-efficacy, which might refer to self-belief in terms of organisation and time management abilities (Betz 2007). The domains in which content- or task-specific self-efficacy have been measured include computing (Lent et al. 2008), engineering (Lent et al. 2007), leadership and public speaking (Bieschke 2006), academic self-efficacy (Gore 2006; Lindley and Borgen 2002; Pajares 1996; Rottinghaus, Betz and Borgen 2003) and, most recently, musical self-efficacy (Ritchie and Williamon 2011).

The other important feature of SCCT relevant to this chapter is experiential/ learning factors. Since music researchers are likely to have been influenced by their musical experiences, it is worth noting the observation that 'surprisingly little attention has been paid to operationalizing [learning experiences generally] for research purposes' (Betz 2007, p. 407), other than in relation to mathematics (for example, Lent et al. 1996) and social skills (Anderson and Betz 2001), although Schaub and Tokar (2005) provide evidence from their study using path analysis in support of the hypothesis that learning experiences contribute to self-efficacy. These in turn contribute to the six 'interests' postulated by Holland's (1997) theory of vocational personalities and work environments: Realistic, Investigative, Artistic, Social, Enterprising and Conventional.

On the whole, research on the learning experiences that influence individuals to pursue particular careers focuses on those that occur in the later years of secondary school or early stages of undergraduate study (for example, Lamb et al. 2006; Mariño et al. 2006). Furthermore, these learning experiences are not necessarily related to the work that is actually undertaken in the course of a career. The accountancy students surveyed by Myburgh (2005), for example, were most likely to be influenced by their friends and families – as were those who had chosen careers in the hospitality industry (Wong and Liu 2010) – and the medical students who completed Newton, Grayson and Thompson's (2005) questionnaire intended to choose their specialties on the basis of the extent to which they were associated with more or less 'friendly' lifestyles and, therefore, projected income. In very few of the professions that have been studied, then, is there evidence that individuals make career choices on the basis of early experiences, such as in the hypothetical case of an orthopaedic surgeon inspired by his or her experience of breaking a limb in childhood and being treated in hospital.

Yet Metzler et al. (2008) observe: 'Pre-professional socialisation, identity and values are formulated well before the start of a teacher education programme' (p. 461), citing the apprenticeship of observation metaphor (Lortie 1975), whereby experiences as a pupil are likely to encourage the selection of a career, in turn, as a teacher. Further, according to Besecke and Reilly (2006), female scientists are likely to be inspired by 'contact with a role model and often an intimate involvement with the process that serves as an invitation into the world of scientific inquiry'.

Given the length of time it takes to develop expertise in music (for example, Ericsson, Krampe and Tesch-Römer 1993), performers and composers cannot but choose their profession on the basis of experiences that are likely to occur even earlier than in the lives of prospective teachers and scientists. A wealth of literature on the developmental psychology of music (see, for example, Burland

and Davidson 2002; Davidson et al. 1996; Hargreaves 1986; McPherson 2006; Moore, Burland and Davidson 2003; Sloboda et al. 1996) discusses the many interrelated factors underlying the career choices of musicians and the extent to which they are capable of achieving their aspirations. Less is known about the influences on those who choose the profession of music researcher.

While the professional careers of dancers and athletes are limited by the demands on their bodies over many years, there is no reason why musicians – particularly composers, conductors and most, although not all, types of instrumentalist – should not enjoy many years of music-making. They do not necessarily do so, however. Using data from the United States National Longitudinal Survey of Youth (1979–98), Alper and Wassall (2002) investigated artists' – including performers' (that is, actors, dancers, musicians, composers and announcers) – longevity and career transitions. Among the artists surveyed, architects and designers were most likely and non-visual artists least likely to stay in the same field. As for performers, fewer than 40 per cent had remained in their careers for two decades. Alper and Wassall (2006) suggest that the reasons for career transitions include changes in the market for the performer's art, the search for the 'right' occupational niche as an artist or simply (for those who make early career changes) completing their training. Bennett (2008), in a wide-ranging survey of musicians in Australia, Europe, Asia and the United States, investigated several aspects of musicians' careers besides longevity, including the balance of performing, teaching and other activities, and how it changes over time; the skills required; and the factors underlying attrition. One such factor is – as for athletes – injury or ill-health. In her analysis of the narratives of freelance professional musicians, Daykin (2005) reveals their experiences of disruption and dissonance, which forced them to rebuild their identities. Similar challenges are reported by Oakland, MacDonald and Flowers (2009) in their study of the effects of redundancy on professional opera choristers' identity. Oakland et al. emphasise that singers – perhaps to an even greater extent than instrumentalists – rely on the social and professional evaluations of others when they form their own sense of self; on the other hand, adaptation to changes in circumstances, such as they experienced when they were made redundant from a company from which some of them had sung for many years, is highly individual.

Advice on career transition has been on offer for some time (for example, Corbin 2002; Rosen and Paul 1998), but the career transitions of people whose first careers began particularly early in life have only recently begun to be documented: Upper (2004) and Jeffri and Throsby (2006) discuss dancers, while Anderson (2009) and Bruner et al. (2009) focus on athletes and More, Carroll and Foss (2009) on performers more generally. A useful comparison of performers' and athletes' career transitions is provided by Hays (2002), who points out that these may occur prematurely as the result of performance anxiety or tension, compounded by overuse, leading to injury. In these circumstances, the distress experienced at and following retirement (Miller and Kerr 2002) may result not only from loss but also from self-blame.

When and Why did Music Researchers Decide to Work in the Field?

In comparison with some of the research outlined above, in which participants had only recently experienced transitions in their working lives, the approach taken by the authors of the present chapter was retrospective. In our online survey (Wöllner et al. 2011), music researchers were asked to look back on their careers. Since music psychology, for example, combines two disciplines, we were interested to find out which one provided the original impetus and to what degree researchers were, and continue to be, involved in music-making (see below). We found that, in many cases, music had been important to researchers from their earliest years. Some had become professional musicians, and most maintained their interest in music. A sub-set of music researchers, then, consists of former musicians, in the sense that music was once their primary occupation or even source of income.

On the whole, music researchers made their decision to enter the profession later than people who work in other disciplines, at a mean age of 30.0 years (standard deviation, SD = 9.13). At this age, most respondents expressed an interest in music psychology or a particular topic; this was among the key reasons they had chosen this direction (see Figure 8.1). They were less likely to mention career opportunities, but these still played an important role, as did general interest and enjoyment in research. A small number of researchers were inspired to work in the field relatively early by 'influential others'. Sample responses relating to career decision-making are provided below.

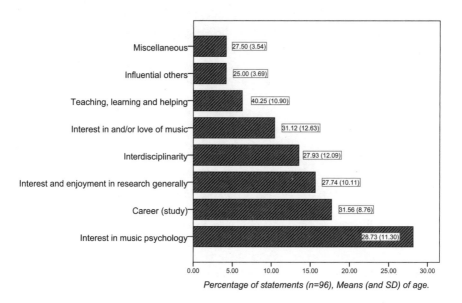

Percentage of statements (n=96), Means (and SD) of age.

Figure 8.1   Reasons why and age when (mean and SD) respondents chose to work in the field

- Interest in music psychology: 'Interested in the application of science to improving and understanding musical performance' (age decided to work in field, 22; female; current age, 27).
- Career (study): 'Job opportunities' (age decided to work in field, 36; female; aged 42).
- Interest and enjoyment in research generally (or no particular topic mentioned): 'I like to reflect and investigate issues that engage me and I like to write' (age decided to work in field, 30; female; aged 43).
- Interdisciplinarity: 'It combines my two passions – music and psychology' (age decided to work in field, 22; male; aged 62).
- Interest in and/or love of music: 'I love music and discovering more how music works just fascinates me' (age decided to work in field, 25; female; aged 27).
- Teaching, learning and helping: 'Because as a teacher and as a [performer] I'm searching for answers, ways to help students, ways to understand performance problems and so on' (age decided to work in field, 47; female; aged 58).
- Influential others: 'Inspiring Masters project supervisor' (age decided to work in field, 22; female; aged 33).
- Miscellaneous: 'no reflected decision, it was my way' (age decided to work in field, not provided; female; aged 28).

Questionnaire data can provide no more than a snapshot of respondents' views on a given topic, but the reflections provided by the present sample on the role of their interest in the subject matter of their research nevertheless suggest some support for the importance of experiential/learning factors put forward in Lent et al.'s SCCT theory (1994, 2000). It may be that musicians make use of the opportunity to become researchers on the grounds that they will benefit from a 'friendlier' lifestyle, in the same way that medical students have been found to choose their career specialties (Newton et al. 2005). We did not find evidence in this study of musicians taking up careers in research because they had been forced to retire from performing, but such evidence exists in the studies cited above (for example, Daykin 2005; Oakland et al. 2009).

Do Music Researchers Still Engage in Making Music?

Motivations for carrying out research in a specific field may be influenced by the nature of the individual researcher's previous experience, career-related and otherwise. Many researchers in our study had accumulated hours of deliberate practice on their musical instrument(s), equivalent in some cases to the amount of practice undertaken over a similar period of time by professional musicians. Some had in fact been professional musicians before switching their focus to research; others were reluctant to define themselves primarily as musicians or researchers, preferring to combine their two professions and be known as performer-researchers. While there is no question that in such instances research and practice are closely related, since these researchers are highly familiar with the subject of their investigations, it is not generally considered to be obligatory for researchers in music psychology and related areas to be actively involved in music making. It was not the case, for example, that all 13 influential music researchers whose autobiographies were obtained by Cohen (2009) reported that their initial musical training, including learning to play an instrument, was of crucial importance to their future careers. They were all driven, however, by their strong interest in and love for music.

In our study, with a larger sample of music researchers, at different stages of their careers, we were able to explore the possibility that the importance of musical performance to researchers can explain their approach to research. We wanted to know the extent to which their skills as performers influence their choice of topic and methods, and if (and in what ways) their findings have an impact on their musical practice. We therefore asked them about their active engagement in music performance. A total of 97 per cent had taken instrumental or vocal lessons in the past, and all (including some who had not had formal music training) reported participating in musical activities such as improvising. About half the respondents had composed or improvised music. Both composition and improvisation require a high level of creativity, which could be related to general creativity in other fields (cf. Kenny and Gellrich 2002; Sawyer 2000). Two-thirds of our sample had taken lessons in two, three or four instruments. Although we have found no other research reporting music-making in a comparable international population of researchers working in fields other than music, we believe that these results indicate a very high level of interest in learning and making music among respondents that may be specific to music researchers. In other words, the motivation to carry out research in music is likely to be – at least partially – influenced by researchers' own practical experiences.

We also wanted to know if the researchers in our sample had continued to perform music. While this was the case for two-thirds of the sample, who were still playing one or more of the instruments they had studied in the past, about 10 per cent of the respondents had stopped playing, and reported that they were currently not making music at all. The most commonly mentioned reasons for this were perceived lack of time and shift of interest from practising to other activities. Another reason was self-criticism: some respondents were no longer satisfied with their skill levels. Others were forced to stop playing following performance-related injuries, subsequently deciding to change the direction of their careers. Respondents tended to spend less time on active music making when their professional work as researchers became more time-consuming and important to them. Because our study was cross-sectional, we were able to compare responses from researchers at different stages of their careers. The more established researchers were in their field (the most established were full professors), the less time they spent practising their instruments. By contrast, researchers at the outset of their careers, such as PhD students and research assistants, were able to spend the most time on instrumental practice. Although there was some variation in reported weekly practice hours attributable to individual differences, the data suggest that, as researchers climb the academic ladder, so to speak, their involvement in active music-making is likely to decrease. An alternative explanation is that – since experienced musicians have accumulated so many hours of practice over time that they need less practice and rehearsal time than students to produce similarly polished performances – practice hours per se may not indicate performance quality or the total time researchers spend making music. Indeed, we found no significant correlation between academic position and number of public performances given. Our respondents took part in a mean of about eight concerts in a typical year, again with considerable variation between individuals.

How important did our respondents believe it is for music researchers to continue making music in daily life? On a seven-point scale, mean ratings were slightly above the neutral centre position (mean = 4.9, SD = 1.9). The more respondents were convinced that it is important for music researchers to be engaged in performance, the more they believed that research findings (not necessarily their own) had influenced their musical practice. In response to the request to say how research influenced them, respondents typically referred to findings relating to strategies for practice and performance. Sample responses are given below.

- 'Research on the parameters of expressive performance helps me think and be more conscious of how I shape the performance expressively' (male, aged 59).
- 'A lot of my research now focuses on effective practice, looking at concepts such as self-regulated learning and uses of imagery within practice, which I now use a tremendous amount in my own practising' (male, aged 29).

Most of the researchers we surveyed had been or continued to be highly engaged in practising and performing music. Even if some of them currently play fewer instruments or practise less than they did before, the experience gained from numerous hours of practice undoubtedly influences the work of many researchers, particularly in applied fields of music research.

What Kinds of Music do Music Researchers Listen to, and How do they Listen to it?

It would seem obvious that music researchers must enjoy listening to music. Nevertheless anecdotal evidence suggests that in some cases professional engagement with music diminishes the desire to listen to music outside the context of work. In our study (Wöllner et al. 2011), we asked how much time researchers spent listening to music, what kind of music they like best, and what they perceive to be the functions of music listening, for them. As shown in the previous section, most researchers engage in active music making. When we asked them to rate their enjoyment of performing and listening to music, so that we could compare their ratings for the two activities, respondents rated listening as more enjoyable. About three-quarters of respondents reported listening to music actively for up to 4 hours each week; 'active' was defined as mainly concentrating on the music itself. Five per cent said they did not listen actively at all, while 7 per cent reported active listening for more than 10 hours each week. 'Passive' listening was defined as having music on in the background, and the proportions of respondents reporting such listening were very different from those reporting active listening. Almost 20 per cent claimed that they never engaged in passive listening, either because they prefer silence or, as one respondent suggests, as soon as they become aware of background music it becomes their focus of attention and they find themselves listening actively to it. About one-third of respondents listened passively for up to 8 hours, and a further third between 8 and 20 hours each week. There was a significant correlation between hours of passive listening and age, such that the younger the respondent, the more he or she engaged in passive listening. Older respondents preferred active listening and spent more time on it than younger respondents. Longitudinal studies and in-depth interviews could clarify the extent to which the relationship between listening behaviour and age of researcher could be explained by changes in attitude. It may be that attitudes to listening are influenced by increasing engagement with research, or perhaps they simply corroborate general age-related trends in listening habits and use of listening devices (cf. Rösing 1984).

As to the musical genres preferred by researchers, classical music genres and jazz/blues/RnB were favoured over other kinds of music. Although ratings for pop, rock and folk were significantly lower, the results of our study showed that musical preferences were distributed across various genres. In short, the majority of respondents seemed relatively 'open-eared' (a term coined by Hargreaves 1982). Only rap, hip-hop, dance and techno were clearly disliked by most respondents. Nevertheless, significant negative correlations indicated a relationship between age and preference ratings: the older respondents were, the less they liked pop, rap and hip-hop. This finding supports the traditional view of music researchers as having been trained predominantly in classical music. This view, in turn, is reflected in the results of a recent investigation of articles published in the journal *Music Perception*. Tirovolas and Levitin (2010) collated all the music used as stimuli in experiments reported over a 26-year time period. They found that – for experimental purposes, at least – researchers preferred classical music to pop. More than 70 per cent of the musical examples used were 'classical' – typically Bach, Mozart and Beethoven – in comparison with the 2 per cent of examples representing pop music. In recent years researchers have begun to use stimuli from an increasing variety of musical genres. To what extent is it likely, we wonder, that researchers' own musical backgrounds and their individual preferences for music influence their research?

Finally, we investigated listening styles and the functions attributed to music listening (see Sloboda 1999). The responses of our sample of researchers were no different from those we would expect from the general population. Our respondents indicated that they listen to music mainly as a source of pleasure and enjoyment and that music can motivate, excite or calm. 'Emotional' listening styles were significantly more prevalent than 'analytical' listening styles, whereby listeners concentrate on the structure of the music. In this regard, the notion of expert listening as postulated by Adorno (1962/2003) and others does not mirror the habits of the researchers who took part in our study. Yet several respondents indicated that they listen to music in ways that reflect their professional engagement with music, such as 'imagining the performance' (male, aged 33) or '[listening] critically for aspects of performance error such as intonation' (male, aged 35).

These data provide an overview of the different ways in which music researchers listen to music, a topic which – in comparison with that of the listening behaviours of populations such as adolescents (for an overview, see North and Hargreaves 2008) – has been largely neglected in research to date. Although familiarity with musical genres is to some extent related to musical preferences and vice versa, researchers also commonly investigate and analyse music in genres that they may not listen to otherwise, that is, in non-professional contexts. If a particular musical genre such as classical music is over-represented in research (as suggested by the findings of Tirovolas and Levitin 2010), then the question arises as to whether this is a reflection of researchers' musical preferences or should be attributed merely to the fact that researchers are simply less familiar with other genres. An increasing number of researchers have begun, nevertheless, to use more popular genres as stimuli in their research, taking in the musical preferences of the wider population.

Conclusions

The process of research – including the undertaking of studies, analysis and interpretation of findings – can no longer been seen as 'objective' and independent of its context. Biographical accounts of researchers and reflections on their roles, attitudes and personal familiarity with the topics of their research have been published for more than 50 years (see Boring et al. 1952). The consideration of context is particularly pertinent to music psychology, performance science and related fields, since so many researchers originally trained as musicians or even had first careers as professional performers. There is increasing interest among music researchers in the topics of reflexivity (for example, the conferences entitled *The Reflective Conservatoire* held at the Guildhall School of Music and Drama, London in 2007, 2009 and 2012, and associated publications; see Odam and Bannan 2005) and career transitions (for example, the SEMPRE seminar held at the same institution in 2011, entitled *Stepping Out of the Shadows: Practitioners becoming Researchers*). While many researchers have specific insights into the field based on their musical training, these insights may potentially bias directions in research or influence the selection of musical examples in research studies.

Even if researchers could achieve what used to be thought of as the ideal, objective, critical distance from the subject of their research, would this be desirable? Musical experiences are ubiquitous, and for many, the appeal of music lies in its power to move us, its potential to influence our moods and behaviours, and its capacity to represent states and feelings that are difficult to express verbally. To be *familiar* with music is, in many ways, to be human. It is therefore important to address the extent to which familiarity undermines or enhances music research, particularly for those who work in fields such as music psychology and performance science, which employ empirical methods and aspire to some extent toward objectivity. Yet, on the basis of the results of our study, reported above, it is almost impossible to answer the question about critical distance, for it is very difficult indeed to find researchers who are neither performers nor listeners and, therefore, who are unfamiliar with music. A majority of our respondents reported listening *actively* to music for several hours per week, and some spent considerably more time *passively* listening. In terms of making music, 97 per cent had taken instrumental or vocal lessons in the past and only 10 per cent reported having stopped playing and performing altogether.

In short, music researchers are by and large 'musical' in their pursuits, and the potential for their experiences and preferences to influence their work output is indeed very high. Is this a problem? If it restricts – consciously or otherwise – their openness to research on music from genres and cultures with which they are less familiar or performance traditions that are not their own, then yes. It does seem to be the case that the label 'within the Western art music tradition' could be added accurately as a suffix to the vast majority of phenomena described in the extant music psychology and performance science literature. Although there are notable forays into multi-cultural and multi-genre research in music (for an overview see, for example, Thompson and Balkwill 2010), one could argue that more of such research could enrich our understanding of fundamental properties of music perception, cognition and performance (cf. Huron 2008).

There are, no doubt, areas in which intimate familiarity can give a researcher particular insight into musical processes that would otherwise not be available. For instance, knowledge or personal experience of performance anxiety symptoms or physical ailments among performers can both motivate researchers to conduct work in a specific area and give them a head start in defining research questions and employing appropriate methodological approaches. Similar examples can be provided for experiences in learning, teaching and listening to music, as well as in music therapeutic settings. The importance of reflexivity in this respect – as well as the self-knowing and self-declaration that this entails – cannot be understated. The extent to which current researchers are aware of how their individual familiarity with music influences their investigations of musical phenomena is at present unclear and ripe for future investigation.

References

Adorno, T.W. (1962/2003). *Dissonanzen. Einleitung in die Musiksoziologie.* [*Introduction to the Sociology of Music.*] Frankfurt/Main: Suhrkamp.

Alper, N.O. and Wassall, G.H. (2002). Artists' Careers – Preliminary Analysis from Longitudinal Data for the US. *Paper presented at the 12th International Conference on Cultural Economics*, Rotterdam, 13–15 June.

Alper, N.O. and Wassall, G.H. (2006). Artists' Careers and their Labor Markets. In V. Gisburgh and D. Throsby (eds), *Handbook of the Economics of Art and Culture*, Volume 1 (pp. 813–64). Amsterdam: Elsevier.

Anderson, D. (2009). A Balanced Approach to Excellence: Life-skill Intervention and Elite Performance. *Paper presented at the International Symposium on Performance Science*, Auckland.

Anderson, S.L. and Betz, N.E. (2001). Sources of Social Self-efficacy Expectations: Their Measurement and Relation to Career Development. *Journal of Vocational Behavior* 58/1: 98–117.

Arthur, M.B. and Rousseau, D.M. (2006). *The Boundaryless Career: A New Employment Principle for a New Organizational Era.* Oxford: Oxford University Press.

Ashmore, M. (1989) *The Reflexive Thesis: Wrighting Sociology of Scientific Knowledge.* Chicago, IL: University of Chicago Press.

Bandura, A. (1977). *Self-efficacy: The Exercise of Control.* New York, NY: Freeman.

Bandura, A. (1986). *Social Foundations of Thought and Action: A Social Cognitive Theory.* Englewood Cliffs, NJ: Prentice-Hall.

Bandura, A. (2009). Social Cognitive Theory of Mass Communication. In J. Bryant and M.B. Oliver (eds), *Media Effects: Advances in Theory and Research* (2nd edn) (pp. 94–124). Mahwah, NJ: Lawrence Erlbaum.

Barz, G.F. and Cooley, T.J. (eds) (2008). *Shadows in the Field: New Perspectives for Fieldwork in Ethnomusicology* (2nd edn). New York: Oxford University Press.

Bayer, M., Brinkkjaer, U., Plauborg, H. and Rolls, S. (2009). *Teachers' Career Trajectories and Work Lives: An Anthology.* Dordrecht: Springer.

Bennett, D.E. (2008). *Understanding the Classical Music Profession: The Past, The Present and Strategies for the Future.* Aldershot: Ashgate.

Bennion, A. and Locke, W. (2010). The Early Career Paths and Employment Conditions of the Academic Profession in Seventeen Countries. *European Review* 18/1: S7–S33.

Besecke, L.M. and Reilly, A.H. (2006). Factors Influencing Career Choice for Women in Science, Mathematics, and Technology: The Importance of a Transforming Experience. *Advancing Women in Leadership Online Journal* 21: 1–8.

Betz, N.E. (2007). Career Self-efficacy: Exemplary Recent Research and Emerging Directions. *Journal of Career Assessment* 15/4: 403–22.

Betz, N.E. and Fitzgerald, L.F. (1987). *The Career Psychology of Women*. San Diego, CA: Academic Press.

Bieschke, K.J. (2006). Research Self-efficacy Beliefs and Research Outcome Expectations: Implications for Developing Scientifically Minded Psychologists. *Journal of Career Assessment* 14/1: 77–91.

Bordin, E.S., Nachmann, B. and Segal, S.J. (1963). An Articulated Framework for Vocational Development. *Journal of Counselling Psychology* 10: 107–16.

Boring, E.G, Langfeld, H.S., Werner, H. and Yerkes, M. (eds) (1952). *A History of Psychology in Autobiography*. Worcester, MA: Clark University Press.

Bruner, M.W., Erickson, K., McFadden, K. and Côté, J. (2009). Tracing the Origins of Athlete Development Models in Sport: A Citation Path Analysis. *International Review of Sport and Exercise Psychology* 2/1: 23–37.

Burland, K. and Davidson, J.W. (2002). Training the Talented. *Music Education Research* 4/1: 121–40.

Carter, N.M., Gartner, W.B., Shaver, K.G. and Gatewood, E.J. (2003). The Career Reasons of Nascent Entrepreneurs. *Journal of Business Venturing* 18/1: 13–39.

Clases, C. and Wehner, T. (2009). Situated Learning in Communities of Practice as a Research Topic. In F. Rauner and R. Maclean (eds), *Handbook of Technical and Vocational Education and Training Research* (pp. 708–25). Berlin: Springer.

Cohen, A.J. (2009). Autobiography and Psychomusicology: Introduction to the Special Volume *A History of Music Psychology in Autobiography*. *Psychomusicology* 20/1–2: 10–17.

Coogan, P.A. and Chen, C.P. (2007). Career Development and Counselling for Women: Connecting Theories to Practice. *Counselling Psychology Quarterly* 20/2: 191–204.

Cools, E. and Vanderheyden, K. (2009). *The Search for Person–Career Fit: Do Cognitive Styles Matter*: Vlerick Leuven Gent Management School.

Corbin, K. (2002). *The Career Transition Pocketbook*. Alresford: Management Pocketbooks Ltd.

Davidson, J.W. (ed.) (2004). *The Musical Practitioner: Research for the Music Performer, Teacher and Listener*. Aldershot: Ashgate.

Davidson, J.W., Howe, M.J.A., Moore, D.G. and Sloboda, J. (1996). The Role of Parental Influences in the Development of Musical Ability. *British Journal of Developmental Psychology* 14: 399–412.

Davies, B., Browne, J., Gannon, S., et al. (2004). The Ambivalent Practices of Reflexivity. *Qualitative Inquiry* 10: 360–89.

Daykin, N. (2005). Disruption, Dissonance and Embodiment: Creativity, Health and Risk in Music Narratives. *Health: An Interdisciplinary Journal for the Social Study of Health, Illness and Medicine* 9/1: 67–87.

Dunne, L., Goddard, G. and Woolhouse, C. (2008). Mapping the Changes: A Critical Exploration into the Career Trajectories of Teaching Assistants who Undertake a Foundation Degree. *Journal of Vocational Education and Training* 60/1: 49–59.

Edmondson, A.C., Bohmer, R.M. and Pisano, G.P. (2001). Disrupted Routines: Team Learning and New Technology Implementation in Hospitals. *Administrative Science Quarterly* 46: 685–716.

Ericsson, K.A. (ed.) (1996). *The Road to Excellence: The Acquisition of Expert Performance in the Arts and Sciences, Sports, and Games*. Mahweh, NJ: Erlbaum.

Ericsson, K.A., Krampe, R.Th. and Tesch-Römer, C. (1993). The Role of Deliberate Practice in the Acquisition of Expert Performance. *Psychological Review* 100/3: 363–406.

Fitzgerald, L.F. and Crites, J.O. (1980). Toward a Career Psychology of Women: What Do We Know? What Do We Need to Know? *Journal of Counselling Psychology* 27: 44–62.

Fournier, G.V. and Bujold, C. (2005). Nonstandard Career Trajectories and their Various Forms. *Journal of Career Assessment* 13/4: 415–38.

Ginzberg, E., Ginsburg, S.W., Axelrad, S. and Herma, J.L. (1951). *Occupational Choice*. New York: Columbia University Press.

Goldacre, M.J., Laxton, L. and Lambert, T.W. (2010). Medical Graduates' Early Career Choices of Specialty and their Eventual Specialty Destinations: UK Prospective Cohort Studies. *British Medical Journal*, 340; doi: 10.1136/bmj/c3199.

Gore, P.A. (2006). Academic Self-efficacy as a Predictor of College Outcomes: Two Incremental Validity Studies. *Journal of Career Assessment* 14/1: 92–115.

Gottfredson, L.S. (1981). Circumscription and Compromise: A Developmental Theory of Occupational Aspirations. *Journal of Counselling Psychology Monograph* 28: 545–79.

Gottfredson, L.S. (2002). Gottfredson's Theory of Circumscription, Compromise and Self-creation. In D.B.A. Associates (ed.), *Career Choice and Development* (4th edn) (pp. 85–148). San Francisco, CA: Jossey-Bass.

Gottfredson, L.S. and Lapan, R. (1997). Assessing Gender-based Circumscription of Occupational Aspirations. *Journal of Career Assessment* 5/4: 419–41.

Hackett, G. and Betz, N.E. (1981). A Self-efficacy Approach to the Career Development of Women. *Journal of Vocational Behavior* 18/3: 326–39.

Hackett, G., Lent, R.W. and Greenhaus, J.H. (1991). Advances in Vocational Theory and Research: A 20-year Retrospective. *Journal of Vocational Behavior* 38/1: 3–38.

Hanson, E.J. (1994). Issues Concerning the Familiarity of Researchers with the Research Setting. *Journal of Advanced Nursing* 20: 940–42.

Hargreaves, D.J. (1982). The Development of Aesthetic Reactions to Music. *Psychology of Music* (special issue), 51–4.

Hargreaves, D.J. (1986). *The Developmental Psychology of Music*. London: Cambridge University Press.

Harper, B. (2002). Workplace and Health: A Survey of Classical Orchestral Musicians in the United Kingdom and Germany. *Medical Problems of Performing Artists* 17/2: 83–92.

Hays, K.F. (2002). The Enhancement of Performance Excellence among Performing Artists. *Journal of Applied Sport Psychology* 14/4: 299–312.

Hershenson, D.B. and Roth, R.M. (1966). A Decisional Process Model of Vocational Development. *Journal of Counselling Psychology* 13: 368–70.

Holland, J.L. (1959). A Theory of Vocational Choice. *Journal of Counselling Psychology* 6: 35–45.

Holland, J.L. (1985). *Making Vocational Choices*. Englewood Cliffs, NJ: Prentice Hall.

Holland, J.L. (1997). *Making Vocational Choices* (3rd edn). Lutz, FL: Psychological Assessment Resources.

Huron, D. (2008). Science and Music: Lost in Music. *Nature* 453: 456–7.

Ilgen, D.R., Lloyd, J.W., Morgeson, F.P., et al. (2003). Personal Characteristics, Knowledge of the Veterinary Profession and Influences on Career Choice among Students in the Veterinary School Applicant Pool. *Journal of the American Veterinary Medical Association* 223/11: 1587–94.

Inal, G. and Karatas-Ozkan, M. (2007). A Comparative Study on Career Choice Influences of Turkish Cypriot Restaurateurs in North Cyprus and Britain. In M. Özbilgin and A. Malach-Pines (eds), *Career Choice in Management and Entrepreneurship* (pp. 484–508). Aldershot: Edward Elgar.

Irving, R. (2009). *Career Trajectories of Women in Policing in Australia*. Canberra: Australian Institute of Criminology.

Jeffri, J. and Throsby, D. (2006). Life after Dance: Career Transition of Professional Dancers. *International Journal of Arts Management* 8/3: 54–63.

Johns, R. and Johnson, C. (2006). Building a Career in Building: Career Trajectories Among Women in the Australian Construction Industry. *Paper presented at the 14th International Employment Relations Association Conference*, Hong Kong, 5–8 July.

Jootun, D., McGhee, G. and Marland, G.R. (2009). Reflexivity: Promoting Rigour in Qualitative Research. *Nursing Standard* 23: 42–6.

Kaufman, S.B. and Kaufman, J.C. (2007). Ten Years to Expertise, Many More to Greatness: An Investigation of Modern Writers. *Journal of Creative Behavior* 41/2: 114–24.

Kenny, B.J. and Gellrich, M. (2002). Improvisation. In R. Parncutt and G.E. McPherson (eds), *The Science and Psychology of Music Performance: Creative Strategies for Teaching and Learning* (pp. 117–34). New York: Oxford University Press.

Kornhauser, Z., Lamm, J., Goldschmiedt, E., et al. (2008). The Road to Becoming a Researcher: The Path from Self-efficacy and Outcome Expectations to Research Career Interests. *Proceedings of the Columbia University Spring Undergraduate Research Symposium* 3/1: 39, http://cusj.columbia.edu/sym posium/viewissue.php (accessed 5 November 2011).

Lamb, G., Evans, N. and Baillie, D. (2006). A Career in Child and Adolescent Psychiatry? Survey of Trainees' Views. *Psychiatric Bulletin* 30/2: 61–4.

Lent, R.W. and Brown, S.D. (2006). On Conceptualising and Assessing Social Cognitive Constructs in Career Research: A Measurement Guide. *Journal of Career Assessment* 14: 12–35.

Lent, R.W., Brown, S.D. and Hackett, G. (1994). Toward a Unifying Social Cognitive Theory of Career and Academic Interest, Choice, and Performance. *Journal of Vocational Behavior* 45/1: 79–122.

Lent, R.W., Brown, S.D. and Hackett, G. (2000). Contextual Supports and Barriers to Career Choice: A Social Cognitive Analysis. *Journal of Counselling Psychology* 47/1: 36–49.

Lent, R.W., Brown, S.D., Gover, M.R. and Nijjer, S.K. (1996). Cognitive Assessment of the Sources of Mathematics Self-efficacy: A Thought-listing Analysis. *Journal of Career Assessment* 4/1: 33–46.

Lent, R.W., Lopez Jr, A.M., Lopez, F.G. and Sheu, H.-B. (2008). Social Cognitive Career Theory and the Prediction of Interests and Choice Goals in the Computing Disciplines. *Journal of Vocational Behavior* 73/1: 52–62.

Lent, R.W., Singley, D., Sheu, H.-B., et al. (2007). Relation of Social-cognitive Factors to Academic Satisfaction in Engineering Students. *Journal of Career Assessment* 15/1: 87–97.

Levinson, D.J., Darrow, C.N., Klein, E.B., et al. (1978). *Seasons of a Man's Life*. New York: Knopf.

Lindley, L.D. and Borgen, F.H. (2002). Generalized Self-efficacy, Holland Theme Self-efficacy and Academic Performance. *Journal of Career Assessment* 10/3: 301–14.

Linstead, S. (1994). Objectivity, Reflexivity, and Fiction – Humanity, Inhumanity, and the Science of the Social. *Human Relations* 47: 1321–46.

Lortie, D.C. (1975). *Schoolteacher*. Chicago, IL: Chicago University Press.

Loui, M.C. (2005). Ethics and the Development of Professional Identities of Engineering Students. *Journal of Engineering Education* 94/4: 383–90.

Malach-Pines, A. and Kaspi-Baruch, O. (2008). The Role of Culture and Gender in the Choice of a Career in Management. *Career Development International* 13/4: 306–19.

Mariño, R.J., Morgan, M.V., Winning, T., et al. (2006). Sociodemographic Backgrounds and Career Decisions of Australian and New Zealand Dental Students. *Journal of Dental Education* 70/2: 169–78.

McPherson, G. (ed.) (2006). *The Child as Musician. A Handbook of Musical Development*. Oxford: Oxford University Press.

Metzler, M.W., Lund, J.L. and Gurvitch, R. (2008). Chapter 2: Adoption of Instructional Innovation across Teachers' Career Stages. *Journal of Teaching in Physical Education* 27: 457–65.

Miller, P.S. and Kerr, G.A. (2002). Conceptualizing Excellence: Past, Present, and Future. *Journal of Applied Sport Psychology* 14/3: 140–53.

Moore, D., Burland, K. and Davidson, J. (2003). The Social Context of Musical Success: A Developmental Account. *British Journal of Psychology* 94: 524–49.

More, E., Carroll, S. and Foss, K. (2009). Knowledge Management and the Performing Arts Industry: The Case of Australia's SCOPE Initiative. *Asia-Pacific Journal of Business Administration* 1/1: 40–53.

Moreland, R.L., Argote, L. and Krishnan, R. (1998). Training People to Work in Groups. In R.S. Tindale and L. Heath (eds), *Theory and Research on Small Groups: Social Psychological Applications to Social Issues* (pp. 37–60). New York: Plenum Press.

Myburgh, J. (2005). An Empirical Analysis of Career Choice Factors that Influence First-year Accounting Students at the University of Pretoria: A Cross-racial Study. *Meditari: Research Journal of the School of Accounting Sciences* 13/2: 35–48.

Newton, D.A., Grayson, M.S. and Thompson, L.F. (2005). The Variable Influence of Lifestyle and Income on Medical Students' Career Specialty Choices: Data from Two U.S. Medical Schools, 1998–2004. *Academic Medicine* 80/9: 809–14.

North, A.C. and Hargreaves, D.J. (2008). *The Social and Applied Psychology of Music*. Oxford: Oxford University Press.

Oakland, J., MacDonald, R. and Flowers, P. (2009). The Meaning of Redundancy for Opera Choristers: An Investigation of Musical Identity in the Context of Job Loss. *Paper presented at the 7th Triennial Conference of European Society for the Cognitive Sciences of Music (ESCOM)*, University of Jyväskylä, 12–16 August.

Odam, G. and Bannan, N. (eds) (2005). *The Reflective Conservatoire: Studies in Music Education* (Guildhall Research Studies). Aldershot: Ashgate.

Pajares, F. (1996). Self-efficacy Beliefs in Academic Settings. *Review of Educational Research* 66/4: 543–78.

Parncutt, R. and McPherson, G.E. (eds) (2002). *The Science and Psychology of Music Performance*. New York: Oxford University Press.

Parsons, F. (1909). *Choosing a Vocation*. Boston, MA: Houghton Mifflin.

Pasquini, M.W. and Olaniyan, O. (2004). The Researcher and the Field Assistant: A Cross-disciplinary, Cross-cultural Viewing of Positionality. *Interdisciplinary Science Reviews* 29: 24–36.

Quesenberry, J.L. and Trauth, E.M. (2007). What Do Women Want? An Investigation of Career Anchors among Women in the IT Workforce. *Paper presented at the Proceedings of the ACM SIGMIS Conference on Computer Personnel Research*, St Louis, MO, 19–21 April.

Ritchie, L. and Williamon, A. (2011). Measuring Distinct Types of Musical Self-efficacy. *Psychology of Music 39/3*: 328–44; doi: 10.1177/0305735610374895.

Rosen, S. and Paul, C. (1998). *Career Renewal: Tools for Scientists and Technical Professionals*. London: Academic Press.

Rösing, H. (1984). Listening, Behaviour and Musical Preference in the Age of 'Transmitted Music'. *Popular Music* 4: 119–49.

Rottinghaus, P.J., Betz, N.E. and Borgen, F.H. (2003). Validity of Parallel Measures of Vocational Interests and Confidence. *Journal of Career Assessment* 11/4: 355–78.

Sawyer, R.K. (2000). Improvisational Cultures: Collaborative Emergence and Creativity in Improvisation. *Mind, Culture and Activity* 7: 177–9.

Schaub, M. and Tokar, D.M. (2005). The Role of Personality and Learning Experiences in Social Cognitive Career Theory. *Journal of Vocational Behavior* 66/2: 304–25.

Schreuder, A.M.G. and Coetzee, M. (2006). *Careers: An Organisational Perspective* (3rd edn). Claremont, Capetown: Juta.

Sellers, R.D. and Fogarty, T.J. (2010). The Making of Accountants: The Continuing Influence of Early Career Experiences. *Managerial Auditing Journal* 25/7: 701–19.

Simonton, D.K. (1997). Creative Productivity: A Predictive and Explanatory Model of Career Trajectories and Landmarks. *Psychological Review* 104/1: 66–89.

Simonton, D.K. (2000). Creative Development as Acquired Expertise: Theoretical Issues and an Empirical Test. *Developmental Review* 20/2: 283–318.

Simonton, D.K. (2009). Creative Genius in Classical Music: Biographical Influences on Composition and Eminence. *The Psychologist* 22: 1076–79.

Sloboda, J.A. (1999). Everyday Uses of Music Listening: A Preliminary Study. In S.W. Yi (ed.), *Music, Mind and Science* (pp. 354–69). Seoul/Korea: Western Music Research Institute.

Sloboda, J.A., Davidson, J.W., Howe, M.J.A. and Moore, D.G. (1996). The Role of Practice in the Development of Expert Musical Performance. *British Journal of Psychology* 87: 287–309.

Stevenson, H. (2006). Moving Towards, Into and Through Principalship: Developing a Framework for Researching the Career Trajectories of School Leaders. *Journal of Educational Administration* 44/4: 408–20.

Super, D.E. (1953). A Theory of Vocational Development. *American Psychologist* 8: 185–90.

Super, D.E. (1980). A Life-span, Life-space Approach to Career Development. *Journal of Vocational Behavior* 16: 282–98.

Super, D.E. (1990). A Life-span, Life-space Approach to Career Development. In D. Brown and L.B. Associates (eds), *Career Choice and Development* (pp. 197–261). San Francisco, CA: Jossey-Bass.

Super, D.E., Crites, J.O., Hunnel, R., et al. (1957). *Vocational Development: A Framework for Research*. New York: Bureau of Research, Teachers' College, Columbia University Press.

Thompson, W.F. and Balkwill, L.-L. (2010). Cross-cultural Similarities and Differences. In P.N. Juslin and J.A. Sloboda (eds), *Handbook of Music and Emotion: Theory, Research, Applications* (pp. 755–88). New York: Oxford University Press.

Tiedeman, D.V. and O'Hara, R.P. (1963). *Career Development: Choice and Adjustment*. New York: College Entrance Examination Board.

Tirovolas, A.K. and Levitin, D.J. (2010). 26 years of Music Perception: Trends in the Field. *Paper presented at the 11th International Conference on Music Perception and Cognition*, University of Washington, Seattle, WA, 23–27 August.

Twyman, C., Morrison, J. and Sporton, D. (1999). The Final Fifth: Autobiography, Reflexivity and Interpretation in Cross-cultural Research. *Area* 31/4: 313–25.

Upper, N. (2004). *Ballet Dancers in Career Transition: Sixteen Success Stories*. Jefferson, NC: McFarland.

Vuust, P., Gebauer, L., Hansen, N.C., et al. (2010). Personality Influences Career Choice: Sensation Seeking in Professional Musicians. *Music Education Research* 12/2: 219–30.

Williamon, A. (ed.) (2004). *Musical Excellence. Strategies and Techniques to Enhance Performance*. Oxford: Oxford University Press.

Wöllner, C., Ginsborg, J. and Williamon, A. (2011). Music Researchers' Musical Engagement. *Psychology of Music* 39/3: 364–82.

Wong, S.C.-k. and Liu, G.J. (2010). Will Parental Influences Affect Career Choice? Evidence from Hospitality and Tourism Management Students. *International Journal of Contemporary Hospitality Management* 22/1: 82–102.

Zwaan, K., ter Bogt, T.F.M. and Raaijmakers, Q. (2010). Career Trajectories of Dutch Pop Musicians: A Longitudinal Study. *Journal of Vocational Behavior* 77/1: 10–20.

Chapter 9

Familiarity in Music Education

Susan Hallam

Dictionary definitions of familiarity refer to it as the state of 'being well known' or having close relationships. In these terms, 'familiarity' is relevant to almost every aspect of music education. All learning involves developing familiarity with a body of knowledge. In music this may include the tonal, rhythmic or structural forms of a particular musical system or genre, specific pieces of music, or a wide range of musical skills including performing, improvisation, composition or analysis and critique. Extensive familiarity with knowledge in a domain supports the development of expertise. This is acquired in relation to particular communities of practice, each of which operates at a number of different levels. Within such communities the process of increasing familiarity within the domain may be enhanced by familiarity with the approaches to learning and teaching that are adopted, and with the relationships developed with others within the community which support the learning process. This chapter will explore issues of familiarity in relation to music education through consideration of existing research with particular reference to enculturation of musical language, developing musical skills and teachers and teaching, including the impact of educational and life transitions.

Listening, Enculturation and Liking

The musical knowledge that adults acquire over time is built on structures present at birth. These provide the basis for the perception of sound and musical enculturation, the process by which the child develops internal schemata of the music of its culture. The newborn has very well developed systems for processing music and is predisposed to attend to melodic contour, rhythmic patterning and consonant sounds. Infants possess similar sensitivity to the pitch and rhythmic grouping of sounds as adults. What they need to learn during early childhood is the tonal (or other) music language system of their culture. The process of learning this begins in the womb. The human auditory system is functional 3–4 months before birth and familiarity with the tonal system predominant in the infant's environment begins then, foetuses reliably reacting to external sounds after 28–30 weeks of gestation (Woodward 1992). Infants recognise music that they have heard in the womb both before and immediately after birth (Hykin et al. 1999; Shahidullah and Hepper 1994) and react positively to music that their mothers have listened to regularly in the last 3 months of pregnancy (Feijoo 1981; Hepper 1988).

Developing understanding of particular tonal systems takes time and depends on the type and extent of exposure to music. The greater the exposure, the more fully and speedily this knowledge will be acquired. It is not necessary for children to focus on listening to music for this to occur. Musical schemata are acquired without conscious awareness. This means that the environment of children in their early years can enhance their implicit knowledge of music even while they are undertaking other activities. Children are likely to be exposed to music in the home through the radio, recordings or television. Globalisation has meant that worldwide most children are exposed to Western popular music, which may be in addition to other genres or world musics. Children exposed to Western tonality, generally, by the age of five can organise songs around stable tonal keys, but still do not have a stable tonal scale system that can be used to transpose melodies. This develops later (Lamont and Cross 1994). Individuals may develop an understanding of and become proficient in the technical requirements and stylistic nuances of two musical systems; in the same way that they can become bilingual (O'Flynn 2005), they can become bi-musical (Hood 1960). To achieve this they have to commit the time necessary for enculturation into two systems to take place.

For music to become familiar children must be able to remember it. Implicit knowledge of the tonal system provides a framework that facilitates the memorisation of particular songs or pieces of music. Infants initially become accustomed to the particular structures of the music to which they are exposed. By the age of 12 months they prefer patterns that conform to those structures (Trehub, Hill and Kamenetsky 1997). By 6 months they can remember the tempo and timbre of music with which they have become familiar, but when pieces are played at new pitches or in a different timbre they do not recognise them (Halpern 1989). Overall, infant representations in memory are closely related to the form that they took on initial presentation. However, this changes over time. Gradually, highly familiar melodies come to be recognised when they are played in a different key (Halpern 1989; Trehub, Thorpe and Morrongiello 1987). The ability to transform representations depends on the extent of familiarity (Schellenberg and Trehub 1996, 1999). By the time that children enter pre-school, they can recognise familiar tunes across many different types of transformations. Some children develop what has come to be known as absolute pitch. This depends on familiarity with particular sounds, the naming system for those sounds and making connections between them. Most possessors of absolute pitch acquire it naturally as they engage with formal music tuition (Levitin 1994). The exact nature of the absolute pitch that individuals develop depends on the particular pitches and timbres with which they have become familiar, and accuracy increases with greater familiarity (Bahr, Christensen and Bahr 2005; Takeuchi and Hulse 1993).

Increased familiarity with a specific piece of music also changes individuals' understanding of the structure of the music and how themes within it are related (Pollard-Gott 1983). There seems to be an intuitive recognition of this. A recent study exploring individuals' beliefs about the way that musical understanding is acquired showed that 63 per cent of a sample of adults and children believed that it was developed through listening. This increased to 75 per cent in a sample of professional musicians (Hallam 2009; also see Chapter 2).

Familiarity with specific musical systems is also crucial for interpreting the emotions portrayed through music. Where music is from cultures with which the individual is unfamiliar, understanding its emotional meaning can be difficult. The recognition of emotion in music develops in young children, although studies vary in identifying the ages when this occurs. What is clear is that children in their early years begin to recognise musical depictions of emotion and respond to them. The specific age at which this occurs is likely to depend on the extent of the child's exposure to music within and across genres, and their general cognitive and emotional development (see Hallam 2006).

Familiarity plays an important role in liking for music. This is illustrated particularly well in infants, who prefer music and voices with which they are familiar (Panneton 1985; Satt 1984), particularly the voices of their mothers (Lecanuet 1996). Despite this, over-familiarity with particular pieces of music can lead to boredom and dislike. The complexity of music, the degree of variability or uncertainty, is critical in determining whether it is liked. The relationship between complexity and liking can be conceptualised as an inverted U-shaped curve (Berlyne 1974). A moderate level of complexity elicits the maximum liking for the music. The effect of exposure to music, repetition, training or practice is to lower its perceived complexity. The relationship between familiarity and listening to music suggests that musical preferences can be changed through prolonged exposure (see Chapter 1), a phenomenon that is important for music education and has been illustrated, for instance, by Shehan (1985), who engaged sixth-grade students over a period of 5 weeks in studying Japanese, Indian and Hispanic music in their cultural contexts and found a marked change in liking.

Young children are open-eared and respond positively to a wide range of different kinds of music (Kopiez and Lehmann 2008), but as they develop, social factors increasingly play a role in musical preferences. Musical taste is not acquired in a vacuum. It is an integral part of the lifestyle of the individual, reflecting their identity and their cultural, historical, societal, familial and peer group background (North and Hargreaves 2007a, b). Current negative attitudes towards 'classical' music may be more related to these factors than the music itself. Where 'classical' music is presented in contexts outside the concert hall, for instance, in advertisements, television programmes, film or sporting contexts, it is not only accepted but becomes part of popular culture. This, in part, may be because it has been allowed to become familiar.

Partly in response to adolescents' liking for popular music, there have been suggestions that it should be included in the school curriculum as a means of increasing motivation. In the UK, the Musical Futures approach has introduced informal learning into the classroom with a focus on popular music, students learning to copy a popular CD track, moving on to composing their own songs

(Green 2008). However, students are reluctant to move beyond popular music to other genres (Hallam, Creech and McQueen 2009b). Although the approach has had considerable success, some teachers who have experienced a more formal classical training themselves find the change problematic.

Developing familiarity with music through listening in everyday life leads to the acquisition of sophisticated listening skills over time. Those with no formal musical training can recognise melodies as well as trained musicians (Dowling 1978) and in some cases the skills of non-musicians are greater than those who have received professional training. For instance, Mito (2004) asked young people to memorise the same Japanese pop song, by ear, over four 10 minute practice sessions and reproduce it after each session. Those without formal music training, particularly those who performed regularly at karaoke sessions, memorised the song better than those who had received formal training. The latter reported particular difficulties in learning the songs without notation to support the process. This demonstrates the extent to which the development of any musical skill depends on familiarity and repetition of activities. Having high levels of musical expertise in a related area only partially supports performance when different skills are required. To develop new skills requires considerable time and effort focused on activities designed to develop those skills (Sudnow 1978).

Developing Musical Skills

There are three main phases in developing musical knowledge and skills (see Table 9.1). Initially, the learning is largely under cognitive, conscious control. In skill development, the learner has to understand what is required to undertake the task and carries it out while consciously providing self-instruction. In acquiring a new body of declarative knowledge, in the acclimation phase, there is development of an understanding of the scope of the knowledge domain where the learner is introduced to specific facts, rules, terminology or conventions, definitions, simple concepts and principles. Simple links between the component parts of the knowledge domain are made. As procedural skills are acquired the learner begins to put together a sequence of responses, which become more fluent over time. Errors are detected and eliminated. Feedback from the sounds produced and the teacher play an important role. In acquiring a body of knowledge in the competence phase, more complex interrelationships are established as the knowledge base is expanded and refined so that it can be used to solve problems. In the final phase, the knowledge domain is secure and the aim is to improve speed and accuracy (proficiency/expertise), while in acquiring skills automisation develops and the skills can be used fluently and quickly (Alexander 1997; Fitts and Posner 1967).

In music, many skills are acquired simultaneously so new skills are constantly added to the repertoire. As mastery of more advanced skills is acquired, skills learned earlier are continuously practised so they achieve greater automaticity. As one set of skills is becoming increasingly automated, others will be at the associative and cognitive stages. The development of the knowledge base and the skills needed to work with that knowledge in most domains are inextricably intertwined. There is an interplay between two types of knowledge: declarative (knowing something) and procedural (knowing how).

Table 9.1   Stages of procedural and declarative skill acquisition

Procedural knowledge (knowing how)	Declarative knowledge (knowing something)
Cognitive–verbal–motor stage	Acclimation
Associative stage	Competence
Autonomous stage	Proficiency/expertise

Knowledge, whether declarative or procedural, once learned is stored in long-term memory and can be retained over many years, motor skills being particularly resistant to forgetting. Remembering depends on familiarity and draws on a range of different resources. Musical memory relies primarily on aural representations, that is, remembering the sound. When playing an instrument, kinaesthetic schemata are acquired. These are based on memory for movement. Where music is learned from notation, visual memory may play a part, some musicians being able to visualise the music on the page as they are playing. If the music is long and complex, the musician may need to support what has been learned automatically through practice with knowledge of the structure of the piece, the automated sections being fitted into this structure (Chaffin, Imreh and Crawford 2002; Hallam 1997).

To develop musical skills, therefore, the learner must become familiar with those skills over a long period of time, regularly engaging with their use. This enables changes to occur in the brain. When we learn, many neurons in the brain are active simultaneously, and if we persist in particular activities, there are changes in the growth of axons and dendrites and the number of synapses connecting neurons related to those activities. Over time, as the individual persists, changes in the efficacy of existing connections are made through a process known as myelinisation. This involves an increase in the coating of the axon of each neuron, which makes the established connections more efficient, thus reinforcing learning and supporting the development of automaticity. In response to these changes, as learning occurs, the cerebral cortex is re-organised. This means that every individual has a unique brain pattern that reflects their life experiences (Pantev et al. 2003).

In learning to play an instrument or develop vocal skills for performance, the existing schemata that have already been developed relating to the sound of music play a crucial role (Dowling 1993). Familiarity with the tonal, rhythmic and structural systems predominant in a particular musical culture facilitates the development of musical skills, providing learners with templates against which to judge their learning. This is important across a range of pursuits including performing, improvisation, composition, analysis and critique. Without well-developed aural schemata, individuals are unable to detect errors or create music that is recognisable within the tonal system of their culture (Hallam 2001a). The ways in which learners spend the required time to develop expertise are varied. There may not be any 'best' way providing that the learner has sufficient familiarity with the body of knowledge or the skill to recognise when they are making errors. Feedback from someone with greater expertise plays an important role in this process. This does not need not be a teacher working in a formal learning environment; it can be a peer or a member of a community of practice.

Learners also need to acquire a range of strategies to support their learning. Learning is more effective when the learner is familiar with and can utilise a range of learning strategies. Being able to assess the nature of a particular task, being aware of personal strengths and weaknesses, having a range of learning strategies, being familiar with the nature of the required outcomes and being able to monitor progress towards the goal are all important (Hallam 1995, 2001b). Developing these meta-cognitive skills is crucial in the acquisition of high levels of expertise in terms of both performing and creating music.

Although historically, creating music through composition has tended to be considered in terms of individual innate abilities, in fact, those acknowledged as among the greatest composers acquired their skills over long periods of time through familiarity and engagement with music from a very early age. While composers differ in their individual characteristics, they do share common experiences, including: the opportunity to become involved in music; commitment; extensive knowledge of music; and extensive experience of working in the musical domain. Those acknowledged as among the greatest also began composing when they were very young, made their first contributions to the repertoire at a very young age and continued to be prolific in their writing throughout their lives (Simonton 1997). Similarly, 'enabling skills' have been identified in young people engaged in composing, including musical expertise, conceptual understanding and aesthetic sensitivity. These are influenced by a range of factors including motivation and environment (Webster 1988, 1991). Familiarity with a wide range of music and the skills associated with it are as important then for composers as for performers.

To acquire the level of musical expertise needed to become an international soloist in the West takes up to 16 years with many hours of solitary practice (Ericsson, Krampe and Tesch-Romer 1993). Extensive familiarity with particular content and skills is needed, although there are differences between instruments, with some having greater technical requirements and a larger repertoire to be mastered (Jorgensen and Hallam 2009). While there are ongoing debates relating to the actual time required to be spent in individual 'deliberate practice' (Ericsson et al. 1993) as opposed to other musical activities, for instance, playing in a range of musical groups, playing more than one instrument, improvising, composing or listening, there is powerful evidence that a considerable time investment in active engagement in music making is necessary to develop high levels of expertise. The specific amount of time is influenced by the demands of particular genres and instruments; for instance, jazz guitarists (Gruber, Degner and Lehmann 2004) and singers tend to begin formal training later (Kopiez 1998), while different instrumental groups in the conservatoire practise for different amounts of time (Jorgensen 2002).

In learning to play an instrument, a range of sub-skills may also have to be acquired, for instance, reading notation and sight reading (Goolsby 1994; Lehmann and Ercisson 1996). These develop through high levels of familiarity with notation and become automated in the same way as other musical skills. This is illustrated in the way that players can be misled into making 'proof reader's errors' (playing notes that are not really there) because they identify familiar patterns of notation rather than reading individual notes (Sloboda 1976). Generally, written notation provides insufficient information for the music to be reproduced. For instance, in playing a Viennese Waltz the performance of the underlying rhythm is distorted considerably, with the first crotchet being shortened and the second lengthened. This is not indicated in notation, so musicians need to be familiar with this tradition. Musicians learn to develop expressive patterns (how to manipulate timing, dynamics, articulation and other parameters in their playing) in accordance with contemporary performance trends that are also linked to styles of playing. Some of these trends are culturally specific – hence there are 'schools of playing', such as the Russian school of piano playing in the twentieth century, or the virtuosic violin school led by Paganini. Similarly, in ensemble performance, Davidson and Good (2002) refer to the coordination of process and content that involves reconciliation of entrances, exits, cues and so on (process) as well as negotiation of shared knowledge and cultural conventions (content). Equally, listeners can detect expressive variations in performance and interpret them. The most effective performers exaggerate expression and are more consistent in its use (Sloboda 1983). To achieve this requires extensive familiarity with musical conventions.

There are two further issues that merit attention about the role of familiarity and the development of musical skills in relation to performance: anxiety and communication. In most areas of music education in the West, learners are required to perform. Anxiety or arousal about performance is common among musicians whatever their age. While this can be advantageous – some degree of tension can energise and enhance the performance (Caldwell 1990; Salmon 1991) – it can become debilitating and disrupt musical skills (Brotons 1994). More recently, Valentine (2002) distinguishes reactive, adaptive and mal-adaptive types of anxiety, that is, anxiety caused by inadequate preparation (reactive), beneficial (adaptive) and deleterious (mal-adaptive). For those aspiring to become professional musicians, anxiety poses a serious threat. Most psychological approaches to reducing performance anxiety have focused on changing familiar negative and task-irrelevant behaviours or cognitions and substituting optimistic task-oriented self-talk (Kendrick et al. 1982), that is, changing familiar debilitating thought patterns to those that are more adaptive.

Musicians who have not felt the need to seek professional advice about reducing anxiety refer to the importance of being technically and musically well prepared (Bartel and Thompson 1994; Valentine 2002), hence being extremely familiar with the material to be performed. Becoming familiar with a range of strategies to cope with anxiety is also important, including warming up, over-learning, slow practice, listening to recordings, practising more difficult materials than those required, sight reading new works and ensuring that instruments are in excellent condition. Another strategy that young musicians sometimes adopt is playing informally for friends to familiarise themselves with how they will feel during performance. Developing familiarity through imagining performance in a range of different acoustical environments can also be useful (Edlund 2000).

A key element of public performance is communication with fellow musicians and the audience. Communication depends on shared meanings, understandings and intentions on the part of performers and audience, as indicated previously. This relies on a mutual knowledge of the underlying musical system and the alternative ways in which music can be performed. For ensemble performers, playing in small or large groups requires team work. Familiarity with co-workers influences the development of musical and social relationships (see Chapter 12). To be successful in the long term, music rehearsals have to be underpinned by strong social frameworks (for a full review see Davidson and King 2004), and ensemble performers have to develop important social skills, such as trust and respect (Young and Colman 1979).

Teachers and Teaching

The length of time required to develop high levels of expertise in music, particularly through learning to play an instrument, means that the same teacher may work with a student over many years and that long-term relationships are frequently developed between teachers and pupils. Most instrumental lessons are undertaken on a one-to-one basis, within the apprenticeship model or in very small groups. To support the development of high-level musical skills, instrumental and vocal teachers need high levels of musical expertise. They also have to be able to develop positive interpersonal relationships with their students. Early motivation to engage with and continue making music requires teachers to be relatively uncritical, encouraging and enthusiastic (Sosniak 1990), although as students progress, the relationship with the teacher changes from one of liking and admiration to respect for their expertise. For those who are committed to becoming a musician, teachers become role models (Davidson et al. 1998) and may become idealised (Abeles 1975), with learners adopting their opinions and attitudes (Gaunt 2010). This aspect of familiarity may be counter-productive in that it can prevent the individual from becoming independent and developing their own musical ideas.

Relationships with teachers have a significant impact on the level of expertise attained (Creech and Hallam 2009, 2010; Manturzewska 1990; Sosniak 1990). Creech (2009) researched the contribution of interpersonal interaction to teaching and learning and found that psychological remoteness within pupil–teacher relationships had a detrimental effect on learning. Mutual respect, common purpose and the establishment of child-centred rather than teacher-centred goals were associated with positive outcomes. The characteristics that seem to support this process include emotional sensitivity, being caring and empathetic, encouraging, friendly, people-oriented, interested in students and having a sense of humour (Pembrook and Craig 2002).

Where teachers are not sufficiently familiar with the musical knowledge and skills that they need to carry out their teaching, they tend to lack confidence. This is particularly the case in primary schools, where generalist teachers are frequently expected to teach music (Hallam and Creech 2008; Hallam et al. 2009a). At secondary school the majority of music teachers have been trained in the classical performance tradition, which may make them reluctant to engage their pupils in performing or creative activities in genres with which they are unfamiliar and may even dislike. Teachers' backgrounds are a significant factor in their own work. The potential to engage with new ideas is often undermined by existing beliefs and expectations. Unfamiliarity with other genres or particular activities, for instance, composing and improvising, may lead to these activities being included less often in music lessons or indeed not at all (MacDonald and Miell 2000).

The teaching styles that teachers develop over time are heavily influenced by their own life histories, and in particular past relationships with their own teachers (Morgan 1998). These experiences may lead to specific interpersonal dynamics in lessons that are based on unconscious processing. Defence mechanisms, for instance, projection or turning passive into active may be used by teachers to avoid unpleasant memories being recalled relating to their own experiences as learners. Past problems experienced by the teacher may be projected onto the pupil (Gustafson 1986).

Issues relating to levels of attachment may be particularly important in teacher–learner relationships. In a small-scale study of the relationships between teachers and students in singing lessons, de Sa Serra Dawa (2010) explored differences in interactions between those with secure or dismissive-avoidant attachment styles. The latter were less receptive to the needs to their students, less interested in their general well-being and more rigid in their teaching approaches. Students with anxious, dismissive-avoidant behaviour were more dependent on their teachers, particularly in relation to career advice, were less likely to ask questions and tended to raise relationship problems in the lesson (see Table 9.2). Where teacher and student both had secure styles the relationship was more productive and provided the necessary elements for the singing activities to progress smoothly. In mixed dyads the difficulties in relationships tended to interfere with learning, particularly where the teacher had a dismissive-avoidant style.

Table 9.2   Teacher and learner attachment styles

Secure attachment style	*Dismissive-avoidant attachment style*
Teacher's style of attachment	
Care for students: enquiring about students' well-being at the start of the lesson and asking about their activities in the previous week	**Little personal involvement:** little interest in students' external life
Teaching adaptability: able to adopt a range of strategies for the benefit of a student's development	**Teacher rigidity:** maintaining the same approach regardless of its success
Adaptation to the physical condition of the student's voice: teachers are willing to respond to the physical condition of the student's voice	**Maintenance of the same methods regardless of condition of voice:** the same methods are used regardless of physical condition
Balanced justification of teaching strategies: singing interspersed with explanations when they are needed	**Over-justification or no justification of teaching options:** teachers adopt one of these extreme behaviours
Available: regular lessons given and contact maintained between lessons if necessary	**Little availability:** lessons are difficult to schedule and there is little communication between lessons
Student's style of attachment	
Questioning: learner questions teacher more than other styles	**Fears rejection:** does not question teacher's choices or suggestions
Independent: makes own decisions, depends less on the teacher	**Dependent:** depends on teacher for decisions, particularly about career
Non-problematic: few problems with other individuals in the institution	**Problematic:** student brings to lessons relational problems and disagreements with other people
Confident: confident in singing	**Worried:** fear of failure

Source: Adapted from de Sa Serra Dawa (2010).

Music teaching (and learning) occurs in a wide range of different ways and in different contexts, not all of which involve formal educational environments. For instance, in the UK, within the brass band tradition, new entrants to the band learn as they participate in the ensemble by being shown the necessary technical (and other) skills by older, more experienced members. Within such communities of practice, players progress through a sequence of stages and the senior players act as mentors or (informal) tutors. Equally, observations of amateur bluegrass banjo players revealed that, as inexperienced participants became more familiar with the community, they progressed as auditors, listeners and inceptors until becoming competent banjoists. The inexperienced players thus developed familiarity with the music and the instrument via role models in an informal context that was highly educational (Adler 1980).

In schools, teachers may encourage learning through peer interaction; indeed, group music making is common in performance as well as other activities, such as composition and improvisation. Pupils develop skills through listening, trial and error, repetition, watching and taking advice from other group members. Where groups are dysfunctional little is achieved, much time being spent in attempting to resolve interpersonal difficulties. Pupils stress the importance of groups being able to work productively together and indicate that they cannot always resolve interpersonal difficulties (Hallam et al. 2009b). There can be advantages in learners working with their friends as relationships do not have to be worked on and effort can be focused on the task. For example, the compositions produced by friendship groups have been rated as superior to those of children working with someone who was only an acquaintance. The interaction, both verbal and musical, between best friends is characterised as being more conducive to good quality collaboration for learning (MacDonald, Hargreaves and Miell 2002; Miell and MacDonald 2000).

The importance of familiarity in educational contexts is illustrated by the disruptive impact of transitions. Transitions can involve changes in: learning environments, either between or within institutions; teachers or facilitators; pedagogical practices; peer and friendship groups; expectations of performance and types of assessment; required skills; conceptual understanding; perceptions of the nature of knowledge itself; ways of thinking about learning; and ways of thinking about self and identity. All of these require a process of cognitive restructuring and the negotiation of change. While some individuals are able to embrace change and the unfamiliar easily, many are not (Hallam 2010).

The move from a familiar to an unfamiliar educational environment can lead to individuals dropping out, particularly if they have low levels of self-efficacy and commitment related to an activity (Hallam 1998). Even when they persist, transition can disrupt learning. For instance, in the transition to the conservatoire, students often experience a dip in self-esteem, self-efficacy and motivation, with an increase in anxiety compared with pre-entry, although there is some partial recovery during the second year of study (Creech, Gaunt and Hallam 2009; Long et al. 2010). Learners also need to become familiar with new ways of learning to derive the most benefit from them. For instance, perceptions of the value and purpose of masterclasses is different between students with prior masterclass performing experiences and those lacking such experience. The former are more likely to value learning by listening to their peers in a masterclass setting and appraise the performance themselves rather than focusing on the comments and demonstrations made by the master. They also regard masterclasses as more motivational (Long et al. 2011).

Both familiarity and life transitions may also play a role in the extent to which people are motivated to engage with music in later life. Typically, amateur musicians are involved in music in their early years at home and at school, disengage while developing their career and focusing on family matters, and return in their 50s, although some continue an engagement with music throughout the transitions of their lives (Hallam et al. 2010; Taylor 2009). Taylor (2009) in a qualitative study with 21 adults playing either the piano or the keyboard found that participants had very early memories of positive family activities involving music. These contributed significantly to their motivation to make music as adults. Those who had been motivated to learn the piano when they were children were encouraged to do so by teachers with a friendly demeanour, focussed attention, clear teaching and praise. This contributed to their positive musical identity construction as pianists in childhood and, reconstructed as memories, contributed to their enjoyment of and intrinsic motivation for music as adults. Some participants were unable to take piano lessons as children because their families could not afford lessons, but had a strong desire to learn that was satisfied when they were older adults. Piano lessons in teenage years were common but were not always positive experiences and many gave up playing through boredom, lack of enjoyment, perceived irrelevance or the pressure of school work. Many had positive memories of music making in the community, usually of singing in a choir and taking part in music festivals. Parental musical interest and/or positive experiences of music making at home, school and elsewhere provided the foundation for the respondents' lifelong interest in musical participation and learning in general and for playing a keyboard instrument in particular.

Conclusion

Familiarity plays a key role in music education. Coming to know is crucial in relation to the enculturation of musical language, the development of aural schemata, and the declarative and procedural knowledge that underpins the development of performing and creative expertise. The process of actively engaging with music, whether through formal tuition, informal learning in a community of practice or rehearsing with others is enhanced where relationships are well established and based on mutual respect. Familiarity with musical engagement in childhood – inside and outside of school – also seems to have an impact on re-engagement in later life. However, the way that we tend to like what is familiar can be problematic when we need to make transitions and engage with new ideas. In these cases, the roles of teachers and peers in educational settings are vital in supporting individuals so as to ensure that change can be handled successfully.

References

Abeles, H.F. (1975). Student Perceptions of Characteristics of Effective Applied Music Instructors. *Journal of Research in Music Education* 23: 147–54.

Adler, T. (1980). *The Acquisition of a Traditional Competence: Folk-musical and Folk-cultural Learning among Bluegrass Banjo Players*. Unpublished PhD dissertation, Indiana University, *Dissertation Abstracts International* 41/3: 1165A.

Alexander, P.A. (1997). Mapping the Multidimensional Nature of Domain Learning: The Interplay of Cognitive, Emotional and Strategic Forces. In M.L. Maehr and P.R. Pintrich (eds), *Advances in Motivation and Achievement*, Volume 10 (pp. 213–50). Greenwich, CT: JAI Press.

Bahr, N., Christensen, C.A. and Bahr, M. (2005). Diversity of Accuracy Profiles for Absolute Pitch Recognition. *Psychology of Music* 33/1: 58–93.

Bartel, L.R. and Thompson, E.G. (1994). Coping with Performance Stress: A Study of Professional Orchestral Musicians in Canada. *The Quarterly Journal of Music Teaching and Learning* 5/4: 70–78.

Berlyne, D.E. (ed.) (1974). *Studies in the New Experimental Aesthetics: Steps toward an Objective Psychology of Aesthetic Appreciation*. New York: Halsted.

Brotons, M. (1994). Effects of Performing Conditions on Music Performance Anxiety and Performance Quality. *Journal of Music Therapy* 31: 63–81.

Caldwell, R. (1990). *The Performer Prepares*. Dallas, TX: PST.

Chaffin, R., Imreh, G. and Crawford, M. (2002). *Practicing Perfection: Memory and Piano Performance*. Mahwah, NJ: Erlbaum.

Creech, A. (2009). Teacher–Parent–Pupil Trios: A Typology of Interpersonal Interaction in the Context of Learning a Musical Instrument. *Musicae Scientiae* 13/2: 163–82.

Creech, A. and Hallam, S. (2009). Interaction in Instrumental Learning: The Influence of Interpersonal Dynamics on Outcomes for Parents. *International Journal of Music Education (Practice)* 27/2: 93–104.

Creech, A. and Hallam, S. (2010). The Influence of Interpersonal Dynamics on Outcomes for Violin Teachers. *Psychology of Music* 38/4: 403–22.

Creech, A., Gaunt, H. and Hallam, S. (2009). Plans and Aspirations of Young Musicians: An Investigation into Aspirations and Self-perceptions in the Conservatoire. *Paper presented at the 2nd International Reflective Conservatoire Conference: Building Connections*, Guildhall School of Music and Drama, London, 28 February to 3 March.

Davidson, J.W. and Good, J.M.M. (2002). Social and Musical Co-ordination between Members of a String Quartet: An Exploratory Study. *Psychology of Music* 30: 186–201.

Davidson, J. and King, E.C. (2004). Strategies for Ensemble Practice. In A. Williamon (ed.), *Musical Excellence: Strategies and Techniques to Enhance Performance* (pp. 105–22). Oxford: Oxford University Press.

Davidson, J.W., Moore, J.W., Sloboda, J.A. and Howe, M.J.A. (1998). Characteristics of Music Teachers and the Progress of Young Instrumentalists. *Journal of Research in Music Education* 46: 141–60.

de Sa Serra Dawa, A.S.A. (2010). *The Teacher–Student Relationship in One-to-one Singing Lessons: A Longitudinal Investigation of Personality and Adult Attachment*. Unpublished PhD dissertation, University of Sheffield.

Dowling, W.J. (1978). Scale and Contour: Two Components of a Theory for Memory for Melodies. *Psychological Review* 85: 341–54.

Dowling, W.J. (1993). Procedural and Declarative Knowledge in Music Cognition and Education. In T.J. Tighe and W.J. Wilding (eds), *Psychology and Music: The Understanding of Melody and Rhythm* (pp. 5–18) Hillsdale, NJ: Erlbaum.

Edlund, B. (2000). Listening to Oneself at a Distance. In C. Woods, G. Luck, R. Brochard, et al. (eds), *Proceedings of the 6th International Conference on Music Perception and Cognition*, Keele University.

Ericsson, K.A., Krampe, R.T. and Tesch-Romer, C. (1993).The Role of Deliberate Practice in the Acquisition of Expert Performance. *Psychological Review* 100/3: 363–406.

Feijoo, J. (1981). Le Feotus, Pierre et Le Coup. In E. Herbinet and M-C. Busnel (eds), *L'Aube des Sens* (pp. 192–209). Cahiers du Nouveau-né, Paris: Stock.

Fitts, P.M. and Posner, M.I. (1967). *Human Performance*. Belmont, CA: Brooks Cole.

Gaunt, H. (2010). One-to-one Tuition in a Conservatoire: The Perceptions of Instrumental and Vocal Students. *Psychology of Music* 38/2: 178–208.

Goolsby, T.W. (1994). Profiles of Processing: Eye Movements during Sight reading. *Music Perception* 12: 97–123.

Green, L. (2001). *How Popular Musicians Learn: A Way Ahead for Music Education*. Aldershot: Ashgate.

Green, L. (2008). *Music, Informal Learning and the School: A New Classroom Pedagogy*. Aldershot: Ashgate.

Gruber, H., Degner, S. and Lehmann, A.C. (2004). Why Do Some Commit Themselves in Deliberate Practice for Many Years – and So Many Do Not? Understanding the Development of Professionalism in Music. In M. Radovan and N. Dordevic (eds), *Current Issues in Adult Learning and Motivation* (pp. 222–35). Ljubljana: Slovenian Institute for Adult Education.

Gustafson, R.I. (1986). Effects of Interpersonal Dynamics in the Student–teacher Dyads in Diagnostic and Remedial Content of Four Private Violin Lessons. *Psychology of Music* 14/2: 130–39.

Hallam, S. (1995) Professional Musicians' Orientations to Practice: Implications for Teaching. *British Journal of Music Education* 12/1: 3–20.

Hallam, S. (1997). The Development of Memorization Strategies in Musicians: Implications for Instrumental Teaching. *British Journal of Music Education* 14/1: 87–97.

Hallam, S. (1998). Predictors of Achievement and Drop Out in Instrumental Tuition. *Psychology of Music* 26/2: 116–32.

Hallam, S. (2001a). The Development of Metacognition in Musicians: Implications for Education. *The British Journal of Music Education* 18/1: 27–39.

Hallam, S. (2001b) The Development of Expertise in Young Musicians: Strategy Use, Knowledge Acquisition and Individual Diversity. *Music Education Research* 3/1: 7–23.

Hallam, S. (2006). *Music Psychology in Education*. London: Institute of Education, University of London.

Hallam, S. (2009). In What Ways do we Understand Music? *Keynote presentation at the Australian Society of Music Education 17th Conference: Understanding Music*, University of Tasmania, Launceston, 10–14 July.

Hallam, S. (2010). Transitions and the Development of Expertise. *Psychology Teaching Review* 16/2: 3–32.

Hallam, S. and Creech, A. (2008). *EMI Music Sound Foundation: Evaluation of the Impact of Additional Training in the Delivery of Music at Key Stage 1*. London: Institute of Education, University of London.

Hallam, S., Burnard, P., Robertson, A., et al. (2009a). Trainee Primary School Teachers' Perceptions of their Effectiveness in Teaching Music. *Music Education Research* 11/2: 221–40.

Hallam, S., Creech, A. and McQueen, H. (2009b). *Musical Futures: A Case Study Investigation: Interim Report December 2010*. London: Institute of Education, University of London.

Hallam, S., Creech, A., Gaunt, H., et al. (2010). Promoting Social Engagement and Welbeing in Older People through Community Supported Participation in Musical Activities. *Paper presented at the International Society of Music Education*, Beijing, 1–6 August.

Halpern, A.R. (1989). Memory for the Absolute Pitch of Familiar Songs. *Memory and Cognition* 17: 572–81.

Hepper, P.G. (1988). Fetal 'Soap' Addiction. *Lancet* 1: 1147–8.

Hood, M. (1960). The Challenge of Bi-musicality. *Ethnomusicology* 4/2: 55–9.

Hykin, J., Moore, R., Duncan, K., et al. (1999). Fetal Brain Activity Demonstrated by Functional Magnetic Resonance Imaging. *Lancet* 354: 645–6.

Jorgensen, H. (2002). Instrumental Performance Expertise and Amount of Practice among Instrumental Students in a Conservatoire. *Music Education Research* 4: 105–19.

Jorgensen, H. and Hallam, S. (2009). Practising. In S. Hallam, I. Cross and M. Thaut (eds), *Oxford Handbook of Music Psychology* (pp. 265–73). Oxford: Oxford University Press.

Kendrick, M.J., Craig, K.D., Lawson, D.W. and Davidson, P.O. (1982). Cognitive and Behavioural Therapy for Musical Performance Anxiety. *Journal of Consulting and Clinical Psychology* 50: 353–62.

Kopiez, R. (1998). Sangerbiographien aus Sicht der Expertiseforschung: Ein Schwachstellenanalyse [Singers are Late Beginners: Singers Biographies from the Perspective of Research on Expertise. An Analysis of Weaknesses]. In H. Gembris, R. Kraemer and G. Maas (eds), *Singen als Gegenstand der Grundlagenforschung* (pp. 37–56). Augsberg: Wissner.

Kopiez, R. and Lehmann, M. (2008). The 'Open-earedness' Hypothesis and the Development of Age-related Aesthetic Reactions to Music in Elementary School Children. *British Journal of Music Education* 25: 121–38.

Lamont, A. and Cross, I. (1994). Children's Cognitive Representations of Musical Pitch. *Music Perception* 12/1: 27–55.

Lecanuet, J.P. (1996). Prenatal Auditory Experience. In I. Deliege and J.A. Sloboda (eds), *Musical Beginnings: Origins and Development of Musical Competence* (pp. 3–25). Oxford: Oxford University Press.

Lehmann, A.C. and Ericsson, K.A. (1996). Structure and Acquisition of Expert Accompanying and Sight-reading Performance. *Psychomusicology* 15: 1–29.

Levitin, D. (1994). Absolute Memory for Musical Pitch: Evidence for the Production of Learned Memories. *Perception and Psychophysics* 56: 414–23.

Long, M., Gaunt, H., Creech, A. and Hallam, S. (2010). Beyond the Conservatoire: A Socio-cognitive Perspective on the Development of Professional Self-concept among Advanced Music Students. *Proceedings from the Students' Ownership of Learning Symposium*, Royal College of Music in Stockholm, 15–17 September.

Long, M., Hallam, S., Creech, A., et al. (2011). Do Prior Experience, Gender or Level of Study Influence Music Students' Perspectives on Masterclasses? *Psychology of Music* 40/6: 683–99; doi: 10.1177/0305735610394709.

MacDonald, R.A.R. and Miell, D. (2000). Creativity and Music Education: The Impact of Social Variables. *International Journal of Music Education* 36: 58–68.

MacDonald, R., Hargreaves, D. and Miell, D. (eds) (2002). *Musical Identities*. Oxford: Oxford University Press.

Manturzewska, M. (1990). A Biographical Study of the Life-span Development of Professional Musicians. *Psychology of Music* 18/2: 112–39.

Miell, D. and MacDonald, R.A.R. (2000). Children's Creative Collaborations: The Importance of Friendship when Working Together on Musical Composition. *Social Development* 9/3: 348–69.

Mito, H. (2004). Role of Daily Musical Activity in Acquisition of Musical Skill. In J. Tafuri (ed.), *Research for Music Education: The 20th Seminar of the International Society for Music Education Research Commission*, Las Palmas, 4–10 July.

Morgan, C. (1998). *Instrumental Music Teaching and Learning: A Life History Approach*. Unpublished PhD dissertation, University of Exeter.

North, A.C. and Hargreaves, D.J. (2007a). Lifestyle Correlates of Musical Preference: 1. Relationships, Living Arrangements, Beliefs and Crime. *Psychology of Music* 35/1: 58–87.

North, A.C. and Hargreaves, D.J. (2007b). Lifestyle Correlates of Musical Preference: 3. Travel, Money, Education, Employment and Health. *Psychology of Music* 35/3: 473–97.

O'Flynn, J. (2005). Re-appraising Ideas of Musicality in Intercultural Contexts of Music Education. *International Journal of Music Education* 23/3: 191–203.

Panneton, R.K. (1985). *Prenatal Auditory Experience with Melodies: Effects on Postnatal Auditory Preferences in Human Newborns*. Unpublished DPhil dissertation, University of North Carolina at Greensboro, NC.

Pantev, C., Engelien, A., Candia, V. and Elbert, T. (2003). Representational Cortex in Musicians. In I. Peretz and R. Zatorre (eds), *The Cognitive Neuroscience of Music* (pp. 382–95). Oxford: Oxford University Press.

Pembrook, R. and Craig, C. (2002).Teaching as a Profession. In R. Colwell (ed.), *Handbook of Research on Music Teaching and Learning.* (pp. 786–817). New York: Schirmer.

Pollard-Gott, L. (1983). Emergence of Thematic Concepts in Repeated Listening to Music. *Cognitive Psychology* 15: 66–94.

Salmon, P.G. (1991). A Primer on Performance Anxiety for Organists: Part I. *The American Organist* May: 55–9.

Satt, B.J. (1984). *An Investigation into the Acoustical Induction of Intra-uterine Learning.* Unpublished DPhil dissertation, Californian School of Professional Psychologists, Alliant International University.

Schellenberg, E.G. and Trehub, S.E. (1994). Processing Advantages for Simple Frequency Ratios: Evidence from Young Children. In I. Deliege (ed.), *Proceedings of the 3rd International Conference on Music Perception and Cognition* (pp. 129–30). Liege: ESCOM.

Schellenberg, E.G. and Trehub, S.E. (1996). Children's Discrimination of Melodic Intervals. *Developmental Psychology* 32: 1039–50.

Schellenberg, E.G. and Trehub, S.E. (1999). Culture-general and Culture-specific Factors in the Discrimination of Melodies. *Journal of Experimental Child Psychology* 74: 107–27.

Shahidullah, S. and Hepper, P.G. (1994). Frequency Discrimination by the Fetus. *Early Human Development* 36: 13–26.

Shehan, P.K. (1985). Transfer of Preference from Taught to Untaught Pieces of Non-Western Music Genres. *Journal of Research in Music Education* 33/3: 149–58.

Simonton, D.K. (1997). Products, Persons and Periods: Historiometric Analyses of Compositional Creativity. In D. Hargreaves and A. North (eds), *The Social Psychology of Music* (pp. 107–22). Oxford: Oxford University Press.

Sloboda, J.A. (1976). The Effect of Item Position on the Likelihood of Identification by Inference in Prose Reading and Music Reading. *Canadian Journal of Psychology* 30: 228–36.

Sloboda, J.A. (1983). The Communication of Musical Metre in Piano Performance. *Quarterly Journal of Experimental Psychology* 35A: 377–96.

Sosniak, L.A (1990). The Tortoise and the Hare and the Development of Talent. In M.J.A. Howe (ed.), *Encouraging the Development of Exceptional Skills and Talents* (pp. 149–64). Leicester: The British Psychological Society.

Sudnow, D. (1978). *Ways of the Hand: The Organisation of Improvised Conduct.* London: Routledge and Kegan Paul.

Takeuchi, A.H. and Hulse, S.H. (1993). Absolute Pitch. *Psychological Bulletin* 113: 345–61.

Taylor, A. (2009). *Understanding Older Amateur Keyboard Players: Music Learning and Mature Adult Musical Identity*. Unpublished PhD dissertation, Institute of Education, University of London.

Trehub, S.E., Hill, D.S. and Kamenetsky, S.B. (1997). Parents' Sung Performances for Infants. *Canadian Journal of Experimental Psychology* 51: 385–96.

Trehub, S.E., Thorpe, L.A. and Morrongiello, B.A. (1987). Organizational Processes in Infants' Perception of Auditory Patterns. *Child Development* 58: 741–9.

Valentine, E. (2002). The Fear of Performance. In J. Rink (ed.), *Musical Performance: A Guide to Understanding* (pp. 168–82). Cambridge: Cambridge University Press.

Webster, P.R. (1988). New Perspectives on Music Aptitude and Achievement. *Psychomusicology* 7/2: 177–94.

Webster, P.R. (1991). Creativity as Creative Thinking. In D. Hamann (ed.), *Creativity in the Classroom: The Best of MEJ* (pp. 25–34). Reston, VA: Music Educators National Conference.

Woodward, S.C. (1992). *The Transmission of Music into the Human Uterus and the Response of Music of the Human Fetus and Neonate*. Unpublished PhD dissertation, University of Cape Town.

Young, V.M. and Coleman, A.M. (1979). Some Psychological Processes in String Quartets. *Psychology of Music* 7: 12–16.

PART III
Performance

Elaine King and Helen M. Prior

The final section of this volume examines the role of familiarity in various stages of musical performance, from the learning, memorisation and rehearsal process to a first performance and beyond. This material is strongly influenced by the authors' performance experiences and backgrounds as research practitioners. As in Part II, the chapters adopt analytical, empirical and philosophical perspectives. The section begins with a study by Apostolaki (Chapter 10) of the effects of familiarity with a particular conceptualisation of the musical language – namely, the solfège system – on performers' effectiveness in memorising music. This focus on musical pitch is followed by a consideration of the role of familiarity in the performance of musical rhythm (Chapter 11). Here, Oliver discusses the notion of 'groove', not only as a shared conceptualisation of musical time between performers, but also as familiarity with a temporal frame of reference, exploring the ways in which this works in recordings of popular music.

In Chapter 12, King explores the role of familiarity in shaping socio-emotional behaviour between members of ensembles. She examines such behaviours through a range of theoretical models, before analysing discourse from within chamber music rehearsals. Her findings provide a fascinating insight into the effects of familiarity on musicians' interactions, establishing the basis for further study. Moving into the performance arena in Chapter 13, Doğantan-Dack traces the effects of the unique type of familiarity gained by musicians through the act of musical performance. Her research also focuses on chamber work.

Overall, the section examines the effects of familiarity on a diverse range of performance activities in learning, rehearsal and performance contexts; in individual and group settings; and across classical and popular domains.

Chapter 10

The Significance of Familiar Structures in Music Memorisation and Performance

Artemis Apostolaki

> So this bloke with the dog didn't have a name. I mean, he must have had one at some stage, but he told me he didn't use it any more, because he didn't agree with names. He reckoned they stopped you from being whoever you wanted to be ... he could be like a hundred different people all in one day ... Human beings are millions of things in one day, and his method understands that much better than like the Western way of thinking about it. (N. Hornby, *A Long Way Down*)

In the above quote, the fictional character Nodog, a homeless wanderer, seems to believe that names play a decisive role in shaping our and others' opinions about self and personality; in his decision to not have a name, thus leaving the people around him to use impromptu, temporary referents to invoke him, he is shown to be able to interact with the linguistic status quo and manipulate meta-language, categorical perception and perceptual labelling in a remarkable way. His motivation to do this appears to come mainly from personal and social reasons, but his example is particularly interesting and useful to study from the perspective of the researcher in language and cognition: how do everyday, familiar linguistic labels affect our perception of the world? In order to start answering this question, this chapter will begin by considering the basis upon which the language/perception/reality edifice is based. It will then examine the role of familiarity in memory and how these structures can possibly affect music memorisation; the possibility of an additional mode of familiarisation with a piece will be discussed in relation to the Working Memory Model and relevant findings from research on music memorisation, music cognition and absolute pitch.

Familiarity and Perception

Let us assume that our perception of the world is based on a combination of perceptual universals and perpetuating familiarity patterns; this is not a philosophical statement related to issues of objective or subjective reality, but rather a hypothesis supported by the findings of numerous psychological studies and experiments, some of which will be discussed below. The relative contribution of biological systems and various aspects of the environment to perception is still uncertain, and such questions provoke stimulating debate that is relevant to the core concept of music and familiarity. In this case, the issue of familiarity being the basis of perceptual reality will be discussed from the psychologist's perspective, by looking at research testing human perception using various experimental methods.

Notable examples of research illuminating specific facets of familiarity patterns and perception are related to colour perception universals (Berlin and Kay 1969; Hardin and Maffi 1997), visual perception and grouping (for a review see Bruce, Green and Georgeson 1996; Gordon 1997; Wade and Swanston 1991; Yantis 2001), music perception and grouping (Deliège and Sloboda 1997; Deutsch 1999a; Krumhansl 1990; Sloboda 2005) and other similar pieces of research. The role of language in perceptual encoding, processing, storage and retrieval has been discussed extensively (Goldstone and Barsalou 1998; Lupyan 2008a); I will not, at this point, deal with the possible ways language can shape patterns constructing our perceptual reality, but instead focus first on the importance of language as the principal tool through which referents are constructed.

Speakers use language to construct referents in order to be able to communicate their thoughts to other users effectively; this is achieved through the mechanisms of abstraction and generalisation (Christidis 2007; Vygotsky 1962). Through abstraction, language allows us to use arbitrarily chosen symbols – words – to refer to specific physical objects *in absentia*; through generalisation we can use the same term to refer to objects that we class as similar and hold the same set of properties (the word *table*, for example, is not just used for objects identical to the first table there ever was; instead, the signified can vary in colour, size, height, function, number of legs and so on). The boundaries of object categories are dictated by perceptual and cognitive rules and universals (despite the varying properties and functions of tables, we would readily recognise the object in the middle of the kitchen as a table and the object someone would use to sit on, next to the table, as a non-table; that said, there are certain examples, like the IKEA table/stool, which constitute ingenious exceptions that transcend linguistic and perceptual boundaries, only to confirm the general rule). Going further from these basic principles, the fact that each language holds a different amount of separate words to represent variations within a single category (for example, desk, secretaire, nightstand) reflects the speakers' choices, dictated by social, historical and cultural demands but still always subject, albeit less strictly, to perceptual grouping mechanisms, familiar to all speakers.

Perception, Memorisation and Language

Does the manner of perceiving the world around us, based on both automated and voluntary familiarity categorisations, affect the way we memorise things? Perception and memorisation are naturally very closely connected: whilst in the act of memorising, the target stimuli need to be first encoded in one or more ways and then stored so that they can be easily retrieved on demand. Moreover, familiarity also lies in the heart of every memorising process. Familiarity in memorising procedures can be traced at two different levels:

1. In relation to the perceptual familiarity mechanisms described above. The use of familiar patterns, which are appropriate to the task at hand, serves the need of the finite human memory and storage capacity for space and time-effective management. In this respect, familiarity can be perceived in relation to the 'raw' material to be memorised: the particular methods employed in the memorisation process of a stimulus depend on the amount of prior rehearsal for similar types of stimuli. The familiarity with stimulus type and rehearsal method is a crucial factor, whether the stimulus is modality specific – for example, visual information and cues for the memorisation of an image – or multimodal, as in the case of a piece of music, where aural, visual, kinaesthetic and semantic information need to be combined in order to achieve efficient memorisation.

2. Familiarity, however, is also apparent in the molecular level, as research shows that the brain uses roughly the same patterns and neural pathways in recall as in the original encoding (Craik, Naveh-Benjamin and Anderson 1996; Mishkin and Appenzeller 1987; Moschovich et al. 1995; Squire, Shimamura and Graf 1985); this finding is also extended to the recognition of musical material (Zatorre, Evans and Meyer 1994; Zatorre et al. 1996 on musical imagery processes; and Demorest et al. 2007 on enculturation effects in music recognition). This connection implies that the way the original encoding is formulated is potentially very important for effective recall; the implications are crucial and will be elaborated on later in this chapter.

If we combine the above with research findings pointing to encoding flexibility – the ability to encode a given stimulus in several different ways – and with findings showing that additional associations can enhance the retrievability potential of the encoded stimulus (Craik and Lockhart 1972; Craik and Tulving 1975), the following hypothesis can emerge: the use of an exceptionally familiar medium, like language, can facilitate the process of memorising any non-verbal stimulus.

The role of language as a mediator between reality, perception and human behavioural output has been the subject of numerous philosophical and linguistic analyses (Bartlett and Suber 1987; Fodor 1975; Saussure 1983; Whorf 1956), which have provided invaluable insights to the field and to the discussion led by psychologists and cognitive scientists (Carmichael, Hogan and Walters 1932; Carruthers 2002; Fodor 2008; Gleitman and Papafragou 2005). The exact manner in which verbal labels interact with encoding and retrieval processes is controversial, and many models have been proposed to demonstrate how linguistic categorisation does or does not affect object representations and, if it does, at which stage this happens and what the implications are (Bloom and Keil 2001; Carruthers 2002; Lupyan 2008b). Although these theories are extremely important in the discussion about familiarity and bear direct consequences on the development of any memorisation theory, the present chapter is going to be based on the lower-level assumption that, regardless of the exact mechanism underlying the application of verbal labels, the acquisition of such labels does have an impact on retrieval, even if this is restricted to the effect of having generated an additional – linguistic – representation for the stimulus.

In the case of music, an additional verbal encoding of the musical stimulus could potentially present us with a relative advantage in memorisation and/or recall, as the linguistic mode is substantially more exercised, and therefore more familiar, than any other type of cognition, in relation to memorisation tasks. Two potential problems with such a hypothesis would be:

1. The issue of economy in the management of memory resources seems to be overridden by the presence of multiple encoding options.
2. What about the separation of music and language processing? Can these two elements interact in perception and memory in a way that is beneficial for the stimulus' trace in storage?

The solution to the first problem has already partially been provided by the reference to encoding flexibility (Craik and Lockhart 1972; Eysenck 1977): the issue of encoding multiplicity has been thoroughly researched and research findings show that having more than one encoding option for the original stimulus is not detrimental, but in fact optimal for speed of storage and effectiveness of recall (Craik and Tulving 1975; Baddeley 1999). Any attempt to answer the second argument would ultimately have to be based on Fodor's concept of modularity of the brain (Fodor 1983) as well as on Gardner's multiple intelligences theory (Gardner 1983), both of which have inspired researchers to investigate the possibility of music perception as a completely distinct cognitive function. The links between music and language processing and the extent to which they interact under various conditions have been researched extensively (Bonnel et al. 2001; Hébert et al. 2003; Peretz and Hyde 2003; Peretz and Morais 1989; Peretz, Radeau and Arguin 2004a; Samson and Zatorre 1991; Williamson et al. 2010a; Williamson, Baddeley and Hitch 2010c); however, the argument made in this chapter is not that music and language are unified in perception, but rather that, whether unified or separate, a combination of the two elements can aid and improve performance in music memorisation tasks. Thus, we are ultimately addressing again a case of multiple encoding, where the musical stimulus is encoded as both music and language; a discussion of whether this is possible and how it can be achieved will follow.

Having settled the initial arguments, the next syllogism can be put together, in order to make the previous hypothesis more subject-specific:

- Language, being the most elaborate and well-practised system of human communication, has a strong familiarity factor compared with other human symbolic systems.
- The combination of different perceptual modes during the memorisation process provides a relative advantage for memorisation performance.
- Therefore, the involvement of language as an additional mode in the memorisation of a musical stimulus can improve performance on the memorisation task.

If we accept the above, the emerging question is: what could be the nature of such an involvement? How and at which stage can language be productively incorporated in music memorisation?

Music Memorisation and Memory for Music

Before we start to answer the above questions, we first need to make a distinction between memory for music and music memorisation. On the one hand we have a generic term, which refers to the general ability to store musical information of any kind; it is a property shared by all human beings, apart from, in certain cases, amusic individuals (for more on congenital amusia, see Peretz et al. 2002; and for an overview of case studies on musical cognition disorders, see Stewart et al. 2009), and relevant research is being conducted using both musicians and non-musicians, using mostly melody recognition and pitch discrimination paradigms. Music memorisation, on the other hand, is a deliberate course of action, taken usually by trained individuals towards a specific goal; it is a specialised function of musical memory, normally performed by musicians, regardless of proficiency level. In this chapter, the issue we are going to tackle is the role of familiarity in the process of memorising a piece of music for performance; nevertheless, literature from both the field of music memorisation and the more general field of musical memory is going to be used, as music memorisation techniques are partly determined by music memory capacity, processing and storage, as well as the order and manner in which these techniques are employed.

One of the main focus points for music memorisation research has been the identification of the distinct modes and techniques musicians use in order to memorise music; various studies have examined the processes and mechanisms involved in the memorisation of music by instrumentalists and vocalists (Chaffin and Imreh 1997; Ginsborg 2002) and possible methods of refining memorisation skills (Ginsborg 2002; Ginsborg and Sloboda 2007; Hughes 1915; Matthay 1926). In the largest part of this body of research, four basic types of strategies have been identified for memorising music: aural ('how the music sounds', including pitch, timbral and rhythmic information, pitch contour); visual (for example, remembering images of the printed music, or seeing finger movements); kinaesthetic (such as feeling bodily gestures); and 'conceptual' or analytical (that is, memory based on knowledge of the music's structural characteristics).

The two layers of familiarity, both in the sense of pattern identification and regarding the structural similarity of encoding/retrieval, can be discerned throughout this body of research: expert memorisers are shown to be able to identify more easily the chunks and patterns useful for memorisation and may rely on certain modalities more than others according to individual differences and preferences (Chaffin and Imreh 1997; Ginsborg 2002; Williamon and Valentine 2002). The initial question, however, of whether there is a particular combination of modalities recommended for optimising memorisation performance, although posed by some researchers (Williamon, Valentine and Valentine 2002), has not been answered; in the rest of the chapter we will investigate the possibility that this optimal combination is based on the linguistic mode.

A review of the literature reveals that the linguistic element is not only represented in the list of the acknowledged memorisation techniques, but is, in fact, the one believed to lead to the most effective and durable memorisation: this is achieved through 'conceptual' memorisation (Chaffin and Imreh 1997; Ginsborg 2002). Conceptual memorisation is implicitly linguistic, as language is definitely and necessarily used to perform any kind of analysis; the extent, however, to which linguistic terms, along with any aural, visual and kinaesthetic information, are evoked during recall has not been determined with certainty. In any case, it has been shown that acquisition of this type of verbally mediated meta-knowledge of a piece of music is a method preferred and recommended by professional musicians in order to prepare for a memorised performance (Chaffin and Imreh 1997); this is, naturally, always combined with data from aural, visual and kinaesthetic memory. The reason behind this partiality towards conceptual memorisation could be, besides the analytical value and effectiveness of 'chunking' that the method entails, the fact that conceptual memory links musical input to another, independent and extremely well-rehearsed modality: speech. As mentioned above, this strategy is only implicitly linguistic, the language element only coming in as an extraneous, additional encoding for meta-musical, and possibly extra-musical, information. Is there not a way, then, to exploit the advantages offered by the linguistic element in earlier stages of the memorisation, or even the encoding, process?

What we are effectively looking for is a way to attach a verbal label to the musical stimulus as close to its creation and as intrinsically related to its musical essence as possible. Fortunately, we need not look too far; as unlikely and obsolete it might seem, solfège fits the description perfectly. For the benefit of the readers who are only familiar with solfège through Julie Andrews and the seven youngsters of the von Trapp family, I will use the following paragraphs to introduce it briefly.

A Brief History of Solfège

According to Grove Music Online, solfège is:

> The use of syllables in association with pitches as a mnemonic device for indicating melodic intervals. Such syllables are, musically speaking, arbitrary in their selection, but are put into a conventionalized order. (Hughes and Gerson-Kiwi 2001)

Solfège was devised most probably between 1025 and 1030 by Guido d'Arezzo, an Italian Benedictine monk, well known for his pioneering work on music methodology. Up until his time, songs were mostly taught by rote and continuous repetition; Guido decided to turn his attention from song-learning to *music* learning in general, so that choir singers would be able, first of all, to *hear* consecutive pitches accurately and then also be able to sing them. In order to achieve this, he attributed a different syllable to each pitch, so the piece could also be taught as a series of syllables. There are various theories regarding the origins of the first solfège syllables (Miller 1973); what is important is that the system Guido devised – or, at least, perfected – proved so successful that it gained the approval of Pope John XIX and, having the blessing of the head of the Catholic church, spread easily in Europe and its colonies over the following centuries. The solfège system, of course, has not remained unchanged since Guido's days; on the contrary, many would argue that the Guidonian solfège, based on the gamut and tailored to serve eleventh-century music needs, bears little or no resemblance with solfège as used in the modern world. It should be noted, however, that what is being argued here is not the uninterrupted historical evolution of solfège, which might or might not be true, but rather that Guido's idea to link musical pitches to syllables through singing was the first of its kind in the Western music tradition and lies diachronically in the core of solfège practice.

An important division that emerged during the evolution of solfège in the modern world was the one between the fixed-do and the moveable-do systems: in fixed-do solfège, every syllable always corresponds to a specific pitch; for example 'do' is always C, 're' is always D and so on. Moveable-do, or tonic sol-fa, syllables on the other hand, denote degrees of the scale rather than specific pitches; in this instance, 'do' is always the tonic. The issue of whether fixed- or moveable-do is preferable has been the subject of a long debate with heated arguments on both sides (Bentley 1959; Humphreys 2006; Siler 1956); for the purposes of this chapter, I will be using the term 'solfège' to refer mainly to the fixed-do system, although, in many instances, the assumptions, hypotheses and conclusions can be generalised to moveable-do. Thus, the word 'solfège' in the present text will denote the practice of assigning the verbal labels 'do', 're', 'mi', 'fa', 'sol', 'la', 'si/ti', to particular acoustic frequencies within a musical context.

How is solfège different to assigning letter names or numbers to pitches? The advantages of solfège compared with other systems can be summarised in three ways. Firstly, *solfège is perceptually and functionally appropriate within the linguistic referential system*. Solfège syllables are a part of language: they can be found in any major dictionary of any language and, just as any other word, they have a one-to-one correlation between signifier and signified: contrary to letters and numbers, solfège syllables were specifically created to serve as linguistic labels to musical pitch. Solfège syllables provide the possibility of reference through the mechanisms of abstraction and generalisation; from this viewpoint, solfège is not only linguistically acceptable and relevant, but, being a part of language, actually functions as an 'essential constitutive ingredient of human reality' (Barlett and Suber 1987, p. 5; Whorf 1956) – in this case, musical reality. In this respect, moveable-do represented a natural first step towards the further refinement that took place with fixed-do, where solfège syllables are rendered non-transposable, denoting specific pitches instead of scale degrees.

Secondly, *solfège is musically relevant*. Apart from the fact that they have an all-musical purpose of existence, solfège syllables are also extremely comfortable to sing, compared with both letter names and numbers; this has been proven throughout music history, as solfège has been used to improve the tone and enunciation of singers (Sands 1942–43).

Thirdly, from the previous two points, it can be made clearer why people who have been taught solfège at an early age can automatically associate it with music; for the solfège-trained musician, *solfège syllables are an indispensable aspect of music and music-making*. This is also the difference between pre-existing pitch labelling systems and solfège: solfège can, and does, become intrinsically linked to both music notation and music practice. Despite the fact that a musician trained in letter names automatically thinks of such names when reading a piece of music, (s)he cannot reproduce the piece singing it with letter names. It is not, of course, impossible to do so, but singing a piece with letter names would be extremely rare, simply because music students have not been trained to do this; one reason for the lack of such training is that, as mentioned in the previous point, singing 'C–C–G–G–A–A–G' is rather uncomfortable from a phonetics point of view – thus, understandably, 'twinkle twinkle little star' is the preferred version. Setting lyrics, or a neutral syllable, to a melody is a perfectly acceptable and successful method for learning songs, but is very different from learning music: the transcendence Guido d'Arezzo conceived of by assigning different syllables to each note can only be achieved through using such syllables to *sing* a piece of music. The most important aspect of solfège is that it is *sung*, so that the association between verbal label and pitch is materialised through music; only in this way can verbal labels become an integral part of the note in our perceptual system.

At this point an apparent paradox in the syllogism may have become evident: in making an argument in favour of solfège, claiming that it can potentially provide a higher level of *familiarity* with the musical stimulus, at the same time letter names and numbers are being dismissed as inferior to solfège syllables on the grounds of their polysemy, or, in other words, because of their previously acquired *familiarity*. This apparent logical leap is a result of a play with words more than anything else: what causes the confusion is that there is only one word, 'familiarity', to describe phenomena which, although they share common features, function on distinct levels in perception. In the first instance, familiarity with the musical stimulus refers to the deep structure and thus occurs in a primary level: the musical stimulus (pitch) becomes more familiar because of its acquisition of a unique verbal label, which renders it more functional in terms of referential potential.[1] In the second instance, familiarity refers to surface structures: letters and numbers are characterised as more familiar because we are assuming that prior training, albeit towards a different aim, has taken place; this is no less true for solfège syllables in some countries, as it is only cultural factors and chance that indicate that reading and counting should be learned before music. The crucial familiarity factor lies in the medium through which numbers, letters or solfège syllables are conveyed, which is language, for all three modes; from this aspect, 'do' is no more or less of an arbitrary selection than 'one' or 'C'.

Consequently, it seems that solfège practice constitutes an optimal way to incorporate language in the initial stages of music processing, thus creating a separate perceptual category and enhancing encoding, storage and retrieval possibilities. What is the exact way, though, in which this unique perceptual amalgam would interact with the brain? In order to answer this, we will place the discussion about solfège within the framework of the working memory model (Baddeley 2000; Baddeley and Hitch 1974). The choice of the particular model amongst other alternatives was made primarily due to its multi-component structure, which can accommodate the explanation of music memorisation in general and solfège practice in particular; for similar reasons, the working memory model has repeatedly constituted the basis for music processing investigations (Berz 1995; Williamson 2010a, b).

---

[1] This process should not be confused with the one taking place when learning/ recalling musical material *combined* with verbal elements or words, that is, singing. As mentioned in the previous paragraph, in songs, the linguistic and the musical material have a purely circumstantial connection: the rules of fit between music and lyrics are only consistent within the same piece of music and cannot be applied to a wider paradigm, mainly owing to the semantic connotations of words. Using the same example again, learning the song 'Twinkle twinkle little star' only results in knowing the song 'Twinkle twinkle little star', rather than being able to mix the musical material to produce other pieces like 'Little star twinkle Twinkle' or even 'star star star'. Learning a song with words does make use of intricate connections between music and language learning (Bonnel et al. 2001; Hébert et al. 2003; Peretz et al. 2004b), and solfège is, ultimately, singing, in the sense of vocal production which is simultaneously musical and verbal; the main, and extremely significant, advantages solfège bears compared with a, hypothetical, song-database system is economy and monosemy.

Working Memory

The working memory model, first proposed by Baddeley and Hitch (1974), has been one of the most influential and widely accepted models for explaining interactions between various memory stores, perceptual processes and attention. It is based on the notion that the short-term store is not simply a temporary store with limited capacity, but is itself constituted by separate sub-systems with both processing and storage functions, essential to activities such as comprehension and reasoning in language, and sight-reading in music (Aiello and Williamon 2002). This short-term processing and storage system is not independent from long-term stores: in order to overcome the restrictions posed by the relatively small capacity of short-term stores, individuals have to resort to using previously learned rules and strategies, stored in long-term memory (LTM), in order to chunk and categorise information (Berz 1995; Ericsson 1985; Ericsson and Kintsch 1995).

The constituents of working memory were initially proposed to be three: a central executive component with two slave systems, the phonological loop and the visuo-spatial sketchpad. The phonological loop is responsible for the manipulation of speech and other acoustic information and comprises a phonological store and an articulatory loop, responsible for controlling sub-vocal rehearsal; the visuo-spatial sketchpad deals mainly with visual data. The general supervision, manipulation and regulation of the flow of information between the two slave systems and towards the long-term memory store are performed by the central executive system. In 2000, Baddeley revised this model to include a fourth element: the episodic buffer (Baddeley 2000; Repovs and Baddeley 2006); this buffer functions along the phonological loop and the visuo-spatial sketchpad as a multimodal store for combined information from the other sub-systems and from LTM (see Figure 10.1). One of the most common criticisms for the working memory model has been its inadequacy in explaining the handling of acoustic information other than speech, as well as other sensory information, such as smell and taste; the combinatory nature of the recently added episodic buffer is not always successful in explaining certain specialised cases of memory feats and pathologies.

Berz (1995) provides a review of literature showing that musicians draw on LTM strategies developed through training for recall; the recall process, Berz assumes, must involve a specified musical processing component connected to the memory task, as the phonological loop in the working memory model is, by definition, restricted to verbal coding of acoustic information and thus cannot account sufficiently for music storage and processing. Even with the incorporation of the episodic buffer in the system, music processing seems to place extremely specific demands on storage and retrieval to be accommodated by a general, multimodal, multifunctional and limited capacity store; Salamé and Baddeley (1989) have also acknowledged the possibility of an additional form of acoustic storage capable of dealing with non-speech stimuli.

Berz thus puts forward the proposition to incorporate an additional slave system to the central executive, specific to music, but leaves the question regarding the

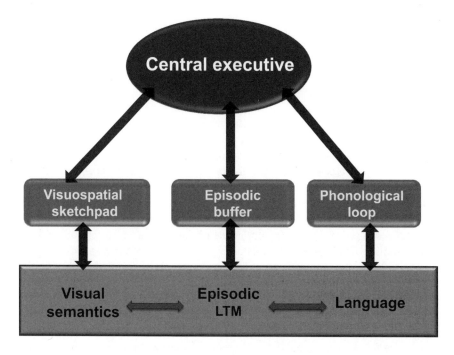

Figure 10.1   The working memory model, including the episodic buffer
Source: Adapted from Baddeley (2000), p. 421.

nature of the relationship between the proposed music-specific sub-component and the phonological loop open-ended; he states that the music sub-system could either be a specialised function of the phonological loop or an entirely different entity. In any case, he argues that the ties between the two systems ought to be extremely loose, given the differences between verbal and musical stimuli and research findings showing differences in the processing, retrieval and decay of these stimuli. Berz also assumes that the nature of the musical component should be very similar to that of the phonological loop, also displaying a musical store, parallel to the phonological store, and a set of control processes based on inner – musical – speech.

According to Berz's model, solfège, as a linguistic element, has a clear-cut place in the phonological loop. If the music sub-component is a specialised function of the phonological loop, then the interaction between solfège's linguistic and musical functions can be justified through the assumed 'loose ties' existing between verbal and musical storage. If the music sub-component is an entirely different entity, then the presence of solfège becomes slightly more problematic, as there is no prediction for possible manners of interaction between sub-components. In either case, the explanation is hardly satisfactory; as valuable as Berz's model is in making the first attempt towards updating music memorisation theories, its deficiencies lie mainly within its weakness to determine specific modes of interaction between the phonological loop and the music sub-component.

The model proposed below does not perform much better in terms of thoroughness; however, it does provide more clear-cut assumptions for the possible ways of interaction between verbal and musical elements. Consequently, it also provides a more accurate description of how solfège would function in working memory; nonetheless, specific testing is necessary in order to either verify or disprove it.

The proposition is that there is, indeed, a distinct, music-specific system, but it is definitely not a part of the phonological loop; it is separate, and, just like the phonological loop, its formulation is dependent on developmental and evolutionary processes, environmental stimulation and appropriate training (Apostolaki 2009). The existence of the phonological loop is indisputable and presumably a result of the continuous presence and predominance of language as a perception, coding, understanding and communication medium, both in developmental terms and with respect to human evolution history. Music, on the other side, although inherent to humans, has numerous functions, expressive as well as performative, whose boundaries have not always been defined with precision; in the cases of musicians, however, there are specialised cognitive functions taking place, which differ depending on the individual and on prior training. In this sense, the same could be true for all other perceptual modalities; we could assume that, depending on extensive practice, one could develop an exceptional olfactory working memory or a specialised memory for taste. This kind of expanded modal organisation of working memory, which will include specialised units for modalities other than the ones designated in working memory models proposed so far, is supported by recent psychological studies. Andrade and Donaldson (2007) support the possible existence in working memory of a sub-component dedicated to the manipulation of olfactory information; Johnson and Miles (2009) also point to the conclusion that working memory is a modally organised system, with specialised components for certain modalities, one of which could possibly be olfaction.

Working memory can be thus viewed as a loosely defined set of functions, which are controlled by a central executive; the potential for development of a language, spatio-temporal, visuo-spatial or music-specific sub-system is inherent as a kind of memory 'competence' in the Chomskyan sense: the dynamic for such functions is present in everyone and the realisation depends largely on individual and environmental differences. If this is the case, a music-specific sub-system would be formed according to available perceptual and coding resources; its relation to the phonological loop would not be one of attachment but rather of sharing from a common pool of processes (see Figure 10.2).

According to this model, working memory is a flexible system, depending on or sharing resources with other cognitive structures. This is also in line with the point made by Ericsson and Kintsch (1995), who argue the existence of a 'long-term working memory', formulated by the skilled use of long-term memory storage (p. 211). This view, combined with the skilled memory theory (Ericsson

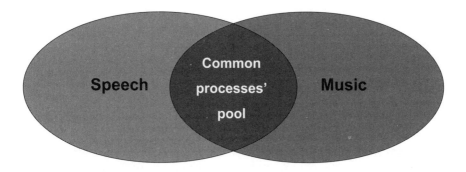

Figure 10.2   Music-specific sub-system of the working memory model:
The central overlapping area demonstrates the cognitive
processes shared by verbal and musical functions

1985) implies the possibility for the development of a domain-specific set of
rapid retrieval skills, which would correspond to the working memory for this
specific domain. In the case of music, if solfège is included in early training, we
can assume that it promotes the use of a set of coding and retrieval processes
common to the phonological loop for verbal information; rapid retrieval can be
made possible through the use of all available encoding resources, through close
tailoring of the manipulation of these resources to serve music-related activities,
as well as through the use of familiar structures common to the verbal and musical
phonological loop.

The proposed link between working memory and solfège is not the first time
working memory has been associated with music memorisation: apart from the
direct activation of short-term processes and stores that is associated with the mental
and physical engagement with a piece whilst performing a music memorisation
task (Palmer 2005; Williams 1975), working memory has been shown to be
actively involved in sight-reading. Sight-reading is one of the most prominent
musical activities to be affected by working memory capacity and processes (Aiello
and Williamon 2002), as it requires rapid online processing of musical material.
The importance of sight-reading in memorisation is supported by Chaffin et al.
(2003), who argue that the initial 'artistic image' of the piece, formulated during
sight-reading, serves as a starting point for the shaping of its final performance and
effectively leads to the desirable memorised performance. The same view is also
advocated empirically by many music teachers and conductors, who put forward
the need for an efficient first sight-reading of any given piece in order to be able to
construct a solid representation of the music, which will be subsequently elaborated
on. What is argued in this chapter is that, if solfège is practised proficiently, the first
sighting of the piece and its subsequent representation can be benefited enormously
by the simultaneous acquisition of verbal cues, which facilitate and speed up the
familiarisation process with the music.

Evidence from Absolute Pitch Research

The model proposed here suggests that the use of solfège promotes a coalescence of musical and linguistic modes, so that the role of familiarity in music memorisation is enhanced beyond the use of previously learned structures and patterns. Although this is still a theoretical account, which is yet to be tested (see Apostolaki 2012), there is a branch of music psychology research whose findings can be used to draw analogies with solfège: absolute pitch research. Absolute pitch (hereon AP) research provides extremely useful insights into the relationship between category labels and musical sound; the verbal nature of solfège and the assumed double encoding it provides allow space for drawing parallels between AP and solfège labels. It should be noted, however, that the main drawback pertaining to AP research itself could pose a problem to such parallels: there is the possibility that findings from AP research might only be relevant to the 0.01 per cent of the general population who happen to possess AP (Takeuchi and Hulse 1993; Ward 1999). Bearing these stipulations in mind, it is worth looking into AP in order to understand better the possible interactions of linguistic labels with acoustic information.

Researchers argue that the association between category label and auditory stimulus is necessary for the emergence of AP (see Levitin and Rogers 2005; Levitin and Zatorre 2003 for a review); in addition, cognitive psychologists report that neural regions recruited in pitch recognition by AP possessors are the same as in conditional associative learning, which is general for verbal labelling of sensory percepts (Zatorre et al. 1998, in Levitin and Zatorre 2003, p. 106). This leads to the assumption that, in the case of AP, specific neural networks are tuned to pitch labelling, a finding also in line with evidence on the modularity of music perception. In such a case, solfège, seen as a category label for musical pitch, is functionally music-specific; specialised neuroimaging studies investigating the localisation of solfège usage in the brain would be necessary to determine whether or not this specificity is also an anatomical one.

The importance of category labels is reflected in Levitin's two-component model for AP, which includes a pitch memory and pitch labelling element (Levitin 1994); this is in accordance with research attributing the superiority of AP possessors to the verbal rather than pitch memory (Takeuchi and Hulse 1993). Takeuchi and Hulse, in reviewing literature on different measurements of AP acuity, report one measure to be the level of decay in memory for pitch. Looking at AP and non-AP possessors' performance in music recall tasks, they suggest that non-AP possessors are solely based on echoic memory, whereas AP possessors have verbal memory readily available. The assumed utility of verbal labels in AP is demonstrated further in experiments where pitch names cannot be used in order to discriminate between pitches, as happens in cases above certain frequencies (over 5000 Herz) or in frequencies falling between tones; in such cases AP and non-AP possessors perform the same (Takeuchi and Hulse 1993; Ward 1999).

Data from research relating solfège to the acquisition of either relative or absolute pitch are quite inconsistent and mostly based on theoretical accounts, observation and anecdotal evidence; there seems to be a general agreement on the fact that solfège in most cases improves relative pitch (Bullis 1936; Smith 1934), but there is also evidence on solfège facilitating the acquisition of AP, with the most prominent example being that of the Yamaha method for music teaching, which explicitly uses fixed-do solfège in order to cultivate AP (Miranda 2000). This brings forward the issue of the importance of early training in the acquisition of AP; Levitin and Zatorre (2003) argue that the pitch labelling networks involved in AP acquisition should be expected to comply with the same rules as other neurodevelopmental events. Apart from the importance of AP training during a critical period in childhood, researchers also stress the importance of the *type* of training: musical training, insofar as it emphasises pitch relations, reinforces relative pitch and can prove detrimental to the acquisition of absolute pitch (Levitin and Zatorre 2003; Takeuchi and Hulse 1993; Ward 1999).

This provides supplementary evidence that music training aiming to achieve the acquisition of AP has to be specifically appropriate; emphasis on category labels, such as solfège, is the first step towards this goal, but the establishment of pitch names has to precede understanding of pitch relations. This is also in tune with the natural developmental sequence of sensory perception in general (Takeuchi and Hulse 1993): children first learn how to perceive stimuli in absolute terms and only later, with the introduction of abstractive thinking, are they made able to form relations between perceptual events. From this viewpoint, fixed-do solfège seems to be developmentally more appropriate in early training than moveable-do in that, regardless of any association with AP, it promotes absolute pitch labelling. Even if solfège syllables do improve the potential for AP acquisition, specific testing is required to determine whether AP relies on the same verbal coding structures as solfège; it should be noted that the verbal nature of solfège renders it suitable for investigation using the same methods as the ones used in verbal memory research, also allowing direct comparisons between the functions of solfège labels and AP pitch labelling.

Conclusions

Music memorisation can be ultimately viewed as a process of continuous familiarisation with a piece; the aim of a musician memorising a piece for performance is to *get to know* all aspects of the piece as thoroughly as possible. Familiarity with a piece of music can be achieved in multiple levels of surface structures: familiarity with the overall auditory result, a visual image, the movements required to produce its performance; also important is familiarity with extramusical information and connotations of the piece, regarding its history and possible meaning. Going one level deeper, the musician can strive to get familiar with distinctive patterns and sequences constituting the piece, whether they are related to pitch, rhythm, timbre or dynamics; the categorisation of such patterns happens according to previously learned rules, which can result from cognitive universals or cultural traditions. At the base of the piece, however, is the essence of the music itself, a sequence of acoustic frequencies that have to be reproduced; familiarity with this material is the necessary basis on which everything else is added in order to produce an aesthetically appropriate result – the potential interactions between the aesthetically pleasing and the familiar have been the basis of numerous philosophical analyses and art schools (for an introduction, see Graham 1997). This chapter has attempted to show the importance of familiarisation with the primary musical material through verbal labels; language, being a highly practised modality, can prove invaluable when used as a medium between sensory perception and cognition.

It is not implied that solfège is the only possible way to link music and language in perception, providing optimal conditions for familiarisation with the musical stimulus; it is, however, a method that has been present in the Western world for almost a millennium and during this course has been tried, altered and improved by musicians so that it best serves its purpose. When Guido d'Arezzo introduced the innovations of staff notation and solfège syllables, he eliminated much of the obscurity of neumatic notation and formulated Western music as we know it; it is plausible to say that, for the Western world, the concept of the note became concrete after solfège, which opened up a new way of learning, thinking about and knowing music. It is in this sense that Hornby's Nodog was right after all: names, be it for people or for pitches, define properties, lead to categorisations and draw borders on what something, or someone, can or cannot be; for better or for worse, names are our way of restricting the infinite possibilities of reality and reducing them to specific, recognisable and predictable aggregates.

References

Aiello, R. and Williamon, A. (2002). Memory. In R. Parncutt and G. McPherson (eds), *The Science and Psychology of Music Performance* (pp. 167–81). New York: Oxford University Press.

Andrade, J. and Donaldson, L. (2007). Evidence for an Olfactory Store in Working Memory. *Psychologia* 50/2: 76–89.

Apostolaki, A. (2009). Solfège Training and Music Memorization: An Empirical Study. *Paper presented at SEMPRE Postgraduate Study Day: Research in Music Psychology and Education*, Hull.

Apostolaki, A. (2012). *The Interplay between Music Memory, Working Memory and Solfège: A Cross-cultural Study of University Students' Aural and Cognitive skills*. Unpublished PhD dissertation, University of Hull.

Baddeley, A.D. (1999). *Essentials of Human Memory*. Hove: Psychology Press.

Baddeley, A.D. (2000). The Episodic Buffer: A New Component of Working Memory? *Trends in Cognitive Sciences* 4/11: 417–23.

Baddeley, A.D. and Hitch, G.J. (1974). Working Memory. In G.H. Bower (ed.), *The Psychology of Learning and Motivation*, Vol. 8 (pp. 47–90). New York: Academic Press.

Bartlett, S.J. and Suber, P. (eds) (1987). *Self-Reference: Reflections on Reflexivity*. The Hague: Martinus Nijhoff.

Bentley, A. (1959). Fixed or Movable Do? *Journal of Research in Music Education*, 7/2: 163–8.

Berlin, B. and Kay, P. (1969). *Basic Color Terms: Their Universality and Evolution*. Berkeley, CA: University of California Press.

Berz, W. (1995). Working Memory in Music: A Theoretical Model. *Music Perception* 12/3: 353–64.

Bloom, P. and Keil, F.C. (2001). Thinking Through Language. *Mind and Language* 16: 351–67.

Bonnel, A.M., Faita, F., Peretz, I. and Besson, M. (2001). Divided Attention between Lyrics and Tunes of Operatic Songs: Evidence for Independent Processing. *Perception and Psychophysics* 63/7: 1201–13.

Bruce, V., Green, P.R. and Georgeson, M.A. (eds) (1996). *Visual Perception: Physiology, Psychology and Ecology*. Hove: Psychology Press.

Bullis, C. (1936). Harmony Study That Correlates with Musical Needs. *Music Educators Journal* 23/3: 26–7.

Carmichael, L.C., Hogan, H.P. and Walters, A.A. (1932). An Experimental Study of the Effect of Language on the Reproduction of Visually Perceived Form. *Journal of Experimental Psychology* 15: 73–86.

Carruthers, P. (2002). The Cognitive Functions of Language. *Behavioral and Brain Sciences* 25: 657–74.

Chaffin, R. and Imreh, G. (1997). 'Pulling Teeth and Torture': Musical Memory and Problem Solving. *Thinking and Reasoning* 3/4: 315–36.

Chaffin, R., Imreh, G., Lemieux, A.F. and Chen, C. (2003). 'Seeing the Big Picture': Piano Practice as Expert Problem Solving. *Music Perception* 20/4: 465–90.

Christidis, A.-F. (2007). The Nature of Language. In A.-F. Christidis (ed.), *A History of Ancient Greek: From the Beginnings to Late Antiquity* (pp. 27–64). Cambridge: Cambridge University Press.

Craik, F.I.M. and Lockhart, R.S. (1972). Levels of Processing: A Framework for Memory Research. *Journal of Verbal Learning and Verbal Behavior* 11/6: 671.

Craik, F.I.M. and Tulving, E. (1975). Depth of Processing and the Retention of Words in Episodic Memory. *Journal of Experimental Psychology: General*, 104/3: 268–94.

Craik, F.I.M., Naveh-Benjamin M. and Anderson D.N. (1996). Encoding and Retrieval Processes: Similarities and Differences. In M.A. Conway, S.E. Gathercole and C. Cornoldi (eds), *Theories of Memory*, Vol. 2 (pp. 61–86). Hove: Psychology Press.

Deliège, I. and Sloboda, J.A. (1997). *Perception and Cognition of Music*. Hove: Psychology Press.

Demorest, S.M., Morrison, S.J., Beken, M.N. and Jungbluth, D. (2007). Lost in Translation: An Enculturation Effect in Music Memory Performance. *Music Perception* 25/3: 213–23.

Deutsch, D. (1999a). The Processing of Pitch Combinations. In D. Deutsch (ed.), *The Psychology of Music* (2nd edn) (pp. 349–411). San Diego, CA: Academic Press.

Deutsch, D. (ed.) (1999b). *The Psychology of Music* (2nd edn). San Diego, CA: Academic Press.

Ericsson, K.A. (1985). Memory Skill. *Canadian Journal of Psychology* 39/2: 188–231.

Ericcson, K.A. and Kintsch, W. (1995). Long-term Working Memory. *Psychological Review* 102/2: 211–45.

Eysenck, M.W. (1977). *Human Memory: Theory, Research and Individual Differences.* Oxford: Pergamon.

Fodor, J.A. (1975). *The Language of Thought.* Cambridge, MA: Harvard University Press.

Fodor, J.A. (1983). *Modularity of Mind: An Essay on Faculty Psychology.* Cambridge, MA: MIT Press.

Fodor, J.A. (2008). *LOT 2: The Language of Thought Revisited.* Oxford: Clarendon Press.

Gardner, H. (1983). *Frames of Mind: The Theory of Multiple Intelligences.* London: Heinemann.

Ginsborg, J. (2002). Classical Singers Learning and Memorizing a New Song: An Observational Study. *Psychology of Music* 30/1: 58–101.

Ginsborg, J. and Sloboda, J.A. (2007). Singers' Recall for the Words and Melody of a New, Unaccompanied Song. *Psychology of Music* 35/3: 421–40.

Gleitman, L. and Papafragou, A. (2005). Language and Thought. In K. Holyoak and B. Morrison (eds), *Cambridge Handbook of Thinking and Reasoning* (pp. 633–61). Cambridge: Cambridge University Press.

Goldstone, R.L. and Barsalou, L.W. (1998). Reuniting Perception and Conception. *Cognition* 65: 231–62.

Gordon, I.E. (1997). *Theories of Visual Perception.* Chichester: Wiley.

Graham, G. (1997). *Philosophy of the Arts: An Introduction to Aesthetics.* London: Routledge.

Hardin, C.L. and Maffi, L. (eds) (1997). *Color Categories in Thought and Language.* Cambridge: Cambridge University Press.

Hébert, S., Racette, A., Gagnon, L. and Peretz, I. (2003). Revisiting the Dissociation between Singing and Speaking in Expressive Aphasia. *Brain* 126: 1838–50.

Hornby, N. (2005). *A Long Way Down.* London: Penguin Books.

Hughes, A. and Gerson-Kiwi, E. (2001). Solmization. *Grove Music Online,* http://www.oxfordmusiconline.com/subscriber/article/grove/music/26154 (accessed 23 March 2006).

Hughes, E. (1915). Musical Memory in Piano Playing and Piano Study. *Musical Quarterly* 1: 592–603.

Humphreys, J.T. (2006). Senior Researcher Award Acceptance Address: Observations about Music Education. *Journal of Research in Music Education* 54/3: 183–202.

Johnson, A.J. and Miles, C. (2009). Single-probe Serial Position Recall: Evidence of Modularity for Olfactory, Visual, and Auditory Short-term Memory. *The Quarterly Journal of Experimental Psychology* 62/2: 267–75.

Krumhansl, C.L. (1990). *Cognitive Foundations of Musical Pitch*. New York: Oxford University Press.

Levitin, D.J. (1994). Absolute Memory for Absolute Pitch: Evidence from the Production of Learned Melodies. *Perception and Psychophysics* 56/4: 414–23.

Levitin, D.J. and Rogers, S.E. (2005). Absolute Pitch: Perception, Coding and Controversies. *Trends in Cognitive Sciences* 9/1: 26–33.

Levitin, D.J. and Zatorre, R.J. (2003). On the Nature of Early Music Training and Absolute Pitch: A Reply to Brown, Sachs, Cammuso, and Folstein. *Music Perception* 21/1: 105–10.

Lupyan, G. (2008a). The Conceptual Grouping Effect: Categories Matter (and Named Categories Matter More). *Cognition* 108: 566–77.

Lupyan, G. (2008b). From Chair to 'Chair': A Representational Shift Account of Object Labeling Effects on Memory. *Journal of Experimental Psychology: General* 137/2: 348–69.

Matthay, T. (1926). *On Memorizing and Playing from Memory and on the Laws of Practice Generally*. London: Oxford University Press.

Miller, S.D. (1973). Guido d'Arezzo: Medieval Musician and Educator. *Journal of Research in Music Education* 21/3: 239–45.

Miranda, M.L. (2000). Developmentally Appropriate Practice in a Yamaha Music School. *Journal of Research in Music Education* 48/4: 294–309.

Mishkin, M. and Appenzeller, T. (1987). The Anatomy of Memory. *Scientific American* 256: 80–89.

Moscovitch, M., Kapur, S., Kohler, S and Houle, S. (1995). Distinct Neural Correlates of Visual Long-term Memory for Spatial Location and Object Identity: A Positron Emission Tomography Study in Humans (Visual Cortex/Hippocampus). *Proceedings of the National Academy of Sciences USA: Psychology* 92: 3721–5.

Palmer, C. (2005). Sequence Memory in Music Performance. *Current Directions in Psychological Science* 14/5: 247–50.

Peretz, I. and Hyde, K.L. (2003). What is Specific to Music Processing? Insights form Congenital Amusia. *Trends in Cognitive Sciences* 7/8: 362–7.

Peretz, I. and Morais, J. (1989). Music and Modularity. *Contemporary Music Review* 4: 279–93.

Peretz, I., Ayotte, J., Zatorre, R.J., et al. (2002). Congenital Amusia: A Disorder of Fine-grained Pitch Discrimination. *Neuron* 33: 185–91.

Peretz, I., Radeau, M. and Arguin, M. (2004a). Two-way Interactions between Music and Language: Evidence from Priming Recognition of Tune and Lyrics in Familiar Songs. *Memory and Cognition* 32/1: 142–52.

Peretz, I., Gagnon, L., Hébert, S. and Macoir, J. (2004b). Singing in the Brain: Insights from Cognitive Neuropsychology. *Music Perception* 21/3: 373–90.

Repovs, G. and Baddeley, A.D. (2006). The Multi-component Model of Working Memory: Explorations in Experimental Cognitive Psychology. *Neuroscience* 139/1: 5–21.

Salamé, P. and Baddeley, A. (1989). Effects of Background Music on Phonological Short-term Memory. *The Quarterly Journal of Experimental Psychology Section A: Human Experimental Psychology* 41/1: 107–22.

Samson, S. and Zatorre, R.J. (1991). Recognition Memory for Text and Melody of Songs after Unilateral Temporal Lobe Lesion: Evidence for Dual Encoding. *Journal of Experimental Psychology: Learning, Memory, and Cognition* 17/4: 793–804.

Sands, M. (1942–43). The Teaching of Singing in Eighteenth Century England. *Proceedings of the Musical Association* 70: 11–33.

Saussure, F. de (1983). *Course in General linguistics*. Trans. R. Harris. London: Duckworth.

Siler, H. (1956). Toward an International Solfeggio. *Journal of Research in Music Education* 4/1: 40–43.

Sloboda, J.A. (2005). *Exploring the Musical Mind: Cognition, Emotion, Ability, Function*. Oxford: Oxford University Press.

Smith, M. (1934). Solfège: An Essential in Musicianship. *Music Supervisors' Journal* 20/5: 16–61.

Squire, L.R., Shimamura, A.P. and Graf, P. (1985). Independence of Recognition Memory and Priming Effects: A Neuropsychological Analysis. *Journal of Experimental Psychology: Learning, Memory and Cognition* 11: 37–44.

Stewart, L., von Kriegstein, K., Dalla Bella, S., et al. (2009). Disorders of Musical Cognition. In S. Hallam, I. Cross and M. Thaut (eds), *Oxford Handbook of Music Psychology* (pp. 184–96). Oxford: Oxford University Press.

Takeuchi, A.H. and Hulse, S.H. (1993). Absolute Pitch. *Psychological Bulletin* 113/2: 345–61.

Vygotsky, L.S. (1986). *Thought and Language*. Trans. A. Kozulin. Cambridge, MA: MIT Press.

Wade, N.J. and Swanston, M.T. (1991). *Visual Perception: An Introduction*. London: Routledge.

Ward, W.D. (1999). Absolute Pitch. In D. Deutsch (ed.), *The Psychology of Music* (2nd edn) (pp. 265–98). San Diego, CA: Academic Press. Whorf, B.L. (1956). *Language, Thought, and Reality: Selected Writings of Benjamin Lee Whorf* (ed. J.B. Carroll). Cambridge, MA: MIT Press.

Williamon, A. and Valentine, E. (2002). The Role of Retrieval Structures in Memorizing Music. *Cognitive Psychology* 44: 1–32.

Williamon, A., Valentine, E. and Valentine, J. (2002). Shifting the Focus of Attention between Levels of Musical Structure. *European Journal of Cognitive Psychology* 14/4: 493–520.

Williams, D.B. (1975). Short-term Retention of Pitch Sequence. *Journal of Research in Music Education* 23/1: 53–66.

Williamson, V.J., Baddeley, A.D. and Hitch, G.J. (2010c). Musicians' and Nonmusicians' Short-term Memory for Verbal and Musical Sequences: Comparing Phonological Similarity and Pitch Proximity. *Memory and Cognition* 38/2: 163–75.

Williamson, V.J., McDonald, C., Deutsch, D., et al. (2010b). Faster Decline of Pitch Memory over Time in Congenital Amusia. *Advances in Cognitive Psychology* 6: 15–22.

Williamson, V.J., Mitchell, T., Hitch, G.J. and Baddeley, A.D. (2010a). Musicians' Memory for Verbal and Tonal Material under Conditions of Irrelevant Sound. *Psychology of Music* 38: 331–50.

Yantis, S. (ed.) (2001). *Visual Perception: Essential Readings*. Hove: Psychology Press.

Zatorre, R.J., Evans, A.C. and Meyer, E. (1994). Neural Mechanisms Underlying Music Perception and Memory for Pitch. *The Journal of Neuroscience* 14/4: 1908–19.

Zatorre, R.J., Halpern A.R., Perry, D.W., et al. (1996). Hearing in the Mind's Ear: A PET Investigation of Musical Imagery and Perception. *Journal of Cognitive Neuroscience* 8/1: 29–46.

Zatorre, R.J., Perry, D., Becket, C., et al. (1998). Functional Anatomy of Musical Processing in Listeners with Absolute and Relative Pitch. *Proceedings of the National Academy of Sciences* 95: 3172–77.

Chapter 11

Groove as Familiarity with Time

Rowan Oliver

To start, I would like to make what may seem to be an obvious point: the more familiar a musician is with the music they make, the better they should be able to engage with key aspects of performance or production associated with that music. This chapter, however, is not concerned with the broad correlation between an individual's practice regime and the resultant level of musical skill that they might attain (although it is written from a performer's perspective). Nor will the discussion stray into explicitly psychological territory, the extent and potential usefulness of this field notwithstanding.[1] Rather, the focus here will be on the musician's relationship with musical time and, more specifically, the way in which the capacity for groove depends upon familiarity with stylistically nuanced conceptions of time. The context in which I consider the link between groove and familiarity will be expanded to include the role of the listener as well as the skewing effect of sampling. Despite the nearly limitless supply of hitherto unexplored musical examples that might be used here, some familiar groovological warhorses will also be re-examined from a new perspective in order to illustrate my thinking. The first part of this chapter explores the concept of groove in the light of recent research in the field and introduces the idea of solo groove as it relates to time, with particular reference to sampling and the role of the producer. The second part defines a model for conceptualising groove and time by defining three categories of interaction, with analyses of funk performances to illustrate the discussion.

'Ain't That a Groove?'

In simple terms, groove in music is a concept which encompasses, in varying proportions, an amalgamation of rhythm, feel, sound, dance and a host of related tangents, so its elusive nature is, unsurprisingly, the subject of ongoing debate. Simon Zagorski-Thomas (2007) provides a recent, succinct overview of the terrain, which is fairly comprehensive and demonstrates the range of conceptual and methodological approaches taken thus far. The diversity of various groovologists' interpretations of the term and its associated concepts seems to reflect an unease with the notion of petrifying the organic and thereby negating

---

[1]   Anyone wishing to explore literature relevant to these approaches could see London (2004), Thaut (2008), Vuust et al. (2009) and Huron (2006) as good starting points.

the vitality of something that is generally agreed to be an embodied, human-centred phenomenon. To arrive at a watertight scholarly definition of groove would be to deny its multiple empirical meanings, to limit its usefulness as an object of multidisciplinary scholarly attention and, ultimately, to miss the point. It is no coincidence that we have not seen a conclusive report from Telif Kfivte's mischievously imagined 'International Groove Definition Committee' (2004, p. 55): if such a body existed, the ideas that follow would only serve to increase their workload!

To date, the growing body of work relevant to groovology has tended to concern itself with grooving as a participatory activity or with groove(s) as the musical result of such activity, although even this consensus is rather loose and is manifested in very different ways according to the disciplinary affiliations of each scholar. Steven Feld, for example, famously considers the participatory dimension of groove from a socio-musical perspective in his seminal work describing the intertwined music and lives of the Kaluli (Keil and Feld 1994). By contrast, Clayton, Sager and Will (2005) utilise approaches drawn from both neuroscience and ethnomusicology in order to examine the role of entrainment in the rhythmic alignment between musicians participating in ensemble performance. Clearly these two examples are anchored by the same underlying themes (indeed Clayton acknowledges the common ground shared with Keil and Feld), but their divergent focuses demonstrate the myriad range of ideas encompassed within groovology. Nevertheless, the principle that groove necessarily involves participation between people making music seems to be a common factor that unites much of the literature.

Solo Groove

Buried deeper in some groovological texts though, there are hints at the possibility that groove exists in the playing of a solitary musician. The terminological water is muddied somewhat by an overlap between this idea and the concept of musical 'feel'.[2] Indeed, if groove is understood to result when groups of more than one musician interact, then perhaps another word might be more appropriate when a single musician is playing, in which case it would seem reasonable to describe one musician's gestural, rhythmic idiosyncrasies as their 'feel'. I propose, however, that because of the individual's awareness of some form of time, the interaction between one musician and time itself could be said to constitute grooving and that this is particularly evident when the time element of this interactive pairing is conceived in stylistically nuanced ways (rather than simply as absolute, scientific or metronomic time, although this too can have a role in groove, as I describe later).

Solo groove, as I will call the model wherein one musician grooves in relation to an abstract sense of time, is mentioned by Joseph Prögler (1995), who himself

---

[2]   Farmelo (1997) notes this and lists several other seemingly interchangeable terms.

cites Allen Farmelo (1997)[3] as a precedent. In fact, the existence of solo groove is implied whenever scholars address an unaccompanied drum excerpt (such as that perennial groovological trope, the 'Funky Drummer' break[4] as played by Clyde Stubblefield in a James Brown recording from 1969). The reason why this often remains a tacit implication may lie in the drum kit's curious organological evolution, in which a collection of instruments that had hitherto been distributed between various musicians – as in an orchestral percussion section for example, or, more pertinently, a New Orleans mardi gras band – gradually fused together and came under the musical control of a single drummer.[5] Thus Charles Keil, for example, is able to refer to the drummer's limbs almost as though they were independent grooving entities (Keil and Feld 1994, p. 98) participating with one another.

Whilst the drum kit *could* be viewed as a collection of instruments and whilst drummers undoubtedly strive for independent rhythmic control of their limbs, there are several arguments for instead seeing the instrument as a whole and therefore hearing the sounds and rhythms that it produces as originating from the same source. Of course this is not a radical suggestion and will only reinforce what is already self-evident to the majority of listeners, but the distinction is relevant to the idea of solo groove in more ways than one, principally because it leads towards the concept of timelines, or what Ives Chor describes as 'temporal frameworks' (Chor 2010, p. 40). This, in turn, relates to the stylistically nuanced conceptions of time that can act as temporal sparring partners for the musician in solo groove, ideas that are explored in more detail later.

Groove, Shared Time and Sampling

I have argued that it is possible to conceive of groove in terms of the solo performer interacting with abstract (musical) time; yet, moving beyond this possibility, groove might operate across dimensions when a particular conceptualisation of time is shared by recording artists, producers, co-performers or listeners, even when they are not co-present. In a work that sets the stage for Christopher Small's (1998) subsequent ideas around musicking, Alfred Schutz (1964) ponders the subject of shared time, envisaging the relationship between performer and listener as one that is based on such sharing. He goes on to accommodate audio technology

---

[3]   Although this 1997 version is later than Prögler's work, it represents the only readily available incarnation of Farmelo's essay and appears to concur with the ideas from the earlier version that Prögler cites.

[4]   I trust that it is no more necessary to preface this section with an explanation of the origins or meaning of the 'drum break' than it would be to always describe the inner workings of the piano prior to discussing Chopin! Rose (1994, pp. 73–4) should help clear up any uncertainty.

[5]   See Stewart (2000) for a discussion of the rhythmic evolution whose genesis was in New Orleans.

within this conceptual framework, suggesting that the 'quasi simultaneity between the [performer] and the listener' that results from listening to a record is of equal validity to the more traditional live music arrangement that takes place without 'the interposition of mechanical devices' (p. 174). Subsequently, a key way in which Small's thinking refines Schutz's ideas is the egalitarian levelling that musicking wreaks on the old hierarchy that privileged composer over performer, and performer (as the mediator of the composer's genius) over listener: in musicking, the sharing becomes at least a two-way process.

Viewed in the light of both Schutz's acceptance of technology as a facilitator of time sharing and Small's assertion that everyone involved in musical activity contributes to the process, I propose that sampling technology enables musicking to occur across time and space. In effect, the producer who samples the Funky Drummer break is collaborating with Clyde Stubblefield, whether that much-sampled but less-frequently-acknowledged drummer likes it or not. (Neither disrespect nor antagonism are intended here, but the ubiquity of this classic breakbeat makes it a high-profile example.) The legal, financial and moral implications which historically dog sampling are in no way addressed or alleviated by my proposed conceptual model – it is intended simply as a way to reassess ideas of groove and musicking as applied to contemporary production practice. The urge to call the process 'transspatiotemporal musicking' is to be resisted I think, on grounds of unwieldiness!

In a passage that discusses listeners' responses to a groove, but which could equally apply to performers themselves, Lawrence Zbikowski (2004, p. 280) argues that 'being a member of a musical culture means knowing how to interact with the musics specific to that culture'. I interpret the term 'musical culture' here in a similarly 'loose and encompassing' sense to that by which Benjamin Brinner (1995, p. 2) understands 'tradition', where distinction occurs primarily by stylistic rather than geographic, ethnographic or other boundaries.[6] Zbikowski's expansive description ('member of a musical culture') can be seen to include performer, producer, listener, dancer and indeed any other participant in the process by which music is made and experienced.

David Brackett echoes Zbikowski and narrows the focus of the listener's interaction with the music, stating that 'something can only be recognised as a groove by a listener who has internalised the rhythmic syntax of a given musical idiom' (Brackett 1995, p. 144). This is a slight shift in perspective (from the performer to the listener), but essentially supports the idea that performed groove derives from the interplay between process and syntax, as Mark Doffman argues when he states that 'we need to think of groove as a dialectic, a continual synthesis of process and structure' (Doffman 2009, p. 296).

Throughout the literature that informs groovology, this duality of process and structure is expressed in many different ways, several of which are drawn

---

[6]   The reasons for drawing the distinction in this way will become clear later when the discussion turns towards groove in the mediated context of sampling in hip hop production.

together in Anne Danielsen's (2006, pp. 46–50) work on funk grooves. Beginning with some stalwart concepts of time and musical rhythm, there is the relationship between figure and gesture, and between metre and rhythm (Hasty 1997). Drawing next on linguistic theory, Danielsen proposes that Mikhail Bakhtin's sentence and utterance, Ferdinand Saussure's *langue* and *parole*, and Louis Hjelmslev's schema and usage are also relevant dualities. Much ethnomusicological work on African rhythm deals with the relationship between sounded and non-sounded patterns (see, for example, Chernoff 1979), with great significance attributed to the effect that the non-sounded patterns have on the way in which musicians, dancers and listeners produce or relate to the sounded patterns. Danielsen finally invokes Gilles Deleuze's version of this duality, discussed in terms of the virtual and the actual (although here a third entity, 'the real' joins the fray).

Whilst each iteration of the duality serves its own conceptual or analytical purpose, they can all be seen, in musical terms, to express a relationship between the abstract and the sounded aspects of music: in this way, they mirror the structure and process duality outlined by Doffman. Returning to Brackett's comment about the internalisation of rhythmic syntax, it is clear that, for the listener to interpret a groove or for the musician to create a groove, all participants need to have a sense of what culturally specific, contextual sense of 'time' is being played against or engaged with.

Contextual Senses of Time

I outline below three broad starting points that might typically form the basis for a contextual sense of musical time in the study of groove, before exploring each in more detail via analysis of selected recordings extrapolated from the funk era. It would, of course, be artificial to expect music in the real world to conform neatly to only one of these categories since there is a degree of overlap between them, particularly when the ideas are manifested in the 'messy reality' of performance, as will be discussed later with reference to a relevant example. Nonetheless, the three categories outlined below begin to help with the characterisation of some typical contextual senses of time.

1. In the first category, the contextual sense of time is contingent upon shared prior knowledge on the part of all musicking participants, as in the reggae 'one-drop' rhythm, for example.
2. In the second category, a musician 'sets up' the contextual sense of time in some way prior to the start of the performance proper for the benefit of the other musicking participants, as in styles based around a stated timeline pattern.
3. The third category relies on a shared sense of metronomic time, although in practice this tends to be more a general feeling of an underlying isochronous pulse rather than a precisely 'metronomic' understanding.

Category 1: Contextual Sense of Time as Shared Prior Knowledge

In reggae drumming, the 'one-drop' rhythm derives its character from the drummer *not* emphasising the first beat of the bar. Instead, the first beat of the bar (the 'one') is 'dropped', typically omitted altogether or marked with only a light stroke on the hi-hat at most, whilst the power of the snare and kick drums are saved for later in the bar. Examples abound, but clear instances can be heard in both The Upsetter's 'One Step Dub' (1976) and Bob Marley and the Wailers' 'Crazy Baldhead' (1976). The one-drop literally turns the standard approach to the drum kit on its head: in non-Jamaican popular music styles, examples of drum kit patterns that do not feature a kick drum on beat one are few and far between. (This is even true in Brazil, where despite the fact that in *samba batucada* the percussive bass emphasis produced by the open tone of the *surdo* generally falls midway through the bar – a close equivalent to the reggae one-drop – by contrast, in *bossa nova*, when the same set of rhythms is transposed from the percussion ensemble to the drum kit, the drummer will still add a kick drum stroke on beat one.)

By de-emphasising the first beat of the bar in this way, often to the point of remaining wholly tacet for a moment, the drummer playing a one-drop rhythm prioritises the non-sounding element of the groove and it is here that the importance of *prior* knowledge becomes important. For drummers playing in most non-Jamaican popular music styles, placing a kick drum stroke on beat one is habitual, a response to the commencement of performance engrained by years of unchanging practice, regardless of what complexity and variation may occur in the subsequent beats of the bar. During a one-drop, other musicians in the reggae band may well place a note on the first beat of the bar, as is heard in the bass part of 'One Step Dub' where the weight of the bass guitar sound compensates in part for the lack of the kick drum. This does not demonstrate a lack of prior knowledge on the part of the bassist but rather an assuredness, based on awareness of the contextual sense of time, that beat one will be an available space, uncluttered by any instrument occupying a similar frequency range. In the case of 'Crazy Baldhead', both bassist and drummer generally remain tacit on beat one, thus emphasising the one-drop pattern even more strongly. Both songs demonstrate that, by knowing that the drummer will be grooving against an idiomatically specific sense of time, the bassist is able to do the same.

If both bassist and drummer are grooving, why not simply describe this as grooving together, in the participatory manner that is most usually attributed to groove? Of course they are grooving together (rather than operating in some artificial, anti-musical isolation), but the point here is that, at the same time as grooving with one another, they are also each grooving individually with the contextual sense of time that is derived from prior knowledge of the one-drop idiom. Thus the contextual sense of time simultaneously becomes the key to their interaction with one another and their solo groove.

Category 2: Contextual Sense of Time as Timeline Pattern

This second category relies on a musician to set up the contextual sense of time in advance of the groove beginning. Depending on the circumstances, the 'set-up' may then continue in some form once the other musicians have begun to play, possibly even throughout the performance, or may recur at key structural moments as appropriate. African diasporic forms, whose musical fabric is woven around a stated timeline pattern (such as Cuban-derived styles based around the clave rhythm), clearly fall into this category too (these are discussed at length by Agawu (1995), Arom (1991), Chernoff (1979) and, more recently, Pressing (2002) and Toussaint (2003)). The category might also include the conductor's role at the start of an orchestral performance, which goes far beyond merely setting the tempo (a task that would seem more appropriate to the third category) and can encapsulate in microcosm the attitude that the orchestra should strive to communicate through their interpretation of the score. A less formal, even seemingly boring, example of this category is found in the bandleader's 'count-in' at the start of a song. In fact, the count-in is a potentially fascinating musical feature, manifested in many nuanced variations and worthy of study in its own right. James Brown's idiosyncratic approach to counting his band in, for example, sheds fresh light on the well-worn quest to get to grips with his groove.

A cursory analysis of the way that James Brown counts the band in for five different songs recorded between 1969 and 1972 reveals a number of trends that characterise his approach and that in turn either influence or relate to each musician's interaction with the contextual sense of time thus created. In terms of cause and effect, the direction of influence between bandleader and musicians here is hard to determine because the band are likely to have rehearsed the songs prior to recording, or at least hastily jammed the rhythm parts in the recording studio during the compositional process. In a filmed interview, Pee Wee Ellis, saxophonist with James Brown and co-writer of 'Cold Sweat' (Brown 1967), describes the series of rhythmic grunts used by Brown to communicate his idea for the song thus

> James took me into his dressing room one night after a gig and mumbled some things like ... [rhythmically] mm-uh-Urrgh-uh-Urrgh-uh-uhhh-ow-Ow-uh-uh-ugh. So I said ... OK! [laughs]. (Knox 1992)

It is clear from this exchange that, even before the song has been fully composed and arranged, Brown wants to convey a specific contextual sense of time that the grooving musicians will be expected to interact with.

To illustrate this, the count-ins of five songs are compared:[7] 'Funky Drummer' (1969), 'Talkin' Loud and Sayin' Nothing (Remix)' (1970), 'Superbad' (1970), 'Hot Pants' (1971) and 'The Boss' (1972). Of these five, 'Funky Drummer' is

---

[7] Data available from the author on request.

unusual in that it contains four different instances of counting altogether, although to hear all four one needs both the 7′01″ edit released as part of the *Star Time* compilation (which features a count-in at the beginning; Brown 1991) and the 9′15″ version from *In The Jungle Groove* (which omits the opening count-in but includes one at the outro that is absent from the shortened *Star Time* edit; Brown 1986). These four counts mark the start of the song (count A), the arrival of the first drum break (count B), the re-entry of the band at the end of this solo (count C), and finally the cue for the closing drum break which functions as a fading outro (count D). Count C occurs in double-time relative to the others, but whether taken at face value or mapped so as to match the tempo of the other counts, it still displays relevant features that relate to the drummer's groove.

Count A sets the scene and demonstrates the first aspect of Brown's approach, namely his fondness for abandoning the use of numbers mid-count and substituting lyrical snatches in their place. (These fragments are sometimes a little indistinct, hence the words used in the following transcriptions are approximate.) So rather than the expected

| One • • • | Two • • • | Three • • • | Four • • • |

the band are instead invited to start playing when they hear:

| One • • • | Two • • • | In • she go | • ah! • • |

This count also demonstrates Brown's avoidance of beat four, which is arguably the most striking feature in terms of the sense of contextual time he aims to communicate here and is common to all of the counts heard in 'Funky Drummer'. In count A, the words 'go' and 'ah!' fall on the nearest semiquaver on either side of beat four, but the beat itself remains unsounded. In terms of the contextual sense of time, these semiquavers are significant to the groove of the drum break, especially the second semiquaver of beat four. During the drum break, each of the eight bars is unique in terms of the pattern played when all the gestural nuances are taken into consideration, but one feature common to all of them is the placement of a combined open hi-hat and kick drum stroke on this second semiquaver of beat four, the hi-hat then closing gradually so that the decay ends on the third semiquaver.

Count B marks the start of the main drum break and is the first instance of another intriguing feature of James Brown's contextual sense of time, in which all of the relevant number names are spoken, but not necessarily in the places that one would expect. So the band drop out when they hear count B thus

| One • • • | Two • • • | Three • • Four | • Get it! • |

but Brown's spoken 'four' actually falls on the last semiquaver of beat 3, rather than on beat four.

Taken at its face-value tempo (that is, in double-time relative to the main tempo of the song), count C only occupies the second half of the bar, serving as a hasty exhortation for the band to re-enter:

| • • • • | • • • a | One • Two • | Three Four • • |

The concertina effect that the double-time count has in relation to the standard-time beats means that the numbers used do not relate to the expected count in any meaningful way, but merely act as a sequence leading towards a structural event – the band's entry. The placement of the word 'four' however, lands on the second semiquaver of beat four again, further emphasising the significance of that moment in the contextual sense of time for both drummer and vocalist, as discussed earlier.

Count D is essentially a re-statement of count B, fulfilling a similar function in that it also prefaces a drum break. The only variation is in the lyrical fragment delivered during the fourth beat

| One • • • | Two • • • | Three • • Four | • The fun-ky | drum-mer

with 'drummer' beginning on the first beat of the next bar, meaning that count D is the only version that spreads across a barline.

The count-ins used for the other four songs are less complex in the way they vary from expectation, but all share the trait whereby Brown substitutes the word 'four', interjecting a lyrical fragment instead, such as 'Hit it!' (or possibly 'Get it!'), 'Uh!' or 'Get down!' Paul Berliner describes the backbeat (the near-ubiquitous emphasis on beats two and four found widely in popular music)[8] as 'an important rhythmic target for improvisers' (Berliner 1994, p. 149), which could explain Brown's tendency to orient his adapted count-ins around beat four. As described above, he treats this rhythmic target in two different ways, either landing on it but using an unexpected word instead of 'four', or avoiding the beat altogether and instead building vocal phrases that cluster in the semiquavers immediately before and after the beat (in effect, creating an unsounded rhythmic target). Both approaches to the rhythmic target fit with an essential aspect of the funk idiom, the concept of 'the one', which describes the need for the various interlocking cyclical patterns played by different members of the band to revolve around, and be anchored by, a strong emphasis of the downbeat in any given bar.[9] So the contextual sense of time that imbues James Brown's approach to counting the band in is specifically geared towards a key musical feature of the

---

[8]   See Mowitt (2002) for an engaging, multilayered discussion of the backbeat.

[9]   Danielsen (2006) deals extensively with 'the one' and it is regularly mentioned by funk musicians in interviews (for example, Bootsy Collin's passionate demonstration in 'Lenny Henry Hunts the Funk' (Knox 1992)).

funk idiom – musicians who interact with the contextual sense of time set up by Brown will therefore be predisposed to appropriate and effective grooving.

Category 3: Contextual Sense of Time as Inferred Isochronous Pulse

Essentially, the distinction between this category and the preceding two is that, since the basis for the sense of time relies on a regular, isochronous pulse, there is no need for context-dependent prior knowledge in order to *feel* the time, as in the first category, nor is there necessarily any need for the time to be set up in advance by a musician, as in the second category, because the regular beat can be inferred. An example of this category in action would be when the audience expresses its understanding of the contextual sense of time by spontaneously clapping along with the main beats of the bar, as an outward expression of the regular, isochronous pulse underlying the music.

In the preceding categories, context has the effect of characterising the timing information (whether stated or unstated). In this third category, given that timing is predetermined by virtue of the isochronous nature of the pulse, context manifests itself instead in a characterisation of the emphasis. So, the underlying pulse may be invariably isochronous, but depending upon the musical context, the majority of the clapping audience may choose to emphasise beats one and three or, alternatively, beats two and four, for example. The following section explores examples in which the contextual senses of time used to groove against overlap the three categories outlined above, as might be expected in a real world of music which eludes narrow categorisation.

Messy Reality versus the Three Categories

As the following example illustrates, most groove interaction is actually based around familiarity with contextual senses of time which can simultaneously fit into more than one of these categories to some extent, or which may seem to fall into one initially then gradually shift towards another category as a result of the (sometimes unwitting) actions of the musicking participants.

Widening the musicking context beyond the confines of the recording studio, a live recording can of course provide evidence of the audience's involvement in groove production and perception. Staying tangentially within the James Brown camp, saxophonist Maceo Parker's live version of 'Shake Everything You've Got' (1992) features a lengthy groove solo by drummer Kenwood Dennard, beginning at 5′44″, in which he utilises the audience's sense of isochronous pulse as a springboard for his own groove interaction. By maintaining a steady snare backbeat from the outset of the solo, and restricting his improvisational gestures to hi-hat flourishes and some varied kick drum patterns, Dennard draws the audience into recognising the prominence of the steady isochronous pulse, so that within the first 20 seconds of the solo audience claps can be heard emphasising beats two and four. He has not needed to set up this contextual sense of time explicitly, as in the second category, but has created an environment within which the underlying pulse of the third category suggests itself strongly to the listeners. Then, as the drumming variations increase, albeit still anchored by the backbeat, there is a sense that Dennard begins to use the audience's clapping as a framework for the greater fluidity in his playing. A new drum pattern emerges in the bar which begins at 7'00'', becoming more playful with cross-rhythms that work against the pulse that the audience is feeling and expressing. Crucially, instead of a snare hit on beat four as expected, this bar culminates in a whole beat of total silence from the drums before Dennard re-enters on the following downbeat. This drum vacuum on beat four is, of course, filled with the audience clap instead and so the groove relationships shift momentarily. The audience's pulse role briefly becomes the actual backbeat, the stated drum pattern becomes unsounding, and expectations are confounded in a way that draws out a sense of shared prior knowledge – in short, it is a simple but magical moment. When the same trick occurs four bars later, the audience really begin to get a sense of sharing Dennard's contextual sense of time, so despite the repetition, this second occurrence is equally thrilling because the perception has changed again.

Some theories occur in response to this passage. Firstly, music that largely conforms to the isochronous pulse category may have the effect of allowing a larger proportion of participants to feel 'in the know', because the rhythmic arcana associated with idiomatic prior knowledge are less prevalent. Secondly, there are moments in musicking when the categorisation of the various contextual senses of time can suddenly shift, and these moments appear to be significant for the participants. Thirdly, groove can perhaps be simultaneously solo *and* participatory in an ensemble context: each musician (or listener, dancer and so on) is grooving in relation to both their own contextual sense of time and to those of the other participants, as manifested in the sounds or movements these other participants make. This last idea hints at a complex, organic, interdependent relationship between individual and shared contextual senses of time which would benefit from more thorough investigation in its own right elsewhere.

Conclusion

I have shown a range of perspectives on the ways in which groove depends upon familiarity with musical time, and particularly such senses of time as are idiomatic, stylistically nuanced or contextually conceived. The three proposed categories that might form the basis for these contextual senses of time have been discussed and exemplified, and it transpires, unsurprisingly, that real music usually defies artificial categorisation. Nevertheless I have shown the categories to be useful tools in understanding the interactive relationship between process and structure in groove, and especially in beginning to address the question of solo groove.

What sampling does to groove is important in the light of the above discussion. First and foremost, by sampling and looping a single bar of a drummer engaged in solo groove, the relationship between process and structure/syntax is shifted, although the nature, degree and direction of the shift is determined by the extent to which the producer manipulates the sample and the character of the sonic setting into which it is recontextualised. Even assuming a simple approach to sampling in which an extract is appropriated wholesale and no manipulation other than looping takes place, the drummer's groove moves from being played as a process in relation to an unsounded structure (whether it be an underlying isochronous pulse or a contextual sense of time supplied previously by another musician), to becoming structure itself when looped as a sample. The contextual sense of time encapsulated both implicitly and explicitly within the sampled breakbeat becomes important following the process/structure shift. It is typically this new contextual sense of time encapsulated by the looped breakbeat that the hip hop MC would then use as a musical entity to interact with, generating a new groove by rapping some lyrics.

Another aspect of groove in relation to sampling (and other production techniques such as programming beats, for example) might begin with Jeff Warren's helpfully catholic definition of the performer as 'the maker of sounds that are considered the performance' (Warren 2008, p. 102). This, and the trends indicated by Danielsen et al. (2010), suggest that the producer can now, in effect, be considered a performer (a position which has arguably been the case since as long ago as the early 1970s when King Tubby first began concocting dub mixes derived from pre-existing multi-track master tapes in Jamaica). I have already discussed the idea that groove need not necessarily occur between interacting musicians, as is seen in solo groove. I have also suggested that sampling allows people to groove together across boundaries of time and space, building interaction onto sampled sections of solo groove.

If we accept the producer as a performer, then it follows that the act of producing – the laborious process by which a track is created through programming, recording, mixing and so on – is all part of performance and could thus be said to contain the potential for grooving too.[10] This suggestion fits with the senses of contextual time defined above. For the producer to create a groove through sampling or programming, familiarity with stylistically nuanced conceptions of time will be just as important as it is for the instrumentalist. Whether this familiarity informs real-time production of sounds from an instrument or gradual electronic manipulation of sounds which will eventually be experienced by other musicking participants in real time, the interaction between the sounded and the unsounded events, between process and structure, functions similarly.

The ideas presented in this chapter demonstrate the relevance of groovological research to our understanding of musicking in the twenty-first century and the role of the contemporary producer in performance, but these ideas are still in the early stages and would benefit from more detailed research in their own right. The possibility that musicking can take place across time and space seems to hold

---

[10]   See Butler (2006) for many more ideas relevant to groove and electronic music.

exciting potential, from the perspectives of both practitioner and researcher, and is undoubtedly of current relevance.

References

Agawu, K. (1995). *African Rhythm: A Northern Ewe Perspective*. Cambridge: Cambridge University Press.

Arom, S. (1991). *African Polyphony and Polyrhythm: Musical Structure and Methodology*. Editions de la Maison des Sciences de l'Homme. Cambridge: Cambridge University Press.

Berliner, P. (1994). *Thinking in Jazz: The Infinite Art of Improvisation*. Chicago, IL: University of Chicago Press.

Brackett, D. (1995). *Interpreting Popular Music*. Cambridge: Cambridge University Press.

Brinner, B. (1995). *Knowing Music, Making Music: Javanese Gamelan and the Theory of Musical Competence and Interaction*. Chicago, IL: University of Chicago Press.

Brown, J. (1967). *Cold Sweat*. King Records (3) 456110.

Brown, J. (1986). *In The Jungle Groove*. Polydor 8296242.

Brown, J. (1991). *Star Time*. Polydor 849108.

Butler, M.J. (2006). *Unlocking the Groove: Rhythm, Meter, and Musical Design in Electronic Dance Music*. Bloomington, IN: Indiana University Press.

Chernoff, J.M. (1979). *African Rhythm and African Sensibility: Aesthetics and Social Action in African Musical Idioms*. Chicago, IL: University of Chicago Press.

Chor, I. (2010). Microtiming and Rhythmic Structure in Clave-based Music. In A. Danielsen (ed.), *Musical Rhythm in the Age of Digital Reproduction* (pp. 37–50). Farnham: Ashgate.

Clayton, M., Sager, R. and Will, U. (2005). In Time with the Music: The Concept of Entrainment and its Significance for Ethnomusicology. *European Meetings in Ethnomusicology* 11: 3–75.

Danielsen, A. (2006). *Presence and Pleasure: The Funk Grooves of James Brown and Parliament*. Middletown, CT: Wesleyan University Press.

Danielsen, A. (ed.) (2010). *Musical Rhythm in the Age of Digital Reproduction*. Farnham: Ashgate.

Doffman, M. (2009). *Feeling the Groove: Shared Time and its Meanings for Three Jazz Trios*. Unpublished PhD dissertation, Open University.

Farmelo, A. (1997). *The Unifying Consequences of Grooving: An Introductory Ethnographic Approach to Unity through Music*, http://musekids.org/UCS. html (accessed 8 August 2011).

Hasty, C.F. (1997). *Meter as Rhythm*. Oxford: Oxford University Press.

Huron, D.B. (2006). *Sweet Anticipation: Music and the Psychology of Expectation*. Cambridge, MA: MIT Press.

Keil, C. and Feld, S. (1994). *Music Grooves: Essays and Dialogues.* Chicago, IL: University of Chicago Press.

Kfivte, T. (2004). Description of Grooves and Syntax/Process Dialectics. *Studia Musicologica Norvegica* 30: 54–74.

Knox, T. (1992). Lenny Henry Hunts the Funk. *The South Bank Show.* Season 15, Episode 14 (television broadcast: 12 January).

London, J. (2004). *Hearing in Time: Psychological Aspects of Musical Meter.* Oxford: Oxford University Press.

Marley, B. and The Wailers. (1976). *Rastaman Vibration.* Island Records 7-900-33-1.

Mowitt, J. (2002). *Percussion: Drumming, Beating, Striking.* Durham, NC: Duke University Press.

Parker, M. (1992). *Life On Planet Groove.* Minor Music 85502324.

Pressing, J. (2002). Black Atlantic Rhythm: Its Computational and Transcultural Foundations. *Music Perception* 19/3: 285–310.

Prögler, J.A. (1995). Searching for Swing: Participatory Discrepancies in the Jazz Rhythm Section. *Ethnomusicology* 39/1 (special issue: Participatory Discrepancies), 21–54.

Rose, T. (1994). *Black Noise: Rap Music and Black Culture in Contemporary America.* Hanover, NH: Wesleyan University Press.

Schutz, A. (1964). Making Music Together: A Study in Social Relationship. In A. Brodersen (ed.), *Collected Papers (Volume 2): Studies in Social Theory* (pp. 159–78). The Hague: Martinus Nijhoff.

Small, C. (1998). *Musicking: The Meanings of Performing and Listening.* Middletown, CT: Wesleyan University Press.

Stewart, A. (2000). Funky Drummer: New Orleans, James Brown and the Rhythmic Transformation of American Popular Music. *Popular Music* 19/3: 293–318.

Thaut, M.H. (2008). *Rhythm, Music, and the Brain: Scientific Foundations and Clinical Applications.* New York: Routledge.

Toussaint, G.T. (2003). Classification and Phylogenetic Analysis of African Ternary Rhythm Timelines. *Proceedings of BRIDGES: Mathematical Connections in Art, Music, and Science* (pp. 25–36).

Upsetters, The (1976). One Step Dub. On Max Romeo, *One Step Forward.* Island Records WIP6305.

Vuust, P., Ostergaard, L., Pallesen, K.J., et al. (2009). Predictive Coding of Music: Brain Responses to Rhythmic Incongruity. *Cortex* 45/1: 80–92.

Warren, J. (2008). Improvising Music/Improvising Relationships: Musical Improvisation and Inter-Relational Ethics. *New Sound* 32: 94–106.

Zagorski-Thomas, S. (2007). The Study of Groove. *Ethnomusicology Forum* 16/2: 327–35.

Zbikowski, L.M. (2004). Modelling the Groove: Conceptual Structure and Popular Music. *Journal of the Royal Musical Association* 129/2: 272–97.

Chapter 12

Social Familiarity: Styles of Interaction in Chamber Ensemble Rehearsal

Elaine King

When two or more musicians first play together in a music ensemble, they begin a process of familiarisation as they get to know each other both musically and socially. There is a complex network of issues to consider in this process as ensemble musicians develop close friendships, cultivate working routines, share and establish musical ideas, explore repertoire and become acquainted with particular rehearsal and performing environments. This chapter examines the styles of social interaction between musicians working at different stages in the familiarisation process through the empirical study of new and established Western chamber ensembles. The purpose is to consider the effects of 'social familiarity', specifically the length of time musicians have known one another, on the discourse produced in ensemble rehearsal.

All human relationships are dynamic because they are in a constant state of flux. Musicians who play together in a small ensemble (with or without a conductor) can be defined as being involved in a *working relationship* because they are engaged in achieving a particular goal together, normally to perform in a concert or gig. Arguably, however, ensemble players develop a certain closeness and friendship in their music-making over a period of time that goes beyond the experiences of a regular working relationship. Indeed, ensemble musicians share a particular bond – a love of music and the desire to play it – which underpins the dynamic relationship between them. In effect, therefore, ensemble musicians, whether amateur, student or professional, are potentially involved in a *close* working relationship that mirrors the experiences in everyday lives among partners, families and friends.[1]

---

[1] According to Berscheid and Peplau (1983), a close relationship is defined by four interdependent criteria: (a) frequency (the individuals have regular impact on each other); (b) strength (the degree of impact on each occurrence is strong); (c) diversity (the impact involves different kinds of action); and (d) duration (the relationship lasts a relatively long time). In the context of a music ensemble, diversity might be relatively constrained as the shared activity – making music – is constant; however, different kinds of repertoire will invariably involve different interpretative ideas, while rehearsals and performances will require contrasting action and spontaneity. Regarding duration, the members of newly formed music ensembles do not fulfil the criteria of a 'close' relationship, even though their

In Western art music practice, the majority of ensembles will spend a high proportion of time in rehearsal as the musicians meet up on a daily or weekly basis to prepare for a particular performance or gig. During rehearsals, ensemble musicians will nurture a major part of their working relationship. For this reason, the aim of this chapter is to examine the ways in which musicians work together in the rehearsal context, specifically to investigate their styles of social interaction. Existing research reveals that musicians interact via verbal and non-verbal discourse in rehearsal alongside playing segments of musical material, normally balancing work on short or long sections of a piece with complete run-throughs (Davidson and King 2004; Goodman 2000; King and Ginsborg 2011; Williamon and Davidson 2002). More talking than playing is likely among newly formed ensembles compared with established ones (Goodman 2000). Verbal discourse facilitates the coordination of work on a particular piece (such as through giving cues, relating starting points and clarifying interpretative ideas; see Davidson and Good 2002) as well as providing explicit information about the musicians' socio-emotional behaviour. It is likely that the conversation taking place between segments of playing will change over time as the group work evolves: John Dalley from the Guarneri String Quartet describes it as a 'constant working-out process' (Blum 1986, p. 7; also see Bayley 2011; Goodman 2002; King and Ginsborg 2011).

So, how do musicians' styles of social interaction change as they become more familiar with one another? In order to address this question, a cross-section of empirical data was selected using transcriptions of verbal and non-verbal discourse arising in rehearsals with musicians working in 'new' and 'established' groups derived from previous studies (Ginsborg and King 2012; Goodman 2000; King and Ginsborg 2011).[2] Two frameworks were used to analyse the data and these will be introduced below prior to evaluation and discussion of the findings. Before moving on, however, there are two important issues that merit attention in the study of chamber ensembles – kinship and survival (longevity). The ensuing section provides brief historical and socio-cultural consideration of these issues as they impacted upon the choice of analytical frameworks as well as the interpretation of the data in this study.

---

first rehearsal might mark the start of a journey towards developing one (and even if the individuals might be close friends already).

[2]   Note: the transcriptions pertain to discourse taking place *between* run-throughs or segments of play in rehearsal. It was beyond the scope of these studies to analyse the audio-video data pertaining to material *within* run-throughs or playing segments, although future studies could usefully contribute knowledge about musical and social interaction in ensemble rehearsal via a study of playing segments.

Kinship and Survival

Over the past few decades there has been a rapid expansion of different types of chamber media alongside traditional duos, trios and quartets: musicians can work together in varying combinations to explore arrangements of existing repertoire, compose and commission new works, and deliver music to different audiences, such as in formal concerts or as incidental background music. Socio-cultural (and other) forces can influence the formation, promotion and success of groups: many of the 'newer' groups adopt the traditional string quartet framework and, like the popular tradition, use concept-led elements such as gender, image and kinship to market the ensemble.[3] Even though kinship seems to be the exception rather than the norm in classical chamber ensembles – there are relatively few groups with members who are blood-related – in some cases, siblings or relatives have formed groups that have enjoyed success over a number of years.[4] Descriptions of experiences by members of both popular and classical ensembles with no blood relations, however, indicate the importance of establishing a 'family feel':

> We do admire each other a lot and it feels like we're one person when we play …
> It's like a family. We are very close and it's given me all these extra people that
> I care about and they care about me. (Interview with New Wave Band Member,
> Kassandra; Bayton 1998, p. 100)

---

[3] Examples of 'newer' quartet ensembles include saxophone quartets, such as Sheffield-based group 'Saxsational' (www.saxsational.co.uk), trombone quartets, such as the award-winning all-female trombone quartet 'Bones Apart' (www.bonesapart.com), and flute quartets, such as the virtuosic all-female group 'London Flutes' (www.londonflutes. com). Equally, there are numerous kinds of vocal quartet ensembles in circulation, some defined by their musical focus (for instance, gospel, barbershop, classical) and others by their specific make-up (for example, all-male, all-female, mixed). The majority of established and new quartets seem to be single-sex, while a recent informal survey of websites promoting trios and quintets indicates that they are predominantly mixed (also see Ford and Davidson 2003; Blank and Davidson 2007). In the popular tradition, gender and image are evident in the promotion of girl and boy bands (see Shuker 1994), while kinship can provide an important recruitment and development tool for groups with family connections (Bayton 1998). Well-known examples of 'family' groups include The Jackson Five, Sister Sledge, The Bee Gees and The Beach Boys.

[4] Aside from the immortalised von Trapp Family Singers, archival evidence points towards the four sisters in the Lucas String Quartet as one of the earliest examples of a 'family' quartet who gave critically acclaimed performances in London, Prague and Vienna in the early twentieth century: 'the association of four sisters to form a string quartet is as interesting as it is remarkable; but when the standard of their performance is as high as that attained by the four Misses Lucas, the phenomenon is quite exceptional' (anonymous critic from *The Musical Times* 1909, p. 469). Other recent examples include the Cann Twins piano duet (www.canntwins.co.uk) and the Daroch Trio (www.darochtrio.com).

It is a dedicated life, a marriage between four people, and if there are spouses,
a doubly dedicated one. (Muriel Nissel describing the Amadeus Quartet; Nissel
1998, p. 164)

These examples highlight the issue of closeness within small musical groups, or
the creation of intimacy through social familiarity.[5]

While bonding through kinship or establishing a 'family feel' might be seen to
be an aspiration for small ensembles, the length of time that a musical group stays
together can be regarded as a benchmark of its success (Murnighan and Conlon
1991). It is widely recognised, however, that musical ensembles are not always
destined to survive; in fact, Bayton (1998) remarks that the average lifespan
of a popular music group is one or two years. She suggests that popular bands
with family members might stay together for longer than other groups because
the family connections serve to 'weld the band' and 'weather the stormy patches'
(Bayton 1998, p. 184). For classical ensembles, a similar one- or two-year span
might be applied, even though players might plan to stay together 'for a long
time' (the most frequent response in Ford and Davidson's 2003 study on wind
quintets). The most remarkable example of a surviving chamber ensemble is the
Amadeus String Quartet, which lasted 40 years from 1947 to 1987 with its original
members (Norbert Brianin, Siegmund Nissel, Peter Schidlof and Martin Lovett).
The group disbanded when Schidlof died because the other members felt that he
was irreplaceable (see Nissel 1998). Other quartets with a relatively long lifespan
include the Alberni (35 years, with several changes of personnel) and the Allegri
(over 50 years with 13 changes of personnel, arguably Britain's longest-surviving
string quartet). Burton-Page (2005) argues that the Allegri survived through
a 'process of slow Darwinian evolution', corroborating the notion of a group
existing as a dynamic entity.[6] Arguably, if surviving chamber music ensembles are
to be understood like 'family' groups, then it is important to consider this in the
analysis of social interaction between players.

---

[5]   Of course, the origin of the word 'familiar' lies in the 'family'; see Chapter 8.

[6]   In his detailed booklet entitled *The Allegri at 50: A Quartet in Five Movements*
(2005; see www.allegriquartet.org.uk/fifty), Piers Burton-Page describes the group's five
'survival mechanisms': (a) a 'youth' policy (allowing younger members to fill places meant
that fresh ideas could be taken on board whilst the younger player had sufficient stamina to
learn a repertoire already known to its senior members); (b) a 'character' policy (the players
needed passion, energy and a sense of humour as well as a deep sense of responsibility
towards the music-making); (c) a 'continuity' policy (whilst the various changes of
personnel worked effectively, some individual members served over 30 years, which
provided important continuity for the group); (d) practical identity (the group established
important links with people or places and undertook educational work in order to ensure its
survival); and (e) musical identity (the group worked on a core repertoire and established a
particular 'Allegri style', which reflected 'maturity and spontaneity' in search for 'musical
truth' over virtuosity).

Two Analytical Frameworks

There are various ways in which musicians' styles of social interaction might be analysed, although the two frameworks drawn from social psychology in this study allowed complementary views of empirical data. The first framework, Robert Bales's Interactive Process Analysis (IPA; Bales 1950, 1999), has been widely applied in the study of small working groups and has been used in previous research with music ensembles (Ginsborg and King 2012; Goodman 2000; King and Ginsborg 2011; cf. Young and Colman 1979). The second draws upon Arnold Sameroff's (2009) transactional theory, which essentially focuses on the interplay of nature and nurture in the study of family relationships, notably child–parent duos, and its application in this research was motivated by the observations made previously about the importance of 'family feel' in chamber music-making. These frameworks will be explained below in turn.

In Bales's IPA, each verbal and non-verbal utterance is coded according to its type as socio-emotional or task-related. The six socio-emotional categories are divided into positive (agree, tension release, solidarity) and negative (disagree, shows tension, antagonism) areas, while the six task-related categories include 'asking' or 'giving' suggestions, opinions and orientation. The frequency of utterances is recorded to enable evaluation of different types of utterances within an exchange and cross-comparison of the interaction arising in different groups; moreover, the frequencies can be compared with the suggested limits for each category according to Bales's accumulated research with other small groups.

Sameroff's ecological transactional theory, originally proposed in 1975, considers long-term behavioural changes between people involved in close relationships; in comparison with IPA, this approach permits a broader view of musicians' social interactional style. Sameroff emphasised the need to monitor bidirectional and interdependent effects on an individual over a period of time in a non-linear fashion: 'development of any process in the individual is influenced by interplay with processes in the individual's context over time' (Sameroff 2009, p. 6). There is a difference, therefore, between transactions and interactions in his work: 'Interactions are documented by finding dependencies in which the activity of one element is correlated with the activity of another … and the correlations are stable over time. Transactions are documented when the activity of one element changes the usual activity of another, either quantitatively by increasing or decreasing the level of the usual response, or qualitatively by eliciting or initiating a new response' (p. 24).

Whilst researchers have applied transactional thinking in a range of studies on family relationships (see Sameroff 2009), its application in the study of working relationships among musicians in a chamber ensemble is entirely novel. Arguably, the concept of transactional thinking is highly relevant as it promotes consideration of the long-term development of a group of musicians. One way in which transactional thinking might be monitored in the music rehearsal context is through coding groups of utterances together as 'frames', or recurring patterns of communication (Fogel 2009, p. 271). In effect, frames are 'dynamically stable transactional patterns in ongoing relationships such as rituals, routines and ways of doing things together' (p. 276). Fogel et al. (2006) identified four different kinds of frames in an observational study of mother–infant communication (see Fogel 2009, pp. 277–8):

- *bridging frames* ('served as an intermediary, a pathway for change, that helped a system move from the old routine into the new one');
- *permeability frames* ('frames blend into each other during the transition period');
- *recapitulation frames* (like regression; communication 'brings back, but only temporarily, an old frame or pattern of frames that was useful and familiar in the past');
- *innovative frames* ('sudden leaps of novelty, new ways of relating, and sparks of inspiration ..., experiment with new ways of doing things').

Even within the context of a single music rehearsal, one might observe these types of frames: *bridging* and *permeability frames* might be evident as musicians become more familiar with one another, perhaps towards the middle or end of a rehearsal; *innovative frames* might be strongest when the musicians are trying to be spontaneous and individual, such as when developing interpretative ideas about a new piece or re-visiting a previously performed work; *recapitulation frames* might characterise behaviour in pre-concert rehearsals as musicians go over familiar repertoire. It is plausible to suggest, therefore, that transactional frames in ensemble music rehearsal might be influenced by a number of factors, including familiarity between players, familiarity with repertoire, familiarity with venue as well as proximity to performance. Furthermore, new and different kinds of frames might emerge in these adult musical groups.

Evaluation of Empirical Data

The data from two previous studies involving the author (Goodman 2000; King and Ginsborg 2011) were selected so as to enable the evaluation of styles of social interaction arising between musicians working in 'new' and 'established' ensembles. The evaluation, like a mini meta-study (see Paterson et al. 2001), involved re-analysing the transcriptions of rehearsals according to the two analytical frameworks outlined above: the frequency of each type of utterance (socio-emotional/task-related) was recorded using IPA and expressed as a percentage of the total number of utterances within each rehearsal; and the transactional behaviour was identified by looking at patterns of communication (Fogel's frames) within each rehearsal. In this case, it was not possible to gather a longitudinal perspective because the data represented a cross-section of groups working in a specific rehearsal scenario (see below); rather, the intention was to identify local transactional frames across a single rehearsal. Here, patterns of communication (frames) were described according to the nature of rehearsal activity, including the lengths of segments of talking and playing, and the frequency of verbal exchanges between musicians in the talking segments. The duration of these frames was recorded and expressed as a percentage of the total rehearsal time.

The data included transcriptions of rehearsals for 11 ensembles: seven newly formed cello–piano duos (Goodman 2000) and four established vocal–piano duos (Ginsborg and King 2012; King and Ginsborg 2011). The participants included university students, music college students and professional musicians, reflecting a range of standards. All of the ensembles were observed in the same scenario, that is, a rehearsal of a set piece (either one movement of a cello sonata or one song by Ivor Gurney) with which they were previously unfamiliar. They were instructed to prepare the piece as if for a performance and were given the opportunity to practise their individual parts in advance of the rehearsal (owing to the design of the original studies, the individual preparation time for the singers and their pianists was limited to 20 minutes, while for the cellists and their pianists, it was unlimited). A summary of relevant background information for each of the ensembles is provided in Table 12.1.

The data were analysed by two independent researchers according to the frameworks described above. Even though other aspects of the rehearsals were examined in the original studies, including the effects of experience and the nature of musical interaction,[7] the primary purpose of this study was to compare the styles of social interaction so as to evaluate explicitly the effects of social familiarity on the rehearsal discourse of ensemble musicians: the seven cello–piano duos represented 'new' ensembles because they were working together for the first time; the four vocal–piano duos represented 'established' ensembles because they were working with a regular partner.[8]

---

[7] For example, in Ginsborg and King (2011) and King and Ginsborg (2012), codes were created to define the content of utterances about the music as basic, expressive or interpretative using the framework established by Chaffin, Imreh and Crawford (2002).

[8] The length of time during which the musicians in the established partnerships had worked together varied from 2 to 15 years. The point or stage at which an ensemble becomes established (or even 'well-established' as may be applicable in the case of older partnerships) will inevitably vary depending on the personalities, aspirations and achievements of the group. For the purpose of this research, a distinction was drawn between 'new' and 'established' based on a straightforward comparison of unfamiliar versus familiar; however, the duration of familiarity (hence the extent to which a partnership is established) might be scrutinised more closely in the future so as to enable greater understanding of this variable in the context of working relationships.

Table 12.1   Summary of background information for the ensembles
              included in the data set

Ensemble	Standard	Mean age	Gender	Piece	Social familiarity
Cello–piano duo A	Professional	28	F/F	Chopin, Op. 65, iii	New
Cello–piano duo B	Professional	38	M/M	Brahms, Op. 38, ii	New
Cello–piano duo C	Professional	44.5	M/M	Debussy, 'Prologue'	New
Cello–piano duo D	University students	20	F/F	Chopin, Op. 65, iii	New
Cello–piano duo E	University students	21	F/F	Chopin, Op. 65, iii	New
Cello–piano duo F	University students	24.5	F/F	Brahms, Op. 38, ii	New
Cello–piano duo G	University students	21.5	F/M	Debussy, 'Prologue'	New
Vocal–piano duo A	Music college students	25.5	F/M	Gurney, 'Up on the Downs'	Established (2 years)
Vocal–piano duo B	Music college students	21.5	F/M	Gurney, 'Up on the Downs'	Established (2 years)
Vocal–piano duo C	Professional	68	F/M	Gurney, 'An Epitaph'	Established (10 years)
Vocal–piano duo D	Professional	57	F/M	Gurney, 'An Epitaph'	Established (15 years)

New Ensembles: Cello–Piano Duos

All of the cello–piano duos began their ensemble rehearsal with a run-through of
the piece that they were preparing for performance. Following this, they worked
through the movement sequentially (or, in the case of two duos, non-sequentially)
so as to familiarise themselves with the music. According to Bales's IPA, the data
showed that talking dominated the non-playing elements of rehearsals (82 per cent
of the utterances were verbal). In general, the opening dialogue of each rehearsal
comprised simple questions about the set piece, such as 'were you happy with the
tempo?' (Duos C, D and F), 'are there any bits that you wanted to do?' (Duos A,
E, G) and 'do you want to tune? (Duos F and G); or ice-breaking 'introductory'
banter, such as comments about the British weather. In the majority of cases (86 per
cent), the first main musical issue for discussion concerned the choice of tempo.

Overall, interaction was predominantly task-related (mean = 72.1 per cent):
the highest category of utterances in each rehearsal was 'giving opinion' (mean
= 32.1 per cent) and the lowest was 'showing antagonism' (mean = 0.2 per cent).
Positive socio-emotional discourse (mean = 25.66 per cent) outweighed negative

socio-emotional discourse (mean = 2.4 per cent). There were high levels of agreement (mean = 19.2 per cent), moments of positive tension release in the form of laughter (mean = 5.36 per cent), yet little praise or showings of solidarity (mean = 1.1 per cent), which reflected individual, unshared behaviour and low levels of dependency (see Table 12.2).[9]

In the initial analysis of the data (Goodman 2000), the 'friendly' tone of the rehearsals was noted along with the high proportion of utterances involving 'giving opinions', although this was described as 'delicate' at times, perhaps symptomatic of the performers' lack of familiarity (p. 96). For example, supportive opinions were often given as a tactic to soften critical suggestions: 'it [the dynamic] makes a really nice contrast, but can we try to ... crescendo really gradually' (Duo F). In some of the partnerships, the pianists tended to *ask* for more orientation and opinions than the cellists (Duos A and B) or give opinions through leading-type questions, such as 'Do you want to put a *tenuto* on that because you've actually got the tune there?' (Duo E); in others, the pianists *gave* more opinions and suggestions than the cellists (Duos F and G). Interestingly, dominant personalities emerged in the partnerships: 'dominance' was attributed to the member of the duo who produced the highest number of opinions and initiated the highest number of verbal utterances following segments of playing, thereby creating a social hierarchy in controlling the direction of the rehearsal. A higher total frequency of verbal utterances did not necessarily reflect the dominant member of the duo, while sex-stereotyping did not appear to influence the musical partnership (see Blank and Davidson 2007).[10]

By comparing the frequency of socio-emotional and task-related utterances with the suggested limits of utterances according to Bales's existing research (see Bales 1950, p. 262; see also Table 12.2), the data indicated that these ensembles manifested less disagreement, tension and orientation in their rehearsals than typical small groups. Interestingly, there were relatively low levels of expressed solidarity and antagonism too. In effect, therefore, the social interaction in these new partnerships was about agreement and giving suggestions (rather than asking for them); it involved creating a positive atmosphere through suggestion-led discourse.

---

[9]   The high standard deviations reflected variability in the frequency of utterances per category between duos, although in comparison with Bales's suggested limits, the trends were highly consistent across the data set (that is, 'gives orientation', 'disagrees' and 'shows tension' were normally below the lower limit).

[10]   According to social dominance theory, individuals are psychologically predisposed towards enhanced or attenuated hierarchies in organisations depending on their personality and socio-political beliefs (see Sidanius and Pratto 1999). A key factor influencing dominance is gender: the assumption is that males tend to be more dominant than females. Blank and Davidson (2007) explored social and musical 'dominance' in a study of 17 piano duos (mixed and single-sex), observing 'musical androgyny' in terms of the organisation, choice of repertoire and processes of communication in partnerships; they noted, however, that sex-stereotyping existed 'outside of the professional relationship' (p. 246).

Table 12.2 Task-related and socio-emotional categories (column 1) with frequency of utterances for the cello–piano duo data (column 2) plus comparison with suggested limits provided by Bales (columns 3 and 4)

Category	Frequency of utterances (%; mean; SD)	Bales's limits (%)	Evaluation of data with Bales's limits
Shows solidarity	19 (1.1%; M = 2.7; SD = 1.7)	0–5	Near lower limit
Tension release	97 (5.36%; M = 13.9; SD = 8.2)	3–14	Near lower limit
Agrees	347 (19.2%; M = 49.6; SD = 27)	6–20	Near upper limit
Gives suggestion	174 (9.6%; M = 24.9; SD = 9.9)	2–11	Near upper limit
Gives opinion	580 (32.1%; M = 82.9; SD = 48.96)	21–40	Within limits
Gives orientation	224 (12.4%; M = 32.1; SD = 13.2)	14–30	*Below* lower limit
Asks for orientation	160 (8.8%; M = 22.8; SD = 11.3)	2–11	Near upper limit
Asks for opinion	151 (8.3%; M = 21.6; SD = 18.4)	1–9	Near upper limit
Asks for suggestion	17 (0.9%; M = 2.4; SD = 1.7)	0–5	Near lower limit
Disagrees	27 (1.4%; M = 3.8; SD = 4.5)	3–13	*Below* lower limit
Shows tension	15 (0.8%; M = 0.8; SD = 2.1)	1–10	*Below* lower limit
Shows antagonism	3 (0.2%; M = 0.4; SD = 0.8)	0–7	Near lower limit

Note: In column 4, the evaluation indicates whether or not these data fell *below* or *above* Bales's ranges, *near* (within 3 per cent) or *within* the upper and lower limits.

Moving on to transactional thinking, given that the musicians in these rehearsals were unfamiliar with one another, their patterns of communication and activity might have been regarded as entirely *permeable*. Yet, it was apparent that pre-existing shared understanding about music making came into play, so their discourse was, to an extent, *recapitulative* and *permeable* (that is, the musicians blended pre-conceived 'familiar' tactics about rehearsal as they worked together for the first time). Two new transactional frames were created in response to these data: (1) the *hesitancy frame* (that is, discourse and rehearsal activity characterised by broken-up conversation, a high frequency of verbal exchanges within talking segments, rapid discussion of musical ideas and short bursts of activity – lots of stops and starts in playing and talking); and (2) the *flowing frame* (that is, discourse and rehearsal activity characterised by relatively long utterances, sustained focus on particular musical issues or longer playing segments).[11] In general, hesitancy dominated the first stages of rehearsal (up to 65 per cent of the duration of the rehearsal), while some of the duos reflected flowing patterns of communication towards the middle and end stages of the rehearsal (at least 35 per cent of the rehearsal). The transactional style of the new duos thus appeared to alter dynamically across the middle to late stages of the rehearsal from hesitant to flowing.

Established Ensembles: Vocal–Piano Duos

The musicians in the established duos seemed to rely upon familiar working habits to achieve the short-term performance goal. Interestingly, based on IPA, analysis of the interactions indicated similarities with the new groups (see Table 12.3): higher levels of task-related discourse were produced (mean = 65.1 per cent) compared with socio-emotional utterances (mean = 34.9 per cent). The highest category of utterances was 'giving opinion' (mean = 31.9 per cent) and the lowest 'showing antagonism' (mean = 0 per cent). Compared with Bales's categories, however, utterances of solidarity were above the norm, indicating that these partnerships produced higher levels of support, praise and closeness towards one another than might be manifested in other small groups (including new ensembles). For example, in response to a recent run-through of material, the singer exclaimed 'that was lovely' and, later, 'sorry, you were right ... I came off the chord', and 'yes, that's together'.[12]

Each of the duos prepared the songs with slightly different strategies: Duo A rehearsed the material under-tempo and phrase-by-phrase ('not too fast to start with'), with the pianist providing cues for the singer's pitches; Duos B and C ran through the material at sight straight away ('let's just do it') before discussing tempi and interpretation; Duo D worked through the material with relatively long discussion of the meaning of the text and how to 'feel' the tempo ('let's see what speed you've got in mind and what speed I've got in mind'). The analyses of the social interactional style of these established partnerships reflected high levels of solidarity, which possibly enabled the musicians to be flexible and spontaneous in their musical collaboration; however, discourse was still predominantly suggestion-led.

---

[11]   The use of the word 'flow' is appropriate in this context as it resonates with the notion of flow used in Csíkszentmihályi's (1990) theory of optimal experience. In this state, the musicians are completely absorbed in the rehearsal process.

[12]   As with the data for the new duos, high standard deviations were evident for some of the individual categories in the established data set, although the overall pattern of responses within each duo was consistent with the overall evaluation against Bales's suggested limits.

Table 12.3   Task-related and socio-emotional categories (column 1) with
              frequency of utterances for the vocal–piano duo data (column 2)
              plus comparison with suggested limits provided by Bales
              (columns 3 and 4)

Category	Frequency of utterances (%; mean; SD)	Bales's limits (%)	Evaluation of data with Bales's limits
Shows solidarity	108 (9.4%; M = 27; SD = 16.9)	0–5	*Above* upper limit
Tension release	71 (6.2%; M = 17.75; SD = 9)	3–14	Near lower limit
Agrees	202 (17.5%; M = 50.5; SD = 19.2)	6–20	Near upper limit
Gives suggestion	100 (8.7%; M = 25; SD = 14)	2–11	Near upper limit
Gives opinion	368 (31.9%; M = 92; SD = 42.8)	21–40	In middle of limits
Gives orientation	163 (14.1%; M = 40.75; SD = 18.2)	14–30	Near lower limit
Asks for orientation	67 (5.8%; M = 16.75; SD = 5.6)	2–11	In middle of limits
Asks for opinion	52 (4.5%; M = 13; SD = 2.2)	1–9	In middle of limits
Asks for suggestion	1 (0.1%; M = 0.25; SD = 0.5)	0–5	Near lower limit
Disagrees	12 (1.0%; M = 3; SD = 5.4)	3–13	*Below* lower limit
Shows tension	10 (0.8%; M = 2.5; SD = 2.4)	1–10	*Below* lower limit
Shows antagonism	0 (0%; M = 0; SD = 0)	0–7	At lower limit

Note: In column 4, evaluation indicates whether or not these data fell *below* or *above* Bales's ranges, *near* (within 3 per cent) or *within* the upper and lower limits.

The transactional style of the established duos was considered in a similar way as for the new duos, although an additional frame was created in the light of the emphasis upon solidarity via the interactional analysis: the *praise frame* (that is, the partners expressed solidarity towards one another by offering support and positive encouragement in their activity).[13] The data revealed that the established duos tended to maintain consistent levels of *praise* across the rehearsal (up to 9

---

[13]   Fogel's *bridging* frame suggests inclusion of praise via 'help', which one would expect in a mother–infant relationship, although given the explicit nature of the praise articulated in these rehearsals, it was considered appropriate to define this frame.

per cent of its duration) with *hesitancy* appearing in the early stages of rehearsal (up to 15 per cent). The majority of the rehearsals involved *flowing* patterns of communication (up to 76 per cent of the rehearsal duration). The transactional style of the established duos reflected a stable kind of *recapitulative* behaviour whereby *hesitant* communication led to *flowing* communication by the first or middle stages of rehearsal.

Discussion: Theories of Group Development

The evaluation of the data highlighted differences in the styles of social interaction between the musicians working together in duo partnerships according to two analytical frameworks (IPA and transactional framing). In the new ensembles, there were high levels of suggestion and low levels of disagreement. Transactions were permeable and recapitulative, characterised primarily by hesitancy. There were comparatively high levels of expressed solidarity (praise) in the established ensembles; transactions were recapitulative and demonstrated predominantly flowing patterns of communication.

The findings corroborate social theories about group development. Typically, a group's development is considered to project along a simple timeline (beginning–lifespan–end) marked by two key events: bonding and breakup. Numerous theoretical models provide descriptions of this timeline according to a series of defined stages (Levinger 1983; Tuckman 1965). For example, Bruce Tuckman's influential model of working relationships reflects the typical developmental sequence of a small group in five stages (Tuckman 1965; Tuckman and Jensen 1977): (1) *forming* (involving group orientation, testing and dependency on key players, such as leaders); (2) *storming* (reflecting resistance to group influence, often characterised by conflict and polarisation); (3) *norming* (moving towards group cohesion, or an 'in-group' feeling, whereby individual roles are adopted and the group identity evolves); (4) *performing* (entailing flexibility among individual players as well as concerted effort towards fulfilling the group's task); (5) and *adjourning* (showing the completion of tasks and the dissolution of the group, which may be planned or unplanned).

Even though different terms can be used to define the stages in developmental models (cf. Levinger 1983), theorists indicate that the first stage is normally characterised by 'individual, unshared behaviour', while the middle to later stages reflect tighter group work as members become 'an effective, integrative, creative social instrument' (Martin and Hill, cited in Douglas 1970, p. 32; also see Atik 1994). Music ensembles will of course work in different ways depending on the levels of experience and expertise of the players (see Ginsborg and King 2012), although these data indicated that the styles of social interaction will differ depending on the extent to which the players are familiar with one another. The cellists and pianists in the new ensembles were effectively 'bonding', manifesting 'unshared' behaviour via low levels of social dependency (lack of disagreement and solidarity). For the singers and pianists in the established ensembles, they were 'bonding' initially, but 'norming' or 'performing' as they expressed solidarity towards one another.[14]

As indicated above (see note 8), the point at which new ensembles become regarded as established will depend upon numerous factors beyond the scope of this discussion. What is important, however, is that the stages advocated in models of group development are regarded as cyclical and non-linear (see Smith 2005): musicians in a new ensemble will need to 'bond' and may progress to the 'norming' stage without experiencing the 'storming' stage; an established ensemble will need to 'bond' (or 're-bond') at the start of each rehearsal regardless of how well the players know one another (the 'hesitancy' frames observed in the established groups' rehearsals indicated this to some extent); all of the ensembles will adjourn (permanently or temporarily) at the end of a rehearsal. Furthermore, 'bonding' might occupy more than one rehearsal, although these data suggested that progression to 'norming' could occur within one rehearsal (as manifested by the flowing transactional frame). The theoretical model – if regarded cyclically and non-linearly – thus explains that chamber ensembles will develop across stages and that different styles of interaction might be captured in these stages. These data fit this model and indicate specifically that social familiarity (new or established partnerships) impacts upon styles of interaction. It is likely, of course, that other factors influenced these musicians' styles of interaction as is typical of everyday rehearsals, such as the moods of individual players (see King 2006), their experiences and expectations of working in duo contexts, their knowledge of particular repertoire, musical styles or rehearsal strategies and so on, yet, arguably, these factors were negotiated within the social parameters of their working relationship, defined in this case according to whether or not they were rehearsing with someone for the first time.

Conclusions

The most important difference between the styles of interaction between the new and established musicians in these rehearsals was in the frequency of utterances of solidarity – the latter exhibited greater solidarity than the former (and compared with other small groups). Otherwise, according to IPA, the nature of discourse was similar: all of the musicians exhibited high levels of agreement and suggestion, with low levels of tension and antagonism. Interestingly, the transactional analysis revealed that the duration and nature of groups of utterances and activity differed among the partnerships according to their social familiarity: the new duos

---

[14]   Interestingly, members of the long-standing Allegri string quartet described their musical identity according to its 'spontaneity and diversity' (Burton-Page 2005), suggesting that their group work was flexible, which adheres to the internal structure of the 'performing' stage in Tuckman's developmental model.

exemplified more *hesitancy* across the first and middle stages of rehearsal than the established duos, while the latter produced more *flowing* patterns of communication across the first, middle and late stages of rehearsal along with *praise*.

Various points merit attention. Firstly, observation of the new duos indicated that three of the partnerships did not really get on: these musicians 'adjourned' quite happily at the end of the study; they did not bond well. However, the social interactional style of these duos was consistent with the other new duos, even though it was not necessarily indicative of their experience or potential to survive – this obviously raises the question of the validity of the scenario: it suggests that, for a limited one-off occasion, one musician can work quite effectively with another so as to create a temporary bond. For these particular duos, however, there were higher levels of *hesitancy* across these rehearsals (up to 70 per cent), suggesting that the transactional style tended to be more disjointed.

Secondly, all of the duos were working on new (unfamiliar) repertoire, which brings into question the influence of the task and other aspects on the creative relationship: the nature of discourse is likely to change as the musicians get to know a particular piece and environment. These data represented a cross-section of the rehearsal process – longitudinal research would enable further insight into the progression of a creative relationship and future studies should involve processing such data. For the cello–piano duos, getting to know the piece was occurring simultaneously with getting to know each other in the set rehearsal: there were thus low levels of social and musical familiarity along with low levels of familiarity with the environment (the prescribed scenario of the study). For the vocal–piano duos, the working relationship was already established even though they were tackling a new piece: there were thus high levels of social familiarity with low levels of musical familiarity along with low levels of familiarity with the environment. The complex ways in which levels of familiarity (social, musical, environmental and so on) operate in rehearsals merit more systematic consideration in future studies.

Thirdly, the personal relationships between the musicians outside of the ensemble rehearsal need to be considered. In the case of three of these partnerships, the musicians were friends (house mates) outside of the music room, while one of the established duos comprised a husband-and-wife team. Personal relationships will inevitably influence professional (musical and social) relationships, although these data do not explore how this might happen. It is plausible to suggest, however, that personal relationships may coincide or create tension with the professional relationship at any point in time (see Blank and Davidson 2007; Doğantan-Dack 2008).

This chapter explored social familiarity in chamber ensembles via analysis of the styles of interaction manifested in discourse between players in rehearsal. It was noted that the working relationships between musicians in chamber groups are potentially close, reflecting a 'family feel', and that social familiarity influences the styles of interaction between players in rehearsal. Two (complementary) analytical approaches were used – IPA and transactional framing – whereby the extension of transactional thinking into this context, that is from a very close family relationship to a working relationship (that happens in some cases to be close), was entirely novel. If developed further, these analytical frameworks could be usefully applied in longitudinal research to explore the growth of chamber ensembles across rehearsals and it is hoped that this study provided the foundation for doing so. This research also complements studies on chamber ensemble performance (for example, see Doğantan-Dack 2008; also Chapter 13) and future work in these areas will allow greater insight into the ways in which individuals make music together.

References

Anon. (1909). Review of Three Chamber Concerts by the Lucas String Quartet. *Musical Times* 50 (July 1)797: 469–70.

Atik, Y. (1994). The Conductor and the Orchestra: Interactive Aspects of the Leadership Process. *Leadership and Organization Development Journal* 13: 22–8.

Bales, R.F. (1950). A Set of Categories for the Analysis of Small Group Interaction. *American Sociological Review* 15: 257–63.

Bales, R.F. (1999). *Social Interaction Systems: Theory and Measurement.* Piscataway, NJ: Transaction.

Bayley, A. (2011). Ethnographic Research into Contemporary String Quartet Rehearsal. *Ethnomusicology Forum* 20/3: 1–27.

Bayton, M. (1998). *Frock Rock: Women Performing Popular Music.* Oxford: Oxford University Press.

Berscheid, E. and Peplau, L.A. (1983). The Emerging Science of Relationships. In H.H. Kelley, E. Berscheid, A. Christensen, et al. (eds), *Close Relationships* (pp. 1–19). New York: Freeman and Company.

Blank, M. and Davidson, J. (2007). An Exploration of the Effects of Musical and Social Factors in Piano Duo Collaborations. *Psychology of Music* 35/2: 231–48.

Blum, D. (1986). *The Art of Quartet Playing: The Guarneri Quartet in Conversation with David Blum.* New York: Cornell University Press.

Burton-Page, P. (2005). *The Allegri at 50: A Quartet in Five Movements,* www.allegriquartet.org.uk/fifty (accessed 20 May 2011).

Chaffin, R., Imreh, G. and Crawford, M. (2002). *Practicing Perfection.* Mahwah, NJ: Erlbaum.

Csíkszentmihályi, M. (1990). *Flow: The Psychology of Optimal Experience.* New York: Harper and Row.

Davidson, J.W. and Good, J.M.M. (2002). Social and Musical Co-ordination between Members of a String Quartet: An Exploratory Study. *Psychology of Music* 30: 186–201.

Davidson, J.W. and King, E.C. (2004). Strategies for Ensemble Practice. In A. Williamon (ed.), *Musical Excellence: Strategies and Techniques to Enhance Performance* (pp. 105–22). Oxford: Oxford University Press.

Doğantan-Dack, M. (2008). Collaborative Processes and Aesthetics of Interaction in Chamber Music Performance: The Alchemy Project. *Proceedings of the 2nd International Symposium on Systems Research in the Arts and Humanities.* Baden-Baden (pp. 7–12). Ontario: International Institute for Advanced Studies.

Douglas, T. (1970). *A Decade of Small Group Theory 1960–1970.* London: Bookstall Publications.

Fogel, A. (2009). What Is a Transaction? In A.J. Sameroff (ed.), *The Transactional Model of Development: How Children and Contexts Shape Each Other* (pp. 271–80). Washington, DC: American Psychological Association.

Fogel, A., Garvey, A., Hsu, H. and West-Stroming, D. (2006). *Change Processes in Relationships: A Relational–Historical Research Approach.* Cambridge: Cambridge University Press.

Ford, L. and Davidson, J.W. (2003). An Investigation of Members' Roles in Wind Quintets. *Psychology of Music* 31: 53–74.

Ginsborg, J. and King, E.C. (2012). Rehearsal Talk: Familiarity and Expertise in Singer–Pianist Duos. *Musicae Scientiae*; doi: 10.1177/1029864911435733.

Goodman, E.C. (2000). *Analysing the Ensemble in Music Rehearsal and Performance: The Nature and Effects of Interaction in Cello–Piano Duos.* Unpublished PhD dissertation, University of London.

Goodman, E.C. (2002). Ensemble Performance. In J. Rink (ed.), *Musical Performance: A Guide to Understanding* (pp. 153–67). Cambridge: Cambridge University Press.

King, E.C. (2006). The Roles of Student Musicians in Quartet Rehearsals. *Psychology of Music* 34: 262–82.

King, E.C. and Ginsborg, J. (2011). Gestures and Glances: Interactions in Ensemble Rehearsal. In A. Gritten and E. King (eds), *New Perspectives on Music and Gesture* (pp. 177–201). Aldershot: Ashgate Press.

Levinger, G. (1983). Development and Change. In H.H. Kelley, E. Berscheid, A. Christensen, et al. (eds), *Close Relationships* (pp. 315–59). New York: Freeman.

Murnighan, J.K. and Conlon, D.E. (1991). The Dynamics of Intense Work Groups: A Study of British String Quartets. *Administrative Science Quarterly* 36: 165–86.

Nissel, M. (1998): *Married to the Amadeus: Life with a String Quartet.* London: Giles de la Mare.

Paterson, B.L., Thorne, S.E., Canam, C. and Jillings, C. (eds) (2001). *Meta-study of Qualitative Health Research: A Practical Guide to Meta-analysis and Meta-synthesis (Methods in Nursing Research).* Thousand Oaks, CA: Sage.

Sameroff, A.J. (ed.) (2009). *The Transactional Model of Development: How Children and Contexts Shape Each Other.* Washington, DC: American Psychological Association.

Shuker, R. (1994). *Understanding Popular Music*. London: Routledge.

Sidanius, J. and Pratto, F. (1999). *Social Dominance: An Intergroup Theory of Social Hierarchy and Oppression*. New York: Cambridge University Press.

Smith, M.K. (2005). Bruce W. Tuckman: Forming, Storming, Norming and Performing in Groups. *The Encyclopaedia of Informal Education*, www.infed.org/thinkers/tuckman.htm (accessed 20 May 2011).

Tuckman, B. (1965). Developmental Sequence in Small Groups. *Psychological Bulletin* 63: 384–99.

Tuckman, B. and Jensen, M.A. (1977). Stages of Small Group Development Revisited. *Group and Organizational Studies* 2: 419–27.

Williamon, A. and Davidson, J. (2002). Exploring Co-performer Communication. *Musicae Scientiae* 6/1: 53–72.

Young, V.M. and Colman, A.M. (1979). Some Psychological Processes in String Quartets. *Psychology of Music* 7: 12–16.

Chapter 13

Familiarity and Musical Performance

Mine Doğantan-Dack

The human capacity to learn from experience is unmatched in the animal world. We are unique in the ways we can transfer and adapt what we have learned to new situations. Take away this capacity, and all aspects of human intelligence, including understanding, classifying, reasoning about, interpreting, evaluating and indeed inhabiting present events, as well as forming expectations and making predictions about future ones, would disappear with it. The remarkable plasticity of the human brain, which continually rewires itself by forming new neural connections, allows us to adjust to changing conditions in the environment by applying what we already know from past experience to current situations, thereby fine-tuning our responses to them. Our surroundings, entrusted to our existing knowledge for its understanding and making, are continually rendered familiar for us. Philosopher Martin Heidegger regarded our sense of *familiarity* with the world as the essence of being human (1962) – or a specifically human way of being – and argued that perception and experience are indeed driven by a 'referential whole' made available by our deep-rooted sense of having-already-been-acquainted with our environment.[1]

The process of becoming familiar with the world starts in the womb, notably through the sense of hearing. Research indicates that hearing is normally fully functional before birth (Lecanuet 1996) and newborns come into a world that is already acoustically familiar, particularly through pre-natal experiences with the prosodic and melodic features of the mother's voice (Welch 2005). It is also an affectively familiar world where the newborn instantly recognises and responds to the greater emotional expressiveness of the mother's speech – evident in its pitch level, contour, tempo and rhythm – and displays a preference for it over women's 'usual' speech (Trehub 2003). Starting at birth, the dynamic between the familiar and the non-familiar shapes all our cognitive and affective relationships

---

[1]   In Heidegger's words, 'My encounter with the room is not such that I first take in one thing after another and put together a manifold of things in order then to see a room. Rather, I primarily see a referential whole … from which the individual piece of furniture and what is in the room stand out. Such an environment of the nature of a closed referential whole is at the same time distinguished by a specific *familiarity*. The … referential whole is grounded precisely in familiarity, and this familiarity implies that the referential relations are *well-known*' (cited in Dreyfus 2001, p. 103).

with our environment and continues to form the basis of all learning, or further acquisition of knowledge.

The concept of familiarity, which defines a psychological phenomenon, is relational in that it involves features of both the environment and the human subject. The contribution of the environment to one's sense of familiarity happens in the form of regularities: stable, coherent and repeating patterns of phenomena are necessary for humans to develop familiarity with the world. The human mind, in turn, actively constructs familiarity by recognising such repeating patterns and similarities, forming and storing mental representations of them, and retrieving what has been stored as and when required so as to mobilise existing knowledge to make sense of – and make decisions about – new contexts and situations. Familiarity is made manifest in one's behaviour through observable signs, such as ease – and confidence – in one's involvement with the world, and entails more than the mere recognition of similarities between past and current experiences: it also implies understanding oneself as part of certain social and cultural practices such that one experiences them as being immediately meaningful, and feels at home within the contexts they provide.[2] In this sense, familiarity merges generic knowledge about the way things usually are in the world and situated knowledge derived from one's exposure to particular social and cultural practices.

Contemporary psychology assigns a basic role to the concept of familiarity in accounting for the cognitive operations of the human mind. For example, the way the cognitive process of attention is structured is typically explained by reference to one's sense of familiarity with objects, events, people and situations: highly familiar stimuli, which concern the most stable, repetitive, constant features of the environment, do not need to be consciously processed at every encounter with them and in this sense they move to the perceptual background, freeing attention to focus on novelties. There is evidence to suggest that familiarity not only organises the allocation of attentional resources but also recruits different kinds of neural networks in comparison to the processing of non-familiar information (Goel, Makale and Grafman 2004).[3] Familiarity further facilitates the achievement of cognitive tasks by rendering many of the steps involved in the processing of new information and the retrieval of stored ones automatic: the more one is familiar with what is currently encountered, the less cognitive effort is required to understand and make decisions about it. Consequently, in a goal-oriented context, familiarity becomes an immensely advantageous – and therefore highly desired –

---

[2]   In this sense, the concept of familiarity is different from that of habituation, which refers to a phenomenon observable in virtually all biological systems and signifies a decrease in behavioural response to repeated stimuli without necessarily invoking conscious awareness of the stimuli or of the response.

[3]   Recent research using data gathered by means of functional magnetic resonance imaging (fMRI) techniques provides evidence that, when humans reason about familiar and unfamiliar situations, they not only employ attentional resources differently, but also utilize different parts of the brain (Goel et al. 2004).

impetus in shaping cognition and behaviour. Familiarity is the primary means by which humans plan their actions and activities, since it affords predictability to the temporal sequence as well as the content of forthcoming events. It is only when we inhabit a familiar world that we can develop expectations, as well as the belief that our expectations will be fulfilled.

Affective experiences are also shaped by the continual interplay between what is familiar and new for us. One of the basic hypotheses put forward in affective psychology is that there is an intimate relationship between cognition and affect in that 'the organism's evaluation of its circumstances (current or remembered or imagined) plays a crucial role in the elicitation and differentiation of its emotions' (Ellsworth and Scherer 2003, p. 573). It appears that the way one cognitively interprets the significance of the events in the environment in terms of one's well-being is a major factor in the arousal of emotions. Environments are never completely constant, but always involve changing features, and one of the fundamental aspects we detect in making appraisals about a situation concerns its degree of familiarity: highly familiar phenomena do not need to be appraised in terms of their significance every time they are encountered, and in that sense the affective tone accompanying their appraisal is often undifferentiated and uniform. Perception of non-familiar components, on the other hand, generate specific affective qualities in the appraisal process depending on whether they are interpreted as posing risks or opportunities for the person. Hence, 'novelty detection in its various forms can be considered as a gateway to the emotion system' (Ellsworth and Scherer 2003, p. 576). There is also evidence to suggest that, because familiarity implies being at ease and thereby a positively valenced affective involvement with our world, it impacts our thought processes and evaluative judgments such that the person in a familiar setting will be cognitively more flexible, experience facility in decision-making, and offer creative solutions to problems (Isen 2004).

The most ubiquitous explanation in contemporary psychology regarding the basis of familiarity is given in terms of *schema*, a theoretical construct used to account for the way the human mind organises knowledge (also see chapters 2 and 9). Introduced into psychology in the 1930s by Frederic Bartlett, the nature and function of schemata have been a dominant interest in cognitive psychology since the 1970s. 'Schema' refers to structured information about particular conceptual categories – including objects, events, people and situations – acquired through experience. As we encounter different situations at different times and notice similarities among them, our minds do not represent and store information about each detail of each particular encounter, but rather average the similarities together into an abstract framework, or schema. Schemata inform us about how the world usually is and serve to contextualise current experience based on prior knowledge, providing the subject with a sense of the degree of familiarity of new events and situations. What makes the world seem familiar to us is the functioning of schematic memories during perception to contextualise it. A listener who has acquired a schema about the social and cultural practices surrounding a classical music concert, and has consequently developed certain expectations about the practice of attending performances in this tradition, does not have to continually evaluate every detail in the concert hall as she takes her seat and waits for the music to start. The relevant schema informing the current experience provides not only the background feeling of familiarity with the kind of event that is about to unfold, that is, a classical music performance, but also about the appropriate ways of acting during the event.

As music in its many diverse manifestations is a universal trait of our species – there has never been any culture or period in human history that lacked music – one would naturally expect familiarity to play a basic role in our musical experiences and behaviour. Over the last several decades music psychology has provided abundant evidence that familiarity is indeed at the root of our cognition of and affective responses to music. A frequently employed term in this connection is the 'experienced listener' (or cf. Chapter 5, where the author uses the term 'qualified listener'). *A Generative Theory of Music* (1983) by Fred Lerdahl and Ray Jackendoff, which is regarded by music psychologist John Sloboda (2005, p. 102) as the work with which music psychology came of age, begins by stating 'the goal of a theory of music to be a *formal description of the musical intuitions of a listener who is experienced in a musical idiom*' (Lerdahl and Jackendoff 1983, p. 1). The authors go on to explain that by 'the musical intuitions of the experienced listener' they mean:

> the largely unconscious knowledge (the "musical intuition") that the listener brings to his hearing – a knowledge that enables him to organize and make coherent the surface patterns of pitch, attack, duration, intensity, timbre, and so forth … A listener without sufficient exposure to an idiom will not be able to organize in any rich way the sounds he perceives. However, once he becomes familiar with the idiom, the kind of organization that he attributes to a given piece will not be arbitrary but will be highly constrained in specific ways. (p. 3)

The hypothesis put forward by Lerdahl and Jackendoff, which received corroborative experimental support (for example, Krumhansl 1990), is that long-term exposure to a musical idiom generates schemata regarding musical syntax – that is, mental representations regarding tonal, rhythmic and metric features of music – such that a new piece of music in the same idiom will be understood spontaneously – and without formal training – by the listener. In this sense, one can speak of one's native music, similar to one's native language, the grammatical rules of which are learned early on during the developmental process without any training.

Familiarity also appears to play a major role in shaping our affective responses to music. Cross-cultural comparative research indicates that listeners who are familiar only with the music of their own culture have difficulty in perceiving the affective content of music from another culture; they also develop unconventional affective responses to music in an unfamiliar idiom, such as a distinct feeling of unease or discomfort in the face of a largely incomprehensible musical syntax (Gregory and Varney 1996).

The possibility of any communication of musical ideas and emotions depends upon the existence of a shared musical tradition that is familiar to composers, performers and listeners alike. Everyone within that tradition would have generic knowledge about how her 'native' music behaves. However, as a highly skilled musical activity requiring extensive physical and mental training, performance in the Western classical tradition involves specialist knowledge, not always shared by non-performing listeners, and acquired in accordance with these particular requirements. Hence, exploring the role of familiarity in shaping the musical activities and experiences related specifically to the making of musical performances necessitates the introduction of further conceptual differentiations.

Some Conceptual Considerations

Musical performance is one of the most complex accomplishments of our species, and 'at its best, is indicative of the upper limits of human physical and mental achievement' (Williamon 2004, p. 7). In order to attain expertise in performance, musicians need to acquire not only high levels of technical mastery in playing a musical instrument – or in applying one's voice musically – but also vast amounts of knowledge to develop meaningful and effective interpretations of the music they play. Research indicates that experience plays a large role in performance expertise in the Western classical tradition (Lehmann and Ericsson 1998), and a successful musical performance depends on the operation of diverse kinds of schemata, representing motor, cognitive, affective and social knowledge that performers develop over many years. Before they become expert performers, musicians acquire a substantial degree of familiarity with the diverse components and conditions involved in performing music.

The kinds of musical behaviour and performances that can be described and explained by reference to the all-embracing umbrella term 'familiarity' display a wide range – both a novice and an expert can be said to be familiar with a musical instrument or a piece of music – and one cannot speak of universally recognisable behavioural or psychological criteria that would define a norm in the application of this term to differing performance contexts and situations. Consequently, conceptualising familiarity such that it retains its explanatory value in relation to musical performance necessitates laying down certain specifications. The first of these is the identification of the *perspective* one wishes to investigate: for example, merely observing the behaviour of performers and the sounds of their performances would not be sufficient to draw any meaningful conclusions regarding their degree of cognitive familiarity with a given piece unless we also note the performers' level of prior experience in performance making. Expert performers who are familiar with a piece of music will be working with significantly different musical schemata compared with beginners who are familiar – at their own level of skill and knowledge – with the same piece; consequently, in order to explore, for example, the impact of cognitive familiarity with a piece on its performance, it is essential to note whose perspective is being investigated.

At the most basic level, in order to be able to perform any kind of music at all, one needs to have *generic knowledge* about how to play an instrument, and *particular knowledge* about the gestural and expressive aspects of a given piece of music. Each musical instrument requires specific technical skills – for example, vibrato in string playing, independence of hands in piano playing (Sloboda 1985, p. 93) – and the know-how that one acquires when learning to play an instrument is the basis of motor familiarity as a pre-condition of performing music with fluency, rhythmic precision and appropriate speed. In the words of Jonathan Dunsby, one of the purposes of practice is to develop a 'fluid "motor" ability to pass through musical time in performance without giving conscious thought to the multiple, vastly complex effort involved: it teaches the musician to "walk", "run", "jump", indeed we might even say "speak music"' (Dunsby 1995, p. 40). It is a well-established fact that, in order to attain expertise in musical performance within the Western classical tradition, one has to start gaining motor familiarity at a very young age, and skills to achieve high levels of performance are acquired through intense practice over a very long time – a minimum of 10 years according to existing research (Ericsson, Krampe and Tesch-Römer 1993). There are great differences in the motor abilities of musicians between the earlier and later stages of this developmental process (ibid.), the implication being that motor schemata undergo substantial transformations throughout the preparatory years: it is not merely the degree of motor familiarity that distinguishes the earlier and later stages, but the musical schemata themselves evolve to acquire different structures, knowledge becoming better organised with increasing expertise. In a real sense, an expert performer's familiarity with a musical instrument is qualitatively different compared with a novice in terms of the knowledge that defines and drives the motor interaction with the instrument.

Another factor that has to be taken into account in exploring the role of familiarity in relation to musical performance is the *cultural context* or the performance tradition within which the performer's experience and knowledge are put into practice. For example, a performer's familiarity with a piece of music in the Western classical tradition would be based upon their knowledge about notational conventions, the historical style of the piece, the cultural codes of musical expression, and so on. In an oral culture, on the other hand, where each performance introduces improvisatory differences into the pitch and rhythmic domains of a piece, familiarity would imply knowledge about the background structural features rather than about each and every surface detail. Each musical culture entails different kinds of knowledge and experience for the making of a successful performance. Even within a given culture, the researcher has to be aware of changing performance traditions and the changing expectations of audiences. For example, contemporary performance practice within the Western classical style obliges performers to give performances of a reproducible kind – an outcome not only of dramatic developments in performance virtuosity over the last century (Lehmann 2006), but also of the technical possibilities offered by recording technology (Philip 2004, pp. 21–2 and 244–50). Consequently, variability or inconsistency in the performance standards of a musician is much less tolerated today than it would have been during the early part of the twentieth century. A present-day performer would spend a lot more time in becoming intimately familiar with the structural and expressive details of a piece while preparing for a concert in comparison to a musician from this era: changing performance traditions mean that the way performers learn a given piece of music also changes in terms of the flexibility of the expressive details that are committed to memory.

Since there are multiple ways of coming to know a piece of music, studying the role of familiarity in performance also requires specifying the nature of *the process* through which familiarity has been attained; there could be significant differences in the details of a performance depending on whether the performer acquires familiarity with a piece only by aural means (for example, through the Suzuki method, by listening intensely to the performance of another musician), by detailed score analysis, or through intuitively driven personal practice. Each of these processes would generate different kinds of cognitive schemata about the piece in question. For performers, coming to know a piece of music is strictly related to how they want it to *sound,* and in this sense the starting point is always the *sounding score.*[4] Performers do not normally analyse a score prior to its production in sound, and if theoretical analysis is undertaken at any stage, this is driven by performance-oriented concerns. Furthermore, while a score-based analysis aims to understand how a piece works – how its structure is created so as to make musical sense – for the performer, the aim of becoming familiar with the music is to *make it work* in sound by creating an effective performance. Consequently, the performer's activity involves a powerful aspect of making something, as distinct from merely observing and contemplating. A performer – *qua* performer – cannot acquire familiarity with the structural and expressive elements of a piece separately from the physical component required to produce them in sound. The kinaesthetic sensations and the physical gestures are a very real part of how a performer comes to know a piece of music, and the cognitive and motor schemata regarding its performance evolve simultaneously. The markers of cognitive familiarity are closely bound up with motor factors, and without the motor component, a performer cannot display cognitive familiarity with a piece *in performance.*

Research in musical performance has largely identified the processes through which a performer becomes familiar with a piece as those taking place in practice sessions and rehearsals. To be sure, practice is an essential part of the process of becoming familiar with a piece of music, and it ensures that the technical

---

[4]   Some of the ideas in the rest of this paragraph were first formulated in a paper entitled 'Practice and Theory: Ways of Knowing Music', presented at the 8th Conference on Systems Research in the Arts, Baden-Baden, Germany, in 2006.

dimension becomes largely automatic so that the performer can direct attention to the interpretative and expressive features of the music in performance. According to a commonly accepted hypothesis, there is a direct relationship between the structure – rather than quantity – of the practice and the quality of the performance (Ericsson 1997; Williamon 2004; Williamon and Valentine 2000). Susan Hallam's research (1997) revealed substantial differences between the practice strategies of experienced musicians and novices such that the former can 'draw on a musical knowledge base which enable[s] them to assess the task, identify task difficulties, recognise errors, monitor progress, and take appropriate action to overcome problems' (Hallam 1997, p. 103). In this sense, performers acquire expertise in practicing as well as in performing music. While practice is the primary activity through which performers become familiar with a piece, learning to perform it effectively does come to an end in the practice room. How performers continue to learn on stage, through live performances, remains a largely neglected area within music performance research.

The Alchemy Project

The practice-based research project entitled *Alchemy in the Spotlight: Qualitative Transformations in Chamber Music Performance*, funded by the AHRC[5] and directed by this author, is to my knowledge the first research undertaken to explore live performance from the perspective of professional performers in the Western classical tradition.[6] The aim of the project was to explore the experience of performing live in the context of a professional piano trio, and to compare and contrast the individual as well as the collaborative cognitive, affective and social processes that shaped a live performance with those that defined rehearsals. The project further aspires to raise awareness of live performance as the ultimate norm or the golden standard in classical performance practice, the articulation of which is particularly important in an age when performances recorded and edited in the studio provide the context for an overwhelming majority of musical experiences. As such, it forms an initial step towards counter-balancing the excessive focus in contemporary performance studies on a 'musicology of recording' (Cook 2010).

One of the most important hypotheses put forward by the Alchemy project was that a prerequisite for an elite performance is the familiarity acquired with a piece

---

[5] The Arts and Humanities Research Council (AHRC) funds postgraduate training and research in the arts and humanities, from archaeology and English literature to design and dance. The quality and range of research supported not only provide social and cultural benefits but also contribute to the economic success of the UK (for further information, see http://www.ahrc.ac.uk).

[6] The project website, designed by Dr Stephen Boyd Davis (Head of the Lansdown Centre for Electronic Arts, University of Middlesex) and maintained by the University of Middlesex, can be visited at http://wwwmdx.ac.uk/alchemy.

through live performances. It is argued that there is a kind of schema for musical performance that can be developed *only* on stage and not in the practice room. Performers do continue to learn on stage, and the kind of knowledge that is thus acquired becomes the basis for future superior performances.

The comparisons between the processes of the rehearsals and the live events undertaken as part of the Alchemy project (documented by means of audio-visual recordings, preparatory and reflective diaries, and interviews) suggest a new conceptualisation of live performance that challenges established notions in musicology. According to the received view, the purpose of practice and rehearsal sessions is to develop and fix an interpretation that is then unfolded for audiences during a live performance: in other words, performing live is more or less a repetition of what has been determined and achieved in the practice room. Such a conceptualisation is, however, inadequate in that it fails to explain the true nature of a live performance: for elite performers, the aesthetic aim of a live performance is to *surpass* what has been achieved in rehearsals and to bring about a certain qualitative transformation that is recognised as 'magical' by both themselves and audiences. During a live performance the cognitive-affective world of the performers and consequently the interpretation of the music they play often undergo certain positive qualitative transformations, which involve processes peculiar to live contexts as distinct from those involved in rehearsals and practice sessions. Among such transformations are an increasing sense of expressive freedom, increasing affective involvement with the music leading to an experience of flow (Csíkszentmihályi 1990), the making of unplanned creative interpretative choices, and the experiencing of certain alterations in time-consciousness. Positive qualitative transformations of this kind are most likely to transpire under optimal psychological, social and acoustical conditions, although they can also be achieved in less favourable circumstances. Being familiar with the processes involved in performing live is an important factor that facilitates the emergence of performance magic.[7]

There are two kinds of knowledge acquired specifically on stage that come to determine a performer's schema for live performance. One is the knowledge concerning particular pieces of music: anecdotal evidence by performers supports the hypothesis of the Alchemy project that a significant part of learning to perform a given piece effectively, convincingly and indeed exquisitely takes place on stage. For instance, pianist Sviatoslav Richter revealed that it was only at his fourth public performance of Mozart's Piano Sonata in A minor that he was able to achieve what he considered a satisfactory interpretation (in Neuhaus 1993, p. 206). Although the number of live performances required to make a given piece

---

[7] In the context of chamber music, the positive qualitative transformations in question do not necessarily happen simultaneously as the three musical parts do not involve the same kind of technical and/or expressive demands at every point: it is, therefore, perfectly possible for the co-performers to experience the transformations at different points along the piece.

work aesthetically vary from one performer to another, Dunsby articulates the general consensus about the necessity of repeat performances by writing that:

> Many performers experience the fact that learning and performing a new piece of music does not happen all at once. The first performance, as good as it may be, is not as satisfactory as the second, and – though I can prove this only anecdotally – it is the third performance where magically everything seems to come together … What is going on during those two gaps that may be hours apart or months? Clearly some kind of unconscious thinking, often called "assimilation". (Dunsby 1995, p. 10).

The other kind of knowledge that performers attain on stage concerns the particular nature of performing live. Acquiring familiarity with the conditions of live performance is necessary in order to develop performance expertise. In a live context, the performer has to work within an environment that involves two kinds of constraints: one of these is the inherent indeterminacy of the event, and the other is the necessity of uninterrupted flow. The indeterminacy is related to the 'living' nature of the performance environment such that at any moment an acoustical, psychological or social incident in the performance venue could displace the attention of the performer as a novel stimulus away from the music and from a focus on performance making – an unexpected noise, the room temperature rising or an audience member falling asleep can all disrupt the performer's concentration. As Dunsby argues, unpredictability plays a fundamental role in musical performance, and 'although the performer can seek to eliminate some uncertainties, there will always be others' (Dunsby 1995, p. 12). The aesthetics of live performance in the Western tradition obliges performers to keep going in the face of disruptive occurrences, and the more familiar performers are with the constraints of live performance, the better they will be able to keep themselves focused on their music making, and smooth out any potential discontinuities. Expert performers are able to create the illusion – by mobilising a remarkable range of skills and experiences – that they are in total control of the performance situation physically, mentally and aesthetically, even though in practice this is never the case. The received musicological notion that a live performance is essentially a repetition of what has been achieved in the practice room is, therefore, erroneous in that it does not take into account the restrictions that the live context imposes on the performer. The cognitive-affective processes involved in the preparatory phase do not lead to those that shape the live event through the logic of linear causality, and consequently it is not possible to understand all that happens in a live performance by reference to the preparatory processes alone.

An important consequence of the unusual nature of live music making is that there is *no guarantee* that the performer will be able to make 'magic' on stage even if earlier practice sessions involve, and promise, high levels of achievement: in the art of musical performance past success does not necessarily secure future accomplishment. Aside from the constraints of performing live, an important reason behind this fact is that part of the performer's familiarity with a piece of music – particularly, though not exclusively, its motor component – that makes its performance possible involves a continuous process of *renewal*, and not just *retrieval* of knowledge and experience. Performers are always at risk of losing their thorough familiarity with the various components that drive a musical performance (Dunsby 1995) and falling back to an earlier stage in their development of expertise – a condition that lies at the heart of the emotional peculiarities of Western performance practice. In this sense, familiarity, if it is to have any material impact on the performance of music, needs to be actively retained through regularly repeating encounters with the conditions of music making both in the practice room and on stage.

While the original aims and objectives of the Alchemy project were designed with Western classical chamber music practice in mind, implying an investigation of ensemble music making in general, soon after the beginning of the project it quickly became apparent that a piano trio is a specific kind of group that functions in accordance with its own criteria of ensemble identity and dynamics. It is different from, say, a cello–piano duo because of the increased size of the group and the attendant complexity of managing interactions between co-performers. It is also different from a string quartet owing to the contrasting nature of the two kinds of instruments involved, namely the piano and the strings. Nevertheless, a piano trio functions in accordance with some of the basic principles that drive any successful chamber music ensemble. In addition to familiarity with the music to be performed, ensemble performance also involves the component of social familiarity between co-performers. Since the Marmara Piano Trio, whose practice forms the research arena of the Alchemy project, was specifically founded for this particular investigation, the performers did not know each other at a personal level prior to the beginning of the project. Psychological research provides abundant evidence that, for any collaborating group of people to work well, there needs to be a sense of trust, support and openness between them, and a sense of belonging in the group (Douglas 1993). Research by Young and Colman on string quartets (1979), for example, indicates the vital role of group closeness, cultivated through trust and respect, for the healthy functioning and longevity of the ensemble (also see Chapter 12).

The general consensus in collaborative practice and research is that '[u]nless collaborators know each other well at a personal as well as meaningful level, the collaboration is almost inevitably headed for trouble at some point' (John-Steiner 2000, p. 278). In a very significant sense, table-fellowship is an integral part of musical fellowship in chamber music. The processes that build trust, mutual respect and support, and constructive ways of criticising when things go wrong, take place not only during rehearsals, but also during other kinds of social activities such as travelling together for performances. While social familiarity in an ensemble context is thus a significant factor leading to performance achievement, it is difficult, if not impossible, to specify *the amount* of social familiarity that is required for success on stage as a group. The most vital criterion seems to be one's subjective evaluation of being comfortable and happy in one's role within the ensemble: this can happen very early on during the life-time of the group, or not at all even after substantial acquaintance with one's musical partners, in which case the *longevity* of the ensemble would be seriously jeopardised. As far as the success of a particular performance is concerned, however, the Alchemy project provided evidence that a high degree of social familiarity among expert performers in a piano trio is not a necessary condition: the very first concert given as part of the project happened shortly after the trio was established, when the co-performers had not yet had time to develop strong social bonds, yet the performance was judged to be successful both by the performers and by the audience.[8]

Another indication of the Alchemy project was that the degree of social familiarity among the members of the trio influenced the way the rehearsal sessions were structured and conducted: rehearsal patterns changed with increasing social familiarity. In the case of the Marmara Piano Trio, while earlier rehearsals displayed more verbal communication, and more caution in making critical comments, later ones involved less talk but increasing ease in the making and receiving of criticism, which supports the findings of other research on communication between co-performers in practice sessions (for example, Williamon and Davidson 2002). Commenting on one's musicianship is a sensitive issue as it implicates the self and how it is perceived, and only increasing social familiarity can build the trust that facilitates the expression of concerns, doubts and self-criticism. It appears that, once the musical skills and competence of the co-performers are established and confirmed through consistently achieved musical standards in various performance situations, the risk of interpreting critical remarks personally rather than as a means to improve the overall performance quality of the ensemble decreases steadily. Interestingly, in the case of the Marmara Trio, it was also found that, during both the earlier and the later stages of the project, verbal praise for each other's musicianship was given sparingly during rehearsals, although there would be clearly understood non-verbal communication of admiration following a beautifully played phrase, an expressive intonation or a singing piano line. Based on the dynamics of the Marmara Trio, it also became apparent that a piano trio, unlike a string quartet, does not necessarily function around an agreed leadership, and that the music itself dictates which instrument will lead at any moment: this is in part due to the fact that there is no natural instrumental hierarchy in a piano trio, which is composed of two different kinds of instruments, and also due to the smaller size of the group in comparison to the string quartet.

Developing cognitive familiarity with a new piece of chamber music involves processes that are different from those concerning the learning of a solo repertoire. In the context of chamber music, the process of becoming familiar has an essential collaborative dimension: in learning and internalising a new piece of music, performers in a chamber ensemble simultaneously internalise a system of shared

---

[8]   Excerpts from this concert, which featured Beethoven's Piano Trio Op. 1 No. 1 in E-flat major, can be heard on the project website (see http://www.mdx.ac.uk/alchemy).

activity, and separating familiarity with the music from group interaction is very difficult, if not impossible. One of the important pedagogical questions is whether performers in a chamber ensemble should learn their individual parts in detail, together with the desired expression. It became clear early on during the Alchemy project that learning and developing an interpretation of one's own part individually can actually hinder putting the parts together musically with a unified ensemble voice:[9] an important part of the process of becoming familiar with a new piece has to be achieved collaboratively as a trio, and a more or less common representation of all three parts of the music needs to be developed together. The more complex a work is, the more urgent such collaborative learning becomes. A vital aspect of chamber music making is listening intensely to the other performers while playing, and one can develop a schema for this kind of interaction only by working within an ensemble.[10] Owing to the different manners of sound production involved by the keyboard and the strings, and the necessity of bringing the three instruments into accord technically and expressively, becoming familiar with a work for piano trio requires collaborative learning. Example 13.1 shows bars 104–08 from the first movement of Beethoven's Piano Trio Op. 70 No. 2 in E-flat. On the downbeat of bar 106, the violinist technically needs extra time to start the theme with a grace note, having ended the previous phrase with a note preceded by its grace note; the pianist cannot predict the particular timing required at the beginning of bar 106 based on familiarity with only her part, which involves a rapid descending scale to be played *leggiermente*. The pianistic intuition here would be to reach the downbeat of bar 106 in one single impulse without any delay, whereas ensemble unity requires adjusting the timing of the piano part to the timing of the violinist.

With regard to the relationship between cognitive familiarity with a piece and its performance, the Alchemy project indicated that, during rehearsals, performers reach a saturation point beyond which attempting to increase the familiarity by further detailed rehearsals does not improve the music making until the live

---

[9] Historically, a unified ensemble voice has not always been an aesthetic aim in chamber music making: Robert Philip (2004, pp. 104–39) discusses numerous examples from recorded history that attest to a casual attitude, a 'general informality' towards ensemble coordination, balance and expressive styles. 'Generally speaking', he writes, 'the best ensembles of today rehearse so that everyone agrees, not just about tempo but also about details … In early twentieth-century recordings, there are sometimes startling contrasts between two or more musicians playing together' (p. 105).

[10] Nevertheless, there is a basic asymmetry in the way the pianist and the string players in a piano trio learn their respective parts in that the former has access to the complete score right from the beginning of the learning process, while the latter have only their own parts. Furthermore, while evidence indicates that solo performers can replicate the timing and dynamic nuances of a performance even over long periods of time, there is no research to show that these expressive parameters can thus be replicated in chamber music performance. Based on my experiences with the Marmara Trio, I would propose that chamber music has a more variable dynamic, and fewer expressive details are retained in long-term memory compared with those of solo pieces of music.

Example 13.1   Beethoven, Piano Trio in E-Flat, Op. 70 No. 2, I, bars 104–08

performance actually takes place. In this sense, knowing when to stop rehearsing is an important factor for performance achievement. To be sure, the amount of pre-concert rehearsal and practice each performer requires is different: some prefer to practice almost until they go on stage, while others do not play at all on the day of the concert. However, there is evidence to suggest that 'extensive practice can reduce the ability to control the expression of sound in performance, for ideas become fixed and harder to adjust' (Goodman 2002, p. 157). As Dunsby writes '[m]any musicians believe that it is possible to, as it were, know too much about a piece, for instance, to have such a clear and preconceived idea of the structure that spontaneity of interpretation becomes impossible' (Dunsby 1995, p. 33).

While ample anecdotal evidence by musicians attest to the importance of live performance in solidifying the knowledge of a piece, there has not been any research on the actual psychological mechanisms through which performers continue to learn through live performances. The Alchemy project indicated that there is a strong affective dimension to learning on stage: the kind of familiarity that is acquired with a piece through live performance is highly emotional, and I would hypothesise that one of the mechanisms at work is the representation of the details of a live performance for long-term memory similarly to other emotional experiences, and differently from the experiences performers have during rehearsals. Indeed, the chemistry of an ensemble in live performance is different from the group dynamics experienced during rehearsals. Live performance is the site where the trust and support between co-performers are tested, confirmed and re-confirmed, and acquire their true practical meaning; furthermore, the excitement and the concentration of the live event, which is shared by the co-performers, creates a unique social interaction and further affective closeness among them. The term 'familiarity' is particularly befitting for explaining the learning that happens on stage in chamber music: one of the definitions of 'familiar' is 'exhibiting the manner of a close friend or pertaining to family' (*New Webster's Dictionary*, 1984) and living through an intense, even peak experience together – as in a successful musical performance – creates an emotional comfort zone for the ensemble, which becomes part of their schema for live performance guiding future music making.

Example 13.2   Beethoven, Trio in E-Flat, Op. 70 No. 2, I, bars 121–30

The other psychological mechanism that leads to learning through the experience of performing live is retrospective: the increasing familiarity acquired through repeat performances throws new light on the musical object and opens the door to further creative solutions to interpretative problems, allowing the performers to experiment with alternatives. It was only after learning and first performing Beethoven's Trio Op. 70 No. 2 with the Marmara Trio that the performers were able to conceive a more creative and effective interpretation of the passage leading to the recapitulation in the first movement. Example 13.2 shows this passage: the recapitulation starts with the upbeat to bar 129 in the piano part, but is preceded by a false recapitulation in bar 127 initiated by the cello.

Our first live performance of this passage involved a straightforward interpretation of what is written in the score. Later, we realised that it could be much more effective if the music in bars 121–7 gradually got calmer with a perceptible *ritardando*, leading to a significantly slower tempo in bars 128–9, showing the moment of true recapitulation as if under a magnifying glass. While this interpretation did not occur to us during the six months when we learned and rehearsed the piece, it took one live performance to establish sufficient affective familiarity with, and a sense of ownership of, the piece to be able to introduce a creative alternative to the shaping of this passage, which has a significant influence on the overall effect on the interpretation of the movement, and its potential to become magical in performance.

Concluding Remarks

The aesthetic impact and power of a musical performance lies in its capacity to intensify the attention of the listeners and their consciousness of the present moment by de-automatising their relationship to the music: a performance at its best breaks down the automatised response that familiar pieces of music invite. Similar to Shelley's poet, the elite performer 'strips the veil of familiarity from the aural world'.[11] A magical performance makes us aware of our own act of listening, and of its temporality. Performers can achieve such magic – the expressive flexibility to de-familiarise the familiar – only after intimate familiarity with a piece is developed through many live encounters with it.[12]

Research on live musical performance is in its earliest stages, and one of the important challenges for contemporary performance studies is to devise scrupulous research methods to understand and theorise about the processes of live music making. Some of the most significant insights regarding what it means and what it takes to perform music will come from systematically exploring the cognitive, affective, physical, aesthetic and social basis for performers of making music on stage. Such research promises to shed light on one of the most interesting aspects of being a performer, namely the existential basis – the desire – that drives the activity of performing live, by revealing the secrets of 'the infinite satisfaction' – to use the poetic words of the organist Dame Gillian Weir – 'of playing a work one has lived with for [many] years, that [fits] like a kid glove, that [lies] contentedly in its warm familiarity deep within the consciousness' (Weir 1995, p. 352).

---

[11]   In his *A Defence of Poetry* (1819), Percy Bysshe Shelley famously wrote that the poet 'strips the veil of familiarity from the world'.

[12]   Two other factors are crucial in preparing for a live performance and, owing to lack of space, they are mentioned here only briefly: one is the familiarity with the acoustics of the actual concert venue, and the other is the necessity for pianists to adjust to an unfamiliar instrument every time they go on stage. Depending on the acoustics of the concert venue, performers often have to make adjustments in tempo, dynamics, articulation and so on within a very short period of time and research is needed to understand the mechanisms responsible for these adjustments. Also, how pianists encode the music they play so that their existing knowledge can be applied to new instruments for successful performances is an important but entirely neglected research area.

References

Bartlett, F.C. (1932). *Remembering: An Experimental and Social Study.* Cambridge: Cambridge University Press.

Cook, N. (2010). The Ghost in the Machine: Towards a Musicology of Recordings. *Musicae Scientiae* XIV/2: 3–21.

Csíkszentmihályi, M. (1990). *Flow: The Psychology of Optimal Experience.* New York: Harper and Row.

Douglas, T. (1993). *Theory of Groupwork Practice.* New York: Macmillan.

Dreyfus, H.L. (2001). *Being-in-the-world: A Commentary on Heidegger's 'Being and Time', Division I.* Cambridge, MA: MIT Press.

Dunsby, J. (1995). *Performing Music: Shared Concerns.* Oxford: Oxford University Press.

Ellsworth, P.C. and Scherer, K.R. (2003). Appraisal Processes in Emotion. In R.J. Davidson, K.R. Scherer and H.H. Goldsmith (eds), *Handbook of Affective Sciences* (pp. 572–95). Oxford: Oxford University Press.

Ericsson, K.A. (1997). Deliberate Practice and the Acquisition of Expert Performance: An Overview. In H. Jørgensen and A.C. Lehmann (eds), *Does Practice Make Perfect? Current Theory and Research on Instrumental Music Practice* (pp. 9–51). Oslo: Norges Musikhogskole.

Ericsson, K.A., Krampe, R.T. and Tesch-Römer, C. (1993). The Role of Deliberate Practice in the Acquisition of Expert Performance. *Psychological Review* 100/3: 363–406.

Goel, V., Makale, M. and Grafman, J. (2004). The Hippocampal System Mediates Logical Reasoning about Familiar Spatial Environments. *Journal of Cognitive Neuroscience* 16/4: 654–64.

Goodman, E. (2002). Ensemble Performance. In J. Rink (ed.), *Musical Performance: A Guide to Understanding* (pp. 153–67). Cambridge: Cambridge University Press.

Gregory, A.H. and Varney, N. (1996). Cross-cultural Comparisons in the Affective Response to Music. *Psychology of Music* 24: 47–52.

Hallam, S. (1997). Approaches to Instrumental Music Practice of Experts and Novices: Implications for Education. In H. Jørgensen and A.C. Lehmann (eds), *Does Practice Make Perfect? Current Theory and Research on Instrumental Music Practice* (pp. 179–231). Oslo: Norges Musikhogskole.

Heidegger, M. (1962). *Being and Time.* Trans. J. Macquarrie and E. Robinson. Oxford: Blackwell. [First published in 1927 as Sein und Zeit. In E. Husserl (ed.), *Jahrbuch für Philosophie und Phänomenologische Forschung* 8.]

Isen, A.M. (2004). Positive Affect and Decision Making. In M. Lewis and J.M. Haviland-Jones (eds), *Handbook of Emotions* (pp. 417–35). New York: Guilford Press.

John-Steiner, V. (2000) *Creative Collaboration.* Oxford: Oxford University Press.

Krumhansl, C.L. (1990). *Cognitive Foundations of Musical Pitch.* New York: Oxford University Press.

Lecanuet, J.P. (1996). Prenatal Auditory Experience. In I. Deliège and J. Sloboda (eds), *Musical Beginnings: Origins and Development of Musical Competence* (pp. 3–34). Oxford: Oxford University Press.

Lehmann, A.C. (2006). Historical Increases in Expert Music Performance Skills: Optimising Instruments, Playing Techniques and Training. In E. Altenmüller, M. Wiesendanger and J. Kesselring, *Music, Motor Control and the Brain* (pp. 3–24). Oxford: Oxford University Press.

Lehmann, A.C. and Ericsson K.A. (1998). Preparation of a Public Piano Performance: The Relation Between Practice and Performance. *Musicae Scientiae* 2: 69–94.

Lerdahl, F. and Jackendoff, R. (1983). *A Generative Theory of Tonal Music*. Cambridge, MA: MIT Press.

Neuhaus, H. (1993). *The Art of Piano Playing*. Trans. K.A. Leibovitch. London: Kahn and Averill.

Philip, R. (2004). *Performing Music in the Age of Recording*. New Haven, CT: Yale University Press.

Sloboda, J. (1985). *The Musical Mind: The Cognitive Psychology of Music*. New York: Oxford University Press.

Sloboda, J. (2005). *Exploring the Musical Mind: Cognition, Emotion, Ability, Function*. Oxford: Oxford University Press.

Trehub, S.E. (2003). Musical Predispositions in Infancy: An Update. In I. Peretz and R. Zatorre (eds), *The Congitive Neuroscience of Music* (pp. 3–20). Oxford: Oxford University Press.

Weir, G. (1995). Organ Music II. In P. Hill (ed.), *The Messiaen Companion*. London: Faber and Faber.

Welch, G.F. (2005). Singing as Communication. In D. Miell, R. Macdonald and D.J. Hargreaves (eds), *Musical Communication* (pp. 239–60). Oxford: Oxford University Press.

Williamon, A. (2004). A Guide to Enhancing Musical Performance. In A. Williamon (ed.), *Musical Excellence: Strategies and Techniques to Enhance Performance* (pp. 3–18). Oxford: Oxford University Press.

Williamon, A. and Davidson, J.W. (2002). Exploring Co-performer Communication. *Musicae Scientiae* 6: 53–72.

Williamon, A. and Valentine, E. (2000). Quantity and Quality of Musical Practice as Predictors of Performance Quality. *British Journal of Psychology* 91: 353–76.

Young, V.M. and Colman, A.M. (1979). Some Psychological Processes in String Quartets. *Psychology of Music* 7: 12–6.

Index

Bold numbers = table; *italic* numbers = figure; ms = music example; n = footnote.